Current Topics in Behavioral Neurosciences

Volume 36

Current Topics in Behavioral Neurosciences provides critical and comprehensive discussions of the most significant areas of behavioral neuroscience research, written by leading international authorities. Each volume offers an informative and contemporary account of its subject, making it an unrivalled reference source. Titles in this series are available in both print and electronic formats.

With the development of new methodologies for brain imaging, genetic and genomic analyses, molecular engineering of mutant animals, novel routes for drug delivery, and sophisticated cross-species behavioral assessments, it is now possible to study behavior relevant to psychiatric and neurological diseases and disorders on the physiological level. The *Behavioral Neurosciences* series focuses on "translational medicine" and cutting-edge technologies. Preclinical and clinical trials for the development of new diagnostics and therapeutics as well as prevention efforts are covered whenever possible.

More information about this series at http://www.springer.com/series/7854

Adam L. Halberstadt · Franz X. Vollenweider
David E. Nichols
Editors

Behavioral Neurobiology of Psychedelic Drugs

Springer

Editors
Adam L. Halberstadt
Department of Psychiatry
University of California San Diego
La Jolla, CA
USA

David E. Nichols
Division of Chemical Biology and Medicinal Chemistry
University of North Carolina
Chapel Hill, NC
USA

Franz X. Vollenweider
Psychiatric University Hospital, Heffter Research Center Zurich
University of Zurich
Zurich
Switzerland

ISSN 1866-3370 ISSN 1866-3389 (electronic)
Current Topics in Behavioral Neurosciences
ISBN 978-3-662-55878-2 ISBN 978-3-662-55880-5 (eBook)
https://doi.org/10.1007/978-3-662-55880-5

Library of Congress Control Number: 2018931899

Cover illustration: Artistic representation of oscillatory synchrony and timing of neurons in networks by Gyorgy Buzsaki

Printed on acid-free paper

This Springer imprint is published by the registered company Springer-Verlag GmbH, DE part of Springer Nature
The registered company address is: Heidelberger Platz 3, 14197 Berlin, Germany

Preface

To fathom Hell or soar angelic,

Just take a pinch of psychedelic.

—Humphry Osmond, in a letter written to Aldous Huxley, March 30, 1956

Scientific interest in psychedelic drugs has increased exponentially over the past decade. Just in the last few years, small controlled clinical trials examining the potential therapeutic utility of psilocybin-assisted therapy in patients suffering from anxiety, depression, and substance abuse have reported extremely positive outcomes (Grob et al. 2011; Bogenschutz et al. 2015; Carhart-Harris et al. 2016a; Griffiths et al. 2017; Johnson et al. 2017; Ross et al. 2016). Several human research studies with LSD have also been reported (Gasser et al. 2014, 2015; Carhart-Harris et al. 2016b; Liechti et al. 2017; Kraehenmann et al. 2017). Hence, we believe it is now an appropriate time for Springer to publish a volume about research with psychedelic drugs as part of their prestigious *Current Topics in Behavioral Neurosciences* (CTBN) series. Our objective for the volume has been to summarize the current state-of-the-art of psychedelics research. As editors, we faced the seemingly impossible task of distilling the entire breadth and depth of this rapidly expanding field into a single volume. It proved difficult to narrow the scope down to a reasonable level. We selected topics that span a wide range of areas, including pharmacological interactions occurring at the molecular level, changes in signaling pathways, effects on individual neurons and neural networks, and reaching all the way up to the level of the whole brain and the mind. We believe the 14 peer-reviewed chapters that were selected for inclusion provide up-to-date information about the pharmacology, neurobiological effects, subjective experience, and therapeutic effects of psychedelic drugs.

The editors of this volume represent three generations of scientists working in the field of psychedelics research. One of us (DEN) entered the field in 1969, at a time when research with LSD, psilocybin, and indeed all psychedelics had become very controversial. Strict legal controls were placed on research projects, most funding sources had dried up, and work with these substances acquired a stigma.

When Dave founded the Heffter Research Institute in 1993 to help promote legitimate scientific research with psychedelics, the possibility that psilocybin would ever again be widely studied in human clinical trials seemed remote, at best. Around that time, however, there were several promising developments. For example, in the early 1990s, Dr. Rick Strassman received permission from the FDA to conduct a research study with intravenous DMT, demonstrating that it was again possible to administer psychedelic drugs to humans. The results of those DMT studies were published in 1994 (Strassman and Qualls 1994; Strassman et al. 1994).

The second editor of this volume (FXV) was one of the first scientists to receive funding from the Heffter Institute. At the time, Franz was a very promising young psychiatrist who was motivated to conduct basic research with psychedelics in humans. With support from Heffter, Franz was able to create a Heffter Research Center in his hospital at the University of Zürich in 1999, enabling him to use state-of-the-art neurophysiological and imaging techniques to investigate the actions and effects of psychedelics. His groundbreaking work continues to demonstrate the high value that psychedelics possess as tools to understand brain function and consciousness (Geyer and Vollenweider 2008; Vollenweider and Kometer 2010).

The third editor (ALH) is a member of the new generation of psychedelics researchers who entered the field after the turn of the century, just as this topic was moving toward the scientific mainstream. Adam completed his postdoctoral training in the translational behavioral neuroscience group headed by Dr. Mark Geyer at UCSD, another founding member of Heffter, and is now running his own laboratory with independent research funding from NIDA.

We are greatly indebted to the corresponding authors and their coauthors for all of their hard work; this volume could not have been completed without their help and expertise. The staff at Springer deserves special recognition for their patience, as well as for guiding us through the publication process. We also owe a debt of gratitude to Mark Geyer, Bart Ellenbroek, Charles Marsden, and Thomas Barnes for their willingness to include this volume in the CTBN series.

Finally, we should provide a comment about our use of the term *psychedelic* in the title of this volume. Despite the recent rapid advancement of this field of research, the terminology used to classify LSD, mescaline, psilocybin, and related substances still remains controversial. Many names have been proposed for these substances over the years, including *delusinogenics, entheogens, hallucinogens, illusinogenics, misperceptionogens, mysticomimetics, oneirogens, phanerothymes, phantasticants, psychedelics, psychodysleptics, psycholytics, psychotaraxics, psychoticants, psychotomimetics, psychotoxins, and schizogens*. Unfortunately, most of these terms are overly specific for one aspect of the drug experience or are nonneutral terms reflecting their perceived utility. The term *psychotomimetic*, which was introduced by Gerard (1956), is a pejorative that emphasizes the ability of these substances to induce a psychosis-like state. However, the effects of these drugs approximate only some of the symptoms of schizophrenia. Similarly, *entheogen*, proposed by several prominent ethnobotanists in 1979 (Ruck et al. 1979), is a narrow term with religious connotations that focuses on the mystical or religious effects that can be produced by this drug class. *Hallucinogen* has perhaps been most widely used in the scientific literature and is the legal designation used to classify these substances in many countries. Unfortunately, it also is a misnomer because these compounds rarely evoke true hallucinations and do not normally impair reality testing. Furthermore, a wide range of psychoactive substances, including cannabinoids, dissociative anesthetics, anticholinergics, and entactogens such as MDMA, can produce "hallucinogenic" effects. Hence, LSD-like drugs are often referred to as *classical hallucinogens* or *serotonergic hallucinogens* in order to distinguish them from other drug classes.

The psychiatrist Humphrey Osmond first proposed the term *psychedelic* at a meeting of the New York Academy of Sciences in 1957 (Osmond 1957). Osmond coined this term (which means "mind manifesting") to highlight the ability of these substances to facilitate exploration of the mind by exposing latent mental states. The word psychedelic was constructed to avoid the negative connotation associated with words such as psychotomimetic, which had been adopted by many medical professionals. This classification did not take hold among many researchers, likely because it carried a stigma due to its association with the counterculture in the 1960s and 1970s. It effectively disappeared from the *scientific literature* for many decades, although it has probably remained the most widely used name in popular culture for more than half a century. These issues, however, have faded with time, and the recent resumption of research with these substances has been accompanied by a reintroduction of the word psychedelic into the scientific vocabulary. During the past two years, papers specifically about "psychedelics" have appeared in several respected scientific journals, for example, *Lancet Psychiatry, Neuropsychopharmacology, Pharmacological Reviews, and Journal of Psychopharmacology* (Carhart-Harris et al. 2016a; Griffiths et al. 2017; Ross et al. 2016; Carhart-Harris and Goodwin 2017; Nichols 2016). Recently, when many prominent investigators gathered together to discuss the therapeutic effects of psilocybin at a conference organized by the Usona Institute, there was much reflection about the need for a professional society devoted to research with these substances. The attendees agreed to form a society, and the unanimous decision was made to name it the International Society for Research on Psychedelics. Hence, there is now a consensus among researchers in the field that these substances should be referred to as psychedelic drugs, and this volume has been titled accordingly.

La Jolla, CA, USA — Adam L. Halberstadt
Chapel Hill, NC, USA — David E. Nichols
Zurich, Switzerland — Franz X. Vollenweider

References

Bogenschutz MP, Forcehimes AA, Pommy JA, Wilcox CE, Barbosa PC, Strassman RJ (2015) Psilocybin-assisted treatment for alcohol dependence: a proof-of-concept study. J Psychopharmacol 29:289–299

Carhart-Harris RL, Goodwin GM (2017) The therapeutic potential of psychedelic drugs: past, present, and future. Neuropsychopharmacology 42:2105–2113

Carhart-Harris RL, Bolstridge M, Rucker J, Day CM, Erritzoe D, Kaelen M, Bloomfield M, Rickard JA, Forbes B, Feilding A, Taylor D, Pilling S, Curran VH, Nutt DJ (2016a) Psilocybin with psychological support for treatment-resistant depression: an open-label feasibility study. Lancet Psychiatry 3:619–627

Carhart-Harris RL, Muthukumaraswamy S, Roseman L, Kaelen M, Droog W, Murphy K, Tagliazucchi E, Schenberg EE, Nest T, Orban C, Leech R, Williams LT, Williams TM, Bolstridge M, Sessa B, McGonigle J, Sereno MI, Nichols D, Hellyer PJ, Hobden P, Evans J, Singh KD, Wise RG, Curran HV, Feilding A, Nutt DJ (2016b) Neural correlates of the LSD experience revealed by multimodal neuroimaging. Proc Natl Acad Sci USA 113:4853–4858

Gasser P, Holstein D, Michel Y, Doblin R, Yazar-Klosinski B, Passie T, Brenneisen R (2014) Safety and efficacy of lysergic acid diethylamide-assisted psychotherapy for anxiety associated with life-threatening diseases. J Nerv Ment Dis 202:513–520

Gasser P, Kirchner K, Passie T (2015) LSD-assisted psychotherapy for anxiety associated with a life-threatening disease: a qualitative study of acute and sustained subjective effects. J Psychopharmacol 29:57–68

Gerard RW (1956) Neuropharmacology: transactions of the second conference. Josiah Macy, Jr. Foundation., New York

Geyer MA, Vollenweider FX (2008) Serotonin research: contributions to understanding psychoses. Trends Pharmacol Sci (TIPS) 29:445–453

Griffiths RR, Johnson MW, Richards WA, Richards BD, Jesse R, MacLean KA, Barrett FS, Cosimano MP, Klinedinst MA (2017) Psilocybin-occasioned mystical-type experience in combination with meditation and other spiritual practices produces enduring positive changes in psychological functioning and in trait measures of prosocial attitudes and behaviors. J Psychopharmacol (in press). https://doi.org/10.1177/0269881117731279

Grob CS, Danforth AL, Chopra GS, Hagerty M, McKay CR, Halberstadt AL, Greer GR (2011) Pilot study of psilocybin treatment for anxiety in patients with advanced-stage cancer. Arch Gen Psychiatry 68:71–78

Johnson MW, Garcia-Romeu A, Griffiths RR (2017) Long-term follow-up of psilocybin-facilitated smoking cessation. Am J Drug Alcohol Abuse 43:55–60

Kraehenmann R, Pokorny D, Vollenweider L, Preller KH, Pokorny T, Seifritz E, Vollenweider FX (2017) Dreamlike effects of LSD on waking imagery in humans depend on serotonin 2A receptor activation. Psychopharmacology (Berl) 234:2031–2046

Liechti ME, Dolder PC, Schmid Y (2017) Alterations of consciousness and mystical-type experiences after acute LSD in humans. Psychopharmacology (Berl) 234:1499–1510

Nichols DE (2016) Psychedelics. Pharmacol Rev 68:264–355

Osmond H (1957) A review of the clinical effects of psychotomimetic agents. Ann NY Acad Sci 66:418–434

Ross S, Bossis A, Guss J, Agin-Liebes G, Malone T, Cohen B, Mennenga SE, Belser A, Kalliontzi K, Babb J, Su Z, Corby P, Schmidt BL (2016) Rapid and sustained symptom reduction following psilocybin treatment for anxiety and depression in patients with life-threatening cancer: a randomized controlled trial. J Psychopharmacol 30:1165–1180

Ruck CAP, Bigwood J, Staples D, Ott J, Wasson RG (1979) Entheogens. J Psychedelic Drugs 11:145–146

Strassman RJ, Qualls CR (1994) Dose-response study of *N,N*-dimethyltryptamine in humans. I. Neuroendocrine, autonomic, and cardiovascular effects. Arch Gen Psychiatry 51:85–97

Strassman RJ, Qualls CR, Uhlenhuth EH, Kellner R (1994) Dose-response study of *N,N*-dimethyltryptamine in humans. II. Subjective effects and preliminary results of a new rating scale. Arch Gen Psychiatry 51:98–108

Vollenweider FX, Kometer M (2010) The neurobiology of psychedelic drugs: implications for the treatment of mood disorders. Nat Rev Neurosci 11:642–651

Contents

Chemistry and Structure–Activity Relationships of Psychedelics

David E. Nichols

Abstract This chapter will summarize structure–activity relationships (SAR) that are known for the classic serotonergic hallucinogens (aka psychedelics), focusing on the three chemical types: tryptamines, ergolines, and phenethylamines. In the brain, the serotonin 5-HT_{2A} receptor plays a key role in regulation of cortical function and cognition, and also appears to be the principal target for hallucinogenic/psychedelic drugs such as LSD. It is one of the most extensively studied of the 14 known types of serotonin receptors. Important structural features will be identified for activity and, where possible, those that the psychedelics have in common will be discussed. Because activation of the 5-HT_{2A} receptor is the principal mechanism of action for psychedelics, compounds with 5-HT_{2A} agonist activity generally are quickly discarded by the pharmaceutical industry. Thus, most of the research on psychedelics can be related to activation of 5-HT_{2A} receptors. Therefore, much of the discussion will include not only clinical or anecdotal studies, but also will consider data from animal models as well as a certain amount of molecular pharmacology where it is known.

Keywords Hallucinogen · Psychedelic · Structure–activity relationships · Serotonin 5-HT_{2A} receptor · Tryptamines · Phenethylamines · Ergolines · LSD

Contents

D.E. Nichols (✉)
Division of Chemical Biology and Medicinal Chemistry, University of North Carolina, Chapel Hill, NC 27514, USA
e-mail: drdave@purdue.edu

Curr Topics Behav Neurosci (2017) 36: 1–44
DOI 10.1007/7854_2017_475
Published Online: 26 March 2017

1 Introduction

Psychedelics (hallucinogens) have remained of high interest for many decades due to their ability to produce unique and dramatic alterations in consciousness. Before they had been pharmacologically classified as 5-HT$_{2A}$ receptor agonists or partial agonists, psychedelic drugs like mescaline, psilocybin, and LSD, were recognized for their powerful effects on the human psyche. They produce such profound effects on perception that it is natural to ask how they work in the brain. What are their biological targets? Where are these targets located in the brain? Are those brain areas recognized to play key roles in perception and cognition? Further, as recent clinical research studies have begun attempts to unravel the basis for human consciousness, it has become apparent that psychedelics offer unique and powerful tools to help to elucidate the basis of consciousness. The age-old questions of who we are and why we are here seem inevitably to arise when people talk about their experiences with psychedelic drugs. Yet, this fascinating class of mind-altering substances has not received significant research attention for more than 50 years, and it is only within the past decade or so that they have been the subject of renewed research interest. A comprehensive review on psychedelics has recently appeared (Nichols 2016).

As modern molecular pharmacology techniques have developed, our understanding has expanded of the roles played by the 5-HT$_{2A}$ receptor in normal brain function, so that studies of 5-HT$_{2A}$ receptor agonist structure–activity relationships (SAR) today take on greater significance, both from a theoretical and practical perspective.

There are three main chemical types of classic hallucinogens: the tryptamines, ergolines related to LSD, which can be considered to be rigidified tryptamines, and phenethylamines. These templates are illustrated in Fig. 1.

Serotonin receptor affinity and functional potency data are not available for many of the known psychedelics. Except for very limited human studies half a century ago, much of the research on hallucinogens involved animal behavioral studies or experiments with a variety of smooth muscle assays (e.g., rat fundus, rat

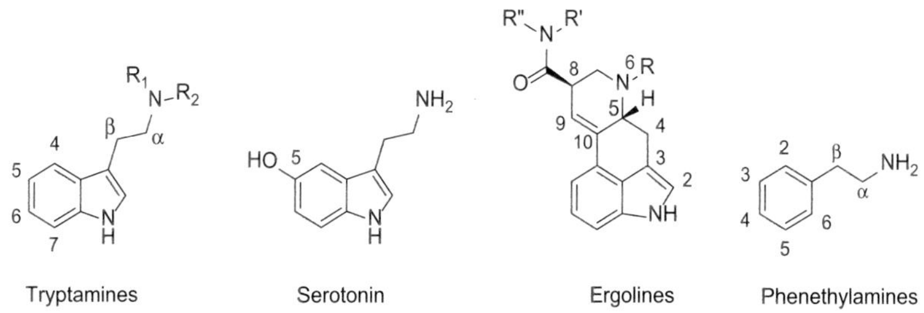

Fig. 1 Comparison of the structure of the neurotransmitter serotonin, with the three basic chemotypes of classic serotonergic hallucinogens

uterus, sheep umbilical artery strips). By contrast, the molecular pharmacology may be known for more recently developed compounds, but these usually lack formal clinical studies, so their human effects can often only be inferred. Fortunately, these substances have been shown to be serotonin 5-HT_{2A} agonists or partial agonists, so there is a sound basis for clinical inferences. Nonetheless, in many cases it is necessary to rely on animal behavioral or smooth muscle data in order to provide a more complete understanding of the SAR of psychedelics. Therefore, reports from early studies that are relevant to a consideration of SAR will largely focus on animal behavior, or in some cases, human hallucinogenic activity. Discussion of more recently developed molecules will include more of the molecular pharmacology, when and where it is known.

There are other types of molecules that are sometimes called hallucinogens, and in some cases they might more properly be called psychotomimetics, but this chapter will address only what are called classic hallucinogens; molecules that activate serotonin 5-HT_{2A} receptors. This chapter will not devote any discussion to 3,4-methylenedioxymethamphetamine (MDMA), salvinorin A (a kappa opioid receptor agonist), ketamine analogues (NMDA receptor antagonists), cannabinoids, or synthetic cannabimimetics. Certainly these latter molecules have become quite popular as recreational drugs, often marketed as "research chemicals," but they differ in their mechanism of action and complete monographs could be devoted to each of them.

2 Tryptamines

Tryptamines are the chemotypes that most closely resemble the natural neurotransmitter serotonin (5-hydroxytryptamine; 5-HT). Ergolines can essentially be considered to be rigidified tryptamines. Although LSD is the most well-known psychedelic, only a very few structural modifications can be made to its structure, and nearly all of those attenuate its activity by about an order of magnitude. In addition, there is a paucity of structure–activity data for ergolines, principally due to the synthetic difficulty inherent in their chemistry. Surprisingly few molecular modifications can be carried out on the tryptamines that allow retention of activity. A number of simple tryptamines, largely *N,N*-substituent variations, have been administered to humans (Shulgin and Shulgin 1997), but their receptor pharmacology remains largely unknown.

2.1 Ring Substituents

The 5-hydroxy group of serotonin (see Fig. 1) stands out as perhaps a key structural feature of this molecule. Serotonin also is a primary amine, and as we shall see psychedelic tryptamine derivatives are generally tertiary amines. High agonist activity at the 5-HT_{2A} receptor, as well as at other serotonin receptor subtypes is also seen in its *O*-methylated derivative, 5-methoxytryptamine. The affinities of 5-HT and 5-methoxytryptamine at the rat 5-HT_{2A} receptor are identical (Gupta et al. 1990; Johnson et al. 1990). Neither serotonin nor 5-methoxytryptamine has activity in vivo if administered orally, presumably as a result of a high first pass effect due side chain deamination by monoamine oxidase A in the liver.

For tryptamines, 5-HT_2 agonist (and psychedelic) activity is generally enhanced by substitution with an oxygen atom at the 4- or 5-position. As an example, *N,N*-dimethyltryptamine (DMT **1**) has a reported K_i of 75 nM in rat brain cortical homogenate (McKenna et al. 1990). Adding a 5-methoxy (**2**) increased the affinity to 14 nM, and the 4-OH compound (psilocin **3**) had a reported affinity of 6 nM.

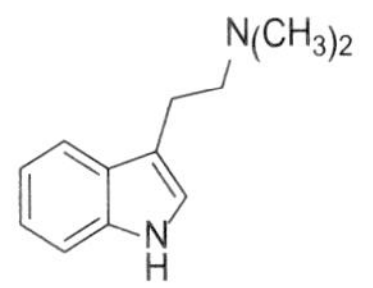

1

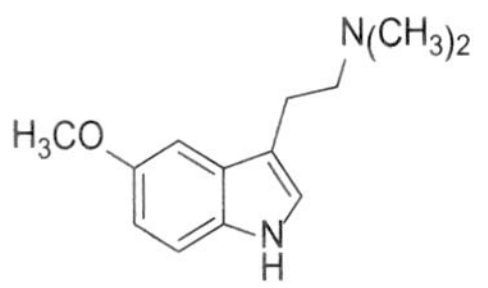

2

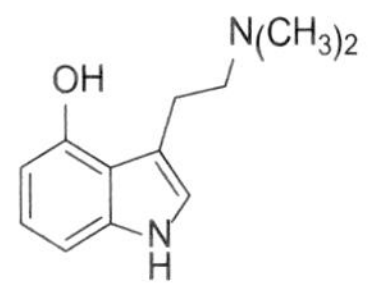

3

Five decades ago it was reported that 6-fluoro-*N,N*-diethyltryptamine (6-F-DET) lacked activity as a hallucinogen (Kalir and Szara 1963). It was recently found that it did not possess LSD- or DOI-like activity in a drug discrimination paradigm in rats (Blair et al. 2000). The affinity of 6-F-DET at the rat 5-HT_{2A} receptor was found to be essentially identical to DET, but its 40 μM EC_{50} in a phosphoinositide (PI) turnover assay was markedly reduced from that of DET (5.4 μM). Further, at a concentration of 100 μM 6-F-DET had an E_{max} of only 63%. The loss of functional efficacy and potency seems the most likely explanation for its absence of significant DET-like activity in man.

The effect of ring fluorination has been studied for four other tryptamines, with comparisons made between 6- and 7-F-psilocin and 4- and 6-fluoro-5-methoxy-DMT, **4**, **5**, **6**, and **7**, respectively, with their nonfluorinated counterparts (Blair et al.

2000). Fluorination of psilocin in the 6- or 7-positions gave compounds with essentially identical affinity at the rat 5-HT_{2A} receptor, and reduced by about one-half compared with psilocin itself. Adding a fluorine to 5-MeO-DMT at either the 4- or 6-position had no significant effect on E_{max}, but the EC_{50} values for the fluorinated compounds were increased to 7.9 and 18.1 μM for the 6-fluoro and 4-fluoro congeners, **6** and **7**, respectively, compared to 2.4 μM for 5-MeO-DMT. Fluorination had almost no effect on affinity at the rat 5-HT_{2C} receptor, but had marked effects on 5-HT_{1A} receptor affinity.

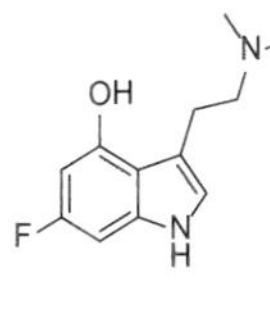

4

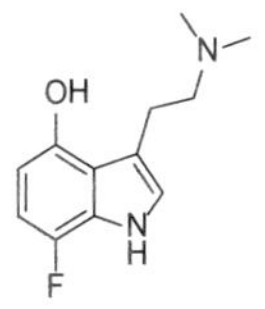

5

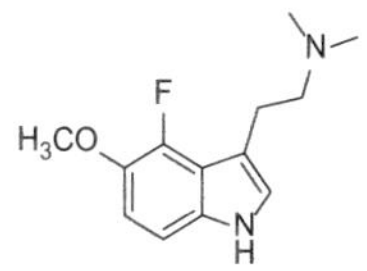

6

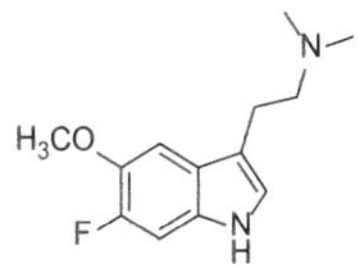

7

The 4-fluoro compound (**6**) had 0.23 nM affinity at the human 5-HT_{1A} receptor, nearly ten-fold greater than 5-MeO-DMT itself (1.7 nM). This substitution pattern was then exploited to create a 5-HT_{1A} ligand by replacing the *N*,*N*-dimethyl substituents with a pyrrolidyl moiety to afford molecule **8**, with 0.12 nM affinity at the human 5-HT_{1A} receptor and in vivo potency in the drug discrimination assay in rats comparable to the 5-HT_{1A} agonist 8-OH-DPAT (Laban et al. 2001).

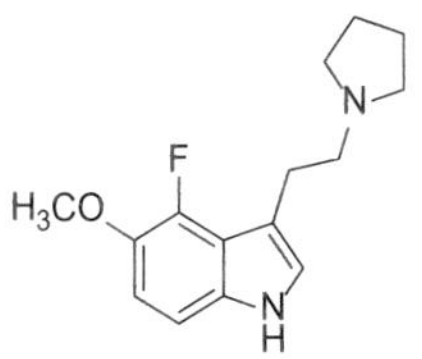

8

Much earlier work with benzo[*b*]thiophenes **9** and 3-indenalkylamines **10** had shown that when the compounds lacked ring substituents, their agonist activity in the rat fundus assay was about comparable to that of tryptamines (Winter et al. 1967). That is, the indole NH was not essential to activate the 5-HT_2 receptor in the rat fundus. The rat fundus receptor was subsequently classified as a 5-HT_{2B} receptor subtype (Baxter et al. 1994), and no recent studies have reported affinity or potency at the 5-HT_{2A} receptor.

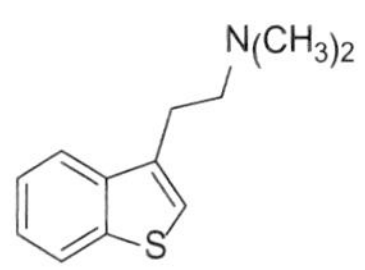

9

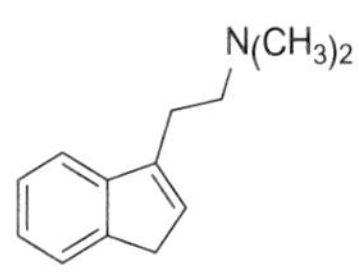

10

Replacing the phenyl ring of DMT with a bioisosteric thiophene was anticipated to lead to molecules that might possess DMT-like activity. The synthesis and biological activity of the thieno[3,2-*b*]-and thieno[2,3-*b*]pyrrole analogues of DMT (**11** and **12**, respectively) were reported by Blair et al. (1999). Both isosteres had lower affinity at the 5-HT$_{2A}$ receptor than DMT, with **12** having greatest affinity (106 nM vs. 65 nM for DMT). Both isomers had somewhat higher affinities than DMT at the 5-HT$_{1A}$ receptor and had higher affinities than DMT at the rat 5-HT$_{2C}$. DMT substituted in a drug discrimination study in rats trained to discriminate LSD from saline, but neither of the thienopyrrole isosteres substituted. Similarly, neither of the isosteres substituted in rats trained to discriminate DOI from saline.

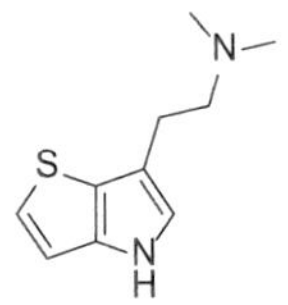

11

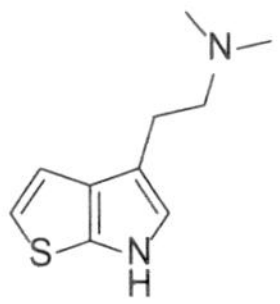

12

In rats trained to discriminate either LSD or DOI, isomer **11** gave the greatest degree of partial substitution, leading to speculation that a hydrogen bond donor in the receptor might be able to engage the sulfur atom in the thienyl ring when it was present in the edge of the molecule that normally carries the oxygen atom of serotonin. Both thiophene isosteres substituted in rats trained to discriminate the 5-HT$_{1A}$ agonist LY293284 from saline, with **11** being about twice the potency of **12**.

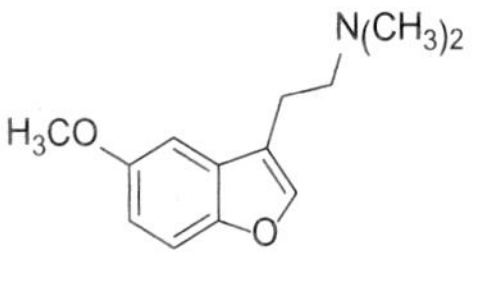

13

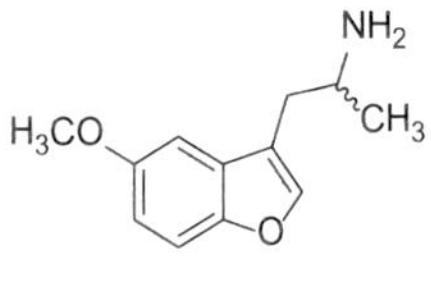

14

Replacing the indole nitrogen of the tryptamines with an oxygen atom affords a benzo[*b*]furan, another potential bioisostere of tryptamines. Compounds **13** and **14** both had about one-sixth the affinity of their indole congeners, using displacement of [^{125}I]DOI from rat frontal cortical homogenate (Tomaszewski et al. 1992). McKenna et al. (1990) reported a similar finding, assessing ability of *N*-methyl-*N*-isopropyltryptamine to displace [^{125}I]-*R*-DOI from rat cortical homogenate, compared with its benzo[*b*]furan isostere. The tryptamine IC$_{50}$ of 38 nM was about 13-fold lower than the benzofuran, which had an IC$_{50}$ of 500 nM.

A variation on ring-substitution patterns was the discovery of indazole ligands with potent 5-HT$_{2A}$ agonist activity (May et al. 2003a, 2006). For example, AL-38022A **15** was developed as a highly potent 5-HT$_{2A}$ agonist that had efficacy in reducing intraocular pressure in glaucoma. Compound **15** was a full agonist at all three 5-HT$_2$ family receptors, with EC$_{50}$ values between 0.5 and 2.2 nM for several functional responses (May et al. 2009). In a drug discrimination assay in rats trained to discriminate the hallucinogen DOM from saline, **15** produced full substitution, with an ED_{50} of 0.05 mg/kg. Similarly, it produced full substitution in monkeys trained to discriminate DOM from saline, with an ED_{50} of 0.04 mg/kg, comparable to the potent 5-$HT_{2A/2C}$ agonist DOI.

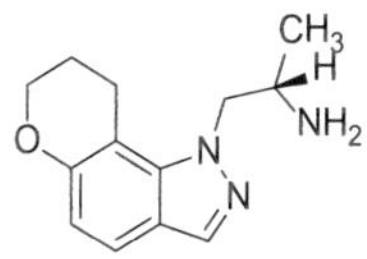

15, AL-38022A

2.2 *N*-Alkylation

Another area for structural modification is the side chain amino group, where *N*-alkylation provides a variety of secondary or tertiary amines. Extensive data have been published for hallucinogenic effects of a number of *N*-substituted tryptamines in humans (Shulgin and Shulgin 1997), but only scant data are available for their receptor affinities or potencies. One of the earliest modifications of the tryptamines to be studied for psychoactive effects was the *N,N*-diethyl analogue of psilocin (CZ-74, **16**). Both CZ-74 and its *O*-phosphoryl derivative CEY 19 (**17**) were studied in humans. Qualitatively, these compounds were very similar to psilocin and psilocybin, respectively, but had somewhat reduced durations of action (Leuner and Baer 1965).

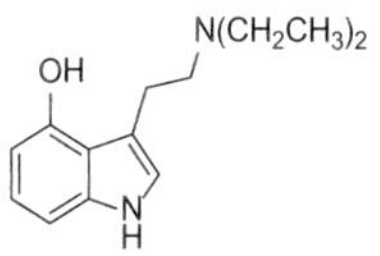

16

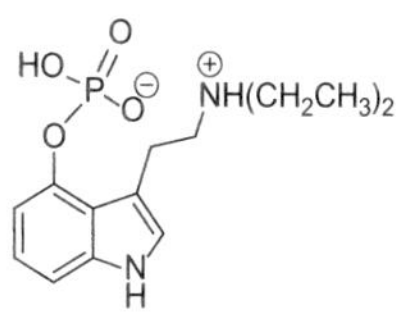

17

A systematic study of the effect of *N*-alkylation on tryptamine receptor affinities was reported by McKenna et al. (1990). *N*-alkylated tryptamines were examined with no ring substituents, a 5-methoxy, or 4-hydroxy group. Highest affinities (4–30 nM) for displacement of [^{125}I]DOI from rat cortical homogenate were observed with *N,N*-dimethyl, *N,N*-diethyl, *N*-methyl-*N*-isopropyl, and *N,N*-diisopropyl substituents. An affinity of 39 nM was reported for 4-OH-*N,N*-di(*sec*-butyl) tryptamine, but the affinity of 4-OH-*N,N*-diisobutyltryptamine was only 260 nM. Tethering the dialkyl groups into a heterocyclic ring gave mixed results; *N*-pyrrolidyl had an affinity similar to *N,N*-dimethyltryptamine (110 vs. 75 nM, respectively), but the affinity for the *N*-piperidyl was much lower, at 760 nM. The *N,N*-disubstituted compound 5-methoxy-*N,N*-diallyltryptamine (5-MeO-DALT **18**) has recently appeared as a new “legal high” on the illicit market (Corkery et al. 2012;

Strano Rossi et al. 2014). Results from broad-based receptor screening led Cozzi and Daley (2016) to conclude that multiple serotonin receptors, as well as several nonserotonergic sites are important for the psychoactive effects of **18** and other *N,N*-diallyltryptamines.

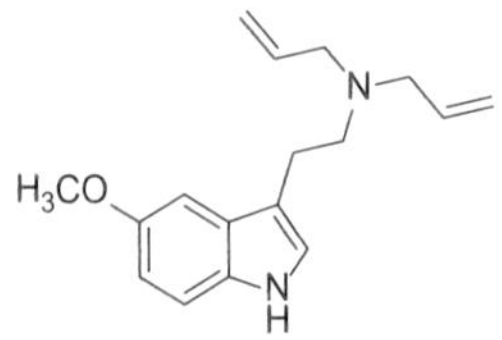

18

Although *N,N*-dimethyltryptamine and its 5-methoxy congener are not orally active, larger *N*-alkyl groups can confer oral activity on the molecules. It was demonstrated *N*-methyl-*N*-isopropyl- and *N,N*-diisopropyltryptamine, as well as the 5-methoxy analogue were both orally active in man, with durations of action of several hours (Shulgin and Carter 1980; Repke et al. 1985).

2.3 Side Chain Alkylation

Alpha-methylation of tryptamine side chains generally renders them orally active, presumably by blocking deamination by liver MAO. For example, 5-methoxytryptamine is inactive in man when given orally, but α-methyl-5-methoxytryptamine is a very potent orally active hallucinogen (Kantor et al. 1980). In mice, racemic α-methyltryptamine (AMT) increased motor activity by a mechanism that apparently involved both dopamine and serotonin (Rusterholz et al. 1979). In man, racemic α-methyltryptamine has been reported to be hallucinogenic (Murphree et al. 1961; Szara 1961; Shulgin and Shulgin 1997). Introduction of the alpha-methyl group also creates a chiral center in the molecule, and tryptamine enantiomers, not surprisingly, have differing biological activities. Affinity of (±)-α-methyltryptamine at the human 5-HT$_{2A}$, 5-HT$_{2B}$, and 5-HT$_{2C}$ receptors was reported to be 164, 58, and 30 nM, respectively (Vangveravong et al. 1998). The affinities of *R*- and *S*-α-methyltryptamine, *R*-**19** and *S*-**19**, were reported as 130 and 46 nM, respectively, for [^{125}I]DOI displacement in rat cortical homogenate (McKenna et al. 1990).

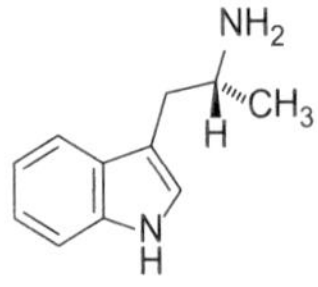

R-(-)-AMT

(*R*)-**19**

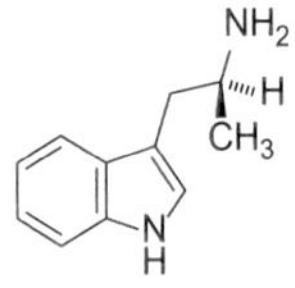

S-(+)-AMT

(*S*)-**19**

The enantiomer with the *S*-(+)-configuration has highest 5-HT_{2A} in vitro agonist activity, at least for molecules with a 5-OH or 5-OCH_3 substituent (Nichols et al. 1988). This in vitro observation is mirrored by human hallucinogenic activity, where 2.4 mg of (*S*)-(+)-5-methoxy-α-methyltryptamine is an effective hallucinogenic dosage in humans, whereas 3.0 mg of the *R* isomer produced no significant effect (Shulgin and Shulgin 1991). The *S*-(+)-enantiomer had 5-HT_{2A} affinity comparable to the non-alkylated 5-methoxytryptamine, whereas the *R*-(−) isomer was less potent. The affinities of (*R*)- and (*S*)-5-MeO-AMT at the agonist-labeled rat 5-HT_{2A} receptor were reported as 47 and 2 nM, respectively (Johnson et al. 1990). By contrast, the *R* enantiomer had higher affinity than the *S* isomer at the rat 5-HT_{1B} receptor (Nichols et al. 1988).

Extension of the α-methyl to α-ethyl afforded a compound named etryptamine (Monase), which was marketed until 1962 as an antidepressant. It appeared in Germany in 1986 as a "designer drug" that was associated with one death (Daldrup et al. 1986). It was found to have "neurotoxic" properties similar to MDMA in rats (Huang et al. 1991) and has been described as having MDMA-like psychopharmacology in humans (Krebs and Geyer 1993; Schechter 1998). Both isomers substituted with nearly equal potency in rats trained to discriminate MDMA from saline (Hong et al. 2001). In the same report, the (+)-enantiomer substituted in rats trained to discriminate the hallucinogenic phenethylamine DOM from saline, whereas the (−)-isomer substituted in rats trained to discriminate (+)-amphetamine. No data have been published on its affinity at 5-HT_2 family receptors.

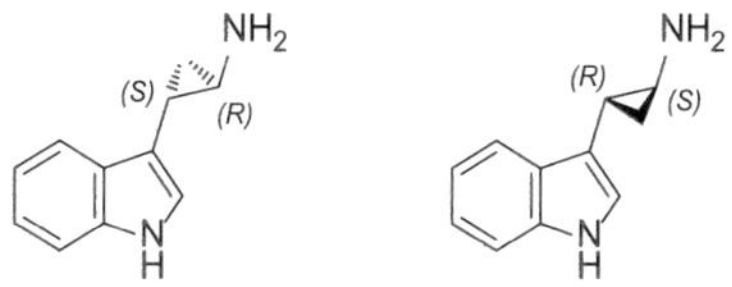

(1*R*,2*S*)-(-)-**20** (1*S*,2*R*)-(+)-**20**

Another variation on side alkylation was provided in a study of *trans*-2-(indol-3-yl)-cyclopropylamines **20** (Vangveravong et al. 1998). Although the (1*R*,2*S*)-(−)- enantiomer of **20** had highest affinity at human 5-HT_{2A} and 5-HT_{2B} sites, the (1*S*,2*R*)-(+)- isomer unexpectedly had higher affinity at the 5-HT_{2C} human receptor. Ring substituents 4-OMe, 5-OMe, and 5-F generally increased affinity over unsubstituted **20**. The difficulty of synthesis and chemical instability of these indolecyclopropylamine compounds precluded preparation of the enantiomeric ring-substituted compounds.

3 Ergolines

The tetracyclic ergoline molecules are ultimately derived from ergot alkaloids, products of the ergot fungus, of the genus *Claviceps*. From the perspective of psychedelic 5-HT_{2A} agonists, the most important one is lysergic acid diethylamide

(LSD **21**), also known as LSD-25. Although LSD is the most potent psychedelic agent in humans, its affinity and potency at the human 5-HT_{2A} receptor is rather unremarkable compared with much simpler molecules such as DOI. Numerous clinical studies of LSD and several of its amide-modified congeners were carried out in the 1950s and 1960s have been reviewed in detail earlier (Brimblecombe and Pinder 1975; Siva Sankar 1975; Shulgin 1982; Nichols 1986). Little new information has been published in the years since, with a few exceptions to be discussed below.

It is only ergolines with the 5*R*,8*R* stereochemistry, as illustrated earlier in Fig. 1 that have biological activity. That isomer is dextrorotatory, so LSD is referred to as (+)-LSD or *d*-LSD. Receptor binding studies by Bennett and Snyder in 1976 first demonstrated that LSD had nanomolar affinity for [^{3}H]LSD-labeled binding sites in rat cortex (Bennett and Snyder 1976). By contrast, its 5*S*,8*S* enantiomer, (−)-LSD, had 2500-fold lower affinity. The 8-position epimerizes readily, particularly at acidic pH, to provide the 5*R*,8*S* epimer (+)-isolysergic acid diethylamide **22**, which has about 30-fold lower receptor affinity and is inactive as a psychedelic.

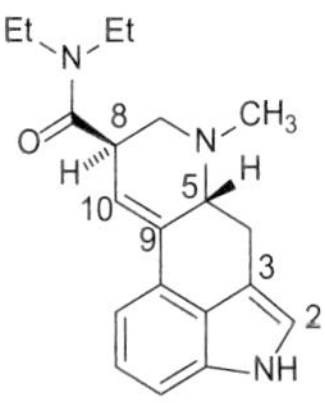

(5*R*,8*R*)-(+)-LSD **21**

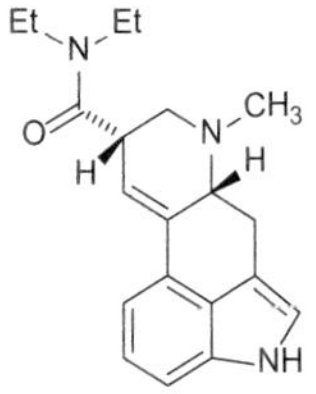

(5*R*,8*S*)-(+)-isoLSD **22**

Because of its structural complexity and tedious approaches to its total synthesis, only a few structural modifications of LSD have been reported. Those principally involved changes to the amide function, reduction of the 2,3- or 9,10-double bonds, a few substitutions on the indole nitrogen, oxidation or halogenation at the 2-position, and replacing the methyl group on the basic nitrogen atom with a small series of other alkyl groups. Unfortunately, only a few of them have been assessed in human psychopharmacology, most being much less active than LSD itself. Although some have been partially characterized for affinity at a few receptors, none of them have been the focus of comprehensive studies using modern molecular pharmacology methods.

If a halogen is introduced at the 2-position of LSD, for example in 2-bromo-LSD (BOL-148) or 2-iodo-LSD, the resulting molecules lack hallucinogenic activity and are antagonists at the 5-HT_{2A} receptor. No work with BOL-148 has been reported since the early 1970s, but it was shown that it could block the effects of LSD in humans.(Ginzel and Mayer-Gross 1956) The radiolabeled 2-iodo congener, [^{125}I] 2-iodo-LSD, has been employed as a radioligand for 5-HT_2 family receptors (Hartig et al. 1983; Nakada et al. 1984; McKenna et al. 1989; Watts et al. 1994). More recently, BOL has shown efficacy in aborting and/or preventing cluster headaches (Karst et al. 2010).

The 9,10-double bond of LSD is apparently crucial for its psychedelic action, and reducing it abolishes hallucinogenic activity (Stoll and Hofmann 1955; Hofmann 1968). Reduced 9,10-dihydro-LSD is still relatively planar, like LSD, so the reason(s) for the loss of activity is unclear (Nakahara et al. 1977). Although 9,10-dihydro-LSD lacks psychedelic effects in humans, there has so far not been a comparison of its receptor activities with those of LSD that might explain its inactivity.

Reducing the 2,3-bond of the indole nucleus results in a compound with about one-eighth the activity of LSD (Gorodetzky and Isbell 1964). It was reported to have a delayed onset of action relative to LSD, and it was speculated that "a metabolic change to a more active substance" might be the explanation. It might be noted that 2,3-dihydroindoles can be fairly readily oxidized to indoles, so such an oxidative transformation might take place in the body, perhaps by action of a mixed function oxidase in the liver.

Replacing the *N*(6)-methyl group of LSD with longer alkyl groups results in compounds that in some cases are more potent than LSD in vivo in rodent behavior and which in some cases have potency comparable to, or slightly greater than LSD in humans (Hoffman and Nichols 1985; Shulgin and Shulgin 1997). Assessment of receptor affinities for some of these analogues has failed to identify any correlation between hallucinogenic potency and nature of the *N*(6) alkyl group.

3.1 Amide Modifications of Lysergic Acid Derivatives

The simplest ergoline with human psychoactive properties is lysergic acid amide (**23**, ergine), reported by Hofmann and Tscherter to be the active component in *Rivea corymbosa* seeds used by the Aztecs in various magical potions and ointments (Hofmann 1971). If the C(8) amide substituent is removed completely to provide the 8-descarboxy **24**, the compound is reported to produce a mouse behavioral profile "remarkably similar to that shown by LSD" (Bach et al. 1974). Unfortunately, no other assays were carried out, nor were human studies carried out that would elucidate whether the presence of an amide substituent is an absolute requirement for activity. That is an important question because even slight modifications to the diethylamide moiety of LSD result in dramatic losses of in vivo activity.

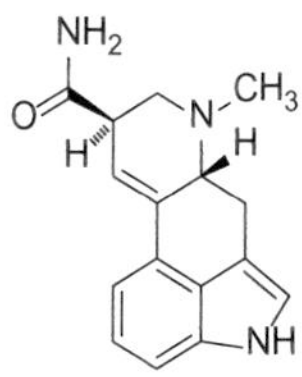

23

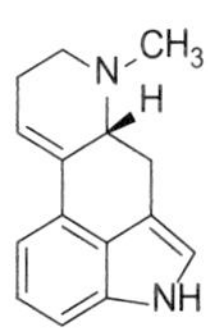

24

With respect to lysergic acid amides, it should be pointed out that the high in vivo potency of LSD seems to depend on the presence of the *N,N*-diethylamide moiety. It has been known for about five decades that any change to the amide moiety, however slight, leads to about an order of magnitude loss in potency. This decreased activity cannot be related simply to hydrophobicity, because compounds such as the *N*-methyl-*N*-propyl, or *N*-methyl-*N*-isopropyl, which are isomers of LSD, are much less potent than LSD itself. It also seems doubtful that it could be related to metabolic stability of the diethyl moiety. Rather, recent work, to be described later, suggests that the 5-HT_{2A} receptor might have a stereochemically defined and sterically constrained region that specifically accommodates the diethylamide moiety.

Evidence that the amide binding region in the receptor might be well defined was provided with the discovery that lysergamides of (*R*)- and (*S*)-2-aminobutane differed in their pharmacological properties (Oberlender et al. 1992). The *R*-configuration in the alkyl of amide **25** was nearly equipotent to LSD in drug discrimination in rats trained to discriminate LSD from saline. By contrast, the lysergamide with the *S*-alkylamide had only one-fourth the potency of LSD in the same assay. Using displacement of [^{125}I]DOI in rat frontal cortical homogenate, the lysergamides with the *R*- and *S*-2-aminobutane amide had affinities of 2.6 and 7.8 nM, respectively, which correlated with their in vivo potencies.

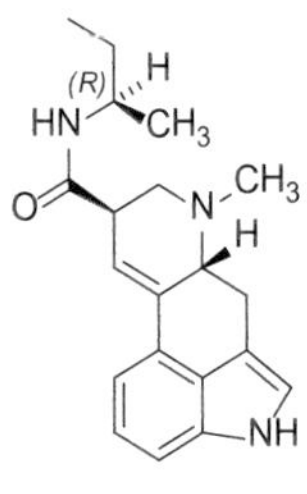

***R*-25**

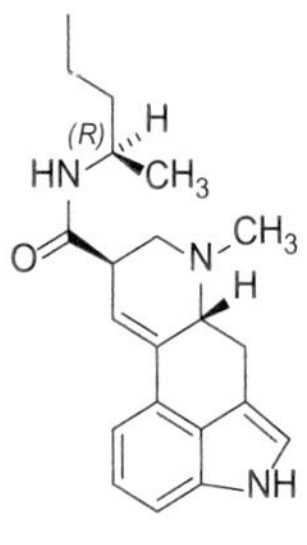

***R*-26**

This approach was extended to study of a series of chiral 2-aminoalkane amides of lysergic acid, with the alkyl group extended from butyl to heptyl (Monte et al. 1995). Using [^{3}H]ketanserin displacement from rat frontal cortex homogenate to measure 5-HT_{2A} receptor affinity, the lysergamide with the *R*-configuration in the secondary alkyl amide group had higher affinity in every case than the one with the *S* configuration. As the chain length increased affinity decreased, with the *R*-2-heptylamide having a K_i of only 80 nM. The pentyl isomers of **26** were the only compounds tested in functional assays, where each isomer proved to be a full agonist in the PI hydrolysis assay, but the *S*-isomer was less potent (see Table 1). Surprisingly, however, extending the length of the 2-alkyl group of the amide *increased* 5-HT_{1A} receptor affinity, with the *R*-2-hexyl substituted amide having a K_i of 0.32 nM! Clearly, the 5-HT_{1A} receptor has greater tolerance for bulk attached to the amide.

Table 1 5-HT_{2A} 5-HT_{2C}, and 5-HT_{1A} receptor affinity and functional effects for selected lysergamides

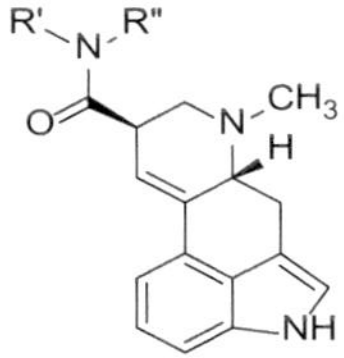

R′	R″	5-HT_{2A}^{a} K_i (nM) (DOI)[b]	5-HT_{2A} K_i (nM) (Ket)[c]	pEC_{50} (G_q)	pEC_{50} (arrestin)	5-HT_{2C}^{a} K_i (nM) (Mes)[d]	5-HT_{2C} K_i (nM) (DOI)[b]	5-HT_{1A} K_i (nM)
Ethyl	Ethyl (LSD)	2.1 ± 0.03	13	6.93	6.69	30	7.8	1.1 ± 0.3
H	*R*-2-Butyl (**25**)	2.6 ± 0.4	18	6.67	7.07	19	15	2.8
H	*S*-2-Butyl	7.8 ± 0.2	21	6.69	6.67	36	8.6	5.9
H	*R*-2-Pentyl (**26**)	4.5 ± 0.5	10	6.28	5.63	18	5.5	1.4
H	*S*-2-Pentyl	34 ± 2	29	6.26	5.44	52	25	7.4
H	*R*-2-Hexyl	16 ± 2	16	5.58	6.03	16	2	1.1
H	*S*-2-Hexyl	55 ± 7	21	5.75	5.29	35	9.7	5.1
H	*R*-2-Heptyl	80 ± 9	13	5.34	7.13	23	5.1	1.4
H	S-2-Heptyl	360 ± 20	35	5.41	5.83	47	17	4.2
H	3-Pentyl	8 ± 0.2	17	6.48	6.02	36	9.6	5.8
H	Isopropyl	1.4	15	7.19	6.94	31	6.4	5.1
H	*tert*-Butyl	33	468	<5	5.65	60	23	163
H	*R*-α-Methylbenzyl	2.3	19	5.86	7.48	16	12	2.8
H	*S*-α-Methylbenzyl	5.5	55	5.58	5.19	82	25	14
Methyl	*R*-2-Butyl	3.2	4.9	6.24	7.02	37	6	8.6
Methyl	*S*-2-Butyl	4.7	6.5	6.06	6.06	81	34	15

(continued)

Table 1 (continued)

R′	R″	5-HT$_{2A}^{a}$ K_i (nM) (DOI)[b]	5-HT$_{2A}$ K_i (nM) (Ket)[c]	pEC$_{50}$ (G$_q$)	pEC$_{50}$ (arrestin)	5-HT$_{2C}^{a}$ K_i (nM) (Mes)[d]	5-HT$_{2C}$ K_i (nM) (DOI)[b]	5-HT$_{1A}$ K_i (nM)
Ethyl	*R*-2-Butyl	2.8	3.3	6.99	6.49	25	7.6	6.9
Ethyl	*S*-2-Butyl	5.2	4.3	6.83	6.56	42	7.9	12
Methyl	Isopropyl	3.2 ± 0.1	6.6	6.71	6.84	53	15	8.5
Ethyl	Isopropyl	10	6.9	6.14	6.71	25	6.2	10
Isopropyl	Isopropyl	9.1	9	5.73	5.94	47	12	35
Allyl	Allyl	8.9	2.85	6.10	6.12	27	11	17
Ethyl	*n*-Propyl	7	4.2	6.20	6.00	48	7.6	11
Ethyl	2,2,2-Trifluoroethyl	1.6 ± 0.03	4.8	5.79	6.46	1.8 ± 0.2	10	21
Ethyl	2-Methoxyethyl	7.1 ± 0.4	9	6.25	5.94	7.8 ± 0.5	7.6	20
cis-2,3-Dimethylazetidide (**27**)		7.9 ± 0.85	10	6.27	6.99	23 ± 2.9	4.4	7.5
R,R-trans-2,3-Dimethylazetidide (**28**)		21 ± 4	10	6.42	6.21	130 ± 11	58	13
S,S-trans-2,3-Dimethylazetidide (**29**)		8.3 ± 1.7	6.2	6.60	7.06	6.5 ± 0.15	2	4.6
cis-2,5-Dimethylpyrrolidide		27 ± 1	6.4	6.11	6.59	11.2 ± 0.5	15	9.4
cis-2,6-Dimethylpiperidide		7.9	5.3	5.81	6.75	31	3.5	5.9
Pyrrolidide		12.2 ± 0.2	57	7.20	6.83	6.1 ± 0.5	29	6.6
Piperidide		2.6 ± 0.1	21	6.66	7.25	2.3 ± 0.1	14	4
Morpholide		16.2 ± 1.8	62	7.14	6.23	51 ± 2.0	16	8.8

PDSP screening data at human receptors unless otherwise specified; [a]Values with SEM are from Parrish (Parrish 2006) [b][^{125}I]DOI; [c][^{3}H]ketanserin; [d][^{3}H] mesulergine

Tests in rats trained to discriminate LSD from saline showed that full substitution occurred with the *R*-2-pentyl lysergamide **26**, but not with the *S*-pentyl, hexyl, or heptyl compounds. In vitro affinities observed at the rat 5-HT_{2A} receptor parallel these in vivo results.

To test the hypothesis that the receptor might have a region that was optimally complementary to the *N,N*-diethylamide, the synthesis and testing of conformationally constrained 2,3-dimethylazetidine amides of lysergic acid was carried out (Nichols et al. 2002). These dimethylazetidines exist in three isomeric forms: the 2,3-*cis* meso isomer **27**, the *R,R-trans* **28,** and the *S,S-trans* **29** isomers. The amide of each of these was prepared from lysergic acid and tested. In the drug discrimination assay in rats trained to discriminate the effects of LSD, *S,S-trans* azetidide **24** had potency most similar to LSD. As shown in Table 1, the *S,S* congener **29** had an affinity and potency profile most comparable to LSD. *R,R* isomer **28** had two–threefold lower affinity at the 5-HT_{2A} receptor and 50–60-fold lower affinity at the 5-HT_{2C} receptor. Cis compound **27** differed from the *S,S*-isomer in that it had about fourfold lower affinity at the 5-HT_{2C} receptor. Although the *S,S*-isomer had about one-half the potency of LSD in activating phosphoinositide hydrolysis through the 5-HT_{2A} receptor, the *R,R* isomer and cis compound were 8–12-fold less potent.

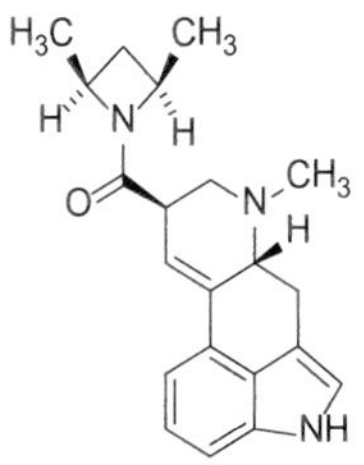

27

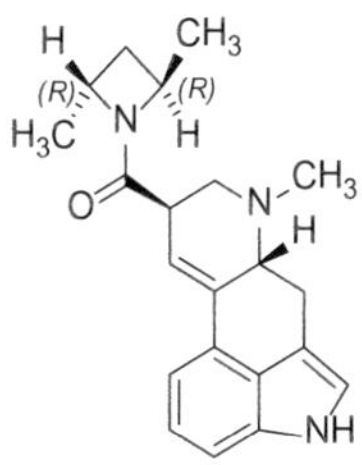

28

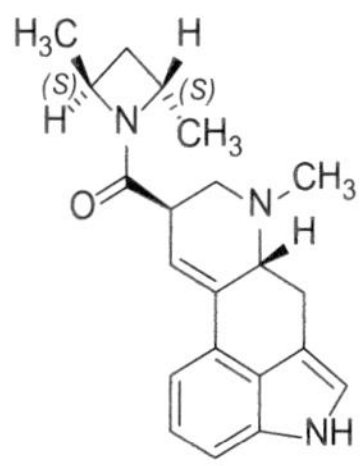

29

Virtual docking of LSD, **28**, and **29** into an in silico agonist-activated model of the 5-HT_{2A} receptor revealed that the diethyl groups of LSD nestle into a region that is bounded by a number of residues near the extracellular face of the receptor (Juncosa 2011). Further, extracellular loop 2 (EL2) was observed to interact with the diethylamide moiety. In particular, Leu-229 in EL2 was found to be critical for this interaction (McCorvy 2012). The conformation of EL2 was very similar after docking either LSD or *S,S*-isomer **29**, whereas EL2 was significantly displaced (ca. 4 Å at Leu-229) by docking of *R,R*-**28**. After docking of LSD, followed by molecular dynamics and minimization, the conformations adopted by the ethyl groups were observed to mirror the configurations in *S,S*-**29**. Curiously, the receptor appears to have evolved to be complementary to the diethyl moiety of LSD in a specific conformation.

3.2 *N(6)-Alkyl Modifications of LSD*

One other structural modification that has led to potent psychedelics is replacement of the *N*(6)-methyl of LSD with a variety of other alkyl groups (Hoffman and Nichols 1985). In a rat drug discrimination assay, in animals trained to discriminate LSD from saline, the *N*(6)-allyl derivative had about twice the potency of LSD itself. The *N*(6)-ethyl was about 1.6-fold more potent than LSD, with the *N*(6)-*n*-propyl being essentially comparable in potency to LSD. The *N*(6)-isopropyl had about 40% of the potency of LSD, with the *N*(6)-*n*-butyl having approximately 10% of the potency of LSD. Neither norLSD (*N*(6)=H), or *N*(6)-2-phenethyl-norLSD gave full substitution in the rats. Anecdotal human experiments then confirmed that the *N*(6)-allyl (AL-LAD) and *N*(6)-ethyl (ETH-LAD) congeners were psychoactive in man at doses that were not all that different from LSD itself, but the two compounds had psychopharmacology that was different from that of LSD (Shulgin and Shulgin 1997). The same source reported that the *N*(6)-n-propyl was much less active, with an oral dose in the range of 100–200 μg. The *N*(6)-propynyl (pargy-LAD) had some activity at 160 μg, and the *N*(6)-*n*-butyl was reported to do "something" at 500 μg. The *N*(6)-2-phenethyl congener was inactive up to 500 μg. These human reports, although anecdotal, do generally parallel the results obtained in the drug discrimination tests.

4 Ergolines as "Research Chemicals"

Interestingly, several LSD analogues have recently appeared on the "research chemical" market. Compound **29** has been distributed as "LSZ," and the (*N*) 1-propionyl derivative of LSD, "1P-LSD" also has been reported. 1P-LSD had never been described in the chemical literature and was an unknown compound prior to its appearance as a new psychoactive substance (NPS). It was hypothesized to be a prodrug of LSD, and when incubated with human serum at 37 °C LSD was detected by LC–MS analysis after a variety of exposure times (Brandt et al. 2016). *N*(6)-ethyl-norLSD (ETH-LAD) also has appeared on the research chemical market, as has *N*(6)-allyl-norLSD (AL-LAD) (Brandt et al. 2017).

5 Phenethylamines and Related Congeners

The phenethylamines are the most extensively explored class of psychedelics largely due to the relatively facile synthesis of phenethylamines. To complement this discussion, the reader is encouraged to read an earlier review on this topic (Nichols 1981), and also a recent review on phenethylamine 5-HT_{2A} agonists (Blaazer et al. 2008).

Mescaline **30**, is the prototype for this class. It is a simple 3,4,5-trimethoxyphenethylamine first isolated from the peyote cactus, *Lophophora williamsii*, at the end of the nineteenth century by chemist/pharmacologist Dr. Arthur Heffter (Heffter 1898). It is an orally active hallucinogen in man, but has very low potency, a typical dose of the sulfate salt being in the range 250–400 mg. The earliest modification to the structure of mescaline was the introduction of an α-methyl into the side chain, giving compound **31**, known as TMA (Hey 1947; Peretz et al. 1955) This compound was the first example of a very large class generically referred to as "substituted amphetamine" hallucinogens. From 1964 to 1969, Dr. Alexander Shulgin carried out an early series of experiments, moving the methoxy ring substituents to different positions. These experiments established that the most potent hallucinogenic amphetamines had the 2,4,5-ring-substitution pattern (Shulgin et al. 1969). Moving the 3-methoxy of TMA to the 2-position afforded TMA-2 **32**.

Additional studies were summarized by Shulgin in 1978 (Shulgin 1978), with a much more comprehensive treatise published on this subject in 1991 (Shulgin and Shulgin 1991). Although no receptor or animal data were reported by the Shulgins in this latter compendium, it does list human dosages and qualitative psychopharmacological effects for a large number of substituted phenethylamines. Studies of many of these compounds in other laboratories have shown that active compounds in man generally have high affinity and are agonists or partial agonists at the 5-HT_{2A} receptor. Much of those data will be cited in the following discussion.

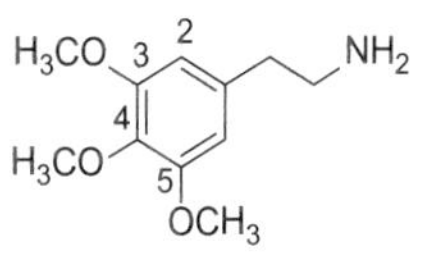

Mescaline

30

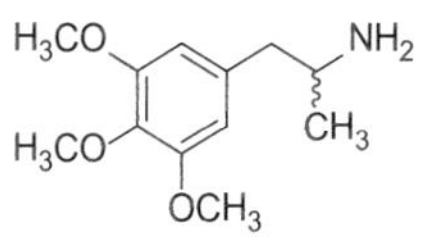

TMA

31

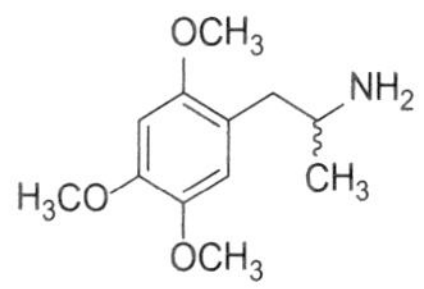

TMA-2

32

The introduction of the α-methyl into the phenethylamine side chain creates a chiral center, and thus the substituted amphetamine type psychedelics have two optical isomers, or enantiomers. An asymmetric synthesis was developed that allowed the facile preparation of the enantiomers of a large number of ring-substituted amphetamines (Nichols et al. 1973). Aldous et al. (1974) later reported a method for chemical resolution of the enantiomers by recrystallization of *N*-benzyloxycarbonyl-L-phenalanine-*p*-nitrophenyl esters. These developments preceded the era of modern molecular biology, and affinity and potency of enantiomers at receptors could not be reported at that time. Some of the assays used then were highly correlated with in vivo hallucinogenic activity in humans, and today we know the effects in those assays are mediated by activation of serotonin 5-HT_{2A} receptors. Thus, one can probably infer that much of the early structure–activity data for hallucinogenic agents reflects agonist activity at that receptor.

R-(−) enantiomers of substituted hallucinogenic amphetamines are most potent in humans, and also are more potent than their *S*-(+) antipodes in activating the human 5-HT_{2A} receptor. This stereochemistry is reversed from that of unsubstituted amphetamine, where the (*S*)-(+)-enantiomer is the more potent psychostimulant. In dog peripheral vasculature, however, the *S*-(+)-isomers of hallucinogenic amphetamines are more potent in producing smooth muscle contraction (Cheng et al. 1974).

5.1 Beta-Oxygenated Phenethylamines

The effect of beta-oxygenation on the 5-HT_{2A} agonist properties of DOB has been studied by Glennon et al. (2004). All four possible stereoisomers of beta-oxygenated amphetamines were studied. As shown below, 1*R*,2*R* stereoisomer **36** had the highest affinity at the 5-HT_{2A} receptor.

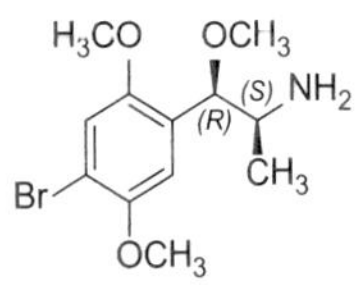
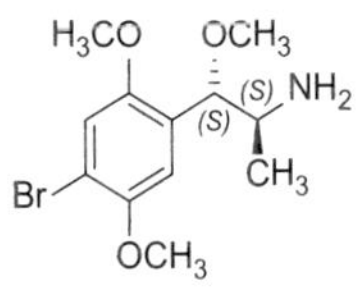
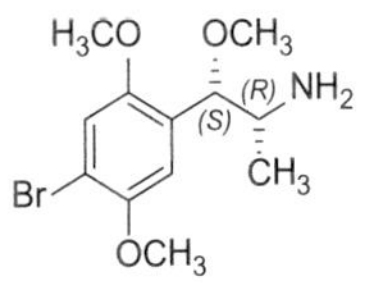
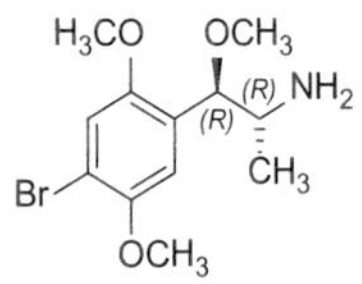

(1*S*,2*R*)	(1*S*,2*S*)	(1*R*,2*S*)	(1*R*,2*R*)
Ki 17.4 nM	Ki 6 nM	Ki 0.8 nM	Ki 0.3 nM
33	**34**	**35**	**36**

Compounds with the *R* stereochemistry at the alpha-carbon have highest affinity in the beta-unsubstituted amphetamines, so it is perhaps not surprising that the highest affinity compounds have the *R* stereochemistry at that position. In a cell-based Ca^{2+} mobilization assay, the 1*R*,2*R* stereoisomer **36** had an E_{max} of 93%, whereas the other isomers were partial agonists, with efficacies varying from 31 to 54%. The analogous beta-hydroxy compounds were both less potent and less efficacious, although the 1*R*,2*R* beta-hydroxy analogue fully substituted in a drug discrimination task in rats trained to discriminate DOM from saline. These data are consistent with an earlier report of analogous beta-oxygenated compounds producing hallucinogen-like effects in man (Lemaire et al. 1985).

5.2 Ring Substituents

After Shulgin had established that the 2,4,5-substitution pattern was optimal for hallucinogenic activity, extensive work followed to establish the range of substituent types that could be tolerated on the ring. It might be noted, however, that 2,4,5-trimethoxyphenethylamine, an isomer of mescaline, lacks mescaline-like effects in man (Shulgin 1978). Although that particular substitution requires an α-methyl in the side chain to be active, we shall see that replacing the 4-methoxy with other groups does afford active compounds, including many that lack the α-methyl group. As a general rule, 2,5-dimethoxy substituents provide optimal hallucinogenic activity, as well as receptor affinity and efficacy. An early drug discrimination study in rats suggests that the 2-methoxy, but not the 5-methoxy, may be replaced by an OH group (Glennon et al. 1982b).

A relatively hydrophobic substituent at the 4-position in 2,4,5- or 3,4,5-substituted molecules affords the most potent compounds. The earliest example of this effect was seen in the potency of the 4-methyl compound, DOM (**37**, STP), which was about ten times more potent than the methoxy congener TMA-2 **32**. The 4-bromo and 4-iodo compounds, **38** and **39**, respectively, had even higher potency.

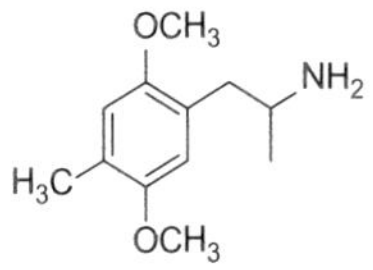

DOM

37

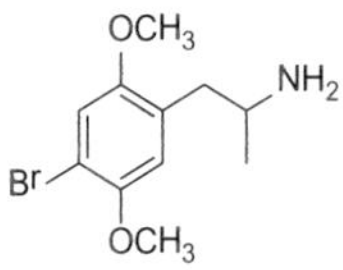

DOB

38

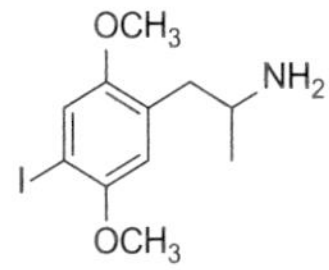

DOI

39

The importance of this substitution pattern is dramatically illustrated by a comparison of the three 2,4,5-substituted isomeric dimethoxy-monoethoxy amphetamines. The 2,5-dimethoxy-4-ethoxy compound (MEM **40**) has good clinical activity, whereas the 2-ethoxy-4,5-dimethoxy **41** 2,4-dimethoxy-5-ethoxy **42** and congeners did not (Shulgin 1968; Nichols et al. 1984b).

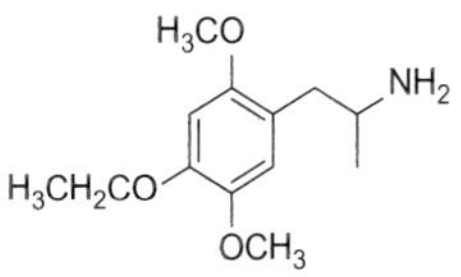

MEM **40**

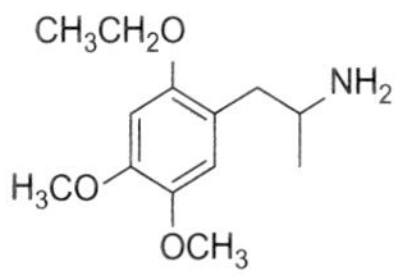

EMM **41**

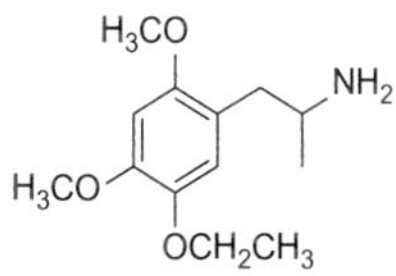

MME **42**

As a similar example, placing a 4-alkylthio substituent at the 4-position gives very active compounds, whereas a methylthio at either the 2- or 5-positions markedly attenuates activity (Jacob et al. 1977). In a rabbit hyperthermia assay, 2,5-dimethoxy-4-methylthio amphetamine **43** had about one-half the potency of DOM (**37**) (Nichols and Shulgin 1976). By contrast, replacing the 2- or 5-methoxy groups with a methylthio afforded much less active compounds (Jacob and Shulgin 1983). Nevertheless, the 5-methylthio analogue had nearly twice the potency of the 2-methylthio compound, indicating some greater receptor tolerance at the 5-position for an atom other than oxygen. Nonetheless, anything other than an oxygen atom at the 2- and 5-positions is very deleterious to activity. The 2- and 5-methoxy groups of DOM (**37**) and its 4-ethyl congener DOET also were individually replaced with methylthio groups. Again, the resulting 2- or 5-thio analogues suffered a dramatic loss of potency (Jacob and Shulgin 1983).

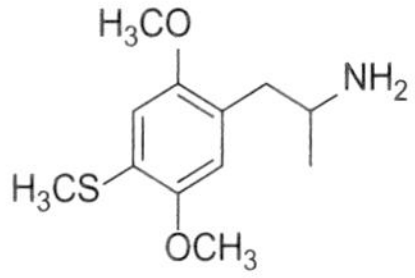

43

Possible explanations for these findings could be that the receptor has hydrogen bond donor residues (e.g., serine or threonine) that interact with the 2- and 5-oxygen atoms (Braden 2007; Braden and Nichols 2007). Sulfur would not provide a good hydrogen bonding partner compared with a methoxy. Another possible explanation could be that the receptor has little tolerance for a group larger than a methoxy in the 2- and 5-positions; the van der Waals radius for oxygen is 1.52 Å, whereas the radius for sulfur is 1.8 Å. Or, both factors could be in play, with the receptor also having less tolerance for steric bulk in the region that interacts with the substituent at the 2-position.

When a parallel approach was applied to mescaline, different results were obtained. In human self-experiments the 3-methylthio compound (**44**) was rated to be about sixfold more potent than mescaline **37** (Jacob and Shulgin 1981). The 4-methylthio compound (**45**) was estimated to be about 12 times more potent than mescaline. Thus, in 3,4,5-trisubstituted compounds, activity increased when either the 3- or the 4-methoxy was replaced with methylthio, suggesting a less critical role for the 3-methoxy in mescaline analogues than the 2-methoxy in 2,4,5-substituted compounds.

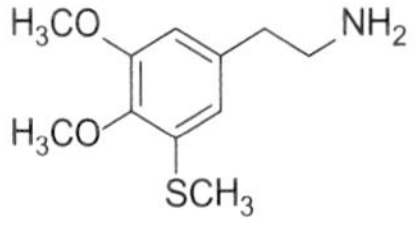

44

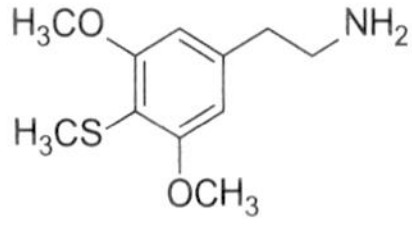

45

Short straight chain alkyl groups in the 4-position of 2,4,5-trisubstituted compounds have been shown to give very active compounds, including ethyl, propyl, and n-butyl, as well as the halogens Cl, Br (**38**), and I (**39**), and a variety of alkylthio substituents (Shulgin et al. 1971; Aldous et al. 1974; Nichols and Shulgin 1976). Branched alkyl groups such as isopropyl or *tert*-butyl are not tolerated (Shulgin and Shulgin 1991). The 4-trifluoromethyl substituent seems to afford a compound with the highest potency (Nichols et al. 1994).

When **39** (DOI), as well as its phenethylamine counterpart 2C-I, were prepared as their radioactive ^{125}I-labeled 4-Iodo congeners, they proved useful as radioligands to label 5-HT$_{2A/2C}$ receptors (Johnson et al. 1987; Glennon et al. 1988; McKenna et al. 1989; Johnson et al. 1990). Indeed, [^{125}I]DOI (**39**) is now widely used as an agonist radioligand to label the serotonin 5-HT$_{2A}$ receptor. Its theoretical specific activity of 2000 Ci/mmol allows the compound to be used to detect low levels of receptor protein. [^{131}I]-labeled **39** also was briefly examined as a potential imaging agent (Braun et al. 1977), and ^{82}Br and ^{77}Br isotopically labeled versions of DOB **38**, were suggested to be useful as brain-scanning agents (Sargent et al. 1975).

What role does a relatively hydrophobic 4-substituent play? It likely contributes in a number of ways to the overall biological activity of these molecules. First of all, it increases the overall hydrophobicity of the molecule so that it partitions better into the central nervous system (Barfknecht and Nichols 1975). That may be the major factor operating for 3,4,5-substituted mescaline analogues that have more hydrophobic substituents in the 4-position (Nichols and Dyer 1977). There is, however, a limitation to the size and bulkiness that this substituent can possess. For 2,4,5-substituted compounds there appears to be a steric limitation for linear alkyl groups of only about three carbon atoms before activity drops off (Nichols et al. 1977). By contrast, a polar moiety such as OH, NH_2, and COOH at the 4-position gives compounds with very low affinity (K_i > 25,000 nM) (Seggel et al. 1990). The latter study also examined compounds with long lipophilic 4-substituents such as *n*-hexyl and *n*-octyl. These longer alkyl groups gave antagonist molecules with high affinities at [^{3}H]ketanserin-labeled sites, whereas smaller alkyl groups gave agonist molecules.

If the 4-substituent is an alkyl group, branching adjacent to the aromatic ring is not tolerated. For example, 2,5-dimethoxy-4-isobutylamphetamine **46** (DOIB) demonstrated significant activity in a rat drug discrimination task, in animals trained to discriminate LSD from saline. DOIB had only about one-third the activity of DOM in humans, with a dose in the 10 to 15 mg range (Shulgin and Shulgin 1991). By contrast, the 2-butyl homolog was about one-third less potent, but also failed to produce full substitution in the rats. The active oral dose in man is reported to be 25–30 mg (Shulgin and Shulgin 1991). In vitro examination of *R* and *S* stereochemistry in the 2-butyl group by displacement of [^{125}I]DOI from rat frontal cortical homogenate revealed identical affinities (K_i values) of 7.8 nM for both isomers (Oberlender et al. 1995). Drug discrimination tests of the two isomers in LSD-trained rats revealed that the *R* isomer (**47**) was only slightly more potent than the *S* (**48**) (ED$_{50}$ of 3.1 vs. 4.8 μmol, respectively). The conclusion is that there is no chiral discrimination by the receptor in the region of the 4-substituent, and that branching in the 4-alkyl group proximal to the aryl ring is detrimental to activity.

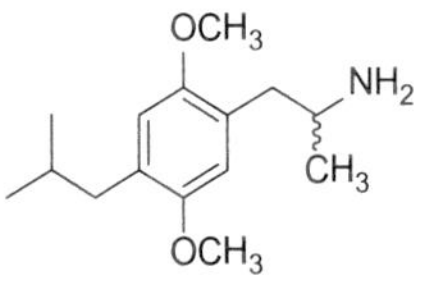

46

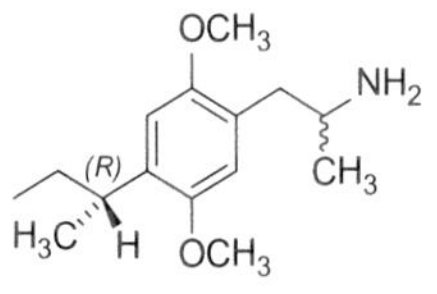

47

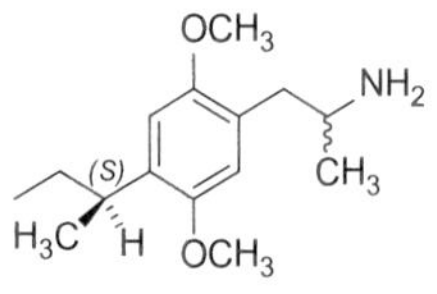

48

Large bulky alkyl groups at the 4-position, such as isopropyl or *tert*-butyl, lead to inactive compounds (Glennon et al. 1981, 1982a; Glennon and Rosecrans 1982; Oberlender et al. 1984). Not surprisingly, therefore, aryl groups attached at the 4-position also gave antagonists, generally with low affinity (Trachsel et al. 2009). Interestingly, however, when a 3-phenylpropyl substituent was introduced at this position, the compound was reported to be a weak partial agonist (Dowd et al. 2000).

The 2,5-dioxygenation pattern coupled with a hydrophobic 4-substituent that meets certain size and hydrophobicity criteria appears optimal to give the most active compounds. Certain structural modifications to this basic pharmacophore also can lead to antagonists with high affinity at the 5-HT_{2A} receptor (Rangisetty et al. 2001).

Although the 4-substituent can have an overall effect on pharmacokinetics and increasing brain penetration, it must be playing other important roles. The correlation between activity and 4-substituent lipophilicity, as well as limitations on the length and bulk of the substituent would be consistent with the presence of a complementary hydrophobic region within the 5-HT_{2A} receptor orthosteric binding domain. The location of this putative region has not yet been elucidated, but it is evident from simulated docking studies (Chambers and Nichols 2002; Isberg et al. 2011) that it should lie somewhere within the vicinity of transmembrane helices 5 and/or 6. Although mutagenesis studies indicate that indoles (i.e., the indole NH) engage Ser-242 in the human 5-HT_{2A} receptor, this residue is apparently not important for binding of 2,5-dimethoxy-substituted phenethylamines (Braden and Nichols 2007).

5.3 Methoxy Mimics—Benzofuran and Benzopyran Analogues

The need for the 2- and 5-oxygen substituents in the phenethylamine hallucinogens raises the question as to what role they may be playing. The most compelling hypothesis is that they serve as hydrogen bond acceptors in the orthosteric ligand binding site. If true, there should then be a dependence on the oxygen unshared electron pair orientations. "Tethering" the 5-methoxy of DOM (**37**) to the 4-position, leading to compound **49**, reduced activity nearly 20-fold in an in vivo rat drug discrimination assay, compared to DOM. (Nichols et al. 1986) When the 5-methoxy of DOB was tethered to the 6-position, however, compound **50** was as potent as DOM (Nichols et al. 1991). Affinity at the [^{125}I]DOI-labeled $5HT_{2A}$ receptor in rat prefrontal cortex was consistent with the in vivo findings, where the affinity of **49** was 488 nM and **50** was 3.1 nM. These results were consistent with the hypothesis that the oxygen electrons needed to project in a specific direction for proper receptor interaction.

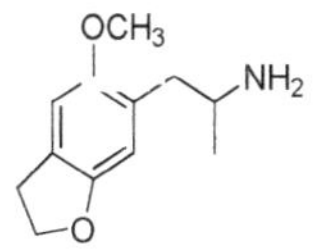

49

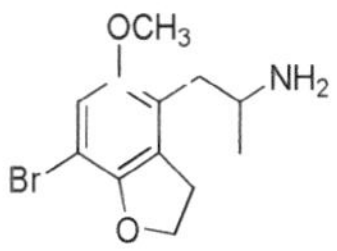

50

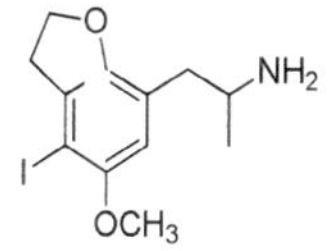

51

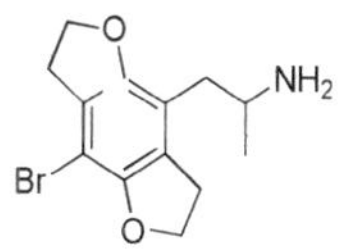

52

Tethering the 2-methoxy into the 3-position to afford compound **51** also led to a potent compound (Monte et al. 1996). Tethering both the 2- and 5-methoxy functions into dihydrofuran rings increased potency even further, as typified by compound **52**, which had an affinity of 0.48 nM at the cloned human 5-HT_{2A} receptor (Monte et al. 1996; Chambers et al. 2001). Although it might be expected that aromatization of the dihydrofuran rings would reduce the hydrogen bonding donor capacity of the furan oxygen atoms, in fact compound **53** was even more potent than its tetrahydrofuranyl congeners, and represents one of the most potent known phenethylamines (Parker et al. 1998). Although the hydrogen bonding potential may be reduced for the oxygen atoms, the large hydrophobic surfaces of the fully aromatic furan rings may be complementary to the relatively hydrophobic interior of the receptor orthosteric binding site.

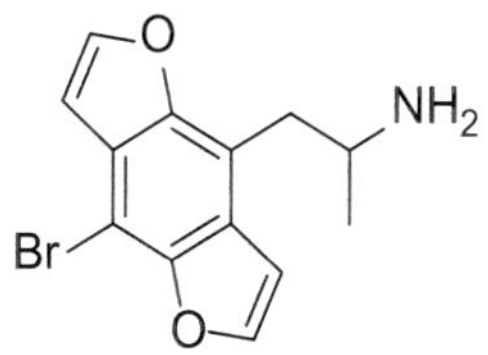

53

These studies naturally led the question: what is the optimal size of the oxygen-containing heterocyclic rings? To address this question, hybrid benzofuran/benzopyran molecules were designed and tested (Schultz et al. 2008). Replacing the 2-methoxy with a dihydrofuran and the 5-methoxy with a dihydropyran gave compound **54**, which had only slightly higher 5-HT_{2A} affinity than did isomer **55** (3.6 nM vs. 5.3 nM, respectively). Compound **54** also was fourfold more potent than **55** in stimulating functional PI turnover. Parallel results were observed in vivo, in rats trained to discriminate LSD from saline, where **54** was three times more potent than **55**. Mutagenesis studies had shown that the 5-methoxy of the phenethylamines likely served as a hydrogen bond acceptor from Ser-239 near the top of transmembrane helix 5 of the receptor (Braden and Nichols 2007). Using that information to guide virtual docking of phenethylamines into a homology model of the 5-HT_{2A} receptor (Chambers and Nichols 2002) directs the 2-methoxy downward toward an intracellular region inside the binding site. Thus, there may well be less space, and hence less steric tolerance for modifications to the 2-methoxy. It should be mentioned that incorporating both oxygen atoms into six-membered dihydropyran rings, to provide hexahydrobenzodipyrans, gave a compound that was about ten-fold less active in a rat drug discrimination task, in rats trained to discriminate LSD from saline. Similarly, the affinity of the dipyrano compound at the rat 5-HT_{2A} receptor was 3.87 nM, whereas the affinity of the difurano homolog was 0.48 nM, approximately paralleling the rat discrimination data (Whiteside et al. 2002).

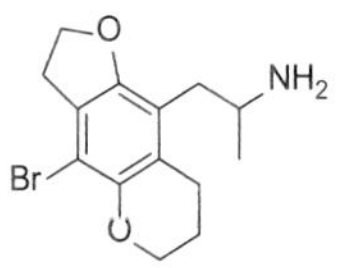

54

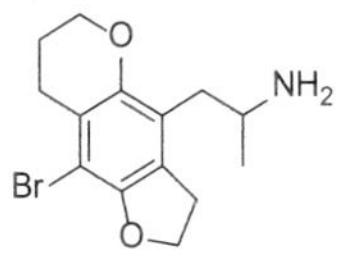

55

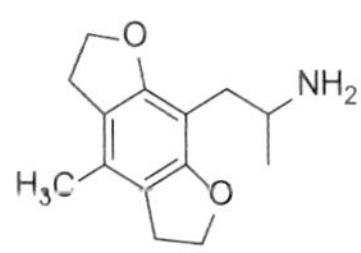

56

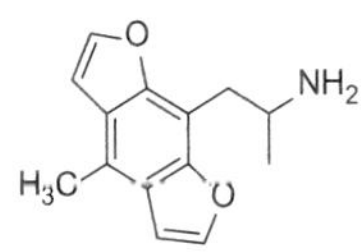

57

Application of a similar approach to 2,6-dimethoxy-4-methylamphetamine likewise resulted in significant potency enhancement in bisdihydrofuran **56**, with a further increase in the fully aromatic **57** (Chambers et al. 2002). For example, the affinity of 2,6-dimethoxy-4-methylamphetamine at the cloned rat 5-HT_{2A} receptor was 49 nM, whereas the affinities of **56** and **57** were 6.3 and 1.8 nM, respectively.

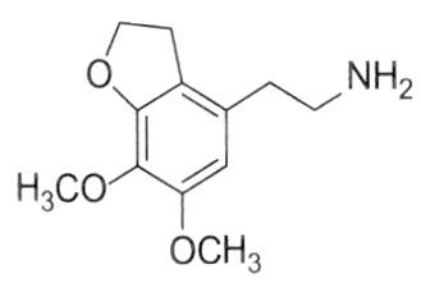

58

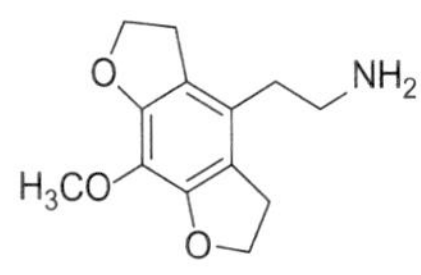

59

When this strategy was applied to 3,4,5-substituted mescaline analogues, however, activity of the tethered compounds was reduced. Although affinity at the 5-HT_{2A} receptor increased compared to mescaline, the monocyclic furano compound **58** lost both efficacy and mescaline-like potency in a rat behavioral model, and difuranyl compound **59** was even less active (Monte et al. 1997). These divergent results suggest that the binding pose of 2,4,5-substituted compounds differs from that of 3,4,5-substituted compounds. Mutagenesis studies support that conclusion, as mutations of polar residues in the orthosteric binding site of the human 5-HT_{2A} receptor have different effects, depending on whether the ligand being examined is a 2,4,5- or a 3,4,5-substituted molecule (McCorvy 2012).

It also has been found that 5-HT_{2A} agonists can reduce intraocular pressure (May et al. 2003b). That finding led to identification of novel 5-HT_{2A} agonists that might be useful to treat glaucoma, but which lacked the hallucinogenic effects that are characteristic of most 5-HT_{2A} agonists. A number of benzodifurans were developed with 4-alkoxymethyl and oxadiazole methyl substituents with high 5-HT_{2A} agonist activity. The purpose of these hydrophilic substituents was to retain 5-HT_{2A} agonist activity, while minimizing penetration into the CNS. Compounds **60**, **61**, and **62** were reported as promising candidates (Feng et al. 2007).

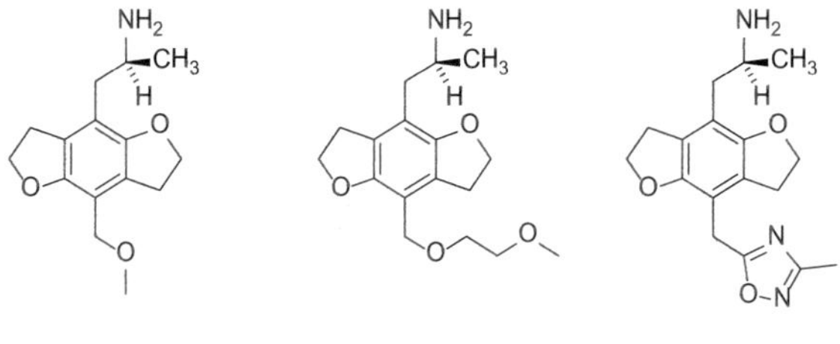

60 **61** **62**

5.4 Effect of Side Chain Alpha-Alkylation

Racemic amphetamines have approximately the same affinity at the human 5-HT_{2A} receptor as their nonmethylated phenethylamine congeners (Johnson et al. 1987; Glennon et al. 1992; Nash et al. 1994; Parrish et al. 2005). At the rat 5-HT_{2A} receptor Parrish et al. (2005) reported that EC_{50}s for stimulating phosphoinositide hydrolysis were virtually identical for phenethylamines and amphetamines, although efficacy for PI hydrolysis was higher for racemic amphetamines. Substituted amphetamine enantiomers, however, have significantly greater affinity, potency, and efficacy in their *R*-(−) enantiomer than in the *S*-(+) antipode. For the enantiomers of DOB (**38**) and DOI (**39**), the difference in EC_{50} was about fourfold higher for the *R* isomer than for the *S*. When the 4-substituent was a CF_3, however, and the two methoxy groups were constrained into dihydrofuran rings (*R*- or *S*-**63**), the difference in *affinity* for the enantiomers was more than 40-fold, but the difference in EC_{50} was less than twofold. In the PI turnover assay *R* enantiomers were about 50% more potent than the *S* enantiomers.

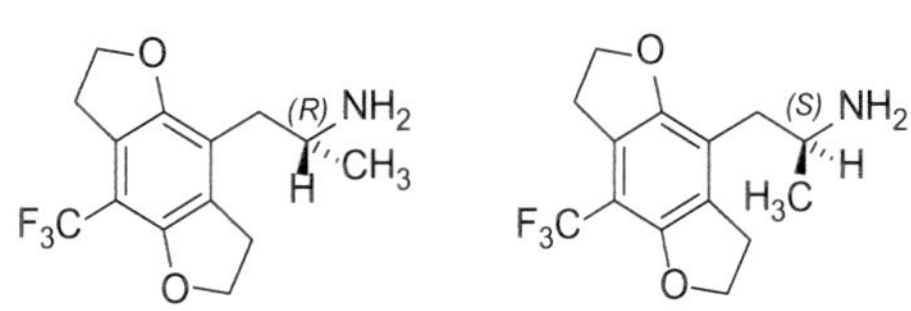

***R*-63** ***S*-63**

Based on molecular modeling studies, Parrish et al. (2005) speculated that only in *R* enantiomers could the alpha-methyl group interact with Phe-340$^{(6.52)}$ in the receptor through van der Waals interactions, possibly explaining the basis for the higher potency and efficacy of the (*R*) enantiomers. It should be mentioned that the SAR of ring-substitution patterns described for substituted amphetamine hallucinogens parallel those for their non-α-methylated phenethylamine congeners. In general, the phenethylamines are less potent and have a shorter duration of action than the alpha-methylated amphetamines with the same ring substitutions. The phenethylamines seem to be considered more pharmacologically benign in vivo than their amphetamine counterparts, and thus are more popular on the illicit drug market than the substituted amphetamines.

For use as a radioligand, a racemic alpha-methyl compound had no advantage over the simpler achiral phenethylamine. That is, [^{125}I]2C-I gave results comparable to [^{125}I]DOI when used as a radioligand to label 5-HT$_{2A/2C}$ receptors (Johnson et al. 1990).

Incorporating the alpha methyl into a cyclopropane ring gives substituted 2-phenylcyclopropylamines with high in vitro and in vivo potency. The cis and trans cyclopropane analogues of mescaline were first reported by Cooper and Walters, who found that trans compound **64** produced an effect in rodents qualitatively resembling mescaline, but with a potency slightly greater than mescaline (Walters and Cooper 1968; Cooper and Walters 1972).

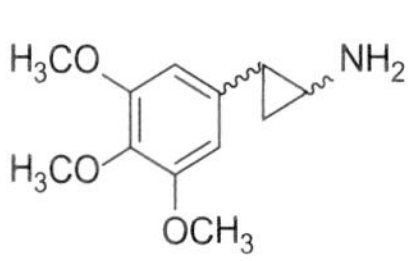

64

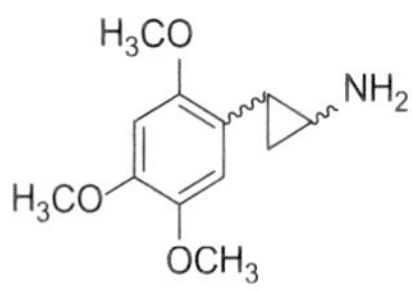

65

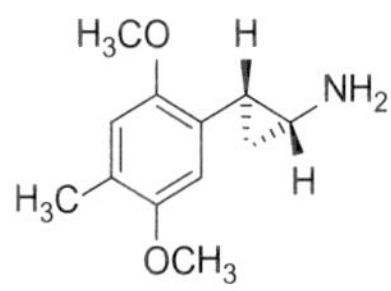

(1*R*,2*S*)-(-)-**66**

Subsequently Aldous et al. (1974) explored cyclopropane analogues of several substituted amphetamines. Using an assay that measured hyperthermia in rabbits, as well as assessing changes in cat EEG, *trans*-2,4,5-trimethoxy compound **65** and *trans*-2,5-dimethoxy-4-methyl compound **66** (DMCPA) had hallucinogen-like activity, with about 20 and 35%, respectively, of the activity of DOM (**37**).

In a later study of the 1*R*,2*S* and 1*S*,2*R* enantiomers of **66**, in three different behavioral responses, DMCPA had activity nearly comparable to DOM (**37**) (Nichols et al. 1978). The (1*R*,2*S*)-(−)-isomer **66** was most potent, with the (1*S*,2*R*)-(+)-antipode being almost inert. Both racemic and the (−) isomer of **66** produced a robust response in the rabbit hyperthermia assay, with the (+) isomer not being different from saline (Nichols et al. 1979). When [^{125}I]2C-I was used to label the

5-$HT_{2A/2C}$ receptor in rat cortical homogenate, affinities of the (1*R*,2*S*)-(−) and (1*S*,2*R*)-(+) enantiomers were about ten-fold different, 2.2 and 21.6 nM, respectively. The K_i of (*R*)-DOI was virtually identical (2.6 nM) to (−)-DMCPA (Johnson et al. 1990). Adding methyl groups to the cyclopropane ring abolished activity (Jacob and Nichols 1982).

The cyclopropyl analogues of DOB (**38**) and DOI (**39**) also have been prepared and receptor affinities and functional potencies measured (Pigott et al. 2012). The two analogues were resolved into their enantiomers and compared with the enantiomers of DOI (**39**). Affinities of the racemic compounds **67** and **68** at the 5-HT_{2A} receptor were five–sixfold higher than those of the open chain analogues **38** and **39**. Assay measuring calcium release at the cloned human 5-HT_{2A} receptor revealed an EC_{50} and E_{max} for (−)-**39** of 3.3 nM and 87%, whereas (−)-**67** had values of 2 nM and 89%, respectively. The bromo compound was somewhat less active, with (−)-**68** having an EC_{50} of 6.3 nM and an E_{max} of 76%.

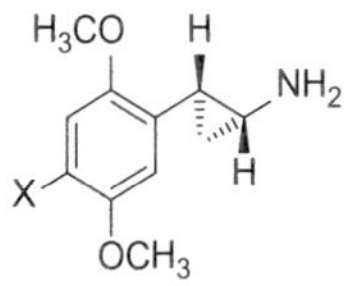

67, X= I
68, X= Br,

Expansion of a cyclopropane to a cyclobutane ring led to a 50–75-fold loss of in vivo activity (Nichols et al. 1984a). Coupled with a variety of studies that have examined conformationally constrained analogues of phenethylamine type hallucinogens, it is reasonable to speculate that the orthosteric binding site within the receptor is sterically restricted around the side chain.

6 Identifying the "Active" Conformation of the Phenethylamines

To understand how phenethylamines and tryptamines may dock in the same receptor site, it can sometimes be useful to identify the active binding conformation through the use of rigid analogues. That is, if a flexible ligand can be conformationally constrained through the use of carefully selected tethers, and if the constrained molecule has high biological activity, it is usually inferred that the constrained conformation is a good representation of how the more flexible ligand binds to the receptor.

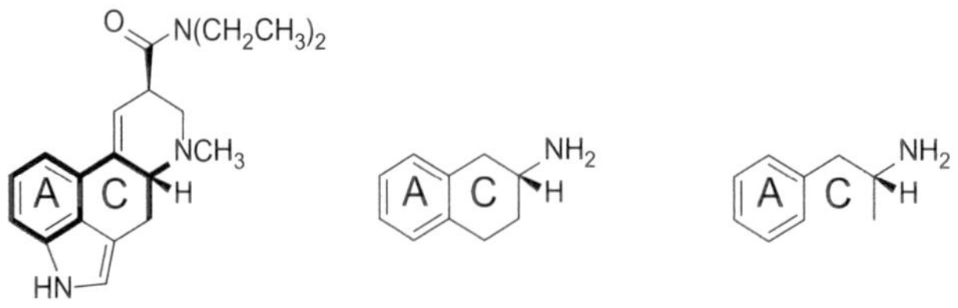

LSD (*R*)-2-aminotetralin (*R*)-amphetamine

A substantial amount of work was carried out in the late 1940s and 1950s to identify the structural element within ergoline molecules (e.g., ergotamine) responsible for their oxytocic and adrenergic activities. The challenges faced by total chemical synthesis of substituted ergoline structures prompted evaluation of much simpler and more synthetically tractable molecules. For example, several early groups focused efforts on structures such as substituted aminotetralins, which they envisioned as representing the A and C rings of the ergolines. When later workers began to study the substituted phenethylamines and amphetamines, and particularly after the more active enantiomer of the amphetamines was shown to have stereochemistry similar to the 5*R* stereochemistry of LSD, this idea was further reinforced. Nonetheless, such a simple analogy is probably not justified, as many studies of rigid analogues of phenethylamines have subsequently shown. The substituted A-ring of the phenethylamines simply mimics, in some unknown way, the indole nucleus of LSD or serotonin. Today we understand that receptors are flexible and dynamic. There is no reason that the structural elements within the orthosteric binding site of the 5-HT$_{2A}$ receptor that engage LSD or the tryptamines must necessarily be the same ones that engage the substituted phenethylamines. Even so, study of rigid analogues of phenethylamines may afford insight into how ligands as structurally different as LSD, tryptamines, and phenethylamines can activate the same receptor.

For phenethylamines (and tryptamines) the side chain has two major rotatable bonds, giving two degrees of conformational freedom: the aryl-Cβ bond, and the Cβ-Cα bond. The trans phenylcyclopropylamines discussed earlier (i.e., **65–68**) lock the Cβ-Cα bond into an approximately trans orientation. Their high activity clearly indicates that the side chain of phenethylamines must exist in a trans extended conformation. Earlier studies had found that 2-aminotetralin and 2-aminoindan derivatives lacked activity, therefore indicating that the side chain probably could not reside in the plane of the aryl ring (Coutts and Malicky 1974; Nichols et al. 1974; Monte et al. 1998). When mescaline was virtually docked into a homology model of an in silico "activated" model of the 5-HT$_{2A}$ receptor (Chambers and Nichols 2001), it was observed that tethering the side chain β carbon back into the aryl ring would afford an aminomethylindan that would closely mimic the docked conformation of mescaline. Interestingly, the modeling also predicted that the active absolute configuration of that molecule would be *R*, because the *S* enantiomer did not provide an acceptable docked pose. Synthesis and subsequent testing of the two enantiomers demonstrated that the *R* enantiomer **69** had 70 nM affinity at the cloned human 5-HT_{2A} receptor. By contrast, the affinity of the *S* enantiomer was only 1120 nM (McLean et al. 2006a). In a functional assay for activation of IP_3 accumulation, the *R* enantiomer had an EC_{50} of 3200 nM, whereas the EC_{50} for the *S* enantiomer was >50,000 nM. The efficacy of the *R* enantiomer was essentially identical to that of mescaline.

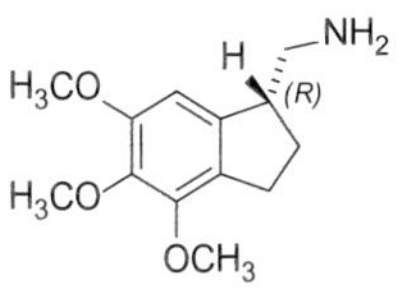

69

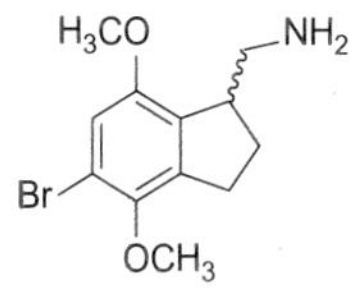

70

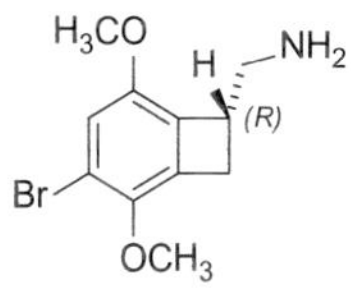

(*R*)-71

When this strategy was applied to a potentially more potent template, a 2,5-dimethoxy-4-bromo substituted ring to afford molecule **70**, its affinity at the human 5-HT_{2A} receptor was slightly greater than that of **69**, but was much lower than the unconstrained phenethylamine. This somewhat surprising finding may indicate that the binding conformation of 3,4,5-substituted compounds differs somewhat from that of 2,4,5-substituted compounds.

When, however, the five-membered ring was contracted to a four-membered benzocyclobutene, the highly potent **71** (TCB-2) was obtained (McLean et al. 2006b). Virtual docking of the enantiomers of **71** into a homology model of the receptor also revealed that only the *R* enantiomer gave an acceptable pose. After resolution and pharmacological evaluation, *R* benzocyclobutene **71** indeed was found to have a K_i of 0.26 nM, whereas the K_i of the *S*-isomer was only 42 nM. In a functional test for IP_3 accumulation, the EC_{50} for the *R* enantiomer was 18 nM, with an E_{max} of 97%. The EC_{50} for the *S* enantiomer was only 460 nM. The contrast between the biological activities of **70** and **71** is not readily explained. The dihedral angle for the aryl-Cβ bond in aminomethylindanes is approximately 100°, whereas for the aminomethylbenzocyclobutene the angle is about 117°. The steric footprint of the five-membered ring is somewhat greater than for the four-membered ring, which could also be a contributing factor. The basis for the difference between the effect of rigidification of mescaline and 2,5-dimethoxy-4-bromophenethylamine is worth further study.

In the two-lever drug discrimination assay in rats trained to discriminate LSD or DOI from saline *R*-**71** had potency comparable to each training drug. LSD had an ED_{50} of 38 nmol/kg in LSD-trained rats, whereas *R*-**71** was slightly more potent than LSD, with an ED_{50} of 24 nmol/kg! No substitution occurred with the *S* enantiomer of **71** at doses up to 250 nmol/kg. These results support the hypothesis that the phenethylamine side chain of hallucinogens binds to the receptor in a conformation displaced out of the aryl ring plane.

6.1 Effect of an N-Benzyl Group

Although the most active tryptamine hallucinogens are *N,N*-dialkylated, the phenethylamines generally cannot tolerate even a single *N*-substitution. Even small groups such as methyl or ethyl (see Table 2) abolish their hallucinogenic activity. It was quite remarkable, therefore, when Heim and coworkers reported that *N*-benzyl groups afforded compounds with remarkable affinity and potency (Heim et al. 1999; Elz et al. 2002; Heim 2003). An oxygen atom at the ortho position of the *N*-benzyl group enhanced activity further (Braden et al. 2006). When an alpha-methyl was introduced into the side chain, affinity dropped about 20-fold. Thus, a side chain alpha-methyl enhances potency of ring-substituted phenethylamines, but is deleterious when an *N*-benzyl is added to the amino group. Compound **72** (25I-NBOMe) is the most studied of these *N*-benzylated analogues. The high potency of this compound, as well as the 4-chloro and 4-bromo analogues has made them very attractive for sale as "research chemicals," often being distributed on blotter papers and deceptively labeled as being LSD. Surprisingly, the NBOMe-type compounds are not orally active, but are typically administered so that they are absorbed through buccal tissues. Sadly, several deaths have occurred as a result of use of NBOMe-type compounds (Poklis et al. 2014; Walterscheid et al. 2014; Nikolaou et al. 2015). In the past, lethality has not been associated with ingestion of hallucinogens, so it is not clear whether death has resulted from toxic amounts of pure drug, or whether there is some inherent toxicity not seen with other hallucinogens. The iodo compound **72**, as well as its 4-bromo and 4-chloro analogues have recently been placed into Schedule 1 of the controlled substances act.

Table 2 Effects of phenethylamine *N*-substituents on affinity at several serotonin receptor subtypes (Braden et al. 2006)

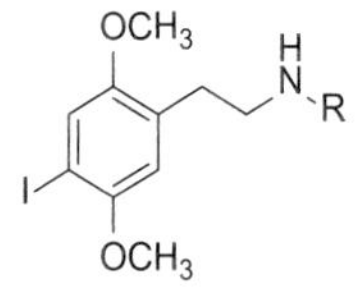

R	h5-HT$_{2A}$ K_i (nM)	r5-HT$_{2A}$ K_i (nM)	h5-HT$_{2C}$ K_i (nM)	r5-HT$_{2C}$ K_i (nM)	h5-HT$_{1A}$ K_i (nM)
H	0.73 ± 0.06	0.65 ± 0.07	1.82 ± 0.20	1.22 ± 0.03	123 ± 24
CH_3	1907 ± 254	1286 ± 64	nd	206 ± 34	247 ± 23
n-Pr	1295 ± 151	734 ± 30	nd	656 ± 127	879 ± 64
Benzyl	0.25 ± 0.05	0.31 ± 0.03	1.08 ± 0.24	1.15 ± 0.90	2205 ± 106
BOMe[a]	0.044 ± 0.006	0.09 ± 0.010	0.43 ± 0.08	0.13 ± 0.02	1696 ± 311

Values are mean ± SEM. *nd* not determined

[a]*N*-(ortho-methoxybenzyl)

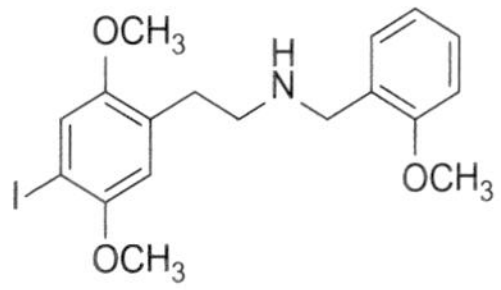

72

The exceptionally high 5-HT_{2A} affinity of **72** made it useful as a radioligand for receptor binding studies (Nichols et al. 2008), and its 4-bromo analogue has found application as a ^{11}C-labeled PET ligand for in vivo imaging studies of the 5-HT_{2A} receptor (Ettrup et al. 2010, 2011). It appears that an aryl group with a hydrogen bond *acceptor*, such as an ether, gives highest activity (see Table 3). Additional modifications of the *N*-benzyl moiety are given in Tables 3 and 4.

Ettrup et al. (2011) examined a series of eight *N*-benzyl substituted phenethylamines for potential as PET imaging agents. The most favorable profile (*i.e.*, largest target-to-background binding ratio) was obtained with the 4-bromo-*N*-(2-methoxybenzyl) compound (25B-NBOMe), which was designated [^{11}C] Cimbi-36. This compound was described as "the most promising candidate for investigation of 5-HT_{2A} receptor binding in the living human brain with PET."

An extensive structure–activity analysis of 48 *N*-benzylphenethylamines by Hansen et al. (2014) identified a highly selective 5-HT_{2A} agonist in the *N*-benzyl series, compound **73** (25CN-NBOH), with the 4-cyano substituent on the ring, and

Table 3 Effect of benzyl group substitution on affinity of *N*-benzylphenethylamines (Braden 2007)

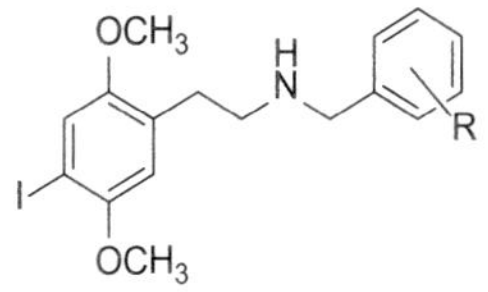

R	h5-HT_{2A} K_i (nM)	r5-HT_{2A} K_i (nM)	h5-HT_{2C} K_i (nM)	r5-HT_{2C} K_i (nM)
2-OCH_3	0.044 ± 0.006	0.09 ± 0.01	0.43 ± 0.08	0.13 ± 0.02
2-OH	0.061 ± 0.012	0.12 ± 0.02	0.13 ± 0.01	0.21 ± 0.02
2-CN	nd	276 ± 65	23.2 ± 4.1	nd
2-$CONH_2$	1.18 ± 0.22	0.84 ± 0.1	nd	0.73 ± 0.09
2-CH_2OH	0.79 ± 0.05	0.44 ± 0.03	nd	0.43 ± 0.01
2-CF_3	1.31 ± 0.15	nd	nd	nd
4-CF_3	205 ± 44	nd	nd	nd
2-F	0.26 ± 0.05	0.28 ± 0.04	2.36 ± 0.41	0.85 ± 0.11
4-F	37.3 ± 6.0	nd	nd	nd
2,3-OCH_2O	0.049 ± 0.008	0.193 ± 0.022	1.7 ± 0.23	0.41 ± 0.07
3,4-OCH_2O	0.69 ± 0.05	nd	nd	nd
2-OH-4,5-OCH_2O	0.82 ± 0.17	nd	nd	nd

Values are mean ± SEM. *nd* not determined

Table 4 Effect of other *N*-aryl groups on phenethylamine affinity (Braden 2007)

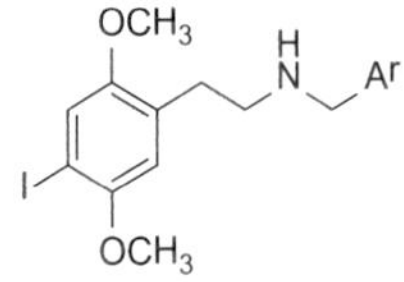

Ar	h5-HT_{2A} K_i (nM)	r5-HT_{2A} K_i (nM)	r5-HT_{2C} K_i (nM)
2-furyl	nd	0.78 ± 0.12	0.99 ± 0.1
2-thienyl	1.02 ± 0.09	0.45 ± 0.09	0.59 ± 0.06
2-pyridyl	nd	3.45 ± 0.7	5.81 ± 1.14
3-indolyl	nd	2.67 ± 0.49	11.8 ± 1.7
1-naphthyl	nd	1.07 ± 0.11	14.0 ± 1.5
2-naphthyl	4.83 ± 0.55	3.74 ± 0.52	176 ± 30
2,3-dihydrobenzofuran-7-yl	0.026 ± 0.006	nd	nd

Values are mean ± SEM. *nd* not determined

the *N*-2-hydroxybenzyl moiety. It had affinity at the human 5-HT_{2A} receptor of 1.3 nM and at the rat 5-HT_{2C} receptor of 132 nM. In functional measures of ability to stimulate inositol phosphate production, **73** had an EC_{50} at the 5-HT_{2A} receptor of 2.1 nM, and at the 5-HT_{2C} receptor of 190 nM. In the mouse head twitch response, an animal behavioral model that has a general correlation with human hallucinogenic activity, **73** elicited the head twitch, but was less potent than DOI (Fantegrossi et al. 2015). In mice trained to discriminate DOI from saline in a two-lever drug discrimination task, **73** engendered 55% generalization to the DOI training dose. These effects were blocked by the selective 5-HT_{2A} antagonist M100907, indicating that the behavioral action of **73** was mediated by 5-HT_{2A} receptor activation.

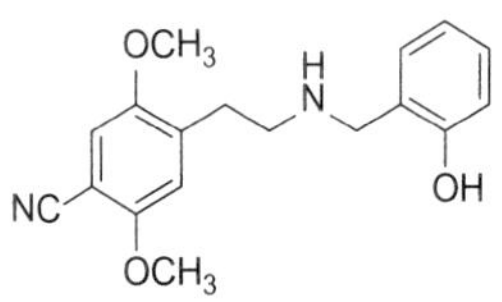

73

The absence of oral activity for the *N*-benzylphenethylamines led Leth-Petersen et al. (2014) to examine the metabolism of a series of NBOMe compounds in human liver microsomes. What they observed was high intrinsic clearance, indicative of high first pass metabolism of the compounds if given orally. In a subsequent study from the same laboratory (Leth-Petersen et al. 2016), an in vivo experiment in pigs confirmed the predicted high first pass metabolism, with the principle phase II metabolism being rapid O-demethylation of the 5-methoxy on the core phenethylamine ring. This hydroxyl metabolite was then rapidly conjugated to provide the 5-*O*-glucuronide **74**. Both metabolic steps (demethylation and glucuronidation) were very fast, with the authors reporting that only minute levels of the intermediate phenolic demethylation product were present at any time point, with the glucuronide being eliminated much slower from the plasma. The authors propose that the glucuronide metabolite may be toxic, and responsible for the adverse effects of NBOMe compounds in humans. Another possibility is that some individuals may lack the P450 isozyme necessary for the rapid *O*-demethylation and that the persistence of the very potent parent compound may lead to toxicity.

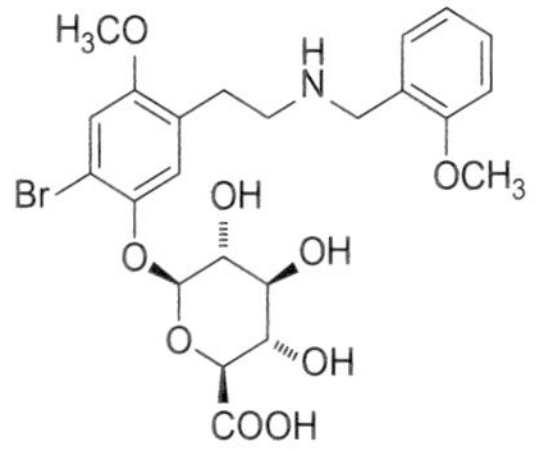

74

An *N*-benzyl moiety also enhances 5-HT_{2A} activity in the tryptamine series, for example compounds such as **75**. A series of 15 analogues of **75** had affinities at the human 5-HT_{2A} receptor ranging from 0.6 to 11 nM (Nichols et al. 2015). Broad receptor screening showed that they had highest affinities for the 5-HT_2 family of receptors, but demonstrated no selectivity for 5-HT_{2A} versus 5-HT_{2C} receptors. In functional assays measuring Ca^{+2} mobilization the compounds generally had an E_{max} between 70 and 90%. Only some of the compounds were active in the mouse head twitch assay, however, but for those that were active, linear regression analysis revealed a significant correlation between the pED_{50} for the head twitch and the pEC_{50} for functional potency in the rat 5-HT_{2A} receptor.

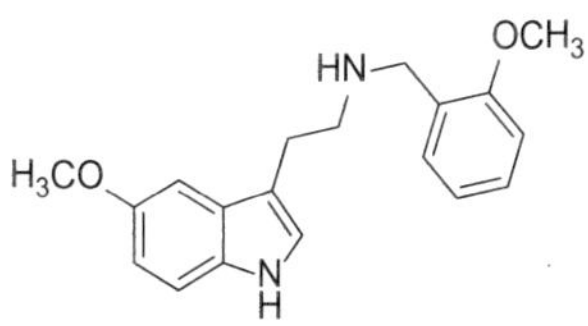

75

6.2 *The Active Conformation of N-Benzylphenethylamines*

In an attempt to identify the active binding conformation of the *N*-benzylphenethylamines, a series of nine of conformationally constrained analogues of **76** was prepared (Juncosa et al. 2013). The most potent of these analogues was *S,S* enantiomer **77**. This compound, as the racemate, had an EC_{50} of 74 nM and an E_{max} of 73% for PI hydrolysis through activation of the human 5-HT_{2A} receptor. In drug discrimination experiments in rats trained to discriminate LSD from saline, **77** had an ED_{50} of 0.41 μmol/kg. Furthermore, (*S*,*S*)-(−)-**77** had 124-fold selectivity for the 5-HT_{2A} receptor versus the 5-HT_{2C} receptor, using antagonist radioligands to measure affinity.

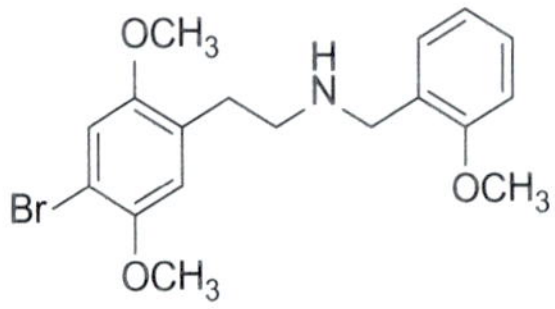

76

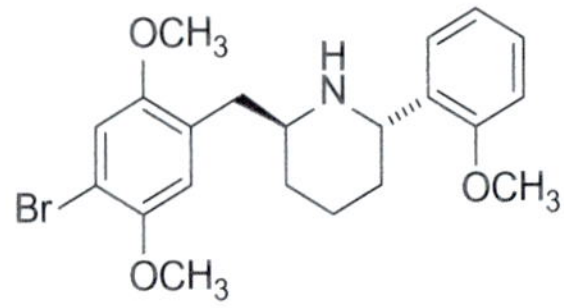

(*S*,*S*)-77

The absolute configuration of (*S*,*S*)-**77** was determined by X-ray crystallography of the molecule with an *N*-acyl chiral auxiliary. If one uses the stereochemistry of (*R*)-**71** discussed earlier, to define the side chain orientation as anti and in a plane approximately perpendicular to the aromatic ring plane, then given its relatively constrained structure, an active binding orientation for (*S*,*S*)-**77** might be envisioned as shown below.

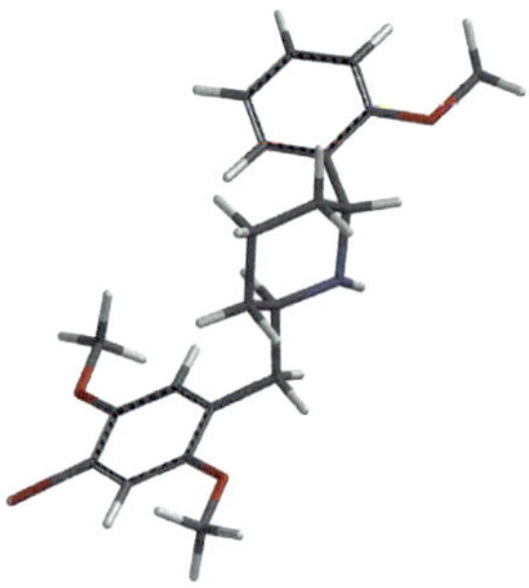

(*S*,*S*)-77

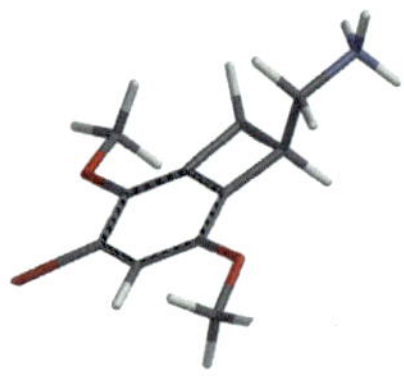

(*R*)-71

7 Modeling the Receptor–Ligand Interaction

The serotonin 5-HT_{2A} receptor is a member of the family A G protein-coupled receptors. It is known that it is comprised of seven membrane-spanning helical segments, and a short helical segment at the C-terminus that lies parallel to the inner leaflet of the membrane. The next step in the evolution of our understanding of SAR for hallucinogens will be the solution of a crystal structure for the 5-HT_{2A} receptor, and of particular importance would be a phenethylamine type ligand.

There have been tremendous strides in receptor protein crystallography over the past 15 years, with nearly two dozen unique structures being characterized

(Costanzi and Wang 2014; Yin et al. 2014). Most recently, the structures of the serotonin 5-HT$_{1B}$ and 5-HT$_{2B}$ receptors with ergotamine bound have been solved (Wacker et al. 2013; Wang et al. 2013). Yet, as of the date of this writing in mid-2016, the structure of the 5-HT$_{2A}$ receptor still has not been solved. One may speculate that the binding of LSD to the 5-HT$_{2A}$ receptor will closely resemble the ergotamine pose in the 5-HT$_{2B}$ receptor crystal structure, but unfortunately there is nothing in that structure that indicates how phenethylamines might bind. If the 5-HT$_{2A}$ receptor could be crystallized with a phenethylamine hallucinogen molecule bound inside, the underlying basis for much of what we know about the SAR of psychedelics should become apparent. In particular, we might finally understand how a phenethylamine can be complementary to a receptor that has evolved to accommodate an indole.

Homology models have so far been the best approach to understanding how hallucinogens bind to their receptors. The earliest in silico-activated homology model of the human 5-HT$_{2A}$ receptor was proposed by Chambers and Nichols (2002) and was developed by in silico activation of the crystal structure of bovine rhodopsin. Virtual docking with various 5-HT$_{2A}$ agonist ligands allowed the formulation of a number of hypotheses about the ligand binding site, which were subsequently validated by site-directed mutagenesis and ligand testing (Braden et al. 2006; Braden and Nichols 2007). Those studies allowed the design of new agonist ligands, including the prediction of their active enantiomers (McLean et al. 2006a, b). Isberg et al. (2011) have recently developed an in silico-activated model of the 5-HT$_{2A}$ receptor from the published structure of the β_2-adrenergic receptor. Nonetheless, these homology models have failed to explain the basis for the binding poses for different ring-substitution patterns in the phenethylamines. In particular, studies have demonstrated that mutations of polar residues in helices 3, 5, and 6 have different consequences for 2,5-dimethoxy versus 3,5-dimethoxy-substituted compounds (McCorvy 2012). Further, homology models have failed to indicate why the 4-position substituent in the 2,4,5-substituted compounds is so important. These, and other aspects of hallucinogen SAR, will likely become apparent when the crystal structure of the 5-HT$_{2A}$ receptor is ultimately solved.

8 Conclusion

The SAR of psychedelics, which are serotonin 5-HT$_{2A}$ receptor agonists or partial agonists, are now well developed for the three major chemotypes of ligands. There were early attempts to identify structural similarities between phenethylamines and tryptamines that might account for the ability of phenethylamines to engage the 5-HT$_{2A}$ receptor, but those were unsuccessful. Despite our current level of understanding, intriguing questions still remain. It is still a mystery why LSD is such a potent hallucinogen when it appears to have rather unremarkable in vitro pharmacological properties. Slight variations in amide substituents of lysergic acid amides have profound effects on the psychopharmacology of lysergamides, and an explanation for this phenomenon is not yet evident.

A complicating factor in the psychopharmacology of hallucinogens is the fact that receptors can couple to multiple effectors, and that different agonists can produce different intracellular biochemical signals. Thus, specific agonists with particular substitution patterns may be able selectively to activate a subset of effectors, a phenomenon now known as functional selectivity (Urban et al. 2007). It seems likely that functional selectivity can at least partially explain some of the differences reported for the human psychopharmacology of hallucinogens. To date there has been no attempt to correlate specific signaling pathways with any aspect of human psychopharmacology of hallucinogens. In addition, Ray (2010) has pointed out that psychedelics often have numerous off-target receptors that could have physiological relevance. A complete and comprehensive study of a particular molecule that involved looking not only at the affinity for each potentially significant brain receptor, but also their functional potency at the most likely signaling pathways would be a herculean task, yet might ultimately be necessary to understand fully the molecular pharmacology of psychedelics.

Clearly, the field of research on psychedelics still offers tremendous challenges. It is a sad fact that there is no significant government funding for this work, despite many emerging clinical studies now demonstrating potential medical utility for these fascinating molecules.

References

Aldous FA, Barrass BC, Brewster K, Buxton DA, Green DM, Pinder RM, Rich P, Skeels M, Tutt KJ (1974) Structure-activity relationships in psychotomimetic phenylalkylamines. J Med Chem 17:1100–1111

Bach NJ, Hall DA, Kornfeld EC (1974) Descarboxylysergic acid (9,10-didehydro-6-methylergoline). J Med Chem 17:312–314

Barfknecht CF, Nichols DE (1975) Correlation of psychotomimetic activity of phenethylamines and amphetamines with 1-octanol-water partition coefficients. J Med Chem 18:208–210

Baxter GS, Murphy OE, Blackburn TP (1994) Further characterization of 5-hydroxytryptamine receptors (putative 5-HT2B) in rat stomach fundus longitudinal muscle. Br J Pharmacol 112:323–331

Bennett JP Jr, Snyder SH (1976) Serotonin and lysergic acid diethylamide binding in rat brain membranes: relationship to postsynaptic serotonin receptors. Mol Pharmacol 12:373–389

Blaazer AR, Smid P, Kruse CG (2008) Structure-activity relationships of phenylalkylamines as agonist ligands for 5-HT(2A) receptors. ChemMedChem 3:1299–1309

Blair JB, Marona-Lewicka D, Kanthasamy A, Lucaites VL, Nelson DL, Nichols DE (1999) Thieno[3,2-b]- and thieno[2,3-b]pyrrole bioisosteric analogues of the hallucinogen and serotonin agonist N, N-dimethyltryptamine. J Med Chem 42:1106–1111

Blair JB, Kurrasch-Orbaugh D, Marona-Lewicka D, Cumbay MG, Watts VJ, Barker EL, Nichols DE (2000) Effect of ring fluorination on the pharmacology of hallucinogenic tryptamines. J Med Chem 43:4701–4710

Braden MR (2007) Towards a biophysical understanding of hallucinogen action. PhD Dissertation

Braden MR, Nichols DE (2007) Assessment of the roles of serines 5.43(239) and 5.46(242) for binding and potency of agonist ligands at the human serotonin 5-HT2A receptor. Mol Pharmacol 72:1200–1209

Braden MR, Parrish JC, Naylor JC, Nichols DE (2006) Molecular interaction of serotonin 5-HT2A receptor residues Phe339(6.51) and Phe340(6.52) with superpotent N-benzyl phenethylamine agonists. Mol Pharmacol 70:1956–1964

Brandt SD, Kavanagh PV, Westphal F, Stratford A, Elliott SP, Hoang K, Wallach J, Halberstadt AL (2016) Return of the lysergamides. Part I: Analytical and behavioural characterization of 1-propionyl-d-lysergic acid diethylamide (1P-LSD). Drug Test Anal 8:891–902

Brandt SD, Kavanagh PV, Westphal F, Elliott SP, Wallach J, Colestock T, Burrow TE, Chapman SJ, Stratford A, Nichols DE, Halberstadt AL (2017) Return of the lysergamides. Part II: Analytical and behavioural characterization of N6 -allyl-6-norlysergic acid diethylamide (AL-LAD) and (2′S,4′S)-lysergic acid 2,4-dimethylazetidide (LSZ). Drug Test Anal 9:38–50

Braun U, Shulgin AT, Braun G, Sargent T III (1977) Synthesis and body distribution of several iodine-131 labeled centrally acting drugs. J Med Chem 20:1543–1546

Brimblecombe RW, Pinder RM (1975) Hallucinogenic agents. Wright-Scientechnica, Bristol

Chambers JJ, Nichols DE (2001) A model of the human 5-HT_{2A} receptor based on constrained homology. Great Lakes/Central Regional American Chemical Society Meeting

Chambers JJ, Nichols DE (2002) A homology-based model of the human 5-HT2A receptor derived from an in silico activated G-protein coupled receptor. J Comput Aided Mol Des 16:511–520

Chambers JJ, Kurrasch-Orbaugh DM, Parker MA, Nichols DE (2001) Enantiospecific synthesis and pharmacological evaluation of a series of super-potent, conformationally restricted 5-HT (2A/2C) receptor agonists. J Med Chem 44:1003–1010

Chambers JJ, Kurrasch-Orbaugh DM, Nichols DE (2002) Translocation of the 5-alkoxy substituent of 2,5-dialkoxyarylalkylamines to the 6-position: effects on 5-HT(2A/2C) receptor affinity. Bioorg Med Chem Lett 12:1997–1999

Cheng HC, Long JP, Nichols DE, Barfknecht CF (1974) Effects of psychotomimetics on vascular strips: studies of methoxylated amphetamines and optical isomers of 2,5-dimethoxy-4-methylamphetamine and 2,5-dimethoxy-4-bromoamphetamine. J Pharmacol Exp Ther 188:114–123

Cooper PD, Walters GC (1972) Stereochemical requirements of the mescaline receptor. Nature 238:96–98

Corkery JM, Durkin E, Elliott S, Schifano F, Ghodse AH (2012) The recreational tryptamine 5-MeO-DALT (N, N-diallyl-5-methoxytryptamine): a brief review. Prog Neuropsychopharmacol Biol Psychiatry 39:259–262

Costanzi S, Wang K (2014) The GPCR crystallography boom: providing an invaluable source of structural information and expanding the scope of homology modeling. Adv Exp Med Biol 796:3–13

Coutts RT, Malicky JL (1974) The synthesis of Analogus of the Hallucinogen 1-(2,5-dimethoxy-4-methylphenyl)-2-aminopropane (DOM). II. Some ring-methoxylated 1-amino-and 2-aminoindanes. Can J Chem 52:381–389

Cozzi NV, Daley PF (2016) Receptor binding profiles and quantitative structure-affinity relationships of some 5-substituted-N, N-diallyltryptamines. Bioorg Med Chem Lett 26:959–964

Daldrup T, Heller C, Matthiesen U, Honus S, Bresges A, Haarhoff K (1986) Etryptamine, a new designer drug with a fatal effect. Z Rechtsmed 97:61–68

Dowd CS, Herrick-Davis K, Egan C, Dupre A, Smith C, Teitler M, Glennon RA (2000) 1-[4-(3-Phenylalkyl)phenyl]-2-aminopropanes as 5-HT(2A) partial agonists. J Med Chem 43:3074–3084

Elz S, Klass T, Heim R, Wamke U, Pertz HH (2002) Development of highly potent partial agonists and chiral antagonists as tools for the study of 5-HT2A-receptor mediated functions. Nauynn-Schmiedeberg's Arch Pharmacol 365(Suppl 1):R29

Ettrup A, Palner M, Gillings N, Santini MA, Hansen M, Kornum BR, Rasmussen LK, Nagren K, Madsen J, Begtrup M, Knudsen GM (2010) Radiosynthesis and evaluation of 11C-CIMBI-5 as a 5-HT2A receptor agonist radioligand for PET. J Nucl Med 51:1763–1770

Ettrup A, Hansen M, Santini MA, Paine J, Gillings N, Palner M, Lehel S, Herth MM, Madsen J, Kristensen J, Begtrup M, Knudsen GM (2011) Radiosynthesis and in vivo evaluation of a series of substituted 11C-phenethylamines as 5-HT (2A) agonist PET tracers. Eur J Nucl Med Mol Imaging 38:681–693

Fantegrossi WE, Gray BW, Bailey JM, Smith DA, Hansen M, Kristensen JL (2015) Hallucinogen-like effects of 2-([2-(4-cyano-2,5-dimethoxyphenyl) ethylamino]methyl)phenol (25CN-NBOH), a novel N-benzylphenethylamine with 100-fold selectivity for 5-HT(2)A receptors, in mice. Psychopharmacology 232:1039–1047

Feng Z, Mohapatra S, Klimko PG, Hellberg MR, May JA, Kelly C, Williams G, McLaughlin MA, Sharif NA (2007) Novel benzodifuran analogs as potent 5-HT2A receptor agonists with ocular hypotensive activity. Bioorg Med Chem Lett 17:2998–3002

Ginzel KH, Mayer-Gross W (1956) Prevention ofpsychological effects of d-lysergic acid diethylamide (LSD 25) by its 2-brom derivative (BOL 148). Nature 178:210

Glennon RA, Rosecrans JA (1982) Indolealkylamine and phenalkylamine hallucinogens: a brief overview. Neurosci Biobehav Rev 6:489–497

Glennon RA, Doot DL, Young R (1981) DOM and related 2,5-dimethoxy-4-alkylphenylisopropylamines: behavioral and serotonin receptor properties. Pharmacol Biochem Behav 14:287–292

Glennon RA, Young R, Rosecrans JA (1982a) A comparison of the behavioral effects of DOM homologs. Pharmacol Biochem Behav 16:557–559

Glennon RA, Young R, Rosecrans JA (1982b) Discriminative stimulus properties of DOM and several molecular modifications. Pharmacol Biochem Behav 16:553–556

Glennon RA, Seggel MR, Soine WH, Herrick-Davis K, Lyon RA, Titeler M (1988) [125I]-1-(2,5-dimethoxy-4-iodophenyl)-2-amino-propane: an iodinated radioligand that specifically labels the agonist high-affinity state of 5-HT2 serotonin receptors. J Med Chem 31:5–7

Glennon RA, Teitler M, Sanders-Bush E (1992) Hallucinogens and serotonergic mechanisms. NIDA Res Monogr 119:131–135

Glennon RA, Bondarev ML, Khorana N, Young R, May JA, Hellberg MR, McLaughlin MA, Sharif NA (2004) Beta-oxygenated analogues of the 5-HT2A serotonin receptor agonist 1-(4-bromo-2,5-dimethoxyphenyl)-2-aminopropane. J Med Chem 47:6034–6041

Gorodetzky CW, Isbell H (1964) A comparison of 2,3-dihydro-lysergic acid diethylamide with LSD-25. Psychopharmacologia 6:229–233

Gupta SK, Diez E, Heasley LE, Osawa S, Johnson GL (1990) A G protein mutant that inhibits thrombin and purinergic receptor activation of phospholipase A2. Science 249:662–666

Hansen M, Phonekeo K, Paine JS, Leth-Petersen S, Begtrup M, Brauner-Osborne H, Kristensen JL (2014) Synthesis and structure-activity relationships of N-benzyl phenethylamines as 5-HT2A/2C agonists. ACS Chem Neurosci 5:243–249

Hartig PR, Kadan MJ, Evans MJ, Krohn AM (1983) 125I-LSD: a high sensitivity ligand for serotonin receptors. Eur J Pharmacol 89:321–322

Heffter A (1898) Ueber pellote. Beitrag zur chemischen und pharmakologischen kenntnis der cacteen. Naunyn Schmiedebergs Arch Exp Path Pharmacol 40:385–429

Heim R (2003) Synthese und Pharmakologie potenter 5-HT2A-Rezeptoragonisten mit *N*-2-Methoxybenzyl-Partialstruktur Entwicklung eines neen Struktur-Wirkungskonzepts. vol. Dr. rer. nat. Freien Universität Berlin, Berlin

Heim R, Pertz HH, Elz S (1999) Preparation and in vitro pharmacology of novel secondary amine-type 5-HT2A receptor agonists: from submillimolar to subnanomolar activity. Arch Pharm Pharm Med Chem 332:34

Hey P (1947) The synthesis of a new homolog of mescaline. Quart J Pharm Pharmacol 20:129–134

Hoffman AJ, Nichols DE (1985) Synthesis and LSD-like discriminative stimulus properties in a series of N(6)-alkyl norlysergic acid N, N-diethylamide derivatives. J Med Chem 28:1252–1255

Hofmann A (1968) Psychotomimetic agents. In: Burger A (ed) Drugs affecting the central nervous system. Edward Arnold, London, p 169

Hofmann A (1971) Teonanacatl and Ololuiqui, two ancient magic drugs of Mexico. Bull Narcotics XXIII:3–14

Hong SS, Young R, Glennon RA (2001) Discriminative stimulus properties of alpha-ethyltryptamine optical isomers. Pharmacol Biochem Behav 70:311–316

Huang XM, Johnson MP, Nichols DE (1991) Reduction in brain serotonin markers by alpha-ethyltryptamine (Monase). Eur J Pharmacol 200:187–190

Isberg V, Balle T, Sander T, Jorgensen FS, Gloriam DE (2011) G Protein- and agonist-bound serotonin 5-HT(2A) receptor model activated by steered molecular dynamics simulations. J Chem Inf Model 51:315–325

Jacob JN, Nichols DE (1982) Isomeric cyclopropyl ring-methylated homologues of trans-2-(2,5-dimethoxy-4-methylphenyl)cyclopropylamine, an hallucinogen analogue. J Med Chem 25:526–530

Jacob P III, Shulgin AT (1981) Sulfur analogues of psychotomimetic agents. Monothio analogues of mescaline and isomescaline. J Med Chem 24:1348–1353

Jacob P III, Shulgin AT (1983) Sulfur analogues of psychotomimetic agents. 2. Analogues of (2,5-dimethoxy-4-methylphenyl)-and (2,5-dimethoxy-4-ethylphenyl)isopropylamine. J Med Chem 26:746–752

Jacob P III, Anderson G III, Meshul CK, Shulgin AT, Castagnoli N Jr (1977) Monomethylthio analogues of 1-(2,4,5-trimethoxyphenyl)-2-aminopropane. J Med Chem 20:1235–1239

Johnson MP, Hoffman AJ, Nichols DE, Mathis CA (1987) Binding to the serotonin 5-HT2 receptor by the enantiomers of 125I-DOI. Neuropharmacology 26:1803–1806

Johnson MP, Mathis CA, Shulgin AT, Hoffman AJ, Nichols DE (1990) [125I]-2-(2,5-dimethoxy-4-iodophenyl)aminoethane ([125I]-2C-I) as a label for the 5-HT2 receptor in rat frontal cortex. Pharmacol Biochem Behav 35:211–217

Juncosa JI (2011) Organic synthesis combined with molecular modeling: a powerful approach to map the functional topography of dopamine and serotonin receptors. PhD, Purdue University

Juncosa JI Jr, Hansen M, Bonner LA, Cueva JP, Maglathlin R, McCorvy JD, Marona-Lewicka D, Lill MA, Nichols DE (2013) Extensive rigid analogue design maps the binding conformation of potent N-benzylphenethylamine 5-HT2A serotonin receptor agonist ligands. ACS Chem Neurosci 4:96–109

Kalir A, Szara S (1963) Synthesis and pharmacological activity of fluorinated tryptamine derivatives. J Med Chem 6:716–719

Kantor RE, Dudlettes SD, Shulgin AT (1980) 5-methoxy-alpha-methyltryptamine (alphs, o-dimethylserotonin), a hallucinogenic homolog of serotonin. Biol Psychiatry 15:349–352

Karst M, Halpern JH, Bernateck M, Passie T (2010) The non-hallucinogen 2-bromo-lysergic acid diethylamide as preventative treatment for cluster headache: an open, non-randomized case series. Cephalalgia 30:1140–1144

Krebs KM, Geyer MA (1993) Behavioral characterization of alpha-ethyltryptamine, a tryptamine derivative with MDMA-like properties in rats. Psychopharmacology 113:284–287

Laban U, Kurrasch-Orbaugh D, Marona-Lewicka D, Nichols DE (2001) A novel fluorinated tryptamine with highly potent serotonin 5-HT1A receptor agonist properties. Bioorg Med Chem Lett 11:793–795

Lemaire D, Jacob P III, Shulgin AT (1985) Ring-substituted beta-methoxyphenethylamines: a new class of psychotomimetic agents active in man. J Pharm Pharmacol 37:575–577

Leth-Petersen S, Bundgaard C, Hansen M, Carnerup MA, Kehler J, Kristensen JL (2014) Correlating the metabolic stability of psychedelic 5-HT(2)A agonists with anecdotal reports of human oral bioavailability. Neurochem Res 39:2018–2023

Leth-Petersen S, Gabel-Jensen C, Gillings N, Lehel S, Hansen HD, Knudsen GM, Kristensen JL (2016) metabolic fate of hallucinogenic NBOMes. Chem Res Toxicol 29:96–100

Leuner H, Baer G (1965) Two new short-acting hallucinogens of the psilocybin group. Neuropsychopharmacol 4:471–474

May JA, Chen HH, Rusinko A, Lynch VM, Sharif NA, McLaughlin MA (2003a) A novel and selective 5-HT2 receptor agonist with ocular hypotensive activity: (S)-(+)-1-(2-aminopropyl)-8,9-dihydropyrano[3,2-e]indole. J Med Chem 46:4188–4195

May JA, McLaughlin MA, Sharif NA, Hellberg MR, Dean TR (2003b) Evaluation of the ocular hypotensive response of serotonin 5-HT1A and 5-HT2 receptor ligands in conscious ocular hypertensive cynomolgus monkeys. J Pharmacol Exp Ther 306:301–309

May JA, Dantanarayana AP, Zinke PW, McLaughlin MA, Sharif NA (2006) 1-((S)-2-aminopropyl)-1H-indazol-6-ol: a potent peripherally acting 5-HT2 receptor agonist with ocular hypotensive activity. J Med Chem 49:318–328

May JA, Sharif NA, Chen HH, Liao JC, Kelly CR, Glennon RA, Young R, Li JX, Rice KC, France CP (2009) Pharmacological properties and discriminative stimulus effects of a novel and selective 5-HT2 receptor agonist AL-38022A [(S)-2-(8,9-dihydro-7H-pyrano[2,3-g] indazol-1-yl)-1-methylethylamine]. Pharmacol Biochem Behav 91:307–314

McCorvy JD (2012) Mapping the binding site of the 5-HT2A receptor using mutagenesis and ligand libraries: insights into the molecular actions of psychedelics. In: Chemistry Medicinal, Pharmacology Molecular (eds) vol. Purdue, PhD West Lafayette, IN

McKenna DJ, Nazarali AJ, Hoffman AJ, Nichols DE, Mathis CA, Saavedra JM (1989) Common receptors for hallucinogens in rat brain: a comparative autoradiographic study using [125I]LSD and [125I]DOI, a new psychotomimetic radioligand. Brain Res 476:45–56

McKenna DJ, Repke DB, Lo L, Peroutka SJ (1990) Differential interactions of indolealkylamines with 5-hydroxytryptamine receptor subtypes. Neuropharmacology 29:193–198

McLean TH, Chambers JJ, Parrish JC, Braden MR, Marona-Lewicka D, Kurrasch-Orbaugh D, Nichols DE (2006a) C-(4,5,6-trimethoxyindan-1-yl)methanamine: a mescaline analogue designed using a homology model of the 5-HT2A receptor. J Med Chem 49:4269–4274

McLean TH, Parrish JC, Braden MR, Marona-Lewicka D, Gallardo-Godoy A, Nichols DE (2006b) 1-Aminomethylbenzocycloalkanes: conformationally restricted hallucinogenic phenethylamine analogues as functionally selective 5-HT2A receptor agonists. J Med Chem 49:5794–5803

Monte AP, Marona-Lewicka D, Kanthasamy A, Sanders-Bush E, Nichols DE (1995) Stereoselective LSD-like activity in a series of d-lysergic acid amides of (R)- and (S)-2-aminoalkanes. J Med Chem 38:958–966

Monte AP, Marona-Lewicka D, Parker MA, Wainscott DB, Nelson DL, Nichols DE (1996) Dihydrobenzofuran analogues of hallucinogens. 3. Models of 4-substituted (2,5-dimethoxyphenyl)alkylamine derivatives with rigidified methoxy groups. J Med Chem 39:2953–2961

Monte AP, Waldman SR, Marona-Lewicka D, Wainscott DB, Nelson DL, Sanders-Bush E, Nichols DE (1997) Dihydrobenzofuran analogues of hallucinogens. 4. Mescaline derivatives. J Med Chem 40:2997–3008

Monte AP, Marona-Lewicka D, Lewis MM, Mailman RB, Wainscott DB, Nelson DL, Nichols DE (1998) Substituted naphthofurans as hallucinogenic phenethylamine-ergoline hybrid molecules with unexpected muscarinic antagonist activity. J Med Chem 41:2134–2145

Murphree HB, Dippy RH, Jenney EH, Pfeiffer CC (1961) Effects in normal man of alpha-methyltryptamine and alpha-ethyltryptamine. Clin Pharmacol Ther 2:722–726

Nakada MT, Wieczorek CM, Rainbow TC (1984) Localization and characterization by quantitative autoradiography of [125I]LSD binding sites in rat brain. Neurosci Lett 49:13–18

Nakahara Y, Niwaguchi T, Ishii H (1977) Studies on lysergic acid diethylamide and related compounds. V. Syntheses of dihydrolysergic acid diethylamides and related compounds. Chem Pharm Bull (Tokyo) 25:1756–1763

Nash JF, Roth BL, Brodkin JD, Nichols DE, Gudelsky GA (1994) Effect of the R(-) and S(+) isomers of MDA and MDMA on phosphatidyl inositol turnover in cultured cells expressing 5-HT2A or 5-HT2C receptors. Neurosci Lett 177:111–115

Nichols DE (1981) Structure-activity relationships of phenethylamine hallucinogens. J Pharm Sci 70:839–849

Nichols DE (1986) Hallucinogens. In: Verderame M (ed) Handbook of CNS agents and local anesthetics. CRC Press, Inc, Boca Raton, pp 303–325

Nichols DE (2016) Psychedelics. Pharmacol Rev 68:264–355

Nichols DE, Dyer DC (1977) Lipophilicity and serotonin agonist activity in a series of 4-substituted mescaline analogues. J Med Chem 20:299–301

Nichols DE, Shulgin AT (1976) Sulfur analogs of psychotomimetic amines. J Pharm Sci 65:1554–1556

Nichols DE, Barfknecht CF, Rusterholz DB, Benington F, Morin RD (1973) Asymmetric synthesis of psychotomimetic phenylisopropylamines. J Med Chem 16:480–483

Nichols DE, Barfknecht CF, Long JP, Standridge RT, Howell HG, Partyka RA, Dyer DC (1974) Potential psychotomimetics. 2. Rigid analogs of 2,5-dimethoxy-4- methylphenylisopropylamine (DOM, STP). J Med Chem 17:161–166

Nichols DE, Shulgin AT, Dyer DC (1977) Directional lipophilic character in a series of psychotomimetic phenethylamine derivatives. Life Sci 21:569–575

Nichols DE, Pfister WR, Yim GK (1978) LSD and phenethylamine hallucinogens: new structural analogy and implications for receptor geometry. Life Sci 22:2165–2170

Nichols DE, Woodard R, Hathaway BA, Lowy MT, Yom KW (1979) Resolution and absolute configuration of trans-2-(2,5-dimethoxy-4- methylphenyl)cyclopropylamine, a potent hallucinogen analogue. J Med Chem 22:458–460

Nichols DE, Jadhav KP, Oberlender RA, Zabik JE, Bossart JF, Hamada A, Miller DD (1984a) Synthesis and evaluation of substituted 2-phenylcyclobutylamines as analogues of hallucinogenic phenethylamines: lack of LSD-like biological activity. J Med Chem 27:1108–1111

Nichols DE, Schooler D, Yeung MC, Oberlender RA, Zabik JE (1984b) Unreliability of the rat stomach fundus as a predictor of hallucinogenic activity in substituted phenethylamines. Life Sci 35:1343–1348

Nichols DE, Hoffman AJ, Oberlender RA, Riggs RM (1986) Synthesis and evaluation of 2,3-dihydrobenzofuran analogues of the hallucinogen 1-(2,5-dimethoxy-4-methylphenyl)-2-aminopropane: drug discrimination studies in rats. J Med Chem 29:302–304

Nichols DE, Lloyd DH, Johnson MP, Hoffman AJ (1988) Synthesis and serotonin receptor affinities of a series of enantiomers of alpha-methyltryptamines: evidence for the binding conformation of tryptamines at serotonin 5-HT1B receptors. J Med Chem 31:1406–1412

Nichols DE, Snyder SE, Oberlender R, Johnson MP, Huang XM (1991) 2,3-Dihydrobenzofuran analogues of hallucinogenic phenethylamines. J Med Chem 34:276–281

Nichols DE, Frescas S, Marona-Lewicka D, Huang X, Roth BL, Gudelsky GA, Nash JF (1994) 1-(2,5-Dimethoxy-4-(trifluoromethyl)phenyl)-2-aminopropane: a potent serotonin 5-HT2A/2C agonist. J Med Chem 37:4346–4351

Nichols DE, Frescas S, Marona-Lewicka D, Kurrasch-Orbaugh DM (2002) Lysergamides of isomeric 2,4-dimethylazetidines map the binding orientation of the diethylamide moiety in the potent hallucinogenic agent N, N-diethyllysergamide (LSD). J Med Chem 45:4344–4349

Nichols DE, Frescas SP, Chemel BR, Rehder KS, Zhong D, Lewin AH (2008) High specific activity tritium-labeled N-(2-methoxybenzyl)-2,5-dimethoxy-4-iodophenethylamine (INBMeO): a high-affinity 5-HT2A receptor-selective agonist radioligand. Bioorg Med Chem 16:6116–6123

Nichols DE, Sassano MF, Halberstadt AL, Klein LM, Brandt SD, Elliott SP, Fiedler WJ (2015) N-Benzyl-5-methoxytryptamines as potent serotonin 5-HT2 receptor family agonists and comparison with a series of phenethylamine analogues. ACS Chem Neurosci 6:1165–1175

Nikolaou P, Papoutsis I, Stefanidou M, Spiliopoulou C, Athanaselis S (2015) 2C-I-NBOMe, an "N-bomb" that kills with "Smiles". Toxicological and legislative aspects. Drug Chem Toxicol 38:113–119

Oberlender RA, Kothari PJ, Nichols DE, Zabik JE (1984) Substituent branching in phenethylamine-type hallucinogens: a comparison of 1-[2,5-dimethoxy-4-(2-butyl)phenyl]-

2-aminopropane and 1- [2,5-dimethoxy-4-(2-methylpropyl)phenyl]-2-aminopropane. J Med Chem 27:788–792

Oberlender R, Pfaff RC, Johnson MP, Huang XM, Nichols DE (1992) Stereoselective LSD-like activity in d-lysergic acid amides of (R)- and (S)-2-aminobutane. J Med Chem 35:203–211

Oberlender R, Ramachandran PV, Johnson MP, Huang X, Nichols DE (1995) Effect of a chiral 4-alkyl substituent in hallucinogenic amphetamines. J Med Chem 38:3593–3601

Parker MA, Marona-Lewicka D, Lucaites VL, Nelson DL, Nichols DE (1998) A novel (benzodifuranyl)aminoalkane with extremely potent activity at the 5-HT2A receptor. J Med Chem 41:5148–5149

Parrish JC (2006) Toward a molecular understanding of hallucinogen action. PhD, Purdue University

Parrish JC, Braden MR, Gundy E, Nichols DE (2005) Differential phospholipase C activation by phenylalkylamine serotonin 5-HT 2A receptor agonists. J Neurochem 95:1575–1584

Peretz DI, Smythies JR, Gibson WC (1955) A new hallucinogen: 3,4,5-trimethoxyphenyl-beta-aminopropane (with notes on a stroboscopic phenomenon). J Mental Sci 101:317–329

Pigott A, Frescas S, McCorvy JD, Huang XP, Roth BL, Nichols DE (2012) trans-2-(2,5-Dimethoxy-4-iodophenyl)cyclopropylamine and trans-2-(2,5-dimethoxy-4-bromophenyl)cyclopropylamine as potent agonists for the 5-HT(2) receptor family. Beilstein J Org Chem 8:1705–1709

Poklis JL, Devers KG, Arbefeville EF, Pearson JM, Houston E, Poklis A (2014) Postmortem detection of 25I-NBOMe [2-(4-iodo-2,5-dimethoxyphenyl)-N-[(2-methoxyphenyl)methyl] ethanamine] in fluids and tissues determined by high performance liquid chromatography with tandem mass spectrometry from a traumatic death. Forensic Sci Int 234:e14–e20

Rangisetty JB, Dukat M, Dowd CS, Herrick-Davis K, Dupre A, Gadepalli S, Teitler M, Kelley CR, Sharif NA, Glennon RA (2001) 1-[2-methoxy-5-(3-phenylpropyl)]-2-aminopropane unexpectedly shows 5-HT(2A) serotonin receptor affinity and antagonist character. J Med Chem 44:3283–3291

Ray TS (2010) Psychedelics and the human receptorome. PLoS ONE 5:e9019

Repke DB, Grotjahn DB, Shulgin AT (1985) Psychotomimetic N-methyl-N-isopropyltryptamines. Effects of variation of aromatic oxygen substituents. J Med Chem 28:892–896

Rusterholz DB, Long JP, Nichols DE (1979) Effect of alpha-methyltryptamine on spontaneous activity in mice. Pharmacol Biochem Behav 10:223–227

Sargent T III, Kalbhen DA, Shulgin AT, Stauffer H, Kusubov N (1975) A potential new brain-scanning agent: 4-77Br-2,5-dimethoxyphenylisoproplamine (4-Br-DPIA). J Nucl Med 16:243–245

Schechter MD (1998) MDMA-like stimulus effects of hallucinogens in male Fawn-Hooded rats. Pharmacol Biochem Behav 59:265–270

Schultz DM, Prescher JA, Kidd S, Marona-Lewicka D, Nichols DE, Monte A (2008) 'Hybrid' benzofuran-benzopyran congeners as rigid analogs of hallucinogenic phenethylamines. Bioorg Med Chem 16:6242–6251

Seggel MR, Yousif MY, Lyon RA, Titeler M, Roth BL, Suba EA, Glennon RA (1990) A structure-affinity study of the binding of 4-substituted analogues of 1-(2,5-dimethoxyphenyl)-2-aminopropane at 5-HT2 serotonin receptors. J Med Chem 33:1032–1036

Shulgin AT (1968) The ethyl homologs of 2,4,5-trimethoxyphenylisopropylamine. J Med Chem 11:186–187

Shulgin AT (1978) Psychotomimetic Drugs: Structure-activity relationships. In: Iversen LL et al (eds) Handbook of psychopharmacology, vol 11. Plenum Press, New York, pp 243–333

Shulgin AT (1982) Chemistry of Psychotomimetics. In: Hoffmeister F, Stille G (eds) Handbook of experimental pharmacology, vol 55/III. Springer, New York, pp 3–29

Shulgin AT, Carter MF (1980) N, N-Diisopropyltryptamine (DIPT) and 5-methoxy-N, N-diisopropyltryptamine (5-MeO-DIPT). Two orally active tryptamine analogs with CNS activity. Commun Psychopharmacol 4:363–369

Shulgin A, Shulgin A (1991) PIHKAL A chemical love story. Berkeley, CA 94701: Transform Press

Shulgin AT, Shulgin A (1997) TIHKAL. The continuation. Transform Press, Berkeley
Shulgin AT, Sargent T, Naranjo C (1969) Structure–activity relationships of one-ring psychotomimetics. Nature 221:537–541
Shulgin AT, Sargent T, Naranjo C (1971) 4-Bromo-2,5-dimethoxyphenylisopropylamine, a new centrally active amphetamine analog. Pharmacology 5:103–107
Siva Sankar DV (1975) LSD: a total study. PJD Publications, Westbury
Stoll A, Hofmann A (1955) Amide der sterreoisomeren lysergsauren und dihydro-lysergsauren. Helv Chim Acta 38:421–433
Strano Rossi S, Odoardi S, Gregori A, Peluso G, Ripani L, Ortar G, Serpelloni G, Romolo FS (2014) An analytical approach to the forensic identification of different classes of new psychoactive substances (NPSs) in seized materials. Rapid Commun Mass Spectrom 28:1904–1916
Szara S (1961) 6-Hydroxylation: an important metabolic route for alpha-methyltryptamine. Experientia 17:76–77
Tomaszewski Z, Johnson MP, Huang X, Nichols DE (1992) Benzofuran bioisosteres of hallucinogenic tryptamines. J Med Chem 35:2061–2064
Trachsel D, Nichols DE, Kidd S, Hadorn M, Baumberger F (2009) 4-aryl-substituted 2,5-dimethoxyphenethylamines: synthesis and serotonin 5-HT(2A) receptor affinities. Chem Biodivers 6:692–704
Urban JD, Clarke WP, von Zastrow M, Nichols DE, Kobilka B, Weinstein H, Javitch JA, Roth BL, Christopoulos A, Sexton PM, Miller KJ, Spedding M, Mailman RB (2007) Functional selectivity and classical concepts of quantitative pharmacology. J Pharmacol Exp Ther 320:1–13
Vangveravong S, Kanthasamy A, Lucaites VL, Nelson DL, Nichols DE (1998) Synthesis and serotonin receptor affinities of a series of trans-2-(indol-3-yl)cyclopropylamine derivatives. J Med Chem 41:4995–5001
Wacker D, Wang C, Katritch V, Han GW, Huang XP, Vardy E, McCorvy JD, Jiang Y, Chu M, Siu FY, Liu W, Xu HE, Cherezov V, Roth BL, Stevens RC (2013) Structural features for functional selectivity at serotonin receptors. Science 340:615–619
Walters GC, Cooper PD (1968) Alicyclic analogue of mescaline. Nature 218:298–300
Walterscheid JP, Phillips GT, Lopez AE, Gonsoulin ML, Chen HH, Sanchez LA (2014) Pathological findings in 2 cases of fatal 25I-NBOMe toxicity. Am J Forensic Med Pathol 35:20–25
Wang C, Jiang Y, Ma J, Wu H, Wacker D, Katritch V, Han GW, Liu W, Huang XP, Vardy E, McCorvy JD, Gao X, Zhou XE, Melcher K, Zhang C, Bai F, Yang H, Yang L, Jiang H, Roth BL, Cherezov V, Stevens RC, Xu HE (2013) Structural basis for molecular recognition at serotonin receptors. Science 340:610–614
Watts SW, Gackenheimer SL, Gehlert DR, Cohen ML (1994) Autoradiographic comparison of [125I]LSD-labeled 5-HT2A receptor distribution in rat and guinea pig brain. Neurochem Int 24:565–574
Whiteside MS, Kurrasch-Orbaugh D, Marona-Lewicka D, Nichols DE, Monte A (2002) Substituted hexahydrobenzodipyrans as 5-HT2A/2C receptor probes. Bioorg Med Chem 10:3301–3306
Winter JC, Gessner PK, Godse DD (1967) Synthesis of some 3-indenealkylamines. Comparison of the biological activity of 3-indenealkylamines and 3-benzo[b]thiophenealkylamines with their tryptamine isosteres. J Med Chem 10:856–858
Yin X, Xu H, Hanson M, Liu W (2014) GPCR crystallization using lipidic cubic phase technique. Curr Pharm Biotechnol 15:971–979

Hallucinogens and Serotonin 5-HT_{2A} Receptor-Mediated Signaling Pathways

Juan F. López-Giménez and Javier González-Maeso

Abstract The neuropsychological effects of naturally occurring psychoactive chemicals have been recognized for millennia. Hallucinogens, which include naturally occurring chemicals such as mescaline and psilocybin, as well as synthetic compounds, such as lysergic acid diethylamide (LSD), induce profound alterations of human consciousness, emotion, and cognition. The discovery of the hallucinogenic effects of LSD and the observations that LSD and the endogenous ligand serotonin share chemical and pharmacological profiles led to the suggestion that biogenic amines like serotonin were involved in the psychosis of mental disorders such as schizophrenia. Although they bind other G protein-coupled receptor (GPCR) subtypes, studies indicate that several effects of hallucinogens involve agonist activity at the serotonin 5-HT_{2A} receptor. In this chapter, we review recent advances in understanding hallucinogen drug action through characterization of structure, neuroanatomical location, and function of the 5-HT_{2A} receptor.

Keywords Serotonin · 5-HT_{2A} receptor · G protein-coupled receptor (GPCR) · Lysergic acid diethylamide (LSD) · Psilocin · Psilocybin · Mescaline · Schizophrenia · Psychosis · Hallucinogen · Psychedelics · Antipsychotics

J.F. López-Giménez
Instituto de Biomedicina y Biotecnología de Cantabria IBBTEC-CSIC,
39011 Santander, Cantabria, Spain
e-mail: jf.lopez.gimenez@csic.es

J. González-Maeso (✉)
Department of Physiology and Biophysics, School of Medicine,
Virginia Commonwealth University, Richmond, VA 23298, USA
e-mail: jgmaeso@vcu.edu

Curr Topics Behav Neurosci (2017) 36: 45–74
DOI 10.1007/7854_2017_478
Published Online: 05 July 2017

Contents

1 Introduction to Receptor Pharmacology

The concept of cellular receptors as we currently know was originally conceived at the end of the nineteenth century and the beginning of the twentieth century by pioneering physiologists such as Langley (1909) and Erlich (1913). At that time, it was speculated that "something" located either at the cell surface or within cells could "read" the chemical information contained in substances present in the environment in such a way that this information is finally converted into observable physiological effects. Later in the 1930s, the British pharmacologist Alfred Joseph Clark introduced the receptor theory as a quantitative discipline. He proposed that the data obtained in experimental assays were the result of unimolecular interactions between the evaluated drug and a type of "substance" found at the cell surface corresponding to what we know now as receptor (Clark 1933).

Currently, it is widely accepted that receptors exist in several forms, including proteins localized at the cell surface, such as enzymes, ion channels, and transporters, as well as nuclear and cytosolic proteins. These receptors are organized in superfamilies according to their protein structure; the largest of these groups includes the seven transmembrane (7TM) receptors, also known as G protein-coupled receptors (GPCRs) because their final response is essentially generated by the activation of heterotrimeric G proteins. From a structural point of view, these receptors are peptides that cross the lipid bilayer through seven transmembrane domains in such a way that they are embedded in the plasma membrane with the amino terminus facing the extracellular space and the carboxyl terminus orientated towards the cytoplasm. At the same time, three extracellular and three intracellular loops interconnect these domains (Katritch et al. 2013). The recent explosion in GPCR crystal structures confirms these biophysical features (Rosenbaum et al. 2009; Audet and Bouvier 2012).

The most extensively characterized mechanism of action of these receptors resides in their coupling to G proteins after binding to a specific activating element known in pharmacology as an agonist. G proteins are heterotrimeric complexes found at the plasma membrane and they are constituted by monomeric Gα subunits and Gβγ dimeric subunits (Oldham and Hamm 2008). Following their interaction with an activated receptor, Gα monomeric proteins dissociate from Gβγ subunits and exchange GDP guanine nucleotide for GTP. Subsequently, the Gα subunits bound to GTP initiate different second messenger cascades at the intracellular level by interacting with other proteins such as adenylyl cyclases or phospholipases. The sequential activation or inhibition of these different elements in a cascade fashion constitutes the biological process known as cell signaling, which may be initiated at the cell surface by the activation of receptors and often concludes in the cell nucleus with modulated transcription of different genes (Ferguson 2001).

All of these cell signaling processes, initiated upon the activation of GPCRs, can be terminated at different points. At the G protein level, bound GTP is hydrolyzed to GDP through the GTPase activity of the Gα subunit, permitting its association with Gβγ subunits to reassemble the heterotrimeric complex, allowing the signaling process to commence again. In the case of GPCRs, receptor activation may be terminated by phosphorylation of different amino acids in the intracellular loops or in the carboxy terminus by specific receptor kinases known as GPCR kinases (GRKs), which leads to receptor desensitization with respect to its G protein coupling. At the same time, phosphorylated receptors are susceptible to interact with β-arrestins, a family of cytosolic proteins involved in the endocytosis of receptors after their activation by agonist molecules (Luttrell 2008). Once receptors are internalized within the different intracellular compartments, the endosomes may either divert to lysosomes by following different degradation pathways, or alternatively may recycle back to the plasma membrane where the receptors will be restored in a state that is susceptible to activation once again by an agonist ligand.

Receptor internalization can be considered as another mechanism that completes the cell signaling processes initiated by receptor activation (Hanyaloglu and von Zastrow 2008). This general paradigm has been widely accepted until the present time and it is based, for the most part, on the investigations carried out over several decades by Robert Lefkowitz and his collaborators who used the β_2 adrenoceptor as an experimental model (Pierce et al. 2002). Nevertheless, new findings discovered in more recent investigations have contributed to a conceptual expanding of this general mechanistic model. Specifically, some recent reports suggest that β-arrestins play new roles in addition to those originally linked to receptor endocytosis. One of these new functions includes the participation of β-arrestins as scaffolding elements that help to facilitate the transmission of signaling from receptors to diverse intracellular effectors—indicating that G proteins, although considered as the canonical pathway for the signal transmission in GPCRs, are not the exclusive elements connecting these receptors to cell signaling pathways (Luttrell and Lefkowitz 2002). Similarly, recent investigations conducted with β_2 adrenoceptors have revealed their capacity to couple to Gα subunits from the lumen of endosomes, meaning that the activation of intracellular effectors continues following endocytosis (Irannejad et al. 2013). Taken together, this intricate amalgam of interacting proteins demonstrates the extreme complexity of the molecular mechanisms involved in the function of these receptors at the cellular level, and makes patent the large number of unanswered questions remaining in this field of knowledge. As evidence of their biological relevance, approximately 900 different types of GPCRs have been identified in the human genome; these receptors participate in the majority of the physiological processes including sense perceptions, as well as neuropsychological, cardiovascular, and endocrine functions, and are the therapeutic target of nearly half of the total number of medicines that are currently prescribed for the treatment of diseases (Overington et al. 2006).

2 Serotonin and Serotonin Receptors: A Historical Perspective

2.1 Serotonin

The discovery of serotonin (5-hydroxytryptamine, 5-HT) dates back to the middle of the nineteenth century when early experimenters recognized that a substance contained in serum was capable of inducing the contraction of smooth muscle. Later in the first third of the twentieth century, Italian scientists extracted a substance from enterochromaffin cells in the gastrointestinal tract, named for this reason "enteramine", that also caused smooth muscle contraction, particularly in stomach and uterus (Mohammad-Zadeh et al. 2008). Nearly at the same time, during the 1940s, Maurice Rapport isolated and characterized a substance from blood that acted as a vasoconstrictor; they called the compound serotonin due to these peculiarities, i.e., *serum* and *tonic* (Rapport et al. 1948). The similar chemical structures observed for enteramine and serotonin, basically defined by the presence of an indole entity, led to conjecture that both substances corresponded essentially to the same compound; this point was corroborated later when serotonin was first synthesized and found to have the same properties as the substances obtained from natural sources, i.e., enterochromaffin cells and serum (Reid and Rand 1952).

The intimate relationship between 5-HT signaling and hallucinogens was envisaged during the period of intense research activity that coincided with the discovery of this biogenic amine. In 1943, the Swiss chemist Albert Hoffman (1906–2008) serendipitously discovered the hallucinogenic properties of lysergic acid diethylamide (LSD) (Hofmann 1979; Nichols 2004). In 1951, John Gaddum at the University of Edinburgh reported the antagonistic action of LSD on the effects induced by 5-HT in rat uterus and rabbit ear preparations (see Green 2008). These results, together with the demonstration in 1968 by George Aghajanian at Yale University that LSD modulates the activity of midbrain neurons containing 5-HT (Aghajanian et al. 1968; Aghajanian 2009), led to conjecture that the psychoactive effects of LSD were mediated by its interaction with the serotonergic system in the CNS.

Currently, it is common knowledge that 5-HT is present in all animal organisms where it participates in numerous physiological functions by acting either as a hormone or as a neurotransmitter. 5-HT can be found in the body at the peripheral level in platelets, mastocytes, and enterochromaffin cells, and in the central nervous system (CNS) in serotonergic neurons located preferentially in the brainstem, which only contains 1–2% of the total amount of 5-HT present in the whole organism (Hannon and Hoyer 2008; Berger et al. 2009).

5-HT is synthesized from tryptophan, an essential amino acid that is obtained from food, following two biochemical reactions. First, tryptophan is hydroxylated to 5-hydroxytryptophan by the enzyme tryptophan hydroxylase. Second, 5-hydroxytryptophan is decarboxylated by an amino decarboxylase, resulting finally in 5-HT. Tryptophan hydroxylase is the rate-limiting step in this sequential reaction. Control of the synthesis of 5-HT is also determined by the availability of oxygen, the heterocyclic compound pteridine, and by the amount of tryptophan present in the bloodstream. The principal route of degradation of 5-HT is by deamination, which is performed by monoamine oxidase (MAO) enzymes. The deamination of 5-HT results in the formation of 5-hydroxyindolacetaldehyde, which in turn is oxidized to the final metabolite 5-hydoxyindolacetic acid or 5-HIAA (Mohammad-Zadeh et al. 2008).

In the CNS, 5-HT plays a fundamental role as a neurotransmitter. However, 5-HT is not able to cross the blood–brain barrier, which means that it must be synthesized within the brain from tryptophan. This synthesis takes place in a cluster of specific neurons with their soma restrictively located in the different raphe nuclei of the brainstem (Dahlström and Fuxe 1964). Axons originating from these neurons innervate practically the entire brain, projecting their terminals to either the forebrain (upper raphe nuclei) or to the spinal cord (lower raphe nuclei) (Tork 1990). 5-HT is released from terminals into the synaptic cleft where it is liberated to interact with specific receptors located mainly in postsynaptic neurons. Once neurotransmission occurs, 5-HT is taken back up into serotonergic neurons by specific transporters found on the plasma membrane of presynaptic terminals and is either stored for future synaptic release or metabolized by MAO enzymes.

5-HT performs its physiological functions by binding to specific cell membrane receptors. Currently, 14 different serotonin receptors, classified into 7 subfamilies according to their primary structure and functional properties, have been described. Excluding 5-HT$_3$, which belongs to the ion channel receptor superfamily, the remainder of the 5-HT receptors are GPCRs (Hoyer et al. 1994).

2.2 Serotonin Receptors

Despite the fact that the existence of at least two different subtypes of 5-HT receptors (initially referred as D and M) had been reported in the 1950s (Gaddum and Picarelli 1957), it was not until the mid-1970s when different 5-HT receptor subtypes were pharmacologically characterized in mammalian brain homogenates using radioligand binding techniques newly developed during that period. Several radioligands developed during that period (such as [^{3}H]5-HT, [^{3}H]LSD and [^{3}H] spiperone) were found to bind to sites suspected to be 5-HT receptors (Peroutka and Snyder 1979). During the course of those early investigations, it was observed that radioligands differentiated between two classes of 5-HT sites; binding sites with high affinity for [^{3}H]5-HT were designated as 5-HT_1 receptors, whereas a second population of sites with high affinity for [^{3}H]spiperone and low affinity for [^{3}H] 5-HT were designated as 5-HT_2 receptors. In this way, by the mid-1980s up to five different 5-HT receptors were described based on their pharmacological profile. The extensive development of new molecular biology techniques that occurred at the end of the 1980s through the beginning of the 1990s permitted cloning of the 14 5-HT receptor subtypes that are currently known. Thus, the contemporary classification of 5-HT receptors has been made according to genetic homologies; consequently, the classification of 5-HT receptors is based on their primary structure, as determined by their amino acid sequence, which in turn is responsible for the functional properties of these cell membrane proteins (for reviews, see Hannon and Hoyer 2008; Hoyer et al. 1994; Millan et al. 2008; Nichols and Nichols 2008; Raymond et al. 2001; Barnes and Sharp 1999).

The 5-HT_2 receptor subfamily comprises three different subtypes, namely 5-HT_{2A}, 5-HT_{2B} and 5-HT_{2C}, which are grouped together due to their high structural homology (their genetic sequences are about 50% identical). After interaction with 5-HT or another agonist, 5-HT_2 receptors couple preferentially to $G_{q/11}$ proteins, promoting the subsequent activation of phospholipase C (PLC). In turn, PLC activation promotes the generation of intracellular inositol phosphates (IP) and promotes the mobilization of intracellular calcium. The 5-HT_{2A} receptor corresponds to the D-type 5-HT receptor described by Gaddum and Picarelli and later characterized by Peroutka and Snyder as a site exhibiting high affinity for [^{3}H] spiperone. There is extensive evidence that the 5-HT_{2A} receptor is responsible for the neuropsychological effects of serotonergic hallucinogens in animal models used for experimentation as well as in human subjects (Halberstadt 2015). This chapter is focused on the biochemical and pharmacological properties of the 5-HT_{2A} receptor in relation to its role in the effects of hallucinogenic drugs of abuse, such as LSD, mescaline, and psilocybin.

3 Chemical Neuroanatomy of 5-HT_{2A} Receptors

The major physiological effects induced by hallucinogens, in particular when evaluated in human subjects, are related to altered states of consciousness, including changes in cognition, mood, and perception. It is widely accepted at the present time that these effects are generated mostly by the interaction of hallucinogens with 5-HT_{2A} receptors as agonists (Halberstadt 2015; Vollenweider et al. 1998; Gonzalez-Maeso et al. 2007; Schmid et al. 2015). Although hallucinogens do not bind exclusively to 5-HT_{2A} receptors (LSD binds to most 5-HT receptor subtypes as well as to dopaminergic and adrenergic receptors), it has been evidenced in both humans and experimental animals that the activation of 5-HT_{2A} receptors is necessary to generate hallucinogenesis and a related behavioral response in animals (Vollenweider et al. 1998; Gonzalez-Maeso et al. 2007). Therefore, the study of the anatomic distribution of 5-HT_{2A} receptor in the CNS is essential to elucidate what brain structures are implicated in the neuropsychological effects elicited by hallucinogenic compounds. As noted above, the development of radiochemicals based on the isotopic labeling of particular ligands permitted the characterization of receptors as binding sites specifically recognized by these radioligands. In addition, by exposure to sensitive autoradiographic films, these radioligands can be used to visualize the distribution of their binding sites in histological sections obtained from the human brain postmortem. The initial receptor autoradiography studies investigating 5-HT_{2A} receptor localization were conducted using either antagonist ([^{3}H]spiperone and [^{3}H]ketanserin) or agonist ([^{3}H]LSD and [^{125}I]DOI) radioligands (Mengod et al. 2006). Although all of these radioligands have high affinity for 5-HT_{2A} receptors, they are not completely selective because they bind to sites corresponding to dopamine receptors ([^{3}H]spiperone), adrenergic receptors ([^{3}H]ketanserin), or to other 5-HT_2 receptor subtypes ([^{125}I]DOI). Because of this nonspecific binding, blockers for those undesired binding sites must be used in order to obtain a specific signal corresponding to 5-HT_{2A} receptors. Another issue that must be taken into consideration when conducting quantitative receptor autoradiography experiments is the pharmacological nature of the radioligand used (in terms of being an agonist or an antagonist). According to the ternary complex model of drug receptor interactions (see below), antagonist radioligands label the entire population of receptors (inactive [R] and active [R^*] states), whereas agonist radioligands selectively bind to the receptors present in their active conformation (R^*)—therefore detecting only a fraction of the total receptor population. The development of [^{3}H]MDL100,907, a highly selective 5-HT_{2A} receptor antagonist, in the mid-1990s, addressed many of the problems associated with other 5-HT_{2A} receptor radioligands (Johnson et al. 1996), particularly in receptor autoradiography studies where [^{3}H]MDL100,907 displayed a remarkably specific signal devoid of nonspecific binding (Lopez-Gimenez et al. 1997, 1998).

The anatomic localization of 5-HT_{2A} receptors in primate brain visualized with [^{3}H]MDL100,907 showed a heterogeneous and wide-ranging distribution throughout different brain areas, and was comparable to that observed in other mammalian species (Lopez-Gimenez et al. 2001). The region containing the highest density of 5-HT_{2A} binding sites was the neocortex, where the autoradiographic signal displayed a banded pattern corresponding predominantly to pyramidal neurons distributed according to the cytoarchitectural organization of different cortical areas. Other regions where 5-HT_{2A} receptors were detected included structures of the hippocampus, thalamic nuclei, the mammillary bodies in the hypothalamus, and different nuclei of the midbrain.

The corpus striatum is a brain structure where 5-HT_{2A} receptors distribution is paradoxical when comparing different mammalian species. Initial investigations of 5-HT_{2A} receptor distribution, performed using [^{3}H]ketanserin as the radioligand, described a homogeneous pattern of labeling throughout caudate and putamen nuclei in both rat and human brain samples, with a high component of nonspecific binding (Pazos et al. 1985, 1987). The nonspecific binding was attributed to the vesicular monoamine transporter since it could be blocked by tetrabenazine.

Interestingly, when equivalent histological sections were incubated with agonist radioligands such as [^{3}H]LSD or [^{125}I]DOI, the autoradiographic signal did not show the abundance of nonspecific binding observed with [^{3}H]ketanserin, and the pattern of distribution was markedly heterogeneous (Appel et al. 1990; Waeber and Palacios 1994). Binding sites were particularly enriched in the nucleus accumbens and in the posterior portion of the caudate-putamen in rat brain, whereas in human samples the distribution displayed a characteristic patchy pattern in both caudate and putamen nuclei corresponding to striosomes.

Striosomes were originally discovered by acetylcholinesterase histochemistry and are anatomical structures differentiated from the rest of the striatum or matrix. In terms of neuronal connections, striosomes and matrix can be differentiated based on their input–output systems (Gerfen 1992). Afferents reaching the matrix originate predominantly from areas related to sensorimotor processing, whereas striosomes receive inputs essentially from limbic regions. On the other hand, the caudate and putamen matrix projects mainly to the pallidum and the substantia nigra pars reticulata, whereas striosomal afferents target dopaminergic nigral neurons. Other radioligands derived from psychoactive compounds, such as benzodiazepines and opioids, also displayed a patchy distribution of labeling in human striatum; suggesting, therefore, the participation of this anatomic structure in neuropsychological effects produced by these types of drugs (Graybiel 1990). Although it was originally speculated that 5-HT_{2A} binding sites in striosomes may correspond to the active conformation of these receptors, where agonist drugs with hallucinogenic effects (such as [^{3}H]LSD and [^{125}I]DOI) bind, this possibility was later disproved by experiments performed with the antagonist radioligand [^{3}H]MDL100,907, which produced the same patchy distribution of striatal labeling that was observed in consecutive histological sections treated with [^{125}I]DOI (Lopez-Gimenez et al. 1999) (Fig. 1).

The peculiar compartmentalized distribution of 5-HT_{2A} receptors in the human corpus striatum is not found in all mammalian species. [^{125}I]DOI labeling of 5-HT_{2A} receptors in striosomes has been clearly observed in human, mouse, and guinea pig brain. By contrast, no striosome labeling has been detected in rat, cat, pig, cow, or monkey brain (Waeber and Palacios 1994). The results reported for cow and monkey

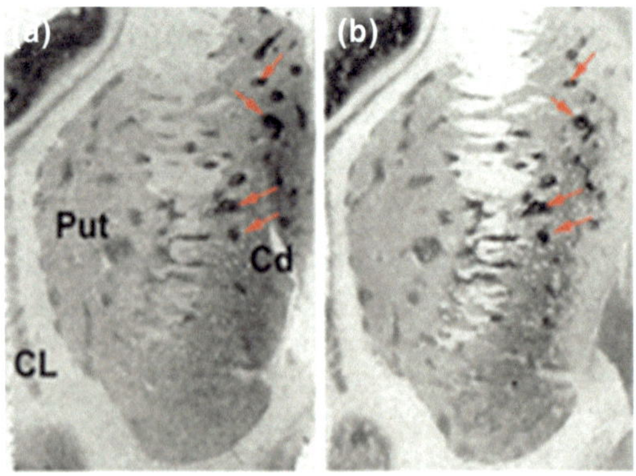

Fig. 1 Consecutive histological sections of human striatum showing the distribution of binding sites labeled by [^{125}I]DOI (**a**) and [^{3}H]MDL100,907 (**b**). *Arrow* heads indicate some of the striosomes visualized with these radioligands. Cd, caudate; Put, putamen; CL, claustrum

are especially remarkable: the autoradiographic signal in cow striatum is high, specific, and homogeneous, whereas only sparse 5-HT$_{2A}$ receptor labeling is detected in monkey caudate, putamen, and accumbens. The cause of these cross-species differences remains to be elucidated. A phylogenetic explanation can be excluded because species closely related in evolutionary terms, such as primates (human and monkey) and rodents (rat, mouse, and guinea pig), often exhibit substantial differences with regard to the chemical neuroanatomy of their striatal 5-HT$_{2A}$ receptors. These differences could have fundamental functional consequences and should be taken into account when evaluating drugs that interact with 5-HT$_{2A}$ receptors, particularly when interpreting and comparing results obtained from behavioral pharmacology studies conducted using common laboratory species such as rat and mouse, or when extrapolating those results to human subjects.

4 Hallucinogenic and Non-Hallucinogenic 5-HT$_{2A}$ Receptor Agonists

The mechanism of action of hallucinogens has attracted the attention of pharmacologists and neuroscientists for decades (Nichols 2004; Aghajanian and Marek 1999; Fantegrossi et al. 2008). These compounds elicit profound alterations of cognition, perception, and mood, and have been used by most human cultures for millennia (Hanks and Gonzalez-Maeso 2013). The role of the 5-HT$_{2A}$ receptor in the mechanism of action of hallucinogens was first proposed by Richard Glennon, Milt Titeler and their teams (Glennon et al. 1984, 1986). However, it was not until the development of 5-HT$_{2A}$ knockout mice in 2003 that the fundamental role of 5-HT$_{2A}$ receptor-dependent signaling in the cellular and behavioral effects of hallucinogens was verified conclusively (Gonzalez-Maeso et al. 2003, 2007). These findings in rodent models are further supported by studies conducted by Franz Vollenweider and his collaborators, which demonstrate that the psychosis-like effects of LSD and psilocybin in healthy volunteers are reversed by the 5-HT$_{2A}$ receptor antagonist ketanserin (Vollenweider et al. 1998; Schmid et al. 2015; Preller et al. 2016, 2017).

From a basic pharmacological perspective, it is particularly interesting that whereas all hallucinogens (such as LSD, mescaline, and psilocin) bind with high affinity and activate the serotonin 5-HT$_{2A}$ receptor, certain closely related 5-HT$_{2A}$ receptor agonists, such as lisuride and ergotamine, do not behave as hallucinogens in humans (Marona-Lewicka et al. 2002; Silbergeld and Hruska 1979; Mokler et al. 1983; Adams and Geyer 1985); indeed, these non-hallucinogenic 5-HT$_{2A}$ receptor agonists are used as therapeutic drugs in the treatment of diseases such as migraine and Parkinson's disease (Fig. 2). Hallucinogenic and non-hallucinogenic 5-HT$_{2A}$ receptor agonists, therefore, serve as attractive pharmacological tools to investigate the molecular and signaling mechanisms that allow chemically related agonists to target the same receptor molecule but induce different neuropsychological effects. What might explain the differences between the neuropsychological effects of hallucinogenic and non-hallucinogenic 5-HT$_{2A}$ agonists?

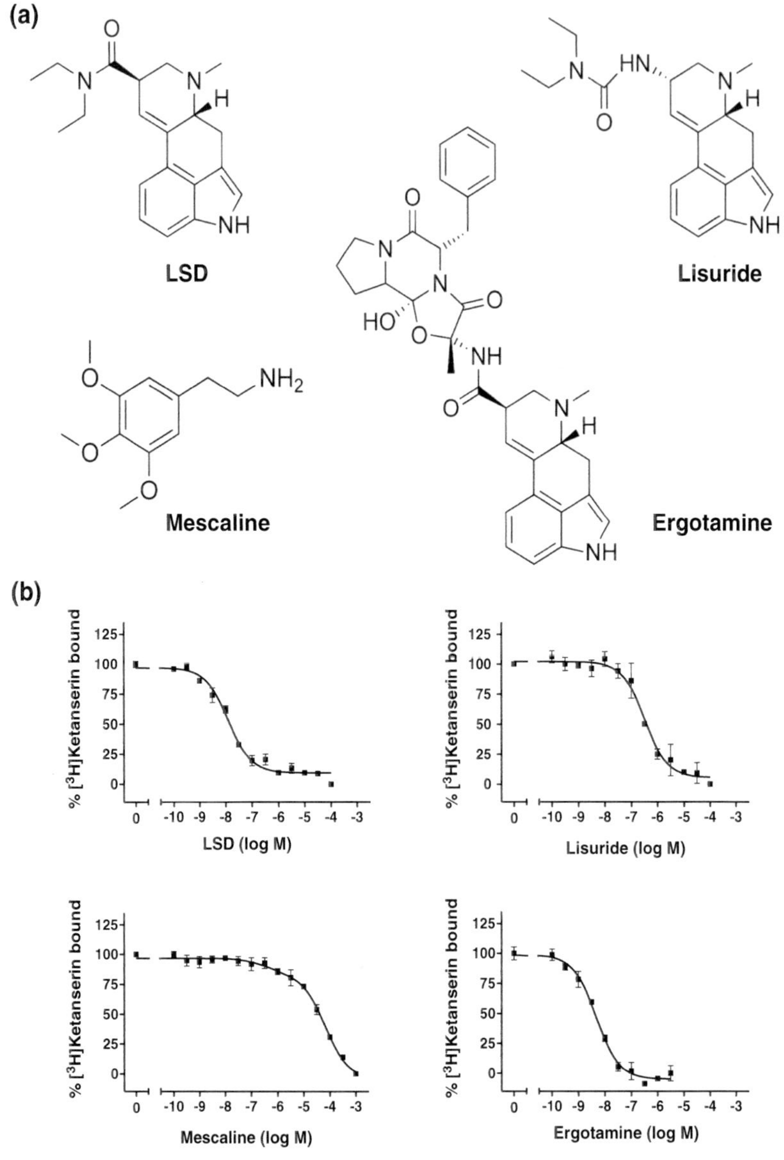

Fig. 2 **a** Chemical structures of hallucinogen (LSD and mescaline) and non-hallucinogen (lisuride and ergotamine) 5-HT_{2A} receptor agonists. **b** [^{3}H]Ketanserin competition curves by LSD, mescaline, lisuride and ergotamine in mouse somatosensory cortex plasma membrane preparations (see Gonzalez-Maeso et al. 2007). Note that the affinity of the ligand does not correlate with its psychoactive potential

Individual GPCRs can couple to multiple signal transduction pathways (Millan et al. 2008; Raymond et al. 2001; Barnes and Sharp 1999; Hoyer et al. 2002). It has been proposed that agonists can stabilize distinct active conformational receptor states (Urban et al. 2007; Kenakin 1995; Kenakin 1997). These active states can differ in their propensity to activate the various signaling proteins coupled to the receptor. This phenomenon of "biased agonism" explains how agonists acting at the same receptor target can elicit different patterns of cellular signaling responses.

As briefly discussed above, the "ternary complex model" (Kenakin 2002; Strange 1998) is the most widely accepted model of GPCR signaling. This model proposes that the receptor is in a dynamic equilibrium between the inactive (R) and the active (R^*) conformational states. Based upon this model, neutral antagonists have identical affinities for the inactive and the active conformational states, whereas agonists exhibit higher affinity for the active state (Fig. 3). Because agonists have higher affinity for the active conformation of the receptor, agonist binding stabilizes GPCRs in their active state, shifting the dynamic equilibrium from R to R^*. The ternary complex model was first proposed by Robert Lefkowitz and his group (De Lean et al. 1980; Lefkowitz et al. 1993) and was based on radioligand binding assays in tissue culture plasma membrane preparations in which they showed that the β_2-adrenergic receptor agonist hydroxybenzylisoproterenol displaced the β_2-adrenergic receptor antagonist [^{3}H]dihydroalprenolol with a biphasic profile, with high-affinity (K_H) and low-affinity (K_L) binding sites (De Lean et al. 1980). They also demonstrated that the fraction of binding sites with high affinity gradually decreased in the presence of increasing concentrations of the non-hydrolyzable GTP analog Gpp(NH)p (De Lean et al. 1980). This model provided a general scheme for the agonist-induced activation of GPCRs and their effectors that has been widely used in pharmacology and drug development (Fig. 3).

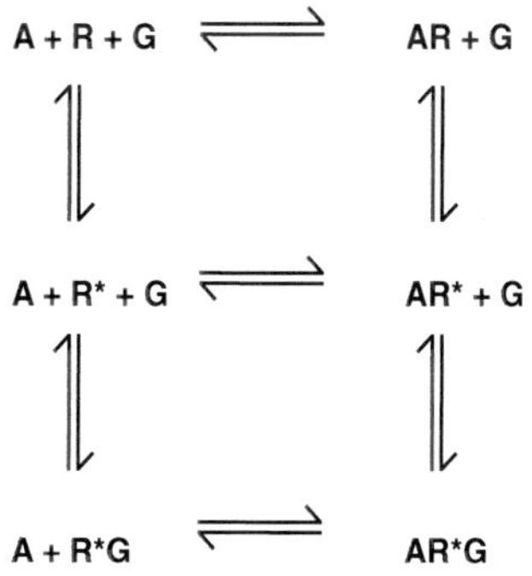

Fig. 3 The ternary complex model. According to this model, the receptor in its inactive state (R) undergoes a conformational transition, which leads to the formation of an active state (R*). This active state in turn interacts with heterotrimeric G proteins (G). Note that according to this model, the relative degree of activation of each effector pathway by the tested agonists must be the same

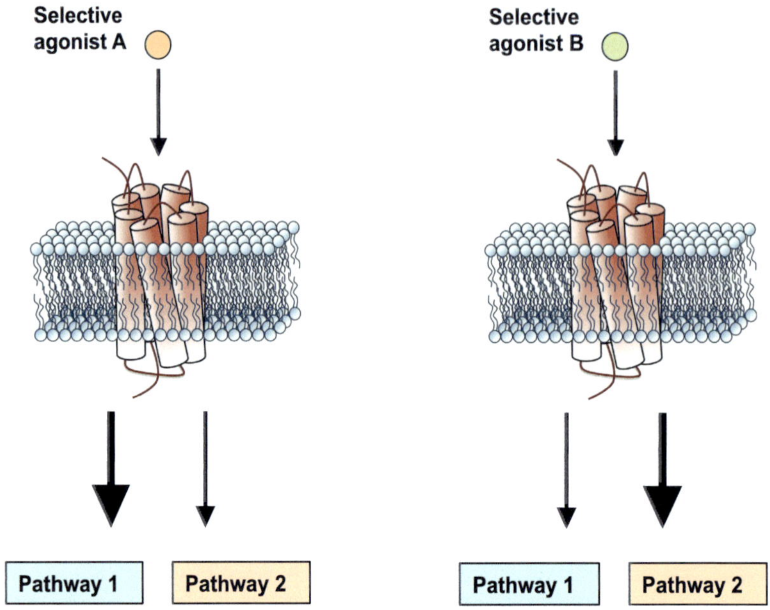

Fig. 4 Model of biased agonism. Selective agonists stabilize a subset of receptor conformations that selectively activates some but not all signaling pathways. Recent findings suggest that biased agonism is involved in the psychoactive differences between hallucinogen and closely related non-hallucinogen 5-HT_{2A} receptor agonists

More recently, the theoretical pharmacologist Terry Kenakin proposed a model where GPCRs adopt multiple conformational states (Kenakin 1995). According to this model, different agonists show a preference (in terms of binding affinity) for a subset of the receptor conformational states. Once these conformational states are stabilized by an agonist, the receptor modulates the activity of some, but not all, of the signaling pathways coupled to the receptor, consequently inducing an agonist-specific functional outcome (Fig. 4). This concept, first termed "agonist-directed trafficking of receptor signaling" and now known as "biased agonism", raised an enormous amount of interest in the molecular pharmacology field because it should theoretically be possible to design new drugs that specifically affect signaling pathways involved in the therapeutic response without inducing unwanted side effects (Urban et al. 2007).

With this background, Kelly Berg, William Clarke, and their team tested the signal transduction pathways activated by a battery of serotonin 5-HT_{2A} receptor agonists in CHO-K1 cells (Berg et al. 1998). Importantly, their findings provided the first demonstration in heterologous expression systems that the relative efficacies of agonists can differ depending upon the signal transduction pathway that is measured (Berg et al. 1998). For example, relative to 5-HT, some agonists, such as 3-trifluoromethylphenylpiperazine (TFMPP), preferentially activate the PLC-IP pathway, whereas other agonists, such as LSD, showed a preference for the PLA_2-AA pathway. These findings provided the basis for further in silico, *in cellulo* and in vivo investigations indicating that the different neuropsychological effects of hallucinogenic and non-hallucinogenic 5-HT_{2A} receptor agonists are likely linked to differences in the 5-HT_{2A} receptor-dependent signaling responses that they elicit.

5 Biased Agonism at the 5-HT_{2A} Receptor

The best characterized 5-HT_{2A} receptor-coupled signaling pathway is $G_{q/11}$-mediated activation of PLC, leading to the formation of inositol phosphates and diacylglycerol, followed by Ca^{2+} release from the endoplasmic reticulum (Hannon and Hoyer 2008; Nichols and Nichols 2008; Raymond et al. 2001; Barnes and Sharp 1999). However, whether this pathway plays a role in mediating the effects of hallucinogens is uncertain. Hallucinogens stimulate PLC-IP signaling with low efficacy, and there is no correlation between behavioral activity and the efficacy of the ligands in vitro and *in cellulo* (Marona-Lewicka et al. 2002; Porter et al. 1999; Egan et al. 1998). It has been demonstrated that head-twitch behavior induced by hallucinogens is absent in 5-HT_{2A} knockout mice (Gonzalez-Maeso et al. 2007) (see also Canal 2012). Importantly, however, it has been shown that the head-twitch response induced by the hallucinogen DOI is only slightly reduced in $G\alpha_q$ knockout mice compared to wild-type controls (Garcia et al. 2007). Together, these findings suggest that although G_q-dependent signaling may contribute to the behavioral effects of hallucinogens in rodents, the G_q pathway is not the only signaling cascade that is responsible for the hallucinogen-induced behavioral response. In addition to $G_{q/11}$-mediated PLC-IP signaling, studies in heterologous expression systems identified several other signaling transduction pathways that are coupled to the 5-HT_{2A} receptor, including phospholipase A_2 (PLA_2) (Kurrasch-Orbaugh et al. 2003), PLA_2-mediated arachidonic acid (AA) release (Berg et al. 1998), and $G_{i/o}$-dependent $G\beta\gamma$-associated activation of ERK1/2 (Kurrasch-Orbaugh et al. 2003). The role of $G_{i/o}$ protein in hallucinogen-specific signaling is further supported by recent findings in heterologous expression systems and in mouse models.

Most signaling pathways ultimately modulate gene expression in response to extracellular stimuli (Hill et al. 2001). It has been shown that GPCR activation can induce concentration-dependent changes in the level of expression of specific genes (Yuen et al. 2002; Wurmbach et al. 2001). To test the hypothesis that hallucinogenic and non-hallucinogenic 5-HT_{2A} receptor agonists modulate specific signaling pathways that are responsible for their behavioral effects, we relied on the quantification of changes in gene expression as a read-out for multidimensional neuronal signaling. This approach was tested in HEK293 cells and in mouse somatosensory cortex, a region that has been implicated in the cellular and behavioral responses induced by hallucinogenic drugs. Importantly, each agonist studied elicited a reproducible and unique response in this signaling assay. Two transcripts (*c-Fos* and *IκBα*) were induced at a similar level by both hallucinogenic and non-hallucinogenic drugs (Gonzalez-Maeso et al. 2003, 2007). However, the transcripts *Egr-1* and *Egr-2* were consistently activated by hallucinogens (DOI, DOM, DOB, mescaline, LSD, and psilocin), but expression of these two genes was unaffected by non-hallucinogenic agonists (*R*-lisuride, *S*-lisuride and ergotamine) (Gonzalez-Maeso et al. 2003, 2007). Additionally, with the exception of $I\kappa B\alpha$, the entire transcriptome fingerprint induced by the hallucinogenic and non-hallucinogenic agonists was abolished in somatosensory cortex of 5-HT_{2A} knock-out mice (Gonzalez-Maeso et al. 2003, 2007). Together, these results indicate that both hallucinogenic and non-hallucinogenic agonists modulate neuronal signaling in somatosensory cortex via the 5-HT_{2A} receptor. Furthermore, the ability to predict behavioral activity based on intrinsic reporter profiles supports the existence of distinct 5-HT_{2A} receptor-dependent signaling responses that characterize the effects of hallucinogens in the brain.

This hypothesis was tested in mouse cortical primary cultures in order to ascertain whether hallucinogens modulate specific 5-HT_{2A} receptor-dependent signaling pathways (Gonzalez-Maeso et al. 2007). Interestingly, it was demonstrated that while hallucinogenic and non-hallucinogenic 5-HT_{2A} receptor agonists activate 5-HT_{2A} receptors that are coupled to PLC, hallucinogen-dependent responses also involve pertussis toxin (PTX)-sensitive heterotrimeric $G_{i/o}$ proteins (Gonzalez-Maeso et al. 2007). These findings revealed that hallucinogen-characteristic transcriptome fingerprint depends on modulation of both $G_{q/11}$ and $G_{i/o}$, and are consistent with the biased agonism model as it explains how distinct neurophysiological responses could be produced by hallucinogenic and non-hallucinogenic 5-HT_{2A} receptor agonists that target the same population of cortical pyramidal 5-HT_{2A} receptors.

Notably, this hypothesis has been recently validated using a quantitative phosphoproteomic approach (Karaki et al. 2014). Thus, it has been demonstrated that hallucinogenic and non-hallucinogenic 5-HT_{2A} receptor agonists induce a distinct pattern of protein phosphorylation in HEK293 cells. Using this heterologous expression system, the authors also demonstrated that pretreatment with PTX abolishes ERK1,2 phosphorylation induced by the hallucinogens DOI and LSD, whereas PTX treatment did not affect the responses induced by the non-hallucinogenic 5-HT_{2A} receptor agonists lisuride and ergotamine (Karaki et al. 2014). Together with previous findings (Gonzalez-Maeso et al. 2007), these observations indicate that hallucinogens selectively activate $G_{i/o}$-dependent signaling in vitro and in vivo, whereas non-hallucinogenic 5-HT_{2A} receptor agonists do not activate $G_{i/o}$ (Karaki et al. 2014).

These findings that were based upon the transcriptome fingerprint induced by hallucinogenic and non-hallucinogenic 5-HT_{2A} receptor agonists are highly relevant to the development of more specific therapeutic drugs (Gonzalez-Maeso et al. 2003, 2007). Indeed, induction of *c-Fos*, *Egr-1,* and *Egr-1* in cortical regions is now widely used to investigate signaling events induced by hallucinogenic 5-HT_{2A} agonists in rodents (Canal et al. 2010; Chiu et al. 2014; Santini et al. 2013; Malkova et al. 2014; Lee et al. 2014; Abbas et al. 2009).

Although heterotrimeric G proteins represent one of the main mechanisms involved in GPCR-dependent responses, it has been demonstrated that G protein-independent mechanisms related to β-arrestin can also play a key role in their signaling properties (see above). The role of β-arrestins in GPCR function is further supported by the recent crystal structure of active β-arrestin-1 bound to a -derived carboxyl terminal (Shukla et al. 2013). Additionally, using the 5-HT precursor 5-hydroxytryptophan (5-HTP) and the hallucinogen DOI, Laura Bohn and others working in her laboratory demonstrated that 5-HT induces the head-twitch response in mice by a β-arrestin-2-dependent mechanism, whereas the effect of DOI on this behavior is independent of β-arrestin-2 (Schmid et al. 2008). They also showed that 5-HT, but not DOI, activates a signaling cascade composed of β-arrestin-2, phosphoinositide 3-kinease, Src, and Akt that is responsible for head-twitch behavior (Schmid and Bohn 2010). Interestingly, more recent findings by the same laboratory suggested that the atypical antipsychotic clozapine induces antipsychotic-like behavioral effects in mice by acting as a 5-HT_{2A} receptor agonist through a β-arrestin-2-independent activation of Akt (Schmid et al. 2014). Further work is needed to delineate the roles of Akt and Src in the behavioral effects of 5-HTP, hallucinogens, and clozapine.

6 Additional Signaling Pathways Modulated by Activation of the 5-HT_{2A} Receptor

Recent findings have elucidated additional neuronal signaling pathways downstream from the 5-HT_{2A} receptor that is potentially involved in the unique behavioral effects induced by hallucinogens (Halberstadt 2015; Hanks and Gonzalez-Maeso 2013). It has been shown that the last four amino acids (VSCV) of the carboxyl terminus of the 5-HT_{2A} receptor constitute a canonical Type I PDZ-binding domain (X-Ser/Thr-X-ϕ) (Kornau et al. 1995; Xia 2003; Xia et al. 2003). PDZ-binding domains are known to physically interact with PSD-95/Discs-large/ZO-1 (PDZ) domain-containing proteins such as postsynaptic density 95 (PSD-95), which is a prototypic member. Co-immunoprecipitation studies suggested that the wild-type 5-HT_{2A} receptor, but not a mutant lacking the last four amino acids of the carboxyl terminus, interacts directly with PSD-95 (Xia et al. 2003). It was also demonstrated that the association with PSD-95 enhanced 5-HT_{2A} receptor-dependent $G_{q/11}$-coupling, and that this augmentation was accompanied by inhibition of agonist-induced 5-HT_{2A} receptor internalization (Xia et al. 2003). 5-HT_{2A} receptor-mediated head-twitch behavior is reduced and the 5-HT_{2A} receptor-dependent induction of genes such as *c-Fos*, *Egr-1* and *Period-1* is disrupted in PSD-95 knockout mice (Abbas et al. 2009). These results suggest that PSD-95 is essential for hallucinogen actions at the 5-HT_{2A} receptor. Additionally, findings based on a proteomic approach that combined affinity chromatography using an immobilized synthetic PDZ ligand with mass spectrometry demonstrated that the 5-HT$_{2A}$ receptor carboxyl terminus interacts with specific PDZ proteins in vitro and in vivo (Becamel et al. 2001, 2002a, 2002b, 2004). These results indicate that the 5-HT$_{2A}$ receptor is associated with protein networks that are important for its synaptic localization and coupling to signaling machinery.

Cellular responses induced by cytokines and growth factors are mediated by the evolutionary conserved Janus kinase/signal transducers and activators of transcription (JAK/STAT) signaling pathway (Kiu and Nicholson 2012). In addition to this integral component of the signaling system, it has been reported that desensitization of 5-HT$_{2A}$ receptor-dependent signaling is accompanied by activation of STAT3 and an increase in RGS7 protein in rat frontal cortex (Singh et al. 2009). It was also found that the 5-HT$_{2A}$ receptor and STAT3 co-precipitate with JAK2, indicating that they are part of the same protein complex (Singh et al. 2009). Because the hallucinogen DOI activates both the MEK-ERK1/2 and JAK2-STAT3 intracellular signaling pathways (Oufkir et al. 2010), it is tempting to speculate that these signal transduction pathways play a role in regulating the effects induced by hallucinogen 5-HT$_{2A}$ agonists. The tyrosine kinase inhibitor genistein (5 μM) decreased the potency of DOI-induced contraction of rat aorta, whereas this effect did not occur with daidzein (5 μM), which is the inactive isomer of genistein (Florian and Watts 1998). In cultured aortic smooth muscle cells, activation of the 5-HT$_{2A}$ receptor stimulated tyrosine-phosphorylation of ERK, and this effect was reduced by the MAPK inhibitor PD098059 (10 μM) (Florian and Watts 1998). Together, these results suggest that hallucinogens cause rat aortic contraction via a pathway that is at least partially independent of the pathways classically associated with the 5-HT$_{2A}$ receptor.

A specific conserved motif, NPxxY, which is found at the junction between TM7 and the carboxyl terminal domains of a number of rhodopsin-like GPCRs, including the 5-HT$_{2A}$ receptor, has been implicated as a determinant of ADP-ribosylation factor (ARF)-mediated signaling because wild-type receptors with an alternative DPxxY motif show selective defects in this pathway (Robertson et al. 2003; Mitchell et al. 1998). Although as described above, the 5-HT$_{2A}$ receptor is known to activate PLC via the heterotrimeric G protein $G_{q/11}$, previous findings have convincingly demonstrated that the 5-HT$_{2A}$ receptor can also signal through the PLD pathway independent of $G_{q/11}$ in an ARF-dependent manner. Both co-immunoprecipitation assays and the effect of negative mutant ARF constructs on 5-HT$_{2A}$ receptor-induced PLD activation demonstrate that ARF1 plays a key role in the function of this receptor (Robertson et al. 2003). The N376PxxY motif in TM7 was shown to be essential for ARF-dependent PLD signaling and co-immunoprecipitation with ARF1. In addition, ARF1 rather that ARF6 participates in this mechanism through a GTP-dependent interaction with the carboxyl terminus of the 5-HT$_{2A}$ receptor (Robertson et al. 2003). It has also been reported that the spatial coordination of the 5-HT$_{2A}$ receptor with transducer and effector proteins into a physical complex is likely to reinforce the impact of receptor activation on G protein-independent signaling pathway (Barclay et al. 2011).

By visualizing GFP-tagged 5-HT$_{2A}$ receptors in living cells, it has been shown that activation of protein kinase C (PKC) by its specific activator phorbol

12-myristate 13-acetate leads to internalization of the receptor in the absence of 5-HT (Bhattacharyya et al. 2002, 2010). Additionally, inhibition of PKC with sphingosine prevents internalization of the 5-HT_{2A} receptor by 5-HT. Because receptors that had been internalized by phorbol 12-myristate 13-acetate exposure in the absence of 5-HT also recycle to the cell surface with a time-course comparable to that seen after activation of the 5-HT_{2A} receptor by 5-HT, these findings suggest that 5-HT_{2A} receptors internalize and return to the cell surface in response to both 5-HT and PKC (Bhattacharyya et al. 2002, 2010). Although the human and rat 5-HT_{2A} receptors differ by only a few amino acids, the human receptor takes longer to recycle to the cell surface after internalization (Bhattacharya et al. 2010). Further investigation based upon the comparison of the primary sequences of human and rat 5-HT_{2A} receptors demonstrated that replacing serine 457 in the carboxyl terminus of the human isoform with alanine resulted in faster recycling (Bhattacharya et al. 2010). By extension, this study also indicates that extrapolating results from non-human receptor isoforms may sometimes lead to misinterpretations.

It is well accepted that protein kinases mediate many of the downstream actions of both ionotropic and metabotropic receptors. Interestingly, relatively recent findings suggest that genetic deletion of p90 ribosomal S6 kinase 2 (RSK2) potentiates 5-HT_{2A} receptor-dependent signaling (Sheffler et al. 2006). Thus, studies of 5-HT_{2A} receptor signaling in fibroblasts obtained from wild-type and RSK2 knockout mice demonstrated that 5-HT_{2A} receptor-dependent phosphoinositide hydrolysis is augmented in RSK2 knockout fibroblasts.

Several lines of evidence have shown that 5-HT_{1A} and 5-HT_{2A} receptors, which are co-expressed in cortical pyramidal neurons (Martin-Ruiz et al. 2001), often show opposite effects on common signaling pathways. It has been demonstrated that activation of 5-HT_{1A} receptors suppresses NMDA receptor function in frontal cortex pyramidal neurons (Yuen et al. 2008). Most importantly, activation of 5-HT_{2A} receptors by hallucinogens significantly attenuates the effect of 5-HT_{1A} receptor on NMDA receptor currents and microtubule depolymerization. Inhibition of the β-arrestin/Src/dynamin signaling was shown to block 5-HT_{2A} receptor-dependent activation of ERK and the counteractive effect of 5-HT_{2A} on 5-HT_{1A}-dependent regulation of NMDA receptor currents. These findings could be important for cognitive control—a function known to be heavily influenced by hallucinogens and the 5-HT_{2A} receptor.

DARPP-32 is a key regulator of kinase-phosphatase signaling cascades modulated by serotonergic, dopaminergic, and glutamatergic neurotransmission (Svenningsson et al. 2004). Four distinct phosphorylation sites determine the function of DARPP-32. Phosphorylation at Thr^{34} converts DARPP-32 into a potent inhibitor of protein phosphatase-1 (PP1) (Svenningsson et al. 2003). Phosphorylated Ser^{97} increases the ability of protein kinase A (PKA) to phosphorylate DARPP-32 at Thr^{34}. Phosphorylation of Ser^{130} prevents the dephosphorylation of Thr^{34} by protein phosphatase-2B (PP2B). Phosphorylation of Thr^{75} converts DARPP-32 into an inhibitor of PKA (Svenningsson et al. 2003). Dopaminergic agonists, such as (+)-amphetamine, serotonergic 5-HT_{2A} agonists such as LSD, and glutamatergic antagonists such as phencyclidine (PCP), have all been shown to induce phosphorylation or dephosphorylation of DARPP-32 at three sites in a pattern predicted to cause a synergistic inhibition of protein PP1 and concomitant regulation of its downstream effector proteins GSK-3, cAMP response element-binding protein (CREB), and c-Fos (Svenningsson et al. 2003). Notably, in mice with point mutations at DARPP-32 phosphorylation sites, the effects of (+)-amphetamine, LSD, and PCP on sensorimotor gating and repetitive movements are strongly attenuated. Thus, three pathways that regulate the state of phosphorylation of Thr^{34}-, Thr^{75}-, and Ser^{130}-DARPP-32 inhibit PP1, which leads to increased phosphorylation of various PP1 substrates (Svenningsson et al. 2003). Further work will be needed to identify the precise PP1 substrates involved in the psychoactive behavioral effects of these drugs.

Brain-derived neurotrophic factor (BDNF) and glial cell line-derived neurotrophic factor (GDNF) are members of the neurotrophin family, a small family of secreted proteins that also includes nerve growth factor (NGF), neurotrophin 3 (NT-3), and neurotrophin 4 (NT-4) (Griebel and Holsboer 2012). Administration of DOI induces a differential regulation *BDNF* mRNA expression in rat hippocampus and neocortex, with downregulation in hippocampus and upregulation in neocortex (Vaidya et al. 1997). This interesting effect is blocked by selective 5-HT_{2A} antagonists but is unaffected by selective 5-HT_{2C} antagonists. Additionally, the mGlu2/3 receptor agonist LY354740 dose-dependently represses the ability of DOI to induce upregulation of *BDNF* mRNA expression in rat frontal cortex, whereas the effect of DOI is enhanced by the mGlu2/3 receptor antagonist LY341495 (Gewirtz et al. 2002). Immobilization stress also decreases the expression of *BDNF* mRNA in rat hippocampus, an effect that is blocked by the 5-HT_{2A} receptor antagonist MDL100,907 (Vaidya et al. 1997). These results suggest that 5-HT_{2A} receptor-dependent signaling is involved in the stress-induced regulation of BDNF expression in the rat hippocampus. Given that the mGlu2/3 receptor agonist LY354740 suppresses the effect of immobilization stress on *BDNF* mRNA expression (Lee et al. 2006), these results are consistent with the hypothesis that mGlu2/3 receptor agonists may modulate 5-HT_{2A} receptor-dependent stress-induced behaviors.

7 Role of mGlu2 Receptor in Hallucinogen Action

Although it is generally accepted that the 5-HT_{2A} receptor expressed in frontal cortical pyramidal neurons represents the main molecular target responsible for the cellular, electrophysiological and behavioral effects of hallucinogens in rodents (Gonzalez-Maeso et al. 2007; Beique et al. 2007; Puig et al. 2003; Celada et al. 2008), effects on sub-cortical regions such as thalamocortical projections may also contribute (Marek et al. 2001; Scruggs et al. 2000). Relatively recent findings demonstrate that the mGlu2/3 receptor agonist LY354740 antagonizes 5-HT_{2A} receptor-induced excitatory postsynaptic potential/currents (EPSPs/EPSCs) in pyramidal neurons (Marek et al. 2000). It was also shown that LY354740 suppresses the head-twitch behavior induced by the hallucinogen DOI (Gewirtz and

Marek 2000). Similar effects have been observed with mGlu2/3 receptor agonists such as LY379268 and LY404030 (Aghajanian and Marek 2000). These findings led to the conclusion that activation of mGlu2 autoreceptors mediates the presynaptic effects of mGlu2/3 agonists in suppressing the electrophysiological effects of hallucinogenic 5-HT_{2A} receptor agonists recorded in cortical pyramidal neurons (Aghajanian and Marek 2000). An alternative (although not mutually exclusive) explanation for the crosstalk between 5-HT_{2A} and mGlu2 receptors is that these two receptors may be expressed in close physical proximity in the postsynaptic density of cortical pyramidal neurons.

Traditionally, GPCRs were thought to function as monomers. The monomeric function of GPCRs is supported by assays that measured agonist binding and G protein coupling of purified receptors reconstituted into a lipid bilayer (including rhodopsin, β_2-adrenergic and μ-opioid receptors) (Whorton et al. 2007, 2008; Kuszak et al. 2009). Nevertheless, many instances of homomerization and heteromerization (macromolecular complexes formed by non-covalent interactions between GPCRs) have recently been reported (Gonzalez-Maeso 2011; Milligan 2013; Ferre et al. 2014). The existence of GPCR heteromers is further suggested by the recent explosion of research elucidating the crystal structures of GPCRs; for example, four crystal structures that were recently reported (CXCR4, μ-opioid, κ-opioid, and β_1-adrenergic receptors) contained receptor dimers (Huang et al. 2013; Manglik et al. 2012; Wu et al. 2010, 2012). Interestingly, previous findings suggest that the $G_{q/11}$-coupled 5-HT_{2A} receptor and the $G_{i/o}$-coupled mGlu2 receptor form a specific GPCR heteromeric complex in heterologous expression systems, as well as in mouse and human frontal cortex (Gonzalez-Maeso et al. 2008; Moreno et al. 2012; Fribourg et al. 2011); these results have been independently confirmed by other groups (Rives et al. 2009). The close molecular proximity between 5-HT_{2A} and mGlu2 receptors does not occur with the closely related $G_{i/o}$-coupled mGlu3 receptor, and is either rescued or disrupted with different mGlu2/mGlu3 chimeric constructs (Gonzalez-Maeso et al. 2008; Moreno et al. 2012, 2016; Fribourg et al. 2011; Baki et al. 2016). The conclusion that 5-HT_{2A} and mGlu2 receptors are co-expressed as a GPCR heteromeric complex in frontal cortex is supported by observations that demonstrate the co-expression and co-immunoprecipitation of these receptors in mouse and human frontal cortex (Gonzalez-Maeso et al. 2008; Fribourg et al. 2011; Moreno et al. 2016), as well as by the close physical proximity of 5-HT_{2A} and mGlu2 receptors at cortical synaptic junctions at the electron microscopy level (Moreno et al. 2012) (Fig. 5).

The role of mGlu2 in the psychoactive effects induced by hallucinogens is supported by the impaired ability of the hallucinogens DOI and LSD to induce head-twitch behavior in mGlu2 knockout mice compared to wild-type littermates (Figs. 6 and 7) (Moreno et al. 2011). It has been demonstrated that the ability of DOI to induce head-twitch behavior in mGlu2 knockout mice is rescued by over-expressing mGlu2 in frontal cortex using a viral (HSV)-mediated transgene

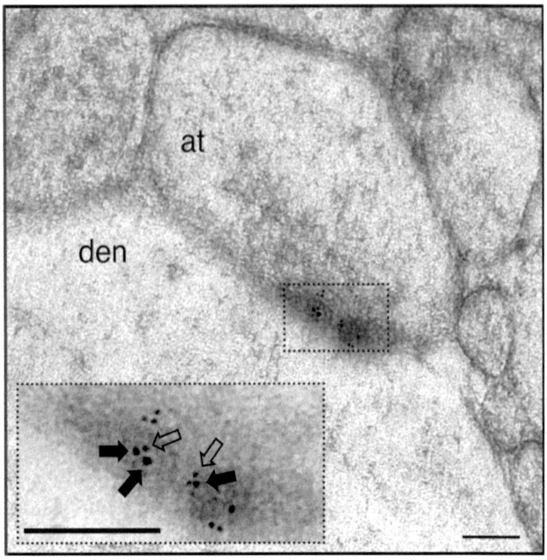

Fig. 5 Immunogold labeling for 5-HT_{2A} and mGlu2 receptors in mouse cortical neurons. Note that the 10-nm gold particles (*filled arrows*) and the 6-nm gold particles (*open arrows*) are located in very close proximity at the synaptic junction (see Moreno et al. 2012). Inset, high magnification view of region delineated in boxed area. *Scale bars*, 100 nm. (*den*, dendrite; *at*, axon terminal)

expression approach (Moreno et al. 2012). Because DOI-induced head-twitch behavior is not rescued in mGlu2 knockout mice over-expressing mGlu2ΔTM4N [a mGlu2/mGlu3 chimeric construct that does not form heteromers with the 5-HT_{2A} receptor (Gonzalez-Maeso et al. 2008; Fribourg et al. 2011)] in frontal cortex (Moreno et al. 2012), these findings suggest that the 5-HT_{2A}-mGlu2 receptor complex is critical for the hallucinogen-like behaviors induced by 5-HT_{2A} receptor agonists (Figs. 6 and 7). The translational potential of these findings is suggested by the alterations in the expression of 5-HT_{2A} and mGlu2 receptors in the frontal cortex of schizophrenic subjects postmortem (Gonzalez-Maeso et al. 2008; Moreno et al. 2012, 2016; Muguruza et al. 2013). However, further investigation of this heteromeric receptor complex is definitely necessary because the functional significance of GPCR homo- and heteromerization remains a controversial topic (Bouvier and Hebert 2014; Lambert and Javitch 2014) [in addition, see: Delille et al. (2012), Frederick et al. (2015)].

8 Future Directions

In conclusion, recent work suggests that hallucinogens may be unique in their ability to modulate the activity of specific 5-HT_{2A} receptor-linked signaling pathways. Hallucinogens are used recreationally; hence, elucidating their biophysical and molecular mechanism of action is an important objective in drug abuse research (Nutt et al. 2013). However, clinical studies also suggest that hallucinogens, when

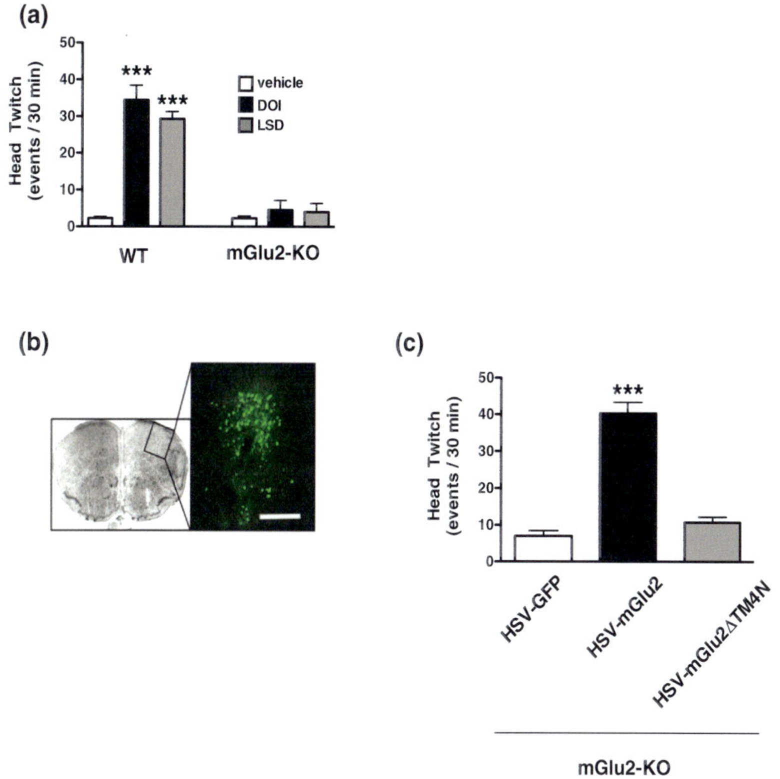

Fig. 6 **a** Head-twitch behavior induced by DOI and LSD is absent in mGlu2 knockout mice (see Moreno et al. 2011). **b**, **c** Expression of 5-HT_{2A} and mGlu2 as a GPCR heteromer is necessary for head-twitch psychosis-like behavior induced by hallucinogenic 5-HT_{2A} agonists in mice (see Moreno et al. 2012). Representative image of HSV-mediated transgene expression in mouse frontal cortex. *Scale bar*, 200 μm (**b**). Virally mediated over-expression of wild-type mGlu2, but not the mGlu2/mGlu3 chimeric construct mGlu2ΔTM4 N that does not form the 5-HT_{2A}-mGlu2 heteromeric receptor complex, rescues the head-twitch behavior induced by the hallucinogenic 5-HT_{2A} receptor agonist DOI (**c**)

administered under medical supervision, may serve as therapeutic drugs that can be used for the treatment of severe psychiatric and neurological disorders such as alcoholism (Krebs and Johansen 2012), obsessive compulsive disorder (Moreno et al. 2006), and cluster headache (Sewell et al. 2006). Further basic and

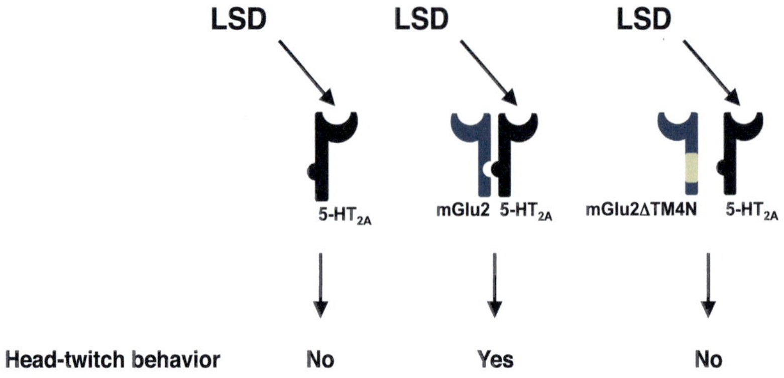

Fig. 7 Model of the mechanism underlying hallucinogen-induced head-twitch behavioral response

translational research is therefore warranted to better define the specific signaling and neuronal circuit mechanisms responsible for their psychoactive effects.

Acknowledgements The preparation of this review article has been supported at least in part by the National Institutes of Health R01MH084894 and R01MH111940 (J.G.M.) and the Spanish Government SAF2010-15663 (J.F.L.G.).

References

Abbas AI, Yadav PN, Yao W-D, Arbuckle MI, Grant SGN, Caron MG, Roth BL (2009) PSD-95 is essential for hallucinogen and atypical antipsychotic drug actions at serotonin receptors. J Neurosci 29:7124–7136

Adams LM, Geyer MA (1985) Patterns of exploration in rats distinguish lisuride from lysergic acid diethylamide. Pharmacol Biochem Behav 23:461–468

Aghajanian GK (2009) Modeling, "psychosis" in vitro by inducing disordered neuronal network activity in cortical brain slices. Psychopharmacology 206:575–585

Aghajanian GK, Marek GJ (1999) Serotonin and hallucinogens. Neuropsychopharmacology 21:16S–23S

Aghajanian GK, Marek GJ (2000) Serotonin model of schizophrenia: emerging role of glutamate mechanisms. Brain Res Brain Res Rev 31:302–312

Aghajanian GK, Foote WE, Sheard MH (1968) Lysergic acid diethylamide: sensitive neuronal units in the midbrain raphe. Science 161:706–708

Appel NM, Mitchell WM, Garlick RK, Glennon RA, Teitler M, De Souza EB (1990) Autoradiographic characterization of (+-)-1-(2,5-dimethoxy-4-[125I] iodophenyl)-2-aminopropane ([125I]DOI) binding to 5-HT2 and 5-HT1c receptors in rat brain. J Pharmacol Exp Ther 255:843–857

Audet M, Bouvier M (2012) Restructuring G-protein-coupled receptor activation. Cell 151:14–23

Baki L, Fribourg M, Younkin J, Eltit JM, Moreno JL, Park G, Vysotskaya Z, Narahari A, Sealfon SC, Gonzalez-Maeso J, Logothetis DE (2016) Cross-signaling in metabotropic glutamate 2 and serotonin 2A receptor heteromers in mammalian cells. Pflugers Arch 468:775–793

Barclay Z, Dickson L, Robertson DN, Johnson MS, Holland PJ, Rosie R, Sun L, Fleetwood-Walker S, Lutz EM, Mitchell R (2011) 5-HT2A receptor signalling through phospholipase D1 associated with its C-terminal tail. Biochem J 436:651–660

Barnes NM, Sharp T (1999) A review of central 5-HT receptors and their function. Neuropharmacology 38:1083–1152

Becamel C, Figge A, Poliak S, Dumuis A, Peles E, Bockaert J, Lubbert H, Ullmer C (2001) Interaction of serotonin 5-hydroxytryptamine type 2C receptors with PDZ10 of the multi-PDZ domain protein MUPP1. J Biol Chem 276:12974–12982

Becamel C, Alonso G, Galeotti N, Demey E, Jouin P, Ullmer C, Dumuis A, Bockaert J, Marin P (2002a) Synaptic multiprotein complexes associated with 5-HT(2C) receptors: a proteomic approach. EMBO J 21:2332–2342

Becamel C, Galeotti N, Poncet J, Jouin P, Dumuis A, Bockaert J, Marin P (2002b) A proteomic approach based on peptide affinity chromatography, 2-dimensional electrophoresis and mass spectrometry to identify multiprotein complexes interacting with membrane-bound receptors. Biol Proced Online 4:94–104

Becamel C, Gavarini S, Chanrion B, Alonso G, Galeotti N, Dumuis A, Bockaert J, Marin P (2004) The serotonin 5-HT2A and 5-HT2C receptors interact with specific sets of PDZ proteins. J Biol Chem 279:20257–20266

Beique JC, Imad M, Mladenovic L, Gingrich JA, Andrade R (2007) Mechanism of the 5-hydroxytryptamine 2A receptor-mediated facilitation of synaptic activity in prefrontal cortex. Proc Natl Acad Sci U S A 104:9870–9875

Berg KA, Maayani S, Goldfarb J, Scaramellini C, Leff P, Clarke WP (1998) Effector pathway-dependent relative efficacy at serotonin type 2A and 2C receptors: evidence for agonist-directed trafficking of receptor stimulus. Mol Pharmacol 54:94–104

Berger M, Gray JA, Roth BL (2009) The expanded biology of serotonin. Annu Rev Med 60: 355–366

Bhattacharya A, Sankar S, Panicker MM (2010) Differences in the C-terminus contribute to variations in trafficking between rat and human 5-HT(2A) receptor isoforms: identification of a primate-specific tripeptide ASK motif that confers GRK-2 and beta arrestin-2 interactions. J Neurochem 112:723–732

Bhattacharyya S, Puri S, Miledi R, Panicker MM (2002) Internalization and recycling of 5-HT2A receptors activated by serotonin and protein kinase C-mediated mechanisms. Proc Natl Acad Sci U S A 99:14470–14475

Bouvier M, Hebert TE (2014) CrossTalk proposal: weighing the evidence for Class A GPCR dimers, the evidence favours dimers. J Physiol 592:2439–2441

Canal CE (2012) Head-twitch response in rodents induced by the hallucinogen 2, 5-dimethoxy-4-iodoamphetamine: a comprehensive history, a re-evaluation of mechanisms, and its …. Drug Testing and Analysis

Canal CE, Olaghere da Silva UB, Gresch PJ, Watt EE, Sanders-Bush E, Airey DC (2010) The serotonin 2C receptor potently modulates the head-twitch response in mice induced by a phenethylamine hallucinogen. Psychopharmacology 209:163–174

Celada P, Puig MV, Diaz-Mataix L, Artigas F (2008) The hallucinogen DOI reduces low-frequency oscillations in rat prefrontal cortex: reversal by antipsychotic drugs. Biol Psychiatry 64:392–400

Chiu HY, Chan MH, Lee MY, Chen ST, Zhan ZY, Chen HH (2014) Long-lasting alterations in 5-HT2A receptor after a binge regimen of methamphetamine in mice. Int J Neuropsychopharmacol 1–12

Clark AJ (1933) The mode of action of drugs on cells

Dahlström A, Fuxe K (1964) Evidence for the existence of monoamine-containing neurons in the central nervous system. I. Demonstration of monoamines in the cell bodies of brain stem neurons. Acta Physiol Scand Suppl SUPPL 232:231–255

De Lean A, Stadel JM, Lefkowitz RJ (1980) A ternary complex model explains the agonist-specific binding properties of the adenylate cyclase-coupled beta-adrenergic receptor. J Biol Chem 255:7108–7117

Delille HK, Becker JM, Burkhardt S, Bleher B, Terstappen GC, Schmidt M, Meyer AH, Unger L, Marek GJ, Mezler M (2012) Heterocomplex formation of 5-HT(2A)-mGlu(2) and its relevance for cellular signaling cascades. Neuropharmacology 1–8

Egan CT, Herrick-Davis K, Miller K, Glennon RA, Teitler M (1998) Agonist activity of LSD and lisuride at cloned 5HT2A and 5HT2C receptors. Psychopharmacology 136:409–414

Erhlich P (1913) Chemotherapeutics: scientific principles, methods and results. Lancet 2:445–451

Fantegrossi WE, Murnane KS, Reissig CJ (2008) The behavioral pharmacology of hallucinogens. Biochem Pharmacol 75:17–33

Ferguson SS (2001) Evolving concepts in G protein-coupled receptor endocytosis: the role in receptor desensitization and signaling. Pharmacol Rev 53:1–24

Ferre S, Casado V, Devi LA, Filizola M, Jockers R, Lohse MJ, Milligan G, Pin JP, Guitart X (2014) G protein-coupled receptor oligomerization revisited: functional and pharmacological perspectives. Pharmacol Rev 66:413–434

Florian JA, Watts SW (1998) Integration of mitogen-activated protein kinase kinase activation in vascular 5-hydroxytryptamine2A receptor signal transduction. J Pharmacol Exp Ther 284: 346–355

Frederick A. L., Yano H., Trifilieff P, Vishwasrao HD, Biezonski D, Meszaros J, Urizar E, Sibley DR, Kellendonk C, Sonntag KC, Graham DL, Colbran RJ, Stanwood GD, Javitch JA (2015) Evidence against dopamine D1/D2 receptor heteromers. Mol Psychiatry

Fribourg M, Moreno JL, Holloway T, Provasi D, Baki L, Mahajan R, Park G, Adney SK, Hatcher C, Eltit JM, Ruta JD, Albizu L, Li Z, Umali A, Shim J, Fabiato A, Mackerell AD Jr, Brezina V, Sealfon SC, Filizola M, Gonzalez-Maeso J, Logothetis DE (2011) Decoding the signaling of a GPCR heteromeric complex reveals a unifying mechanism of action of antipsychotic drugs. Cell 147:1011–1023

Gaddum JH, Picarelli ZP (1957) Two kinds of tryptamine receptor. Br J Pharmacol Chemother 12:323–328

Garcia EE, Smith RL, Sanders-Bush E (2007) Role of G(q) protein in behavioral effects of the hallucinogenic drug 1-(2,5-dimethoxy-4-iodophenyl)-2-aminopropane. Neuropharmacology 52:1671–1677

Gerfen CR (1992) The neostriatal mosaic: multiple levels of compartmental organization. Trends Neurosci 15:133–139

Gewirtz JC, Marek GJ (2000) Behavioral evidence for interactions between a hallucinogenic drug and group II metabotropic glutamate receptors. Neuropsychopharmacology 23:569–576

Gewirtz JC, Chen AC, Terwilliger R, Duman RC, Marek GJ (2002) Modulation of DOI-induced increases in cortical BDNF expression by group II mGlu receptors. Pharmacol Biochem Behav 73:317–326

Glennon RA, Titeler M, McKenney JD (1984) Evidence for 5-HT2 involvement in the mechanism of action of hallucinogenic agents. Life Sci 35:2505–2511

Glennon RA, Titeler M, Young R (1986) Structure-activity relationships and mechanism of action of hallucinogenic agents based on drug discrimination and radioligand binding studies. Psychopharmacol Bull 22:953–958

Gonzalez-Maeso J (2011) GPCR oligomers in pharmacology and signaling. Mol Brain 4:20

Gonzalez-Maeso J, Yuen T, Ebersole BJ, Wurmbach E, Lira A, Zhou M, Weisstaub N, Hen R, Gingrich JA, Sealfon SC (2003) Transcriptome fingerprints distinguish hallucinogenic and nonhallucinogenic 5-hydroxytryptamine 2A receptor agonist effects in mouse somatosensory cortex. J Neurosci 23:8836–8843

Gonzalez-Maeso J, Weisstaub NV, Zhou M, Chan P, Ivic L, Ang R, Lira A, Bradley-Moore M, Ge Y, Zhou Q, Sealfon SC, Gingrich JA (2007) Hallucinogens recruit specific cortical 5-HT (2A) Receptor-mediated signaling pathways to affect behavior. Neuron 53:439–452

Gonzalez-Maeso J, Ang RL, Yuen T, Chan P, Weisstaub NV, Lopez-Gimenez JF, Zhou M, Okawa Y, Callado LF, Milligan G, Gingrich JA, Filizola M, Meana JJ, Sealfon SC (2008) Identification of a serotonin/glutamate receptor complex implicated in psychosis. Nature 452:93–97

Graybiel AM (1990) Neurotransmitters and neuromodulators in the basal ganglia. Trends Neurosci 13:244–254
Green AR (2008) Gaddum and LSD: the birth and growth of experimental and clinical neuropharmacology research on 5-HT in the UK. Br J Pharmacol 154:1583–1599
Griebel G, Holsboer F (2012) Neuropeptide receptor ligands as drugs for psychiatric diseases: the end of the beginning? Nat Rev Drug Discov 11:462–478
Halberstadt AL (2015) Recent advances in the neuropsychopharmacology of serotonergic hallucinogens. Behav Brain Res 277C:99–120
Hanks JB, Gonzalez-Maeso J (2013) Animal models of serotonergic psychedelics. ACS Chem Neurosci 4:33–42
Hannon J, Hoyer D (2008) Molecular biology of 5-HT receptors. Behav Brain Res 195:198–213
Hanyaloglu AC, von Zastrow M (2008) Regulation of GPCRs by endocytic membrane trafficking and its potential implications. Annu Rev Pharmacol Toxicol 48:537–568
Hill SJ, Baker JG, Rees S (2001) Reporter-gene systems for the study of G-protein-coupled receptors. Curr Opin Pharmacol 1:526–532
Hofmann A (1979) How LSD originated. J Psychedelic Drugs 11:53–60
Hoyer D, Clarke DE, Fozard JR, Hartig PR, Martin GR, Mylecharane EJ, Saxena PR, Humphrey PP (1994) International Union of Pharmacology classification of receptors for 5-hydroxytryptamine (Serotonin). Pharmacol Rev 46:157–203
Hoyer D, Hannon JP, Martin GR (2002) Molecular, pharmacological and functional diversity of 5-HT receptors. Pharmacol Biochem Behav 71:533–554
Huang J, Chen S, Zhang JJ, Huang XY (2013) Crystal structure of oligomeric beta1-adrenergic G protein-coupled receptors in ligand-free basal state. Nat Struct Mol Biol 20:419–425
Irannejad R, Tomshine JC, Tomshine JR, Chevalier M, Mahoney JP, Steyaert J, Rasmussen SG, Sunahara RK, El-Samad H, Huang B, von Zastrow M (2013) Conformational biosensors reveal GPCR signalling from endosomes. Nature 495:534–538
Johnson MP, Siegel BW, Carr AA (1996) [3H]MDL 100,907: a novel selective 5-HT2A receptor ligand. Naunyn Schmiedebergs Arch Pharmacol 354:205–209
Karaki S, Becamel C, Murat S, Mannoury la Cour C, Millan MJ, Prezeau L, Bockaert J, Marin P, Vandermoere F (2014) Quantitative phosphoproteomics unravels biased phosphorylation of serotonin 2A receptor at Ser280 by hallucinogenic versus nonhallucinogenic agonists. Mol Cell Proteomics 13:1273–1285
Katritch V, Cherezov V, Stevens RC (2013) Structure-function of the G protein-coupled receptor superfamily. Annu Rev Pharmacol Toxicol 53:531–556
Kenakin T (1995) Agonist-receptor efficacy. II. Agonist trafficking of receptor signals. Trends Pharmacol Sci 16:232–238
Kenakin T (1997) Agonist-specific receptor conformations. Trends Pharmacol Sci 18:416–417
Kenakin T (2002) Drug efficacy at G protein-coupled receptors. Annu Rev Pharmacol Toxicol 42:349–379
Kiu H, Nicholson SE (2012) Biology and significance of the JAK/STAT signalling pathways. Growth Factors 30:88–106
Kornau HC, Schenker LT, Kennedy MB, Seeburg PH (1995) Domain interaction between NMDA receptor subunits and the postsynaptic density protein PSD-95. Science 269:1737–1740
Krebs TS, Johansen PO (2012) Lysergic acid diethylamide (LSD) for alcoholism: meta-analysis of randomized controlled trials. J Psychopharmacol 26:994–1002
Kurrasch-Orbaugh DM, Watts VJ, Barker EL, Nichols DE (2003a) Serotonin 5-hydroxytryptamine 2A receptor-coupled phospholipase C and phospholipase A2 signaling pathways have different receptor reserves. J Pharmacol Exp Ther 304:229–237
Kurrasch-Orbaugh DM, Parrish JC, Watts VJ, Nichols DE (2003b) A complex signaling cascade links the serotonin2A receptor to phospholipase A2 activation: the involvement of MAP kinases. J Neurochem 86:980–991
Kuszak AJ, Pitchiaya S, Anand JP, Mosberg HI, Walter NG, Sunahara RK (2009) Purification and functional reconstitution of monomeric mu-opioid receptors: allosteric modulation of agonist binding by Gi2. J Biol Chem 284:26732–26741

Lambert NA, Javitch JA (2014) CrossTalk opposing view: weighing the evidence for class A GPCR dimers, the jury is still out. J Physiol 592:2443–2445

Langley JN (1909) On the contraction of muscle, chiefly in relation to the presence of 'receptive' substances. J Physiol 39:235–295

Lee Y, Duman RS, Marek GJ (2006) The mGlu2/3 receptor agonist LY354740 suppresses immobilization stress-induced increase in rat prefrontal cortical BDNF mRNA expression. Neurosci Lett 398:328–332

Lee MY, Chiang CC, Chiu HY, Chan MH, Chen HH (2014) N-acetylcysteine modulates hallucinogenic 5-HT receptor agonist-mediated responses: behavioral, molecular, and electrophysiological studies. Neuropharmacology

Lefkowitz RJ, Cotecchia S, Samama P, Costa T (1993) Constitutive activity of receptors coupled to guanine nucleotide regulatory proteins. Trends Pharmacol Sci 14:303–307

Lopez-Gimenez JF, Mengod G, Palacios JM, Vilaro MT (1997) Selective visualization of rat brain 5-HT2A receptors by autoradiography with [3H]MDL 100,907. Naunyn Schmiedebergs Arch Pharmacol 356:446–454

Lopez-Gimenez JF, Vilaro MT, Palacios JM, Mengod G (1998) [3H]MDL 100,907 labels 5-HT2A serotonin receptors selectively in primate brain. Neuropharmacology 37:1147–1158

Lopez-Gimenez JF, Mengod G, Palacios JM, Vilaro MT (1999) Human striosomes are enriched in 5-HT2A receptors: autoradiographical visualization with [3H]MDL100,907,[125I](±)DOI and [3H]ketanserin. Eur J Neurosci 11:3761–3765

Lopez-Gimenez JF, Vilaro MT, Palacios JM, Mengod G (2001) Mapping of 5-HT2A receptors and their mRNA in monkey brain: [3H]MDL100,907 autoradiography and in situ hybridization studies. J Comp Neurol 429:571–589

Luttrell LM (2008) Reviews in molecular biology and biotechnology: transmembrane signaling by G protein-coupled receptors. Mol Biotechnol 39:239–264

Luttrell LM, Lefkowitz RJ (2002) The role of beta-arrestins in the termination and transduction of G-protein-coupled receptor signals. J Cell Sci 115:455–465

Malkova NV, Gallagher JJ, Yu CZ, Jacobs RE, Patterson PH (2014) Manganese-enhanced magnetic resonance imaging reveals increased DOI-induced brain activity in a mouse model of schizophrenia. Proc Natl Acad Sci U S A 111:E2492–2500

Manglik A, Kruse AC, Kobilka TS, Thian FS, Mathiesen JM, Sunahara RK, Pardo L, Weis WI, Kobilka BK, Granier S (2012) Crystal structure of the micro-opioid receptor bound to a morphinan antagonist. Nature 485:321–326

Marek GJ, Wright RA, Schoepp DD, Monn JA, Aghajanian GK (2000) Physiological antagonism between 5-hydroxytryptamine(2A) and group II metabotropic glutamate receptors in prefrontal cortex. J Pharmacol Exp Ther 292:76–87

Marek GJ, Wright RA, Gewirtz JC, Schoepp DD (2001) A major role for thalamocortical afferents in serotonergic hallucinogen receptor function in the rat neocortex. Neuroscience 105:379–392

Marona-Lewicka D, Kurrasch-Orbaugh DM, Selken JR, Cumbay MG, Lisnicchia JG, Nichols DE (2002) Re-evaluation of lisuride pharmacology: 5-hydroxytryptamine1A receptor-mediated behavioral effects overlap its other properties in rats. Psychopharmacology 164:93–107

Martin-Ruiz R, Puig MV, Celada P, Shapiro DA, Roth BL, Mengod G, Artigas F (2001) Control of serotonergic function in medial prefrontal cortex by serotonin-2A receptors through a glutamate-dependent mechanism. J Neurosci 21:9856–9866

Mengod G, Vilaro MT, Cortes R, Lopez-Gimenez JF, Raurich A, Palacios JM (2006) Chemical neuroanatomy of 5-HT receptor subtypes in the mammalian brain. In Roth BL (ed) serotonin T, and therapeutics rFmpth. Humana Press, pp 319–364

Millan MJ, Marin P, Bockaert J, la Cour CM (2008) Signaling at G-protein-coupled serotonin receptors: recent advances and future research directions. Trends Pharmacol Sci 29:454–464

Milligan G (2013) The prevalence, maintenance and relevance of GPCR oligomerization. Mol Pharmacol 158–169 (Epub ahead of print)

Mitchell R, McCulloch D, Lutz E, Johnson M, MacKenzie C, Fennell M, Fink G, Zhou W, Sealfon SC (1998) Rhodopsin-family receptors associate with small G proteins to activate phospholipase D. Nature 392:411–414

Mohammad-Zadeh LF, Moses L, Gwaltney-Brant SM (2008) Serotonin: a review. J Vet Pharmacol Ther 31:187–199

Mokler DJ, Commissaris RL, Warner MR, Rech RH (1983) Blockade of the behavioral effects of lysergic acid diethylamide, 2,5-dimethoxy-4-methylamphetamine, quipazine and lisuride by 5-hydroxytryptamine antagonists. J Pharmacol Exp Ther 227:557–562

Moreno FA, Wiegand CB, Taitano EK, Delgado PL (2006) Safety, tolerability, and efficacy of psilocybin in 9 patients with obsessive-compulsive disorder. J Clin Psychiatry 67:1735–1740

Moreno JL, Holloway T, Albizu L, Sealfon SC, Gonzalez-Maeso J (2011) Metabotropic glutamate mGlu2 receptor is necessary for the pharmacological and behavioral effects induced by hallucinogenic 5-HT2A receptor agonists. Neurosci Lett 493:76–79

Moreno JL, Muguruza C, Umali A, Mortillo S, Holloway T, Pilar-Cuellar F, Mocci G, Seto J, Callado LF, Neve RL, Milligan G, Sealfon SC, Lopez-Gimenez JF, Meana JJ, Benson DL, Gonzalez-Maeso J (2012) Identification of three residues essential for 5-hydroxytryptamine 2A-metabotropic glutamate 2 (5-HT2A.mGlu2) receptor heteromerization and its psychoactive behavioral function. J Biol Chem 287:44301–44319

Moreno JL, Miranda-Azpiazu P, Garcia-Bea A, Younkin J, Cui M, Kozlenkov A, Ben-Ezra A, Voloudakis G, Fakira AK, Baki L, Ge Y, Georgakopoulos A, Moron JA, Milligan G, Lopez-Gimenez JF, Robakis NK, Logothetis DE, Meana JJ, Gonzalez-Maeso J (2016) Allosteric signaling through an mGlu2 and 5-HT2A heteromeric receptor complex and its potential contribution to schizophrenia. Sci Signal 9:ra5

Muguruza C, Moreno JL, Umali A, Callado LF, Meana JJ, Gonzalez-Maeso J (2013) Dysregulated 5-HT(2A) receptor binding in postmortem frontal cortex of schizophrenic subjects. Eur Neuropsychopharmacol 23:852–864

Nichols DE (2004) Hallucinogens. Pharmacol Ther 101:131–181

Nichols DE, Nichols CD (2008) Serotonin receptors. Chem Rev 108:1614–1641

Nutt DJ, King LA, Nichols DE (2013) Effects of Schedule I drug laws on neuroscience research and treatment innovation. Nat Rev Neurosci 14:577–585

Oldham WM, Hamm HE (2008) Heterotrimeric G protein activation by G-protein-coupled receptors. Nat Rev Mol Cell Biol 9:60–71

Oufkir T, Arseneault M, Sanderson JT, Vaillancourt C (2010) The 5-HT 2A serotonin receptor enhances cell viability, affects cell cycle progression and activates MEK-ERK1/2 and JAK2-STAT3 signalling pathways in human choriocarcinoma cell lines. Placenta 31:439–447

Overington JP, Al-Lazikani B, Hopkins AL (2006) How many drug targets are there? Nat Rev Drug Discov 5:993–996

Pazos A, Cortes R, Palacios JM (1985) Quantitative autoradiographic mapping of serotonin receptors in the rat brain. II. Serotonin-2 receptors. Brain Res 346:231–249

Pazos A, Probst A, Palacios JM (1987) Serotonin receptors in the human brain–IV. Autoradiographic mapping of serotonin-2 receptors. Neuroscience 21:123–139

Peroutka SJ, Snyder SH (1979) Multiple serotonin receptors: differential binding of [3H] 5-hydroxytryptamine, [3H]lysergic acid diethylamide and [3H]spiroperidol. Mol Pharmacol 16:687–699

Pierce KL, Premont RT, Lefkowitz RJ (2002) Seven-transmembrane receptors. Nat Rev Mol Cell Biol 3:639–650

Porter RH, Benwell KR, Lamb H, Malcolm CS, Allen NH, Revell DF, Adams DR, Sheardown MJ (1999) Functional characterization of agonists at recombinant human 5-HT2A, 5-HT2B and 5-HT2C receptors in CHO-K1 cells. Br J Pharmacol 128:13–20

Preller KH, Pokorny T, Hock A, Kraehenmann R, Stampfli P, Seifritz E, Scheidegger M, Vollenweider FX (2016) Effects of serotonin 2A/1A receptor stimulation on social exclusion processing. Proc Natl Acad Sci U S A 113:5119–5124

Preller KH, Herdener M, Pokorny T, Planzer A, Kraehenmann R, Stampfli P, Liechti ME, Seifritz E, Vollenweider FX (2017) The fabric of meaning and subjective effects in lsd-induced states depend on serotonin 2a receptor activation. Curr Biol 27:451–457

Puig MV, Celada P, Diaz-Mataix L, Artigas F (2003) In vivo modulation of the activity of pyramidal neurons in the rat medial prefrontal cortex by 5-HT2A receptors: relationship to thalamocortical afferents. Cereb Cortex 13:870–882

Rapport MM, Green AA, Page IH (1948) Serum vasoconstrictor, serotonin; isolation and characterization. J Biol Chem 176:1243–1251

Raymond JR, Mukhin YV, Gelasco A, Turner J, Collinsworth G, Gettys TW, Grewal JS, Garnovskaya MN (2001) Multiplicity of mechanisms of serotonin receptor signal transduction. Pharmacol Ther 92:179–212

Reid G, Rand M (1952) Pharmacological actions of synthetic 5-hydroxytryptamine (serotonin, thrombocytin). Nature 169:801–802

Rives ML, Vol C, Fukazawa Y, Tinel N, Trinquet E, Ayoub MA, Shigemoto R, Pin JP, Prezeau L (2009) Crosstalk between GABAB and mGlu1a receptors reveals new insight into GPCR signal integration. EMBO J 28:2195–2208

Robertson DN, Johnson MS, Moggach LO, Holland PJ, Lutz EM, Mitchell R (2003) Selective interaction of ARF1 with the carboxy-terminal tail domain of the 5-HT2A receptor. Mol Pharmacol 64:1239–1250

Rosenbaum DM, Rasmussen SG, Kobilka BK (2009) The structure and function of G-protein-coupled receptors. Nature 459:356–363

Santini MA, Ratner C, Aznar S, Klein AB, Knudsen GM, Mikkelsen JD (2013) Enhanced prefrontal serotonin 2A receptor signaling in the subchronic phencyclidine mouse model of schizophrenia. J Neurosci Res 91:634–641

Schmid CL, Bohn LM (2010) Serotonin, but not N-methyltryptamines, activates the serotonin 2A receptor via a ß-arrestin2/Src/Akt signaling complex in vivo. J Neurosci official J Soc Neurosci 30:13513–13524

Schmid CL, Raehal KM, Bohn LM (2008) Agonist-directed signaling of the serotonin 2A receptor depends on beta-arrestin-2 interactions in vivo. Proc Natl Acad Sci U S A 105:1079–1084

Schmid CL, Streicher JM, Meltzer HY, Bohn LM (2014) Clozapine acts as an agonist at serotonin 2A receptors to counter MK-801-induced behaviors through a betaarrestin2-independent activation of Akt. Neuropsychopharmacology

Schmid Y, Enzler F, Gasser P, Grouzmann E, Preller KH, Vollenweider FX, Brenneisen R, Muller F, Borgwardt S, Liechti ME (2015) Acute effects of lysergic acid diethylamide in healthy subjects. Biol Psychiatry 78:544–553

Scruggs JL, Patel S, Bubser M, Deutch AY (2000) DOI-Induced activation of the cortex: dependence on 5-HT2A heteroceptors on thalamocortical glutamatergic neurons. J Neurosci 20:8846–8852

Sewell RA, Halpern JH, Pope HG Jr (2006) Response of cluster headache to psilocybin and LSD. Neurology 66:1920–1922

Sheffler DJ, Kroeze WK, Garcia BG, Deutch AY, Hufeisen SJ, Leahy P, Bruning JC, Roth BL (2006) p90 ribosomal S6 kinase 2 exerts a tonic brake on G protein-coupled receptor signaling. Proc Natl Acad Sci U S A 103:4717–4722

Shukla AK, Manglik A, Kruse AC, Xiao K, Reis RI, Tseng WC, Staus DP, Hilger D, Uysal S, Huang LY, Paduch M, Tripathi-Shukla P, Koide A, Koide S, Weis WI, Kossiakoff AA, Kobilka BK, Lefkowitz RJ (2013) Structure of active beta-arrestin-1 bound to a G-protein-coupled receptor phosphopeptide. Nature 497:137–141

Silbergeld EK, Hruska RE (1979) Lisuride and LSD: dopaminergic and serotonergic interactions in the "serotonin syndrome". Psychopharmacology 65:233–237

Singh RK, Dai Y, Staudinger JL, Muma NA (2009) Activation of the JAK-STAT pathway is necessary for desensitization of 5-HT2A receptor-stimulated phospholipase C signalling by olanzapine, clozapine and MDL 100907. Int J Neuropsychopharmacol 12:651–665

Strange PG (1998) Three-state and two-state models. Trends Pharmacol Sci 19:85–86

Svenningsson P, Tzavara ET, Carruthers R, Rachleff I, Wattler S, Nehls M, McKinzie DL, Fienberg AA, Nomikos GG, Greengard P (2003) Diverse psychotomimetics act through a common signaling pathway. Science (New York, N.Y.) 302:1412–1415

Svenningsson P, Nishi A, Fisone G, Girault JA, Nairn AC, Greengard P (2004) DARPP-32: an integrator of neurotransmission. Annu Rev Pharmacol Toxicol 44:269–296
Tork I (1990) Anatomy of the serotonergic system. Ann N Y Acad Sci 600:9–34; discussion 34–35
Urban JD, Clarke WP, von Zastrow M, Nichols DE, Kobilka B, Weinstein H, Javitch JA, Roth BL, Christopoulos A, Sexton PM, Miller KJ, Spedding M, Mailman RB (2007) Functional selectivity and classical concepts of quantitative pharmacology. J Pharmacol Exp Ther 320:1–13
Vaidya VA, Marek GJ, Aghajanian GK, Duman RS (1997) 5-HT2A receptor-mediated regulation of brain-derived neurotrophic factor mRNA in the hippocampus and the neocortex. J Neurosci 17:2785–2795
Vollenweider FX, Vollenweider-Scherpenhuyzen MF, Babler A, Vogel H, Hell D (1998) Psilocybin induces schizophrenia-like psychosis in humans via a serotonin-2 agonist action. NeuroReport 9:3897–3902
Waeber C, Palacios JM (1994) Binding sites for 5-hydroxytryptamine-2 receptor agonists are predominantly located in striosomes in the human basal ganglia. Brain Res Mol Brain Res 24:199–209
Whorton MR, Bokoch MP, Rasmussen SG, Huang B, Zare RN, Kobilka B, Sunahara RK (2007) A monomeric G protein-coupled receptor isolated in a high-density lipoprotein particle efficiently activates its G protein. Proc Natl Acad Sci U S A 104:7682–7687
Whorton MR, Jastrzebska B, Park PS, Fotiadis D, Engel A, Palczewski K, Sunahara RK (2008) Efficient coupling of transducin to monomeric rhodopsin in a phospholipid bilayer. J Biol Chem 283:4387–4394
Wu B, Chien EY, Mol CD, Fenalti G, Liu W, Katritch V, Abagyan R, Brooun A, Wells P, Bi FC, Hamel DJ, Kuhn P, Handel TM, Cherezov V, Stevens RC (2010) Structures of the CXCR4 chemokine GPCR with small-molecule and cyclic peptide antagonists. Science 330:1066–1071
Wu H, Wacker D, Mileni M, Katritch V, Han GW, Vardy E, Liu W, Thompson AA, Huang XP, Carroll FI, Mascarella SW, Westkaemper RB, Mosier PD, Roth BL, Cherezov V, Stevens RC (2012) Structure of the human kappa-opioid receptor in complex with JDTic. Nature 485
Wurmbach E, Yuen T, Ebersole BJ, Sealfon SC (2001) Gonadotropin-releasing hormone receptor-coupled gene network organization. J Biol Chem 276:47195–47201
Xia Z (2003) The PDZ-binding domain is essential for the dendritic targeting of 5-HT2A serotonin receptors in cortical pyramidal neurons in vitro. Neuroscience 122:907–920
Xia Z, Gray JA, Compton-Toth BA, Roth BL (2003) A direct interaction of PSD-95 with 5-HT2A serotonin receptors regulates receptor trafficking and signal transduction. J Biol Chem 278:21901–21908
Yuen T, Zhang W, Ebersole BJ, Sealfon SC (2002) Monitoring G-protein-coupled receptor signaling with DNA microarrays and real-time polymerase chain reaction. Methods Enzymol 345:556–569
Yuen E, Jiang Q, Chen P, Feng J, Yan Z (2008) Activation of 5-HT2A/C receptors counteracts 5-HT1A regulation of NMDAR channels in pyramidal neurons of prefrontal cortex. J Biol Chem

Effects of Hallucinogens on Neuronal Activity

L. Lladó-Pelfort, P. Celada, M.S. Riga, E. Troyano-Rodríguez, N. Santana and F. Artigas

Abstract Hallucinogens evoke sensory, perceptual, affective, and cognitive effects that may be useful to understand the neurobiological basis of mood and psychotic disorders. The present chapter reviews preclinical research carried out in recent years in order to better understand the action of psychotomimetic agents such as the noncompetitive NMDA receptor (NMDA-R) antagonists and serotonergic hallucinogens. Our studies have focused on the mechanisms through which these agents alter cortical activity. Noncompetitive NMDA-R antagonists, such as phencyclidine (PCP) and MK-801 (dizocilpine), as well as the serotonergic hallucinogens DOI and 5-MeO-DMT, produce similar effects on cellular and population activity in prefrontal cortex (PFC); these effects include alterations of pyramidal neuron discharge (with an overall increase in firing), as well as a marked attenuation of the low frequency oscillations (0.2–4 Hz) to which neuronal discharge is coupled in anesthetized rodents. PCP increases *c-fos* expression in excitatory neurons from various cortical and subcortical areas, particularly the thalamus. This effect of PCP involves the preferential blockade of NMDA-R on GABAergic neurons of the reticular nucleus of the thalamus, which provides feedforward inhibition to the rest of thalamic nuclei. It is still unknown whether serotonergic hallucinogens also affect thalamocortical networks. However, when examined, similar alterations in other cortical areas, such as the primary visual cortex (V1), have been observed, suggesting that these agents affect cortical activity in sensory and associative areas. Interestingly, the disruption of PFC activity induced by PCP, DOI and 5-MeO-DMT is reversed by classical and atypical antipsychotic drugs. This effect suggests a possible link between the mechanisms underlying the disruption of

L. Lladó-Pelfort · P. Celada · M.S. Riga · E. Troyano-Rodríguez · N. Santana · F. Artigas
Department of Neurochemistry and Neuropharmacology, Centro de Investigación Biomédica en Red de Salud Mental (CIBERSAM), Institut d'Investigacions Biomèdiques de Barcelona (IIBB-CSIC-IDIBAPS), Barcelona, Spain

F. Artigas (✉)
Department of Neurochemistry and Neuropharmacology, IIBB-CSIC (IDIBAPS), Rosselló, 161, 6th Floor, 08036 Barcelona, Spain
e-mail: fapnqi@iibb.csic.es

Curr Topics Behav Neurosci (2017) 36: 75–106
DOI 10.1007/7854_2017_473
Published Online: 26 February 2017

perception by multiple classes of hallucinogenic agents and the therapeutic efficacy of antipsychotic agents.

Keywords 5-HT_{2A} receptors · Antipsychotic drugs · NMDA receptors · Prefrontal cortex · Thalamus

Contents

1 Hallucinogens as Pharmacological Models in Schizophrenia Research

Schizophrenia is one of the most severe psychiatric conditions, affecting approximately 1% of the population. It has an early onset (typically late adolescence or early adulthood) and shows a chronic and deteriorating course. Many affected individuals suffer from a lifelong disability, with a poor quality of life, and nearly 10% commit suicide. Schizophrenia is characterized by a variety of symptoms, including positive (hallucinations, delusions, disorganized thought and speech, aberrant behavior, etc.) and negative symptoms (emotional blunting, social withdrawal, depression, anxiety, etc.) together with cognitive deficits. Anatomical, cellular and neurochemical alterations have been reported in various brain areas of schizophrenic patients and notably in the prefrontal cortex (PFC) (Harrison 1999a; Selemon and Goldman-Rakic 1999; Lewis and Lieberman 2000; Harrison and Weinberger 2005; Lewis and Gonzalez-Burgos 2006) (see below for additional information).

Historically, schizophrenia has been associated with derangements of multiple neurotransmitter systems, due to the ability of pharmacological agents acting on those systems to mimic symptoms of the illness in healthy individuals and to aggravate symptoms in patients. Hence, more classical views on the pathophysiology of schizophrenia indicated a hyperactivity of midbrain dopaminergic systems, suggested by the ability of dopamine-releasing agents to evoke psychotic symptoms (Carlsson 1977). This early view was later supported by positron emission tomography (PET) scan studies showing a larger striatal dopamine release in schizophrenic patients compared to healthy controls after receiving an amphetamine challenge (Laruelle et al. 1996) as well as a higher basal occupancy of dopamine D_2 receptors (Abi-Dargham et al. 2000). Subsequently, the dopamine hypothesis of schizophrenia was refined, suggesting the function of the ascending mesocortical dopamine system was reduced, which might account for the negative symptoms—and possibly, cognitive deficits—of the illness (Weinberger 1987). This view received support from PET studies showing alterations of postsynaptic dopamine D_1 receptor availability consistent with a low dopaminergic tone (Abi-Dargham et al. 2002).

In addition to dopamine, alterations of other neurotransmitter systems in schizophrenia have been proposed, including serotonin (5-hydroxytryptamine, 5-HT), glutamate, and GABA. In particular, alterations of 5-HT and glutamate receptors have been reported (Rasmussen et al. 2010, 2016; Harrison 1999b). For example, noncompetitive antagonists of NMDA receptors (NMDA-R) are extensively used as pharmacological models of schizophrenia due to the ability of these compounds to mimic the symptoms of the illness (Krystal et al. 2003). Additionally, atypical (second and third generation) antipsychotic drugs mainly target 5-HT receptors (Meltzer and Massey 2011).

In the present chapter, we review some electrophysiological and histological observations suggesting a common pattern of action of serotonergic and glutamatergic agents in the prefrontal cortex (PFC) of experimental animals that may account for the psychotomimetic properties of these agents. We also provide evidence that the alterations of PFC activity evoked by serotonergic hallucinogens and noncompetitive NMDA-R antagonists can be countered by classical (haloperidol) and atypical (clozapine) antipsychotic drugs.

1.1 Serotonergic Hallucinogens

Most 5-HT_{2A} receptor agonists, including *N,N*-dimethyltryptamine (DMT), 5-methoxy-*N,N*-dimethyltryptamine (5-MeO-DMT), lysergic acid diethylamide (LSD), 1-(2,5-dimethoxy-4-iodophenyl)-2-aminopropane (DOI), mescaline and psilocybin, produce hallucinogenic effects, altering perception, thought, emotion and mood in a manner that is similar to endogenous psychosis (Glennon 1991, 1994; Nichols 2004; Vollenweider and Kometer 2010). Chemically, these agents belong to two main families: indoleamines (such as DMT, 5-MeO-DMT, psilocybin, and LSD), which have some structural similarity to the 5-HT molecule, and phenethylamines [such as mescaline, DOI, 1-(2,5-dimethoxy-4-bromophenyl)-2-aminopropane (DOB), and 1-(2,5-dimethoxy-4-methylphenyl)-2-aminopropane (DOM)] (Nichols 2004). One important reason for investigating psychedelic agents lies in their capacity to model certain aspects of psychosis in experimental research

(Geyer and Vollenweider 2008), as well as their ability to facilitate the identification of brain areas/circuits altered in psychiatric disorders (Vollenweider et al. 1997a, b). Moreover, some of these agents, such as LSD and psilocybin, have shown therapeutic efficacy in the treatment of psychiatric disorders [reviewed in Vollenweider and Kometer (2010)]. LSD was marketed by Sandoz in the 1950s to facilitate psychotherapeutic treatment of psychiatric disorders, due to its ability to increase introspection and recall of emotional memories. The widespread recreational use of LSD and its marked effects on perception, mood and personality had a profound influence on young people in the Western world, leading to the development of so-called psychedelic art in the 1960s and 1970s. The hallucinogenic properties of LSD and the role it played in countercultural activities led to its classification as a controlled substance in the USA and most other Western countries.

DOI is another synthetic compound with strong psychedelic effects that has been extensively used in pharmacological research. It evokes long-lasting alterations of consciousness and perception (Nichols 2004). DOI acts by stimulating 5-HT_{2A} receptors, since most of its behavioral, neurochemical and electrophysiological effects can be blocked by the selective 5-HT_{2A} receptor antagonist M100907 (Schreiber et al. 1994; Martin-Ruiz et al. 2001; Puig et al. 2003).

DMT and 5-MeO-DMT are natural components of *ayahuasca*, a hallucinogenic beverage used in religious ceremonies and for healing purposes in the upper Amazon (McKenna 2004; McKenna et al. 1984; Schultes et al. 1991) and recently investigated for potential clinical uses (McKenna 2004). The psychedelic effects induced by *ayahuasca* include visual and auditory hallucinations, synesthesia, and deep psychological introspection. In addition to the tryptamines, *ayahuasca* contains β-carbolines such as harmine, which act as reversible inhibitors of monoamine oxidase-A (MAO-A). The β-carbolines prevent the catabolism of the tryptamines in the gut, thus allowing the tryptamines to be active orally and prolonging their psychedelic effects (McKenna et al. 1984; Agurell et al. 1968). DMT and 5-MeO-DMT are also synthesized and distributed for recreational purposes (Yu 2008) and intoxications have been reported (Sklerov et al. 2005; Brush et al. 2004). Early reports identified 5-MeO-DMT as a possible endogenous psychotoxin and several groups theorized about its potential involvement in schizophrenia (Benington et al. 1965; Gillin and Wyatt 1976; Angrist et al. 1976).

Interestingly, not all 5-HT_{2A} receptor agonists produce hallucinogenic effects, raising questions about the neural mechanisms responsible for their effects, such as differences in signaling pathways (Kurrasch-Orbaugh et al. 2003; Gonzalez-Maeso et al. 2007, see also Chap. 2). Regardless of the cellular mechanisms that are responsible for the effects of hallucinogens, it is important to note that their main target—the 5-HT_{2A} receptor—is expressed at very high levels in the neocortex, including sensory and associative areas (Pazos et al. 1985, 1987; Jakab and Goldman-Rakic 1998; Hall et al. 2000; Lopez-Gimenez et al. 2001; Santana et al. 2004). The 5-HT_{2A} receptor is densely expressed in intermediate and deep layers of the frontal, temporal, parietal and occipital lobes (Pazos et al. 1987; Santana et al. 2004; de Almeida and Mengod 2007), suggesting that their activation may have a strong influence on the processing of information in cortical areas as well as between cortical and subcortical brain regions. 5-HT$_{2A}$ receptors are expressed in midbrain-projecting PFC pyramidal neurons (Vazquez-Borsetti et al. 2009) and 5-HT$_{2A}$ receptor activation by DOI and other hallucinogens has marked effects on the activity of downstream serotonergic and dopaminergic neurons (Martin-Ruiz et al. 2001; Bortolozzi et al. 2003, 2005).

Activation of 5-HT$_{2A}$ receptors by 5-HT and other agonists elicits several electrophysiological effects in cortical neurons recorded in vitro, including enhancement of the response to excitatory synaptic input (Aghajanian and Marek 1997, 1999), induction of membrane depolarization, and reduction of the afterhyperpolarization that follows a burst of spikes (Araneda and Andrade 1991; Villalobos et al. 2005). In vivo, 5-HT$_{2A}$ receptor activation by endogenous 5-HT moderately increases the firing rate of pyramidal cells (Amargos-Bosch et al. 2004; Puig et al. 2005), whereas systemic administration of DOI evokes a dramatic increase in the firing rate of a subpopulation of PFC pyramidal neurons (Puig et al. 2003). However, despite the increasing knowledge of the cellular actions of hallucinogens, there is a conspicuous lack of information concerning their actions at the network level.

1.2 NMDA Receptors. Noncompetitive Antagonists

Rapid ionic effects of glutamate are mediated by three receptor subtypes, namely AMPA, kainate, and NMDA receptors. These receptors are ion channels, and their activation by glutamate allows extracellular Na^+ and Ca^{2+} ions to enter (and K^+ ions to leave) the cytoplasm, thus evoking rapid and marked changes of the membrane potential (depolarization in most instances), which subsequently results in the generation of action potentials. Glutamate can also act on a family of 8 G-protein coupled metabotropic receptors, analogous to monoamine receptors, which are suitable targets for drug development (Swanson et al. 2005; Niswender and Conn 2010).

NMDA-R are involved in a large number of key physiological functions, such as long-term potentiation and other forms of synaptic plasticity, and play a role in several neurological and psychiatric disorders (Lau and Zukin 2007; Paoletti and Neyton 2007). NMDA-R are tetrameric ion channels composed of two GluN1 and two GluN2 subunits (formerly called NR1 and NR2). A third type of NMDA-R subunits (GluN3) has been identified that changes the ionic sensitivity of the NMDA-R channel (Cavara and Hollmann 2008). The NMDA-R ion channel is voltage-sensitive, and the channel is blocked by Mg^{2+} ions at resting membrane potential. After the depolarization of the cell membrane, Mg^{2+} ions are released from the NMDA-R, allowing the passage of other ions (Na^+, Ca^{2+}, K^+) through the channel. Thus, in general, AMPA-induced depolarization precedes the functional activity of NMDA-R.

The NMDA-R contains several binding sites, including the orthosteric site that binds glutamate and competitive antagonists such as AP5 (also known as APV).

Likewise, the NMDA-R contains several regulatory sites, including a glycine binding site, located outside the channel, as well as binding sites for Mg^{2+} and noncompetitive antagonists such as phencyclidine (PCP), which are located inside the channel. The dissociative anesthetics ketamine and PCP are noncompetitive NMDA-R antagonists. These agents have been used as pharmacological models of schizophrenia due to their ability to mimic the positive (psychotic) and negative symptoms of schizophrenia in healthy individuals and to aggravate illness symptoms in schizophrenic patients (Krystal et al. 2003; Javitt and Zukin 1991). Moreover, PCP, ketamine, and dizocilpine (MK-801; not available for human use) evoke a series of behavioral alterations in experimental animals, including induction of hyperlocomotion and stereotypies, and disruption of sensorimotor gating. These alterations are partially or totally antagonized by antipsychotic drugs (Carlsson and Carlsson 1989; Geyer et al. 2001). However, the cellular elements and brain networks involved in these actions are still not completely understood, although work by different research groups in recent years has started to clarify the actions of NMDA-R antagonists on brain function, including activity and processing in the prefrontal cortex (PFC).

2 The Prefrontal Cortex

The PFC is the highest association cortex and plays an important role in the pathophysiology and treatment of psychiatric disorders, including schizophrenia. Not surprisingly, most research on the cellular actions of psychotomimetic agents has been carried in this cortical area. The PFC has poorly defined anatomical boundaries but it can be identified by its reciprocal connectivity with the mediodorsal (MD) nucleus of the thalamus. The PFC is involved in many higher brain functions, such as perception, attention, memory, and cognition. The dorsolateral PFC is primarily involved in cognitive processes, such as working memory and executive function, as well as action planning and decision making (Fuster 2001, 2008; Miller and Cohen 2001). In addition to cognitive functions, the PFC participates in the control of mood and affect. Hence, the anterior cingulate cortex, including its ventral subdivision, is heavily involved in emotional processing (Devinsky et al. 1995; Davidson and Irwin 1999; Cardinal et al. 2002; Phillips et al. 2003) and certain psychotic symptoms such as hallucinations are associated with hyperactivity in this PFC subdivision (Shergill et al. 2000). Likewise, alterations of energy metabolism in the ventromedial PFC have been reported in depressed patients (Drevets 2001; Seminowicz et al. 2004) and high frequency stimulation of this region evokes immediate antidepressant effects in patients refractory to standard treatments (Mayberg et al. 2005; Puigdemont et al. 2012).

As is the case with other cortical regions, the PFC is composed of $\sim$75–80% of pyramidal projection neurons, which use glutamate as a transmitter, and $\sim$20–25% of local circuit inhibitory interneurons, which use GABA as a transmitter. Pyramidal neurons integrate excitatory glutamatergic afferent input from various thalamic nuclei, including the mediodorsal, centromedial and several midline nuclei, as well as the hippocampus, the amygdala, and other cortical areas (Fuster 2008; Groenewegen and Uylings 2000). Local inhibitory inputs arise from GABAergic interneurons. These interneurons have been classified according to their anatomical and neurochemical characteristics and their synaptic relationships with pyramidal neurons (DeFelipe et al. 2013). Firing patterns also differ markedly between subpopulations of interneurons. Fast-spiking, parvalbumin-containing GABAergic interneurons have been suggested to play a role in schizophrenia, in particular in deficits of cognitive control (Lewis et al. 2005, 2012).

PFC neurons are densely innervated by projections from brainstem monoaminergic nuclei such as the dorsal and median raphe nuclei, locus coeruleus and ventral tegmental area, which employ 5-HT, norepinephrine and dopamine, respectively, as neurotransmitters. These nuclei exert an important modulatory role of the excitatory and inhibitory inputs onto PFC neurons (Puig et al. 2005; Steinbusch 1981; Van Eden et al. 1987; Seamans and Yang 2004; Aston-Jones and Cohen 2005; Celada et al. 2013). A large population of pyramidal and GABAergic neurons in PFC express receptors for monoamine neurotransmitters (Santana et al. 2004, 2009, 2013; de Almeida and Mengod 2007, 2008). Atypical antipsychotic drugs such as clozapine show high affinity for monoamine receptors, suggesting that the PFC may be a key brain structure in their therapeutic action, in addition to the well-known blockade of dopamine D_2 receptors in ventral striatum (Artigas 2010).

3 Effects of Psychotomimetic Agents on Neuronal Activity in PFC

3.1 Serotonergic Hallucinogens

In one study (Puig et al. 2003), we examined the effect of DOI on the firing activity of pyramidal neurons in the rat medial PFC (mPFC). These neurons were identified by antidromic activation from the dorsal raphe or the ventral tegmental area, two areas targeted by projections from pyramidal neurons in deep PFC layers (Gabbott et al. 2005). More recent studies indicated that a large proportion of mPFC pyramidal neurons simultaneously project to both monoaminergic nuclei (Vazquez-Borsetti et al. 2011).

The i.v. administration of DOI (0.05–0.3 mg/kg) evoked three types of responses in mPFC pyramidal neurons. DOI enhanced the firing rate in 38% of the neurons examined (21/56), increasing the mean firing rate from 1.34 ± 0.36 to 6.45 ± 0.97 spikes/s (a 481% increase; $n = 21$). Another 32% of the projection neurons examined (18/56) were unaffected by administration of DOI at doses up to 0.3–0.8 mg/kg (i.v.), whereas the remaining cells (17/56, 30%) showed reductions of their firing rate (to 11 ± 3% of baseline). The baseline firing rate of the neurons that were excited or inhibited by DOI did not differ significantly (1.34 ± 0.36

spikes/s for excitations vs. 1.20 ± 0.33 spikes/s for inhibitions). Summing over all of the neurons examined, DOI significantly increased the output of mPFC projection neurons to 238% of baseline, from 1.23 ± 0.18 to 2.93 ± 0.53 spikes/s ($n = 56$; $p < 0.002$, paired t-test).

The DOI-induced excitations were reversed by M100907 (0.1–0.5 mg/kg i.v.) in 6 out of 9 cases examined. Because DOI is also an agonist at 5-HT_{2C} receptors, effects on some pyramidal neurons may also be mediated by this receptor subtype. Alternatively, M100907 may not be able to reverse the pyramidal excitation once the neurons become extensively depolarized, likely by activation of ionotropic glutamate receptors. The inhibition of firing elicited by DOI was also antagonized by M100907 (0.1–0.5 mg/kg i.v.; $n = 6$). In addition, the inhibitory effect of DOI was also reversed by the $GABA_A$ antagonist picrotoxinin (1–3 mg/kg i.v.), indicating that the pyramidal neuron inhibition is mediated by activation of 5-HT_{2A}-R located on GABAergic interneurons (Santana et al. 2004) (Fig. 1).

In another study (Wood et al. 2012), the effects of DOI on neuronal activity were examined in slightly different PFC subdivisions (orbitofrontal cortex (OFC) and anterior cingulate cortex (ACC)) in freely moving rats. At a lower dose (1 mg/kg i.p.), DOI mainly increased population activity in ACC (as was observed in mPFC), whereas a larger proportion of the neurons were inhibited by a higher dose (5 mg/kg i.p.). In contrast, in the OFC, DOI evoked an excitation/inhibition ratio of ca. 1 at 1 mg/kg i.p., and produced a dose-dependent inhibition of neuronal activity at higher doses (Wood et al. 2012). The differences in the relative proportion of excited versus inhibited neurons in the two studies may be a consequence of the distinct PFC subdivisions examined as well as the use of choral hydrate anesthesia in the first study.

The effect of 5-MeO-DMT administration on pyramidal discharge was examined in chloral hydrate anesthetized rats pretreated with the MAO-A inhibitor clorgyline to mimic the effects of *ayahuasca*. The i.v. administration of 5-MeO-DMT (0.1 mg/kg) evoked a response in pyramidal neurons similar to that of DOI, increasing the firing rate of 51% of the neurons (to 406% of baseline), reducing the firing of 35% of the neurons (to 31% of baseline), and leaving the rest of the neurons (14%) unaffected (Fig. 2). Overall, 5-MeO-DMT increased pyramidal firing rate to 215% of baseline ($p < 0.001$, Student's t test; $n = 37$).

3.2 Noncompetitive NMDA Receptor Antagonists

Neuroimaging studies indicate that noncompetitive NMDA-R antagonists increase metabolic activity in the PFC (Breier et al. 1997). We therefore examined the actions of the noncompetitive NMDA-R antagonist PCP on the firing activity of pyramidal neurons in the mPFC of choral hydrate anesthetized rats. As with the studies of serotonergic hallucinogens, pyramidal neurons were identified by antidromic activation from midbrain nuclei, so the results are representative of the same neuronal population that was examined in the studies with DOI and 5-MeO-DMT.

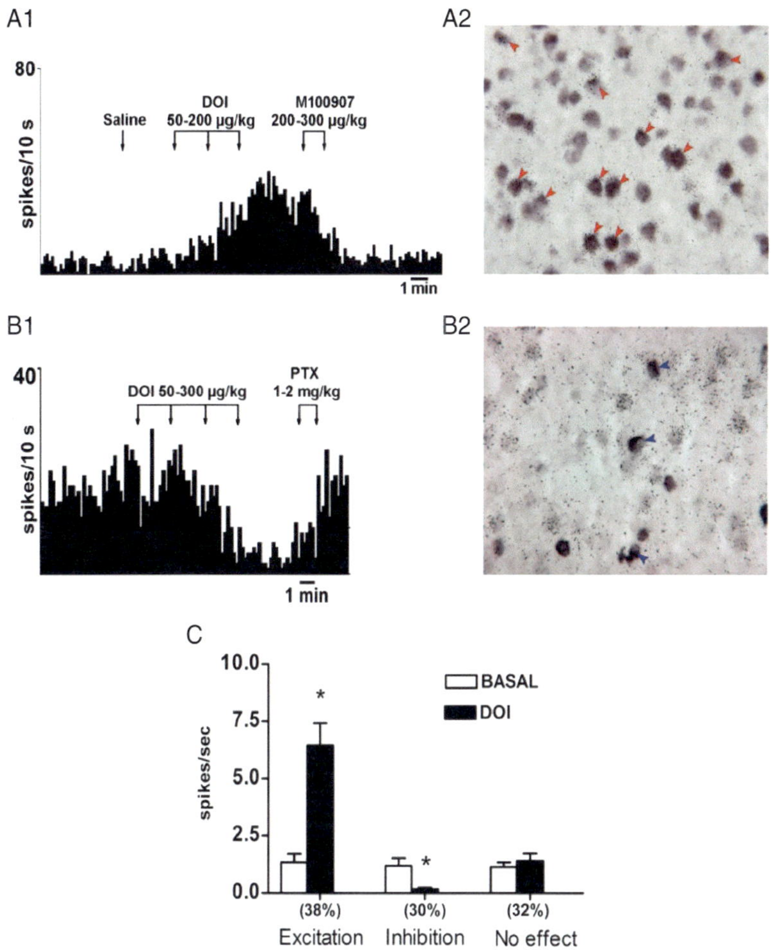

Fig. 1 Effect of 1-(2,5-dimethoxy-4-iodophenyl)-2-aminopropane (DOI) on the firing rate of pyramidal neurons in rat medial prefrontal cortex. **A1** Neuron that responded to DOI with an increase in firing rate. The excitatory effects of DOI were reversed by the 5-HT_{2A}-R antagonist M100907. **A2** Photomicrograph showing the presence of 5-HT_{2A}-R mRNA (^{33}P-labeled oligonucleotides) in pyramidal cells (*red arrows*) identified by the presence of vesicular glutamate transporter 1 (vGluT1) mRNA (dig-labeled oligonucleotides). **B1** Neuron that responded to DOI with a reduction in firing rate. The inhibitory effects of DOI were reversed by M100907 (data not shown) as well as by the $GABA_A$-R antagonist picrotoxinin (PTX). **B2** Photomicrograph showing the presence of 5-HT_{2A}-R mRNA (^{33}P-labeled oligonucleotides) in GABAergic cells (*blue arrows*) identified by the presence of glutamate decarboxylase (GAD) mRNA. **C** *Bar graph* showing the average effect of DOI according to the type of response. *$p < 0.05$ versus the basal firing rate. Adapted from Puig et al. (2003)

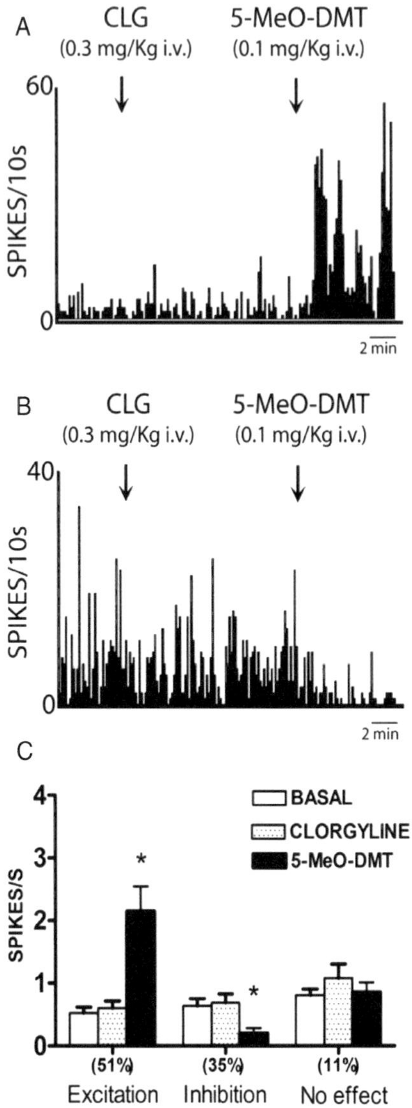

Fig. 2 Effect of 5-methoxy-*N,N*-dimethyltryptamine (5-MeO-DMT) on the firing rate of pyramidal neurons in rat medial prefrontal cortex. Rats treated with 5-MeO-DMT were pretreated with the MAO-A inhibitor clorgyline (CLG) in order to mimic the effects of *ayahuasca*, which contains both β-carbolines (MAO-A inhibitors) and tryptamines. **A** Neuron that responded to CLG and 5-MeO-DMT with an increase in firing rate. **B** Neuron that responded to CLG and 5-MeO-DMT with a reduction in firing rate. **C** *Bar graph* showing the average effect of CLG and 5-MeO-DMT according to the type of response. $^{*}p < 0.05$ versus the basal firing rate. Adapted from Celada et al. (2008)

The i.v. administration of PCP evoked complex effects on the discharge rate of mPFC pyramidal neurons (Kargieman et al. 2007). PCP (0.25 mg/kg i.v.) increased the discharge rate of 45% of the recorded neurons (to 286% of baseline) and reduced the discharge rate of 35% of the neurons (to 43% of baseline), with the remainder of the neurons (22%) unaffected (Fig. 3).

In a similar study, Katayama et al. (2007) also reported that PCP increased pyramidal neuron activity in rats anesthetized with pentobarbital followed by

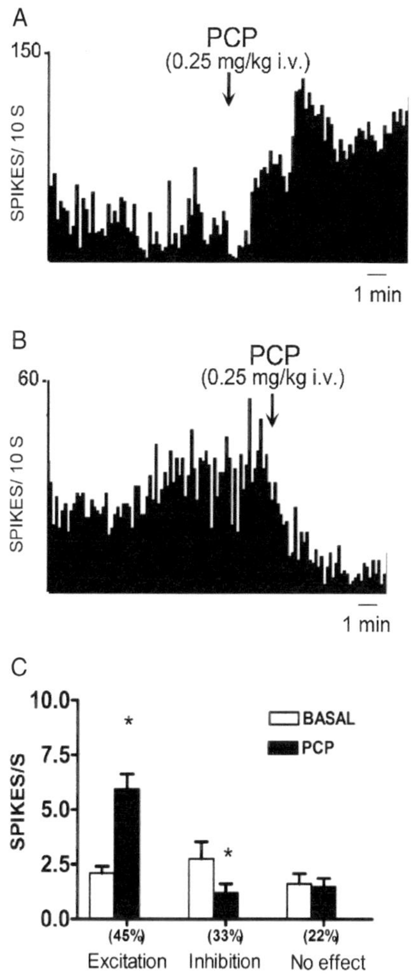

Fig. 3 Effect of phencyclidine (PCP) on the firing rate of pyramidal neurons in rat mPFC. **A** Neuron that responded to PCP with an increase in firing rate. **B** Neuron that responded to PCP with a reduction in firing rate. **C** *Bar graph* showing the average effect of PCP according to the type of response. $*p < 0.05$ versus the basal firing rate. Adapted from Kargieman et al. (2007). Note that the effects of PCP on pyramidal neuron firing are very similar to those produced by DOI (see panel **C** in Fig. 1) and 5-MeO-DMT (see panel **C** in Fig. 2)

urethane; this effect was antagonized by the local application of the AMPA/kainate receptor antagonist 6-cyano-7-nitroquinoxaline-2,3-dione (CNQX).

4 Role of Thalamocortical Inputs

Despite the marked alteration of PFC activity produced by PCP, it is still not clear whether this action is essentially driven by intrinsic mechanisms (e.g., blockade of NMDA-R in the PFC) or whether other cortical and subcortical areas connected with the PFC—in particular the hippocampal formation and the thalamus—are also involved. The fact that systemic MK-801 administration has countervailing effects on the activity of putative pyramidal and GABAergic neurons (increased and decreased discharge, respectively) led to the proposal that the main mechanism of action of this agent involves preferential blockade of NMDA-Rs on cortical GABAergic interneurons (Homayoun and Moghaddam 2007). However, the local application of PCP or MK-801 in the mPFC reduced the discharge rate of putative pyramidal neurons (Jodo et al. 2005; Suzuki et al. 2002) and, in the latter study, it was reported that the local application of PCP in the hippocampal formation facilitated pyramidal neuron activity in mPFC. Moreover, systemic—but not local—administration of noncompetitive NMDA-R antagonists increased neurotransmitter release in PFC (Amargos-Bosch et al. 2006; Lopez-Gil et al. 2007), suggesting that NMDA-R blockade in other brain areas may also contribute to the increase in PFC activity. In particular, interhemispheric projections may play a role since the bilateral (but not unilateral) application of MK-801 and ketamine in mPFC evoked neurochemical and behavioral changes similar to those produced by the systemic MK-801 administration (Lopez-Gil et al. 2012).

Double in situ hybridization experiments revealed that PCP (10 mg/kg i.p.) markedly increased *c-fos* expression in glutamatergic (vGluT1-positive) neurons from several cortical regions (prefrontal, somatosensory, retrosplenial, and entorhinal cortex) (Kargieman et al. 2007; Santana et al. 2011). PCP also induced a very marked increase of *c-fos* expression in various thalamic nuclei, in particular the centromedial and mediodorsal nuclei, which are reciprocally connected with the PFC (Gabbott et al. 2005; Berendse and Groenewegen 1991; Kuroda et al. 1998). PCP also increased *c-fos* expression in the amygdala, but only had a small effect in the hippocampal formation. The results of these *c-fos* studies led us to directly examine the effect of PCP on the discharge rate of neurons in the centromedial and mediodorsal thalamic nuclei. The administration of PCP (0.25 mg/kg i.v.) altered the discharge of thalamic neurons, increasing (to 424% of baseline) and decreasing (to 41% of baseline) the activity of 57 and 20% of the recorded neurons, respectively (23% of the neurons were unaffected) (Santana et al. 2011) (Fig. 4).

The activation of thalamocortical network activity by PCP appears to result from the preferential blockade of NMDA-R in the reticular nucleus of the thalamus (RtN). The RtN consists of a thin layer of GABAergic neurons that provide feedforward inhibition to excitatory thalamic nuclei. Hence, in parallel with the *c-fos* increase, systemic PCP administration evoked a dramatic and generalized decrease of neuronal activity in the RtN (Troyano-Rodriguez et al. 2014) at the same doses that increased excitatory neuronal discharge in the PFC and centromedial/mediodorsal nuclei (Kargieman et al. 2007; Santana et al. 2011) (Fig. 4).

Recent evidence from other groups also supports the involvement of thalamic nuclei in the action of NMDA-R antagonists. For example, ketamine administration increased the discharge rate of cells in the nucleus reuniens of the thalamus, and subsequently in the CA1 hippocampal subfield, a region to which the nucleus reuniens projects (Zhang et al. 2012). The same workers also reported that blockade of NMDA-R with the competitive antagonist APV in the reticular thalamic nucleus evoked bursts of delta oscillations (Zhang et al. 2009). Moreover, the effect of systemic MK-801 administration on slow oscillations (see below) in the PFC was mimicked by the local application of lidocaine in the mediodorsal nucleus of the

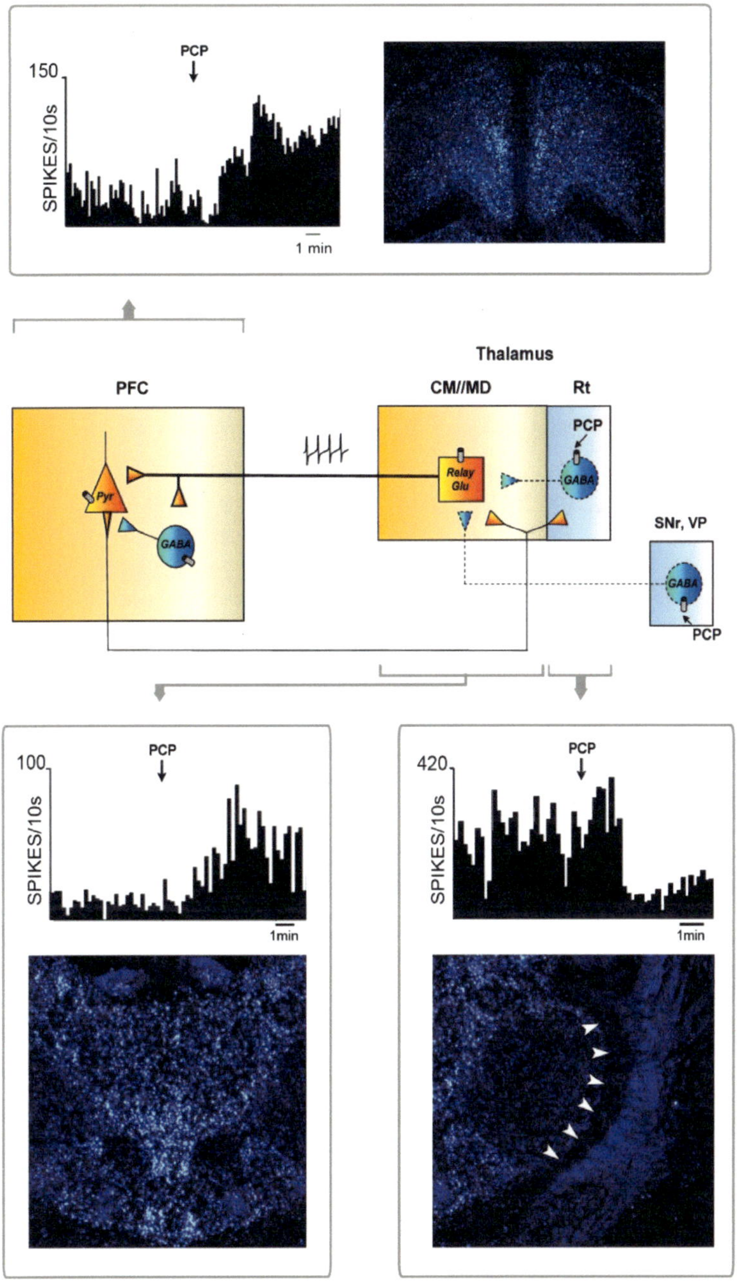

PCP
150
SPIKES/10s
1 min
Thalamus
PFC
CM//MD
Rt
PCP
Pyr
Relay
Glu
GABA
GABA
SNr, VP
GABA
PCP
100
PCP
SPIKES/10s
1min
420
PCP
SPIKES/10s
1min

◂**Fig. 4** Schematic representation of the action of phencyclidine (PCP) on thalamocortical circuits. PCP appears to exert a preferential blockade of NMDA receptors (NMDA-R) in the reticular nucleus (Rt) of the thalamus, which results in disinhibition of the excitatory thalamic nuclei that project to the neocortex (*bottom right panel*). The *upper* trace is an integrated firing rate histogram showing the effect of i.v. administration of PCP (0.25 mg/kg) on the discharge rate of a GABAergic neuron in the Rt. Note the lack of *c-fos* expression in the Rt (indicated by *arrowheads* in the lower image) (*bottom left panel*). In parallel with the reduction of GABAergic neuronal activity in the Rt, a large percentage of excitatory neurons in the mediodorsal (MD) and centromedial (CM) nuclei, reciprocally connected with the prefrontal cortex (PFC), are excited by PCP (0.25 mg/kg i.v.). These nuclei show also a dramatic increase in *c-fos* expression after i.p. administration of PCP (10 mg/kg), as shown in the *lower* histological image (*top panel*). Similar to the effect on the excitatory thalamic nuclei, PCP also increased the firing rate of a large percentage of the pyramidal neurons in the mPFC and increased *c-fos* expression in vGluT1-positive cells (pyramidal neurons). Overall, this combination of electrophysiological and histological evidence supports the view that PCP activates thalamocortical networks through a *bottom-up* process, although additional effects mediated by blockade of NMDA-R in other subpopulations of GABAergic neurons (e.g., fast-spiking cortical interneurons, substantia nigra reticulata (SNr) neurons, etc.) cannot be excluded

thalamus (Kiss et al. 2011a). Furthermore, the motor hyperactivity and the behavioral stereotypies induced by MK-801 in rats were prevented by bilateral application of the $GABA_A$ agonist muscimol directly into the anterior nucleus of the thalamus, indicating that MK-801 reduces $GABA_A$-mediated neurotransmission in the thalamus (Lopez Hill and Scorza 2012). Overall, these observations clearly support the involvement of thalamic nuclei in the behavioral, perceptual, and cognitive alterations induced by noncompetitive NMDA-R antagonists.

Thalamocortical inputs have also been claimed to play a role in the effects of serotonergic hallucinogens. For example, 5-HT and DOI were reported to evoke excitatory postsynaptic potentials in layer V pyramidal neurons in PFC slices, potentially by acting on presynaptic 5-HT_{2A} receptors located on thalamocortical afferents (Marek and Aghajanian 1998; Marek et al. 2001). However, this view has been challenged by several lines of experimental evidence, including the failure of histological studies to identify 5-HT_{2A} receptors located on presynaptic glutamatergic axons (Miner et al. 2003), the fact that DOI has identical effects on pyramidal neuron discharge in normal rats and in rats with extensive thalamic lesions (Puig et al. 2003), and the dependence of 5-HT-mediated facilitation of activity on 5-HT_{2A} receptors intrinsic to the PFC (Beique et al. 2007). However, presynaptic 5-HT_{2A} receptors have been recently identified in thalamocortical networks as modulators of associative learning (Barre et al. 2016).

5 Brain Oscillations: Relevance to Schizophrenia

Higher cognitive functions and executive functions emerge from the coordinated activity of different neuronal networks and brain areas. This coordinated activity is reflected in oscillatory activity, a characteristic feature of cortical dynamics. Brain oscillations can be assessed through electroencephalographic (EEG) recordings, which detect the integrated activity of neuronal networks surrounding the electrodes (Barre et al. 2016). EEG oscillatory activity depends on the synchrony at which local and distal networks operate. Since the discovery of the EEG by Hans Berger (Nunez and Srinivasan 2006) and the first description of the most prominent rhythm (alpha, 8–12 Hz), multiple oscillatory activities have been described: slow (<1 Hz), delta (1–4 Hz), theta (4–7 Hz), beta (12–30 Hz) and gamma (30–80 Hz) rhythms. Brain oscillations are important because they serve to codify neural information and allow for coordinated activity between different neuronal networks in the temporal dimension: information is encoded not only by spiking activity, but also by the time at which the spikes are produced.

The generation of brain oscillations involves a balance between excitatory and inhibitory transmission within a network, which depends on the individual properties of the network components. For example, slow oscillations result from the interaction of cortical, thalamocortical and reticular nucleus oscillators (Berger 1929), whereas delta oscillations depend upon cortical and thalamocortical components (Crunelli and Hughes 2010; Petsche et al. 1984; Leresche et al. 1990; McCormick and Pape 1990; Llinas and Steriade 2006). Low frequency oscillations (slow and delta bands) are involved in several brain functions, including short- and long-term memory (Steriade et al. 1993; Bodizs et al. 2002; Marshall et al. 2006; Basar and Guntekin 2008). Additionally, they are essential for organizing higher frequency activities in sequences of complex oscillations (Binder et al. 2014). Gamma oscillations (30–80 Hz) deserve special attention due to their involvement in multiple cognitive processes (Steriade 2006): sensory and perceptual processing, short and long term memory, attention and executive functions, among others. Parvalbumin-positive GABAergic interneurons are heavily involved in the generation of gamma oscillations. These cells are fast-spiking interneurons, with special electrical properties, and are able to control large populations of pyramidal neurons via large dendritic networks connected by gap junctions (Engel and Singer 2001; Traub et al. 2000, 2001). These properties make them excellent candidates to spread fast oscillations through neuronal networks, although other cells and neurotransmitter systems may also be involved (Galarreta and Hestrin 2001; Belforte et al. 2010; Korotkova et al. 2010).

Because brain oscillations reflect neuronal and network dynamics, they may provide a valuable tool to study the etiology and pathophysiology of mental illnesses such as schizophrenia. Moreover, the study of brain oscillations can be a powerful translational tool, enabling comparisons between patients, healthy individuals and animal models. EEG recordings have been used to identify biomarkers, endophenotypes or prognostic indicators in schizophrenia. Indeed, schizophrenia symptoms may result from impaired connectivity, communication and coordination between brain regions (Carlen et al. 2012; Hoffman and McGlashan 1993; Skelly et al. 2008).

Recent studies show an increase of resting state gamma activity in schizophrenic patients compared to healthy controls (Camchong et al. 2011; Venables et al. 2009; Kikuchi et al. 2011). These findings are consistent with the alterations of GABAergic neurotransmission (especially in fast-spiking interneurons) (Lewis et al. 2005) and the deficits in NMDA-R glutamatergic neurotransmission

(Krystal et al. 2003; Spencer 2011; Konradi and Heckers 2003) that have been reported in schizophrenic patients. Likewise, abnormalities in corticosubcortical (e.g., thalamocortical) communication have been examined in schizophrenic patients via EEG sleep recordings. There are slow wave sleep deficits in schizophrenic patients, changes that are associated with negative symptoms. Specifically, schizophrenic patients show decreased delta wave counts, reductions in delta and theta power, and alterations in the laterality of these measures compared to healthy controls (Woo et al. 2008; Keshavan et al. 1998). Alterations in alpha activity during sleep have also been correlated with positive and negative symptom intensity (Sekimoto et al. 2007). Likewise, enhanced slow wave, delta, theta and beta activity, as well as decreased alpha activity in the resting state, have been associated with schizophrenia (Poulin et al. 2008; Rockstroh et al. 2007; Bates et al. 2009). Alterations of delta band frequency have been associated with negative symptoms of the illness (Begic et al. 2011). Moreover, EEG patterns can be used to distinguish schizophrenia patients from patients with other psychiatric disorders (Camchong et al. 2011; Poulin et al. 2008; Bates et al. 2009), different groups of schizophrenic patients (violent and nonviolent schizophrenic patients (Itoh et al. 2011)), and schizophrenia patients with "positive" versus "negative" symptoms (Schug et al. 2011; Begic et al. 2000).

Finally, different strategies have been used to model schizophrenia in healthy subjects. One such strategy is the administration of psychotomimetic drugs such as NMDA-R antagonists (Krystal et al. 2003). Using this approach, it has been shown that administration of subanesthetic doses of ketamine to healthy subjects augments high frequency oscillations (40–85 Hz) and reduces low frequency oscillations (1–5 Hz), mimicking some of the disturbances of oscillatory activity found in schizophrenia (Begic et al. 2009). Similar results have been reported using serotonergic hallucinogens (Hong et al. 2010; Oughourlian et al. 1971; Riba et al. 2002; Schenberg et al. 2015; Carhart-Harris et al. 2012; Kometer et al. 2015; Riba et al. 2004). After administration of psilocybin, decreases in broadband oscillatory power (Riba et al. 2004; Muthukumaraswamy et al. 2013), as well as reductions of blood flow (Schenberg et al. 2015), have been observed in areas of the default mode network (DMN).

6 Effects of Psychotomimetic Agents on Low Frequency Cortical Oscillations

The effects of hallucinogenic drugs on oscillatory activity have been examined in different animal models (Table 1). Our group has focused on the influence of psychotomimetic agents on low frequency oscillations. In rats and mice anesthetized with chloral hydrate, the aforementioned effects of PCP on pyramidal neuron activity were accompanied by simultaneous and marked reductions of low frequency cortical oscillations (LFCO; 0.3–4 Hz, with the most prominent intensity

Table 1 Effects of hallucinogen drugs in animal models: oscillatory activity in corticothalamic areas

	Area	Hallucinogen	Preparation	Effects	Ref.
NMDA receptor antagonists	mPFC	PCP	Anesthetized rats	↓ power of low frequency oscillations	Kargieman et al. (2007)
	mPFC	PCP	Anesthetized mice	↓ power of low frequency oscillations	Raichle et al. (2001)
	PFC	MK-801	Freely moving rats	↑ gamma power ↓ correlation between gamma power and discharge rate	Wood et al. (2012)
	mPFC	MK-801	Anesthetized rats, subiculum stimulation	Administration of MK-801 systemically or by microinjection into MD changed 2-Hz oscillations to a less regular delta rhythm ↓ paired pulse facilitation	Kiss et al. (2011a), Tseng et al. (2006)
	FC and V1	PCP	Awake macaques during WM task	↑ gamma power in FC and V1 ↓ alpha power in FC ↓ beta power in V1 Prolong P300 in FC	Kiss et al. (2011b)
	FC and sensorimotor cortex	PCP, MK-801	Conscious rats	↑ power of 1–3 Hz oscillations in FC ↑ or ↓ 9–30 Hz power depending on dose	Goonawardena et al. (2016)
	Neocortex	Ket, MK-801	Freely moving rats	↑ gamma power	Sebban et al. (2002)
	Visual cortex	Ket, PCP	Slices, rat	↑ phase coupling between gamma oscillations in layers III and V	Pinault (2008)
	FC-PC	Ket, PCP	Freely moving rats	↑ gamma power	Anver et al. (2011)
	MD and CM	PCP	Anesthetized rats	↓ power of low frequency oscillations	Santana et al. (2011)
	RtN and mPFC	PCP	Anesthetized rats	↓ power of low frequency oscillations ↓ coherence between PFC and RtN	Troyano-Rodriguez et al. (2014)

(continued)

Table 1 (continued)

	Area	Hallucinogen	Preparation	Effects	Ref.
5-HT$_{2A}$ agonists	mPFC	DOI	Anesthetized rats	↓ power of low frequency oscillations	Kargieman et al. (2012)
	PFC	DOI	Freely moving rats	Dose-dependent ↓ of gamma power ↓ correlation between gamma power and discharge rate	Wood et al. (2012)
	mPFC and V1	5-MeO-DMT	Anesthetized rats	↓ power of low frequency oscillations	Celada et al. (2008)
	Various regions of cortex	2C-B	Freely moving rats	↓ beta and gamma power ↑ delta power Effects on coherence depending on the dose (↓ or ↑)	Upton et al. (2014)

↓ decrease; ↑ increase; **2C-B** 4-bromo-2,5-dimethoxyphenethylamine hydrochloride; **5-MeO-DMT** 5-methoxy-*N*,*N*-dimethyltryptamine; **CM** centromedial thalamic nucleus; **DOI** 2,5-dimethoxy-4-iodoamphetamine; **FC** frontal cortex; **Ket** ketamine; **MD** mediodorsal thalamic nucleus; **MK-801** dizocilpine ((5*S*,10*R*)-(+)-5-methyl-10,11-dihydro-5*H*-dibenzo[*a*,*d*]cyclohepten-5,10-imine); **mPFC** medial prefrontal cortex; **PC** parietal cortex; **PCP** phencyclidine; PFC prefrontal cortex; **RtN** reticulotegmental nucleus; **V1** primary visual cortex; **WM** working memory at ~1 Hz) in the mPFC (Kargieman et al. 2007; Santana et al. 2011; Raichle et al. 2001); PCP also reduced delta waves in the thalamus. Hence, PCP administration dramatically reduced the power of LFCO, recorded in parallel with neuronal discharge. The reduction of LFCO power occurred in all experiments, irrespective of whether the recorded pyramidal neuron was excited, inhibited or unaffected by PCP. Figure 5 shows the effect of PCP on LFCO in the mPFC of anesthetized rats. Interestingly, PCP also produced a very marked desynchronization of the neuronal discharge from the active ("up") phases of LFCO. Under normal conditions, spikes are typically fired during the active phases of LFCO, corresponding to "up" states

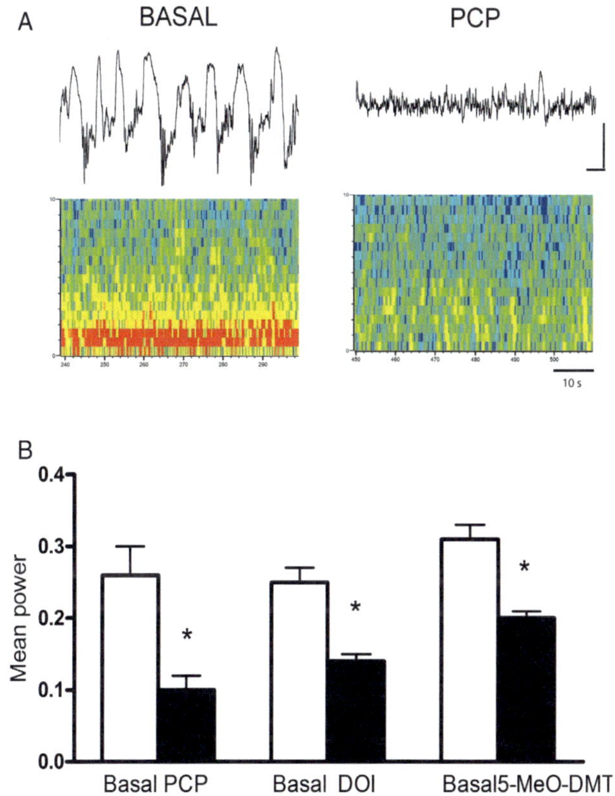

Fig. 5 **A** Representative examples of the local field potential recordings (*upper traces*) and the corresponding power spectrograms (*lower traces*) under basal conditions (*left side*) and after phencyclidine (PCP) administration (*right side*). Note that PCP produced a marked reduction of low frequency cortical oscillations (LFCO, ~1 Hz). The intensity of the power spectrum is color-coded (*red* high intensity; *blue* low intensity). **B** *Bar graph* showing the reduction in LFCO power produced by PCP (0.25 mg/kg i.v.), 1-(2,5-dimethoxy-4-iodophenyl)-2-aminopropane (DOI; 0.25 mg/kg i.v.) and 5-methoxy-*N,N*-dimethyltryptamine (5-MeO-DMT; 0.1 mg/kg i.v.). $^{*}p < 0.05$ versus the basal power level. Adapted from Kargieman et al. (2007, 2012), Celada et al. (2008)

(periods when neurons are depolarized). The percentage of spikes fired during the active phase of LFCO was 90 ± 3% at baseline conditions, whereas PCP reduced this value to 59 ± 11% (note that the maximum possible reduction was to 50% due to the random distribution of active and inactive phases of LFCO) (Kargieman et al. 2007).

The systemic administration of DOI (50–300 μg/kg i.v.) also caused a marked reduction of LFCO in PFC, to 56% of baseline (the effect of DOI on LFCO was slightly less marked than that of PCP). The effect of DOI was antagonized by the subsequent administration of M100907, indicating the exclusive involvement of 5-HT_{2A} receptors (Fig. 5) (Kargieman et al. 2012).

Likewise, the hallucinogen 5-MeO-DMT (0.1 mg/kg i.v.), a component of *ayahuasca*, also reduced the power of LFCO in PFC to 69% of baseline (Figs. 5 and 6) (Celada et al. 2008). Similar to DOI, the effect of 5-MeO-DMT on LFCO was also blocked by subsequent administration of M100907. Interestingly, 5-MeO-DMT also significantly altered neuronal firing and oscillatory activity in primary visual cortex (V1). Administration of 0.1 mg/kg i.v 5-MeO-DMT reduced LFCO power to a similar extent in V1 and PFC (Fig. 6). The reduction of LFCO power evoked by 5-MeO-DMT in PFC and V1 was accompanied by a significant reduction of BOLD signal in parallel fMRI studies (Celada et al. 2008). Interestingly, psilocybin also reduced the BOLD signal in the thalamus and various cortical areas of human volunteers (Schenberg et al. 2015), suggesting that this effect may be a common action of serotonergic hallucinogens. The decrease in BOLD signal may seem paradoxical in view of the fact that serotonergic hallucinogens primarily produce excitatory effects on neuronal activity in anesthetized rodents. However, because generation of LFCO is likely energetically demanding, reduction of oscillatory activity by serotonergic hallucinogens may counteract the energy requirements associated with the increase in neuronal firing, thus resulting in an overall decrease of cerebral blood flow. Additionally, differences in the overall excitatory/inhibitory effects of serotonergic hallucinogens due to the use of anesthesia may also be accountable.

7 Mechanisms Involved in the Reversal by Antipsychotic Drugs

Behavioral and neuronal alterations induced by noncompetitive NMDA-R antagonists and 5-HT_{2A} receptor agonists can be reversed by antipsychotic drugs, in particular by second generation or atypical antipsychotic drugs (Geyer et al. 2001). The similar disruptions of PFC activity produced by the two different models (NMDA-R antagonists and 5-HT_{2A} receptor agonists) indicates that these alterations may underlie their psychotomimetic activity (with the obvious limitation that these observations were carried out in anesthetized rodents). In spite of this limitation—common to most

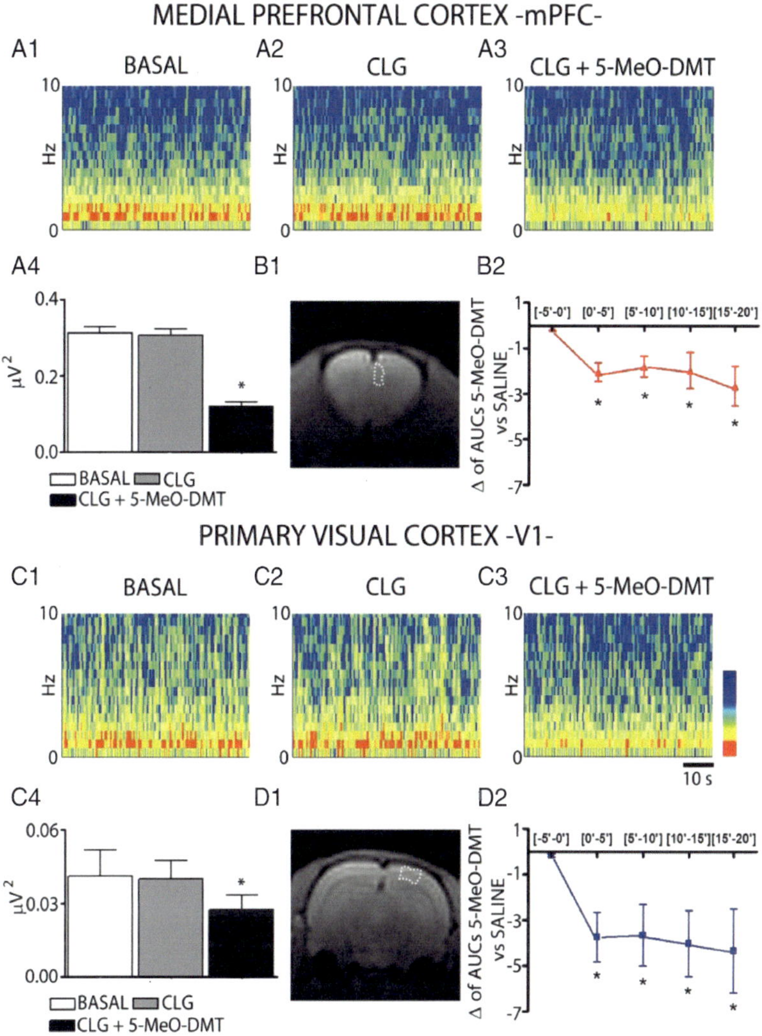

Fig. 6 Effect of the serotonergic hallucinogen 5-methoxy-*N,N*-dimethyltryptamine (5-MeO-DMT) on low frequency cortical oscillations (LFCO) and blood oxygen level dependent (BOLD) response in anesthetized rats. **A1–A3** Spectrograms showing the effect of clorgyline (CLG) and 5-MeO-DMT on local field potentials in medial prefrontal cortex (mPFC). The intensity of the power spectrum is color-coded (*red* high intensity; *blue* low intensity). Note the lack of effect of CLG and the marked reduction of LFCO power induced by 5-MeO-DMT. **A4** Averaged effects of CLG and 5-MeO-DMT on power spectra in mPFC. **B1** Representative coronal section showing the regions of interest (ROI) used to assess effects on BOLD in mPFC. **B2** Kinetics of the effect of 5-MeO-DMT on BOLD signal in mPFC. The effect of 5-MeO-DMT on functional magnetic resonance imaging (fMRI) signal intensity was calculated by subtracting the effect of saline. **C1–C3** Effect of CLG and 5-MeO-DMT on local field potentials in primary visual cortex (V1). **C4** Averaged effects of CLG and 5-MeO-DMT on power spectra in V1. **D1** Representative coronal section showing the ROI used to assess effects on BOLD in V1. **D2** Kinetics of the effect of 5-MeO-DMT on BOLD signal in V1. $^*p < 0.05$ versus the basal level. Adapted from Celada et al. (2008)

pharmacological studies employing in vivo electrophysiology—we attempted to reverse the effects of PCP on LFCO using antipsychotic drugs.

Interestingly, the alterations of PFC activity induced by PCP, DOI and 5-MeO-DMT were reversed by classical (haloperidol) and atypical (clozapine) antipsychotic drugs. The administration of clozapine completely reversed the increase in firing produced by PCP in PFC pyramidal neurons and in thalamic neurons (Kargieman et al. 2007; Santana et al. 2011). Likewise, clozapine pre-treatment prevented the PCP-induced increase in c-*fos* expression in all brain areas examined, including PFC and thalamic nuclei (Santana et al. 2011). Clozapine also reversed the reduction in LFO produced by PCP in PFC and thalamic nuclei (Kargieman et al. 2007; Santana et al. 2011), as well as the effects induced by the serotonergic hallucinogens DOI and 5-MeO-DMT in PFC (Kargieman et al. 2012; Celada et al. 2008) (Fig. 7). Likewise, the administration of haloperidol reversed

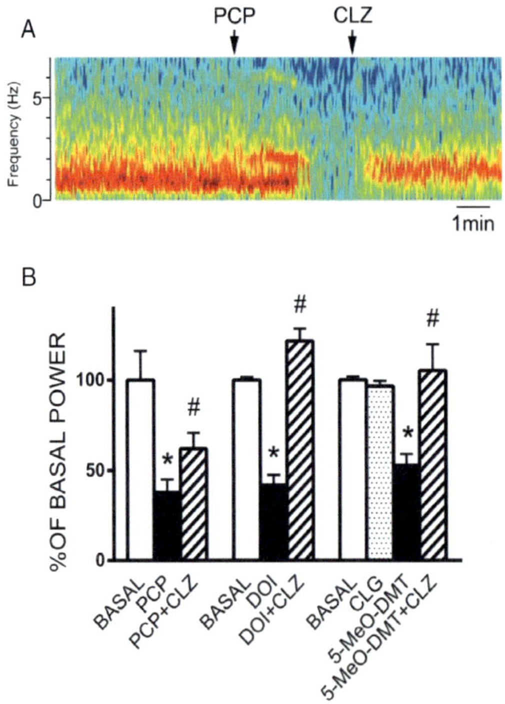

Fig. 7 **A** Representative spectrograms showing the reduction of low frequency cortical oscillations (LFCO) induced by phencyclidine (PCP; 0.25 mg/kg i.v.) in rat medial prefrontal cortex (mPFC) and the reversal induced by subsequent administration of clozapine (CLZ; 1 mg/kg i.v.). **B** *Bar graphs* showing the average effects of PCP, 1-(2,5-dimethoxy-4-iodophenyl)-2-aminopropane (DOI; 0.25 mg/kg i.v.), and 5-methoxy-*N,N*-dimethyltryptamine (5-MeO-DMT; 0.1 mg/kg i.v.) on the power of LFCO, as well as the ability of CLZ (1 mg/kg i.v.) to reverse those effects. $^{*}p < 0.05$ versus basal power; $^{\#}p < 0.05$ versus power in rats treated with PCP, DOI or 5-MeO-DMT alone. Data taken from Kargieman et al. (2007, 2012), Celada et al. (2008)

the effects of PCP, DOI and 5-MeO-DMT on LFCO, and when examined, on neuronal discharge (Kargieman et al. 2007, 2012; Celada et al. 2008).

The mechanisms involved in the reversal of PCP, DOI and 5-MeO-DMT effects have not been fully elucidated. At the cellular level, the reversal of PCP effects by clozapine may depend on increased GABA input to pyramidal neurons, given that the combination of PCP and clozapine had opposing effects on *c-fos* expression in GABAergic neurons (increased vs. PCP alone) and pyramidal neurons (reduced vs. PCP alone) (Kargieman et al. 2007). On the other hand, the reversal of the effects of DOI and 5-MeO-DMT by clozapine may simply involve the displacement of the 5-HT$_{2A}$ receptor agonists from their binding sites by clozapine, which has high affinity and antagonist activity at such sites.

The reversal of PCP effects by clozapine appears to require the participation of 5-HT$_{1A}$ receptors. Hence, PCP was equally effective in reducing LFCO in the PFC of wild-type (WT) mice and of mice lacking 5-HT$_{1A}$ or 5-HT$_{2A}$ receptors (1A-KO and 2A-KO mice, respectively) (Raichle et al. 2001). However, the subsequent administration of clozapine reversed the effects of PCP in WT mice and in 2A-KO mice, but failed to produce an effect in 1A-KO mice (Raichle et al. 2001), indicating the requirement of 5-HT$_{1A}$ receptors. Recent pharmacological studies also support the involvement of 5-HT$_{1A}$ receptors in the reversal of PCP effects on LFCO. Hence, the selective 5-HT$_{1A}$ agonist BAY × 3702 completely reversed the reduction of LFCO power produced by PCP in rat PFC (Lladó-Pelfort et al. 2016; Riga et al. 2014). It does not appear that the reversal of PCP effects by clozapine can be explained by a direct interaction between the latter drug and 5-HT$_{1A}$ receptors. Atypical antipsychotic drugs with low or no in vitro affinity for 5-HT$_{1A}$ receptors—including clozapine—act as functional agonists at 5-HT$_{1A}$ receptors in vivo, as shown by their ability to increase dopamine release in PFC (Meltzer and Massey 2011; Lladó-Pelfort et al. 2016; Rollema et al. 1997; Ichikawa et al. 2001; Diaz-Mataix et al. 2005). The mechanisms involved are unclear but may bear some relationship with the extensive co-expression of 5-HT$_{1A}$ and 5-HT$_{2A}$ receptors in PFC neurons (Amargos-Bosch et al. 2004).

Given the almost exclusive high affinity of haloperidol for dopamine D$_2$ receptors, the reversal of the effects of PCP and serotonergic hallucinogens by haloperidol cannot be explained by a direct competition with PCP, DOI or 5-MeO-DMT at 5-HT$_{2A}$ receptor sites and likely involves network modulation. For example, ventral tegmental area stimulation excites PFC fast-spiking interneurons and concurrently inhibits PFC pyramidal neurons (Bortolozzi et al. 2010). Thus, haloperidol may attenuate PCP-induced excitation of pyramidal neurons via activation of dopamine D$_1$ receptors secondary to an autoreceptor-mediated increase of PFC dopamine release, which adds to the increase of dopamine release induced by PCP.

8 Concluding Remarks

Data obtained in recent years has revealed that psychotomimetic agents markedly disrupt neuronal activity in PFC and in its afferent thalamic areas, such as the centromedial and dorsomedial nuclei. In parallel, these agents reduce the power of low frequency oscillations in PFC and in the thalamus. Given the crucial role of these brain networks in adapting behavioral responses to external sensory input, it is likely that the observed abnormalities partly underlie the psychotomimetic action of these agents. The link between the effects on network activity and the production of schizophrenia-like symptoms is strengthened by the observation that antipsychotic agents reverse the disruption of thalamic and cortical network activity evoked by PCP, DOI and 5-MeO-DMT. Overall, the alterations of thalamocortical activity induced by psychotomimetic agents can be successfully used to gain further insight into the neurobiological basis of psychosis and to examine the potential antipsychotic efficacy of new drugs under development. Likewise, studies of altered PFC network activity may also provide new information on the mechanisms involved in the therapeutic activity of agents targeting NMDA-R and 5-HT_{2A} receptors.

Acknowledgements Supported by the Innovative Medicines Initiative Joint Undertaking (IMI) under Grant Agreement N° 115008 (NEWMEDS). IMI is a public–private partnership between the European Union and the European Federation of Pharmaceutical Industries and Associations. Support from the following grants is also acknowledged: SAF 2015 68346-P (Ministry of Economy and Competitiveness and European Regional Development Fund), PI09/1245 and PI12/00156 (PN de I+D+I 2008–2011, ISCIII-Subdireccion General de Evaluación y Fomento de la Investigación cofinanced by the European Regional Development Fund. "Una manera de hacer Europa") and Centro de Investigación Biomédica en Red de Salud Mental, CIBERSAM (P82, 11INT3). Support from the Generalitat de Catalunya (SGR20093) is also acknowledged. MR is recipient of a IDIBAPS fellowship.

Statement of interest

None.

References

Abi-Dargham A et al (2000) Increased baseline occupancy of D2 receptors by dopamine in schizophrenia. Proc Natl Acad Sci U S A 97(14):8104–8109

Abi-Dargham A et al (2002) Prefrontal dopamine D1 receptors and working memory in schizophrenia. J Neurosci 22(9):3708–3719

Aghajanian GK, Marek GJ (1997) Serotonin induces excitatory postsynaptic potentials in apical dendrites of neocortical pyramidal cells. Neuropharmacology 36(4–5):589–599

Aghajanian GK, Marek GJ (1999) Serotonin, via 5-HT2A receptors, increases EPSCs in layer V pyramidal cells of prefrontal cortex by an asynchronous mode of glutamate release. Brain Res 825(1–2):161–171

Agurell S et al (1968) Identification of two new beta-carboline alkaloids in South American hallucinogenic plants. Biochem Pharmacol 17(12):2487–2488

Amargos-Bosch M et al (2004) Co-expression and in vivo interaction of serotonin1A and serotonin2A receptors in pyramidal neurons of prefrontal cortex. Cereb Cortex 14(3):281–299
Amargos-Bosch M et al (2006) Clozapine and olanzapine, but not haloperidol, suppress serotonin efflux in the medial prefrontal cortex elicited by phencyclidine and ketamine. Int J Neuropsychopharmacol 9(5):565–573
Angrist B et al (1976) Dimethyltryptamine levels in blood of schizophrenic patients and control subjects. Psychopharmacology 47(1):29–32
Anver H et al (2011) NMDA receptor hypofunction phase couples independent gamma-oscillations in the rat visual cortex. Neuropsychopharmacology 36(2):519–528
Araneda R, Andrade R (1991) 5-Hydroxytryptamine2 and 5-hydroxytryptamine1A receptors mediate opposing responses on membrane excitability in rat association cortex. Neuroscience 40(2):399–412
Artigas F (2010) The prefrontal cortex: a target for antipsychotic drugs. Acta Psychiatr Scand 121(1):11–21
Aston-Jones G, Cohen JD (2005) An integrative theory of locus coeruleus-norepinephrine function: adaptive gain and optimal performance. Annu Rev Neurosci 28:403–450
Barre A et al (2016) Presynaptic serotonin 2A receptors modulate thalamocortical plasticity and associative learning. Proc Natl Acad Sci U S A 113(10):E1382–E1391
Basar E, Guntekin B (2008) A review of brain oscillations in cognitive disorders and the role of neurotransmitters. Brain Res 1235:172–193
Bates AT et al (2009) Low-frequency EEG oscillations associated with information processing in schizophrenia. Schizophr Res 115(2–3):222–230
Begic D, Hotujac L, Jokic-Begic N (2000) Quantitative EEG in 'positive' and 'negative' schizophrenia. Acta Psychiatr Scand 101(4):307–311
Begic D, Mahnik-Milos M, Grubisin J (2009) EEG characteristics in depression, "negative" and "positive" schizophrena. Psychiatr Danub 21(4):579–584
Begic D et al (2011) Quantitative electroencephalography in schizophrenia and depression. Psychiatr Danub 23(4):355–362
Beique JC et al (2007) Mechanism of the 5-hydroxytryptamine 2A receptor-mediated facilitation of synaptic activity in prefrontal cortex. Proc Natl Acad Sci U S A 104(23):9870–9875
Belforte JE et al (2010) Postnatal NMDA receptor ablation in corticolimbic interneurons confers schizophrenia-like phenotypes. Nat Neurosci 13(1):76–83
Benington F, Morin RD, Clark LC Jr (1965) 5-methoxy-N, N-dimethyltryptamine, a possible endogenous psychotoxin. Ala J Med Sci 2(4):397–403
Berendse HW, Groenewegen HJ (1991) Restricted cortical termination fields of the midline and intralaminar thalamic nuclei in the rat. Neuroscience 42(1):73–102
Berger H (1929) Electroencephalogram in humans. Archiv fur Psychiatrie und Nervenkrankheiten 87:527–570
Binder S et al (2014) Transcranial slow oscillation stimulation during NREM sleep enhances acquisition of the radial maze task and modulates cortical network activity in rats. Front Behav Neurosci 7:220
Bodizs R et al (2002) Sleep-dependent hippocampal slow activity correlates with waking memory performance in humans. Neurobiol Learn Mem 78(2):441–457
Bortolozzi A et al (2003) In vivo modulation of 5-hydroxytryptamine release in mouse prefrontal cortex by local 5-HT(2A) receptors: effect of antipsychotic drugs. Eur J Neurosci 18(5): 1235–1246
Bortolozzi A et al (2005) The activation of 5-HT receptors in prefrontal cortex enhances dopaminergic activity. J Neurochem 95(6):1597–1607
Bortolozzi A et al (2010) Dopamine release induced by atypical antipsychotics in prefrontal cortex requires 5-HT(1A) receptors but not 5-HT(2A) receptors. Int J Neuropsychopharmacol 13(10):1299–1314
Breier A et al (1997) Association of ketamine-induced psychosis with focal activation of the prefrontal cortex in healthy volunteers. Am J Psychiatry 154(6):805–811

Brush DE, Bird SB, Boyer EW (2004) Monoamine oxidase inhibitor poisoning resulting from Internet misinformation on illicit substances. J Toxicol Clin Toxicol 42(2):191–195
Camchong J et al (2011) Altered functional and anatomical connectivity in schizophrenia. Schizophr Bull 37(3):640–650
Cardinal RN et al (2002) Emotion and motivation: the role of the amygdala, ventral striatum, and prefrontal cortex. Neurosci Biobehav Rev 26(3):321–352
Carhart-Harris RL et al (2012) Neural correlates of the psychedelic state as determined by fMRI studies with psilocybin. Proc Natl Acad Sci U S A 109(6):2138–2143
Carlen M et al (2012) A critical role for NMDA receptors in parvalbumin interneurons for gamma rhythm induction and behavior. Mol Psychiatry 17(5):537–548
Carlsson A (1977) Does dopamine play a role in schizophrenia? Psychol Med 7(4):583–597
Carlsson M, Carlsson A (1989) Dramatic synergism between MK-801 and clonidine with respect to locomotor stimulatory effect in monoamine-depleted mice. J Neural Transm 77(1):65–71
Cavara NA, Hollmann M (2008) Shuffling the deck anew: how NR3 tweaks NMDA receptor function. Mol Neurobiol 38(1):16–26
Celada P et al. (2008) The hallucinogen DOI reduces low-frequency oscillations in rat prefrontal cortex: reversal by antipsychotic drugs. Biol Psychiatry 64(5):392–400
Celada P, Puig MV, Artigas F (2013) Serotonin modulation of cortical neurons and networks. Front Integr Neurosci 7:25
Crunelli V, Hughes SW (2010) The slow (<1 Hz) rhythm of non-REM sleep: a dialogue between three cardinal oscillators. Nat Neurosci 13(1):9–17
Davidson RJ, Irwin W (1999) The functional neuroanatomy of emotion and affective style. Trends Cogn Sci 3(1):11–21
de Almeida J, Mengod G (2007) Quantitative analysis of glutamatergic and GABAergic neurons expressing 5-HT(2A) receptors in human and monkey prefrontal cortex. J Neurochem 103(2):475–486
de Almeida J, Mengod G (2008) Serotonin 1A receptors in human and monkey prefrontal cortex are mainly expressed in pyramidal neurons and in a GABAergic interneuron subpopulation: implications for schizophrenia and its treatment. J Neurochem 107(2):488–496
DeFelipe J et al (2013) New insights into the classification and nomenclature of cortical GABAergic interneurons. Nat Rev Neurosci 14(3):202–216
Devinsky O, Morrell MJ, Vogt BA (1995) Contributions of anterior cingulate cortex to behaviour. Brain 118(Pt 1):279–306
Diaz-Mataix L et al (2005) Involvement of 5-HT1A receptors in prefrontal cortex in the modulation of dopaminergic activity: role in atypical antipsychotic action. J Neurosci 25(47):10831–10843
Drevets WC (2001) Neuroimaging and neuropathological studies of depression: implications for the cognitive-emotional features of mood disorders. Curr Opin Neurobiol 11(2):240–249
Engel AK, Singer W (2001) Temporal binding and the neural correlates of sensory awareness. Trends Cogn Sci 5(1):16–25
Fuster JM (2001) The prefrontal cortex—an update: time is of the essence. Neuron 30(2):319–333
Fuster JM (2008) The prefrontal cortex, 4th edn. Academic Press, Los Angeles, California, USA
Gabbott PL et al (2005) Prefrontal cortex in the rat: projections to subcortical autonomic, motor, and limbic centers. J Comp Neurol 492(2):145–177
Galarreta M, Hestrin S (2001) Electrical synapses between GABA-releasing interneurons. Nat Rev Neurosci 2(6):425–433
Geyer MA, Vollenweider FX (2008) Serotonin research: contributions to understanding psychoses. Trends Pharmacol Sci 29(9):445–453
Geyer MA et al (2001) Pharmacological studies of prepulse inhibition models of sensorimotor gating deficits in schizophrenia: a decade in review. Psychopharmacology 156(2–3):117–154
Gillin JC, Wyatt RJ (1976) Evidence for and against the involvement of N, N-dimethyl-tryptamine (DMT) and 5-methoxy-N, N-dimethyltryptamine (5-MeO-DMT) in schizophrenia. Psychopharmacol Bull 12(4):12–13

Glennon RA (1991) Discriminative stimulus properties of hallucinogens and related designer drugs. NIDA Res Monogr 116:25–44

Glennon RA (1994) Classical hallucinogens: an introductory overview. NIDA Res Monogr 146:4–32

Gonzalez-Maeso J et al (2007) Hallucinogens recruit specific cortical 5-HT(2A) receptor-mediated signaling pathways to affect behavior. Neuron 53(3):439–452

Goonawardena AV et al (2016) Alterations in high-frequency neuronal oscillations in a cynomolgus macaque test of sustained attention following NMDA receptor antagonism. Neuropsychopharmacology 41(5):1319–1328

Groenewegen HJ, Uylings HB (2000) The prefrontal cortex and the integration of sensory, limbic and autonomic information. Prog Brain Res 126:3–28

Hall H et al (2000) Autoradiographic localization of 5-HT(2A) receptors in the human brain using [(3)H]M100907 and [(11)C]M100907. Synapse 38(4):421–431

Harrison PJ (1999a) The neuropathological effects of antipsychotic drugs. Schizophr Res 40 (2):87–99

Harrison PJ (1999b) The neuropathology of schizophrenia. A critical review of the data and their interpretation. Brain 122(Pt 4):593–624

Harrison PJ, Weinberger DR (2005) Schizophrenia genes, gene expression, and neuropathology: on the matter of their convergence. Mol Psychiatry 10(1):40–68; image 5

Hoffman RE, McGlashan TH (1993) Neurodynamics and schizophrenia research: editors' introduction. Schizophr Bull 19(1):15–19

Homayoun H, Moghaddam B (2007) NMDA receptor hypofunction produces opposite effects on prefrontal cortex interneurons and pyramidal neurons. J Neurosci 27(43):11496–11500

Hong LE et al (2010) Gamma and delta neural oscillations and association with clinical symptoms under subanesthetic ketamine. Neuropsychopharmacology 35(3):632–640

Ichikawa J et al (2001) 5-HT(2A) and D(2) receptor blockade increases cortical DA release via 5-HT(1A) receptor activation: a possible mechanism of atypical antipsychotic-induced cortical dopamine release. J Neurochem 76(5):1521–1531

Itoh T et al (2011) LORETA analysis of three-dimensional distribution of delta band activity in schizophrenia: relation to negative symptoms. Neurosci Res 70(4):442–448

Jakab RL, Goldman-Rakic PS (1998) 5-Hydroxytryptamine2A serotonin receptors in the primate cerebral cortex: possible site of action of hallucinogenic and antipsychotic drugs in pyramidal cell apical dendrites. Proc Natl Acad Sci U S A 95(2):735–740

Javitt DC, Zukin SR (1991) Recent advances in the phencyclidine model of schizophrenia. Am J Psychiatry 148(10):1301–1308

Jodo E et al (2005) Activation of medial prefrontal cortex by phencyclidine is mediated via a hippocampo-prefrontal pathway. Cereb Cortex 15(5):663–669

Kargieman L et al (2007) Antipsychotic drugs reverse the disruption in prefrontal cortex function produced by NMDA receptor blockade with phencyclidine. Proc Natl Acad Sci U S A 104 (37):14843–14848

Kargieman L et al (2012) Clozapine reverses phencyclidine-induced desynchronization of prefrontal cortex through a 5-HT(1A) receptor-dependent mechanism. Neuropsychopharmacology 37 (3):723–733

Katayama T et al (2007) Activation of medial prefrontal cortex neurons by phencyclidine is mediated via AMPA/kainate glutamate receptors in anesthetized rats. Neuroscience 150 (2):442–448

Keshavan MS et al (1998) Delta sleep deficits in schizophrenia: evidence from automated analyses of sleep data. Arch Gen Psychiatry 55(5):443–448

Kikuchi M et al (2011) Frontal areas contribute to reduced global coordination of resting-state gamma activities in drug-naive patients with schizophrenia. Schizophr Res 130(1–3):187–194

Kiss T, Hoffmann WE, Hajos M (2011a) Delta oscillation and short-term plasticity in the rat medial prefrontal cortex: modelling NMDA hypofunction of schizophrenia. Int J Neuropsychopharmacol 14(1):29–42

Kiss T et al (2011b) Role of thalamic projection in NMDA receptor-induced disruption of cortical slow oscillation and short-term plasticity. Front Psychiatry 2:14

Kometer M et al (2015) Psilocybin-induced spiritual experiences and insightfulness are associated with synchronization of neuronal oscillations. Psychopharmacology 232(19):3663–3676

Konradi C, Heckers S (2003) Molecular aspects of glutamate dysregulation: implications for schizophrenia and its treatment. Pharmacol Ther 97(2):153–179

Korotkova T et al (2010) NMDA receptor ablation on parvalbumin-positive interneurons impairs hippocampal synchrony, spatial representations, and working memory. Neuron 68(3):557–569

Krystal JH et al (2003) NMDA receptor antagonist effects, cortical glutamatergic function, and schizophrenia: toward a paradigm shift in medication development. Psychopharmacology 169 (3–4):215–233

Kuroda M, Yokofujita J, Murakami K (1998) An ultrastructural study of the neural circuit between the prefrontal cortex and the mediodorsal nucleus of the thalamus. Prog Neurobiol 54(4): 417–458

Kurrasch-Orbaugh DM et al (2003) Serotonin 5-hydroxytryptamine 2A receptor-coupled phospholipase C and phospholipase A2 signaling pathways have different receptor reserves. J Pharmacol Exp Ther 304(1):229–237

Laruelle M et al (1996) Single photon emission computerized tomography imaging of amphetamine-induced dopamine release in drug-free schizophrenic subjects. Proc Natl Acad Sci U S A 93(17):9235–9240

Lau CG, Zukin RS (2007) NMDA receptor trafficking in synaptic plasticity and neuropsychiatric disorders. Nat Rev Neurosci 8(6):413–426

Leresche N et al (1990) Pacemaker-like and other types of spontaneous membrane potential oscillations of thalamocortical cells. Neurosci Lett 113(1):72–77

Lewis DA, Gonzalez-Burgos G (2006) Pathophysiologically based treatment interventions in schizophrenia. Nat Med 12(9):1016–1022

Lewis DA, Lieberman JA (2000) Catching up on schizophrenia: natural history and neurobiology. Neuron 28(2):325–334

Lewis DA, Hashimoto T, Volk DW (2005) Cortical inhibitory neurons and schizophrenia. Nat Rev Neurosci 6(4):312–324

Lewis DA et al (2012) Cortical parvalbumin interneurons and cognitive dysfunction in schizophrenia. Trends Neurosci 35(1):57–67

Lladó-Pelfort L et al (2016) Phencyclidine-induced disruption of oscillatory activity in prefrontal cortex: effects of antipsychotic drugs and receptor ligands. Eur Neuropsychopharmacol 26(3):614–625

Llinas RR, Steriade M (2006) Bursting of thalamic neurons and states of vigilance. J Neurophysiol 95(6):3297–3308

Lopez Hill X, Scorza MC (2012) Role of the anterior thalamic nucleus in the motor hyperactivity induced by systemic MK-801 administration in rats. Neuropharmacology 62(7):2440–2446

Lopez-Gil X et al (2007) Clozapine and haloperidol differently suppress the MK-801-increased glutamatergic and serotonergic transmission in the medial prefrontal cortex of the rat. Neuropsychopharmacology 32(10):2087–2097

Lopez-Gil X et al (2012) Importance of inter-hemispheric prefrontal connection in the effects of non-competitive NMDA receptor antagonists. Int J Neuropsychopharmacol 15(7):945–956

Lopez-Gimenez JF et al (2001) Mapping of 5-HT2A receptors and their mRNA in monkey brain: [3H]MDL100,907 autoradiography and in situ hybridization studies. J Comp Neurol 429(4):571–589

Marek GJ, Aghajanian GK (1998) 5-Hydroxytryptamine-induced excitatory postsynaptic currents in neocortical layer V pyramidal cells: suppression by mu-opiate receptor activation. Neuroscience 86(2):485–497

Marek GJ et al (2001) A major role for thalamocortical afferents in serotonergic hallucinogen receptor function in the rat neocortex. Neuroscience 105(2):379–392

Marshall L et al (2006) Boosting slow oscillations during sleep potentiates memory. Nature 444(7119):610–613

Martin-Ruiz R et al (2001) Control of serotonergic function in medial prefrontal cortex by serotonin-2A receptors through a glutamate-dependent mechanism. J Neurosci 21(24):9856–9866
Mayberg HS et al (2005) Deep brain stimulation for treatment-resistant depression. Neuron 45 (5):651–660
McCormick DA, Pape HC (1990) Properties of a hyperpolarization-activated cation current and its role in rhythmic oscillation in thalamic relay neurones. J Physiol 431:291–318
McKenna DJ (2004) Clinical investigations of the therapeutic potential of ayahuasca: rationale and regulatory challenges. Pharmacol Ther 102(2):111–129
McKenna DJ, Towers GH, Abbott F (1984) Monoamine oxidase inhibitors in South American hallucinogenic plants: tryptamine and beta-carboline constituents of ayahuasca. J Ethnopharmacol 10(2):195–223
Meltzer HY, Massey BW (2011) The role of serotonin receptors in the action of atypical antipsychotic drugs. Curr Opin Pharmacol 11(1):59–67
Miller EK, Cohen JD (2001) An integrative theory of prefrontal cortex function. Annu Rev Neurosci 24:167–202
Miner LA et al (2003) Ultrastructural localization of serotonin2A receptors in the middle layers of the rat prelimbic prefrontal cortex. Neuroscience 116(1):107–117
Muthukumaraswamy SD et al (2013) Broadband cortical desynchronization underlies the human psychedelic state. J Neurosci 33(38):15171–15183
Nichols DE (2004) Hallucinogens. Pharmacol Ther 101(2):131–181
Niswender CM, Conn PJ (2010) Metabotropic glutamate receptors: physiology, pharmacology, and disease. Annu Rev Pharmacol Toxicol 50:295–322
Nunez PL, Srinivasan R (2006) A theoretical basis for standing and traveling brain waves measured with human EEG with implications for an integrated consciousness. Clin Neurophysiol 117(11):2424–2435
Oughourlian JM, Rougeul A, Verdeaux J (1971) Action of hallucinogens on electroencephalograms. Therapie 26(5):953–968
Palenicek T et al (2013) Behavioral, neurochemical and pharmaco-EEG profiles of the psychedelic drug 4-bromo-2,5-dimethoxyphenethylamine (2C-B) in rats. Psychopharmacology 225(1):75–93
Paoletti P, Neyton J (2007) NMDA receptor subunits: function and pharmacology. Curr Opin Pharmacol 7(1):39–47
Pazos A, Cortes R, Palacios JM (1985) Quantitative autoradiographic mapping of serotonin receptors in the rat brain. II. Serotonin-2 receptors. Brain Res 346(2):231–249
Pazos A, Probst A, Palacios JM (1987) Serotonin receptors in the human brain–IV. Autoradiographic mapping of serotonin-2 receptors. Neuroscience 21(1):123–139
Petsche H, Pockberger H, Rappelsberger P (1984) On the search for the sources of the electroencephalogram. Neuroscience 11(1):1–27
Phillips ML et al (2003) Neurobiology of emotion perception I: the neural basis of normal emotion perception. Biol Psychiatry 54(5):504–514
Pinault D (2008) N-methyl d-aspartate receptor antagonists ketamine and MK-801 induce wake-related aberrant gamma oscillations in the rat neocortex. Biol Psychiatry 63(8):730–735
Poulin J, Stip E, Godbout R (2008) REM sleep EEG spectral analysis in patients with first-episode schizophrenia. J Psychiatr Res 42(13):1086–1093
Puig MV et al (2003) In vivo modulation of the activity of pyramidal neurons in the rat medial prefrontal cortex by 5-HT2A receptors: relationship to thalamocortical afferents. Cereb Cortex 13(8):870–882
Puig MV, Artigas F, Celada P (2005) Modulation of the activity of pyramidal neurons in rat prefrontal cortex by raphe stimulation in vivo: involvement of serotonin and GABA. Cereb Cortex 15(1):1–14
Puigdemont D et al (2012) Deep brain stimulation of the subcallosal cingulate gyrus: further evidence in treatment-resistant major depression. Int J Neuropsychopharmacol 15(1):121–133
Raichle ME et al (2001) A default mode of brain function. Proc Natl Acad Sci U S A 98(2):676–682
Rasmussen H et al (2010) Decreased frontal serotonin2A receptor binding in antipsychotic-naive patients with first-episode schizophrenia. Arch Gen Psychiatry 67(1):9–16

Rasmussen H et al (2016) Low frontal serotonin 2A receptor binding is a state marker for schizophrenia? Eur Neuropsychopharmacol 26(7):1248–1250
Riba J et al (2002) Topographic pharmaco-EEG mapping of the effects of the South American psychoactive beverage ayahuasca in healthy volunteers. Br J Clin Pharmacol 53(6):613–628
Riba J et al (2004) Effects of the South American psychoactive beverage ayahuasca on regional brain electrical activity in humans: a functional neuroimaging study using low-resolution electromagnetic tomography. Neuropsychobiology 50(1):89–101
Riga MS et al (2014) The natural hallucinogen 5-MeO-DMT, component of ayahuasca, disrupts cortical function in rats: reversal by antipsychotic drugs. Int J Neuropsychopharmacol 17 (8):1269–1282
Rockstroh BS et al (2007) Abnormal oscillatory brain dynamics in schizophrenia: a sign of deviant communication in neural network? BMC Psychiatry 7:44
Rollema H et al (1997) Clozapine increases dopamine release in prefrontal cortex by 5-HT1A receptor activation. Eur J Pharmacol 338(2):R3–R5
Santana N et al (2004) Expression of serotonin1A and serotonin2A receptors in pyramidal and GABAergic neurons of the rat prefrontal cortex. Cereb Cortex 14(10):1100–1109
Santana N, Mengod G, Artigas F (2009) Quantitative analysis of the expression of dopamine D1 and D2 receptors in pyramidal and GABAergic neurons of the rat prefrontal cortex. Cereb Cortex 19(4):849–860
Santana N et al (2011) Activation of thalamocortical networks by the N-methyl-D-aspartate receptor antagonist phencyclidine: reversal by clozapine. Biol Psychiatry 69(10):918–927
Santana N, Mengod G, Artigas F (2013) Expression of alpha(1)-adrenergic receptors in rat prefrontal cortex: cellular co-localization with 5-HT(2A) receptors. Int J Neuropsychopharmacol 16(5): 1139–1151
Schenberg EE et al (2015) Acute biphasic effects of ayahuasca. PLoS ONE 10(9):e0137202
Schreiber R, Brocco M, Millan MJ (1994) Blockade of the discriminative stimulus effects of DOI by MDL 100,907 and the 'atypical' antipsychotics, clozapine and risperidone. Eur J Pharmacol 264(1):99–102
Schug RA et al (2011) Resting EEG deficits in accused murderers with schizophrenia. Psychiatry Res 194(1):85–94
Schultes RE, Hofmann A (1991) *The botany and chemistry of Hallucinogens*. Charles C Thomas Pub Ltd., Springfield, Illinois, USA
Seamans JK, Yang CR (2004) The principal features and mechanisms of dopamine modulation in the prefrontal cortex. Prog Neurobiol 74(1):1–58
Sebban C et al (2002) Effects of phencyclidine (PCP) and MK 801 on the EEGq in the prefrontal cortex of conscious rats; antagonism by clozapine, and antagonists of AMPA-, alpha(1)- and 5-HT(2A)-receptors. Br J Pharmacol 135(1):65–78
Sekimoto M et al (2007) Reduced frontal asymmetry of delta waves during all-night sleep in schizophrenia. Schizophr Bull 33(6):1307–1311
Selemon LD, Goldman-Rakic PS (1999) The reduced neuropil hypothesis: a circuit based model of schizophrenia. Biol Psychiatry 45(1):17–25
Seminowicz DA et al (2004) Limbic-frontal circuitry in major depression: a path modeling metanalysis. NeuroImage 22(1):409–418
Shergill SS et al (2000) Mapping auditory hallucinations in schizophrenia using functional magnetic resonance imaging. Arch Gen Psychiatry 57(11):1033–1038
Skelly LR et al (2008) Diffusion tensor imaging in schizophrenia: relationship to symptoms. Schizophr Res 98(1–3):157–162
Sklerov J et al (2005) A fatal intoxication following the ingestion of 5-methoxy-N, N-dimethyltryptamine in an ayahuasca preparation. J Anal Toxicol 29(8):838–841
Spencer KM (2011) Baseline gamma power during auditory steady-state stimulation in schizophrenia. Front Hum Neurosci 5:190
Steinbusch HWM (1981) Distribution of serotonin-immunoreactivity in the central nervous system of the rat-cell bodies and terminals. Neuroscience 6(4):557–618

Steriade M (2006) Grouping of brain rhythms in corticothalamic systems. Neuroscience 137(4): 1087–1106

Steriade M, Nunez A, Amzica F (1993) Intracellular analysis of relations between the slow (<1 Hz) neocortical oscillation and other sleep rhythms of the electroencephalogram. J Neurosci 13(8):3266–3283

Suzuki Y et al (2002) Acute administration of phencyclidine induces tonic activation of medial prefrontal cortex neurons in freely moving rats. Neuroscience 114(3):769–779

Swanson CJ et al (2005) Metabotropic glutamate receptors as novel targets for anxiety and stress disorders. Nat Rev Drug Discov 4(2):131–144

Traub RD et al (2000) A model of gamma-frequency network oscillations induced in the rat CA3 region by carbachol in vitro. Eur J Neurosci 12(11):4093–4106

Traub RD et al (2001) Gap junctions between interneuron dendrites can enhance synchrony of gamma oscillations in distributed networks. J Neurosci 21(23):9478–9486

Troyano-Rodriguez E et al (2014) Phencyclidine inhibits the activity of thalamic reticular gamma-aminobutyric acidergic neurons in rat brain. Biol Psychiatry 76(12):937–945

Tseng KY et al (2006) Excitatory response of prefrontal cortical fast-spiking interneurons to ventral tegmental area stimulation in vivo. Synapse 59(7):412–417

Upton N et al. (2014) NMDA receptor antagonist-induced changes in rat EEG power spectra as a model of schizophrenia. Program No. 230.04, in neuroscience 2014 abstracts 2014. Society for Neuroscience, Washington, USA

Van Eden CG et al (1987) Immunocytochemical localization of dopamine in the prefrontal cortex of the rat at the light and electron microscopical level. Neuroscience 22(3):849–862

Vazquez-Borsetti P, Cortes R, Artigas F (2009) Pyramidal neurons in rat prefrontal cortex projecting to ventral tegmental area and dorsal raphe nucleus express 5-HT2A receptors. Cereb Cortex 19(7):1678–1686

Vazquez-Borsetti P et al (2011) Simultaneous projections from prefrontal cortex to dopaminergic and serotonergic nuclei. Int J Neuropsychopharmacol 14(3):289–302

Venables NC, Bernat EM, Sponheim SR (2009) Genetic and disorder-specific aspects of resting state EEG abnormalities in schizophrenia. Schizophr Bull 35(4):826–839

Villalobos C et al (2005) Serotonergic regulation of calcium-activated potassium currents in rodent prefrontal cortex. Eur J Neurosci 22(5):1120–1126

Vollenweider FX, Kometer M (2010) The neurobiology of psychedelic drugs: implications for the treatment of mood disorders. Nat Rev Neurosci 11(9):642–651

Vollenweider FX et al (1997a) Positron emission tomography and fluorodeoxyglucose studies of metabolic hyperfrontality and psychopathology in the psilocybin model of psychosis. Neuropsychopharmacology 16(5):357–372

Vollenweider FX et al (1997b) Metabolic hyperfrontality and psychopathology in the ketamine model of psychosis using positron emission tomography (PET) and [18F]fluorodeoxyglucose (FDG). Eur Neuropsychopharmacol 7(1):9–24

Weinberger DR (1987) Implications of normal brain development for the pathogenesis of schizophrenia. Arch Gen Psychiatry 44(7):660–669

Woo TU, Kim AM, Viscidi E (2008) Disease-specific alterations in glutamatergic neurotransmission on inhibitory interneurons in the prefrontal cortex in schizophrenia. Brain Res 1218:267–277

Wood J, Kim Y, Moghaddam B (2012) Disruption of prefrontal cortex large scale neuronal activity by different classes of psychotomimetic drugs. J Neurosci 32(9):3022–3031

Yu AM (2008) Indolealkylamines: biotransformations and potential drug-drug interactions. AAPS J 10(2):242–253

Zhang Y, Llinas RR, Lisman JE (2009) Inhibition of NMDARs in the Nucleus Reticularis of the Thalamus Produces Delta Frequency Bursting. Front Neural Circuits 3:20

Zhang Y et al (2012) NMDAR antagonist action in thalamus imposes delta oscillations on the hippocampus. J Neurophysiol 107(11):3181–3189

Interactions of Hallucinogens with the Glutamatergic System: Permissive Network Effects Mediated Through Cortical Layer V Pyramidal Neurons

Gerard J. Marek

Abstract Recordings made from layer V (L5) pyramidal cells of the prefrontal cortex (PFC) and neocortex in rodent slice preparations have shown that serotonin (5-hydroxytryptamine, 5-HT) and serotonergic hallucinogens induce an increase in the frequency of spontaneous excitatory postsynaptic currents (EPSCs) in the apical dendritic field by activating 5-HT_{2A} receptors. Serotonergic hallucinogens induce late EPSCs and increase recurrent network activity when subcortical or mid-cortical regions are stimulated at low frequencies (e.g., 0.1 Hz). A range of agonists or positive allosteric modulators (PAMs) for mostly $G_{i/o}$-coupled receptors, including metabotropic glutamate$_2$ (mGlu$_2$), adenosine A_1, or μ-opioid receptors, suppress these effects of 5-HT_{2A} receptor stimulation. Furthermore, a range of mostly $G_{q/11}$-coupled receptors (including orexin$_2$ [OX_2]; α_1-adrenergic, and mGlu$_5$ receptors) similarly induce glutamate (Glu) release onto L5 pyramidal cells. Evidence implicates a number of brain regions in mediating these effects of serotonergic hallucinogens and $G_{q/11}$-coupled receptors including the midline and intralaminar thalamic nuclei, claustrum, and neurons in deep PFC. These effects on 5-HT_{2A} receptors and related GPCRs appear to play a major role in the behavioral effects of serotonergic hallucinogens, such as head twitches in rodents and higher order behaviors such as rodent lever pressing on the differential-reinforcement-of-low rate 72-s (DRL 72-s) schedule. This implies that the effects of 5-HT_{2A} receptor activation on the activity of L5 pyramidal cells may be responsible for mediating a range of behaviors linked to limbic circuitry with connectivity between the PFC, striatum, thalamus, claustrum, striatum, amygdala, and the hippocampal formation.

Keywords In vitro electrophysiology · Layer V pyramidal neurons · DOI-induced head-twitch response · DRL 72-s behavior

---

G.J. Marek (✉)
Global Medical Science, CNS and Pain, Astellas Pharma Global Development, 1 Astellas Way, Northbrook, IL 60062, USA
e-mail: gerard.marek@astellas.com

Curr Topics Behav Neurosci (2017) 36: 107–136
DOI 10.1007/7854_2017_480
Published Online: 26 August 2017

Contents

The present chapter will detail the emergence over the last several decades of evidence demonstrating that hallucinogenic drugs share a common pharmacological action in stimulating cortical 5-hydroxytryptamine$_{2A}$ (5-HT$_{2A}$) receptors, which may provide an important clue to the treatment of psychotic disorders. While focusing on psychosis and potential links to lysergic acid diethylamide (LSD)-induced perceptual changes, we will also discuss hallucinogen-induced changes in mood and cognition as well as potential therapeutic applications. The genesis of this line of work began in many ways with the discovery of LSD by the Sandoz chemist Albert Hoffman in the early 1940s and the observation that this drug, like mescaline, appears to mimic certain symptoms exhibited by patients with schizophrenia. Woolley and Shaw first suggested that the serotonergic properties of LSD might explain its psychotomimetic effects (Shaw and Woolley 1956). By the late 1970s and early 1980s, convergent receptor binding, behavioral and electrophysiological studies suggested that LSD, mescaline, and psilocybin all share a common agonist action at 5-HT$_{2A}$ receptors. While both stimulants like amphetamine and *N*-methyl-D-aspartate (NMDA) receptor antagonists such as phencyclidine (PCP) and ketamine began to overtake serotonergic hallucinogens as models of a broader range of the symptoms of schizophrenia, others continued to use hallucinogen effects on 5-HT$_{2A}$ receptors as a model psychosis with potential links to the family of schizophrenic disorders (Gouzoulis-Mayfrank et al. 2005; Vollenweider et al. 1998). More recently, evidence has accrued for pathophysiologic and therapeutic influences of 5-HT$_{2A}$ receptors in neurodegenerative disease states ranging from Parkinson's disease (PD) psychosis to the psychosis-like symptoms associated with Alzheimer's disease (AD).

The initial clues indicating that serotonergic hallucinogens have effects on glutamate (Glu) release appeared around the time that it was recognized that selective 5-HT$_{2A}$ receptor antagonists are not highly effective antipsychotic drugs in patients with schizophrenia. First, discoveries linking the glutamatergic system and hallucinogens will be discussed. The initial report showing that bath application of serotonin (5-HT) induces spontaneous (i.e., not evoked by electrical stimulation) excitatory postsynaptic potentials/currents (EPSPs/EPSCs) in prefrontal cortical and neocortical layer V (L5) pyramidal neurons in rat brain slice preparations was published in 1997 (Aghajanian and Marek 1997). The basic pharmacological observations were that these spontaneous 5-HT-induced EPSPs/EPSCs were potently blocked by 5-HT$_{2A}$ receptor antagonists at concentrations that were consistent with their K_i values for the 5-HT$_{2A}$ receptor, but were not affected by

5-HT_{1A} or $5\text{-HT}_3/5\text{-HT}_4$ receptor antagonists. The other basic pharmacological finding was that the broad spectrum presynaptic metabotropic glutamate (mGlu) receptor agonist (*1S,3S*)-ACPD also suppressed the frequency of the 5-HT-induced EPSPs/EPSCs. Previously, inhibitory EPSPs were observed in piriform cortical and hippocampal neurons due to excitation of 5-HT_{2A} receptors on GABAergic interneurons. However, the 5-HT-induced synaptic currents in prefrontal and neocortical pyramidal cells were (1) potently and completely blocked by the AMPA receptor antagonist LY293558; (2) only minimally (<15%) suppressed by the $GABA_A$ receptor antagonist bicuculline; (3) completely suppressed by the fast sodium channel blocker tetrodotoxin (TTX); (4) completely suppressed by perfusing the slice with an artificial cerebrospinal fluid solution containing no added calcium ("0" calcium) and high magnesium; and (5) mimicked by iontophoretic application of 5-HT in layers I and Va in a distribution consistent with stimulation of the apical dendrite but not the basilar dendrites. Furthermore, the firing of pyramidal neurons in response to bath applied 5-HT could not be detected even after exhaustive searches, unlike the ready identification of firing interneurons in piriform cortical slices. Taken together, these observations suggested a model where 5-HT, via activation of 5-HT_{2A} receptors, releases Glu from glutamatergic afferents terminating in the layer I and Va dendritic field of L5 pyramidal cells. Subsequently, studies that will be described below suggested that the midline and intralaminar thalamic nuclei are at least one of the sources providing the glutamatergic afferents. Thus, these 5-HT-induced EPSCPs/EPSPs are a novel form of feedforward excitation of glutamatergic afferents to the principle output neurons of the prefrontal cortex (PFC) and neocortex.

Another model paradigm for observing Glu release induced by serotonergic hallucinogens via 5-HT_{2A} receptor activation in PFC slices was subsequently identified (Aghajanian 2009). After applying a serotonergic hallucinogen such as LSD (10 nM) or DOI (3 μM) to a prefrontal cortical slice for 10 min, focal electrical stimulation of the slice in the middle of the cortex at approximately 1 Hz results in an early fast EPSC followed by an UP state and recurrent network activity. More cumbersome protocols to induce what was originally described as "late EPSCs" involved identifying stimulation sites in the white matter below cortical layer VI where stimulation at approximately 0.1 Hz induced these late EPSCs/EPSPs or recurrent network activity during the washout of 5-HT. Subsequent application of a serotonergic hallucinogen for approximately 10 min would then induce a long-lasting (up to several hours) appearance of late EPSCs or recurrent network activity after low-frequency 0.1 Hz stimulation of the white matter. The fact that washout of 5-HT from the slice is required to observe this form of recurrent network activity highlights the likelihood that a primary but relatively brief effect of 5-HT in the PFC is to inhibit Glu release, probably via some combination of effects on 5-HT_{1A}, 5-HT_{1B}, and 5-HT_{1F} receptors (Fig. 1). By contrast, bath application of 5-HT appears to result in a more sustained activation of 5-HT_{2A} receptors. Bath applications of serotonergic hallucinogens such as LSD and DOI, which potently bind to and activate 5-HT_{2A} receptors, together with focal electrical stimulation, combine to produce late synaptic currents or Glu "spillover", which

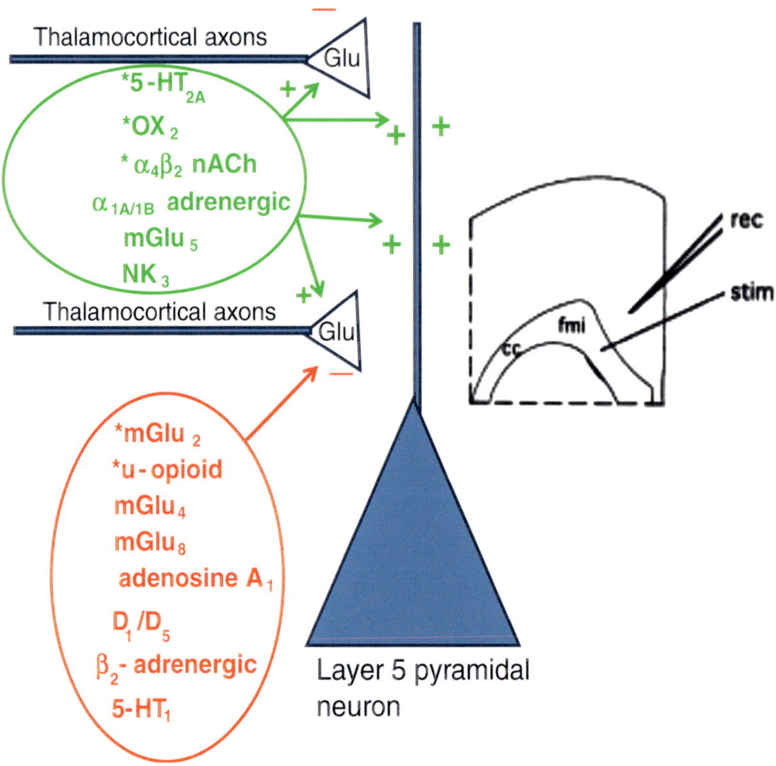

Fig. 1 Model of extrinsic (thalamus and/or claustrum) and intrinsic (medial prefrontal cortex (mPFC) and neocortex) afferents to the apical dendrites of layer 5 pyramidal cells that release glutamate (Glu) in response to activation of 5-HT_{2A}, $\alpha_{1A/1B}$ adrenergic, $mGlu_5$, OX_2, NK_3 or $\alpha_4\beta_2$ nicotinic acetylcholine (nACh) receptors (shown in *green*). A variety of presynaptic receptors suppress the 5-HT_{2A} receptor-induced Glu release including $mGlu_2$, $mGlu_4$, $mGlu_8$, μ-opioid, adenosine A_1, dopamine D_1 and D_5, β_2-adrenergic, and 5-HT_1 receptors. The Glu released in response to 5-HT_{2A} activation activates AMPA receptors on the layer 5 pyramidal neurons. To date, the most convincing evidence suggests that the midline and intralaminar thalamic nuclei are the primary sources of the afferent glutamatergic axons that terminate on layers I and Va of the PFC and neocortex. The inset diagram shows that these experiments were performed under in vitro conditions using slices of the rat brain containing the anterior cingulate and prelimbic regions of the mPFC

can otherwise be described as excitatory feedforward and feedback recurrent network activity.

The 5-HT-induced EPSCs and the recurrent network activity induced by the combination of DOI and focal low-frequency electrical stimulation are suppressed by a range of neurotransmitter receptors that are known to inhibit presynaptic Glu release (Table 1 and Fig. 1). The initial electrophysiological experiments suggested

Table 1 Neurotransmitter receptors suppressing glutamate release induced by 5-HT_{2A} receptor activation

Neurotransmitter	Receptor and Action	5-HT-induced EPSCs	DOI-induced recurrent network activity
Glutamate	$mGlu_{2/3}$ receptor agonists	Suppress	Suppress
Glutamate	$mGlu_2$ receptor PAMs	Suppress	*Not tested*
Glutamate	$mGlu_{4/8}$ receptor agonists	Suppress	*Not tested*
Adenosine	A_1 receptor agonists	Suppress	Suppress
Opioids	μ-opioid receptor agonists	Suppress	*Not tested*
Dopamine	D_1/D_5 receptor agonists	*Not tested*	Suppress
Norepinephrine	β_2 receptor agonists	*Not tested*	Suppress

that mGlu autoreceptors played an important role in modulating the Glu release induced by 5-HT_{2A} receptor activation (Aghajanian and Marek 1997). An autoreceptor role for the $mGlu_2$ receptor was initially suggested based on the fact that the $mGlu_{2/3}$ receptor orthosteric agonists LY354740 and LY379268 could potently suppress 5-HT-induced EPSCs and DOI-induced Glu spillover, whereas the $mGlu_{2/3}$ receptor antagonist LY341495 enhanced EPSC frequency (Marek et al. 2000). LY354740 was approximately threefold more potent in suppressing 5-HT-induced EPSCs and DOI-induced Glu spillover compared to the early evoked EPSP; importantly, the direct effects on LY354740 on membrane potential were minimal, and LY354740 failed to alter the inward currents induced by bath application of AMPA following perfusion of the slice with TTX to block fast sodium channels. Furthermore, $mGlu_{2/3}$ receptor binding sites in layers I and Va of the PFC appeared to overlap with the laminar distribution of 5-HT_{2A} receptor binding sites. Subsequent experiments demonstrated that midline thalamic lesions caused a ~20% reduction in the density of $mGlu_{2/3}$ receptors (measured with [^{3}H] LY354740) in PFC layers I and Va, likely a physiologically significant amount given that most [^{3}H]LY354740 binding sites were subsequently suggested to be $mGlu_3$ receptors (Marek et al. 2001; Wright et al. 2013). An even more definitive identification of a role for $mGlu_2$ receptors in mediating these effects was provided by experiments demonstrating that selective $mGlu_2$ receptor positive allosteric modulators (PAMs) suppressed 5-HT-induced EPSCs in the rat PFC (Benneyworth et al. 2007). These effects have also been replicated in other laboratories and with additional selective $mGlu_{2/3}$ receptor agonists (Wright et al. 2013; Klodzinska et al. 2002; Rorick-Kehn et al. 2007). Evidence for the behavioral salience of these functional 5-HT_{2A} and $mGlu_2$ receptor interactions will be described below. Thus, the $mGlu_2$ receptor appears to function as an autoreceptor on glutamatergic terminals releasing Glu (induced by 5-HT_{2A} receptor activation) from afferents reaching the layer I and Va dendritic field of L5 pyramidal cells in the PFC and neocortex.

Agonists of several other G-protein-coupled receptors (GPCRs) that generally had inhibitory effects on neurotransmitter release or hyperpolarizing electrophysiological effects ($G_{i/o}$-coupled receptors) were also found to suppress the spontaneous 5-HT- or DOI-induced EPSCs (Table 1 and Fig. 1). This included two other mGlu receptors ($mGlu_4$ and $mGlu_8$) that were also well known to decrease Glu release (Palucha-Poniewierg et al. 2008; Slawinska et al. 2013; Zhang and Marek 2007). The adenosine A_1 receptor, another well-known heteroreceptor that decreases Glu release, was also found to suppress spontaneous 5-HT-induced EPSCs and the feedforward and feedback Glu release induced by serotonergic hallucinogens and electrical stimulation (Stutzman et al. 2001). Finally, agonists for the μ-opioid receptor, a $G_{i/o}$-coupled GPCR with enrichment in PFC layers I and Va, were also found to suppress 5-HT-induced EPSCs (Marek and Aghajanian 1998). Thalamic lesions that decreased $mGlu_{2/3}$ receptor binding sites by $\sim$20% also reduced μ-opioid receptor binding sites by $\sim$40%, suggesting that the midline and intralaminar thalamic nuclei are a potential source for the 5-HT-induced EPSCs (Marek et al. 2001). These studies are all consistent with the known localization of $mGlu_2$, $mGlu_4$, adenosine A_1, and μ-opioid receptors in a potential thalamic source of afferents to cortical layers I and V, namely the midline and intralaminar thalamic nuclei (Kolaj et al. 2014). Thus, converging lines of evidence further suggest a model where 5-HT_{2A} receptor activation and a range of $G_{i/o}$-coupled GPCRs have countervailing effects Glu release from at least one pool of subcortical terminals reaching the apical dendritic field of L5 pyramidal cells throughout the PFC and neocortex (Table 1 and Fig. 1).

In addition to the suppression of hallucinogen-induced Glu overflow induced by a number of $G_{i/o}$-coupled GPCRs, several G_s-coupled receptors appear to have similar effects on hallucinogen-induced recurrent network activity (Table 1 and Fig. 1). This effect has been well characterized for the $D_{1\text{-like}}$ class of dopamine receptors (D_1/D_5), which are coupled to G_s. The ability of the D_1/D_5 partial agonist SKF38393 to suppress DOI-induced late EPSCs was blocked by the dopamine D_1/D_5 receptor antagonist SCH23390 but not the dopamine D_2 receptor antagonist raclopride (Lambe and Aghajanian 2007). In contrast, the dopamine D_2 receptor agonist quinpirole failed to suppress DOI-induced recurrent network activity. The critical role played by Glu spillover in the effect of SKF38393 is indicated by mechanistic experiments linking its suppressant effect to the known ability of the G_s pathway to upregulate the excitatory amino acid transporter 3 (EAAT3). Extending the effects observed to levels downstream from receptors, application of the adenylyl cyclase activator forskolin, or the phosphodiesterase-resistant cAMP analogue 8-Br-cAMP suppressed DOI-induced recurrent network activity. The modulatory effects of forskolin and 8-Br-cAMP on DOI-induced recurrent network activity occurred in the absence of effects on fast early EPSCs. A cortical versus thalamic site of action for the D_1/D_5 ligands is consistent with the known cortical versus thalamic localization of these receptors (Ciliax et al. 2000; Huang et al. 1992; Mansour et al. 1992). Both postsynaptic and presynaptic effects relative to L5 pyramidal cells are supported by a range of known mechanisms for dopamine D_1/D_5 receptors (Gao et al. 2001; Rotaru et al. 2007). Preliminary experiments have also suggested that β_2-adrenergic receptor stimulation by epinephrine or clenbuterol can also selectively suppress DOI-induced recurrent network activity (GJ Marek and B Ramos, unpublished observations). The preferential localization of β_2-adrenergic receptors to the midline and intralaminar thalamic nuclei compared to the preferential localization of β_1-adrenergic receptors in the PFC and neocortex, as well as the known physiological effects of β_2-adrenergic receptor function in one of the midline thalamic nuclei, is also consistent with a role of thalamic inputs in DOI-induced Glu release in the mPFC (Nicholas et al. 1993; Rainbow et al. 1984; Zhang et al. 2009). Thus, suppression of DOI-induced Glu release by D_1/D_5 receptor agonists and β_2-adrenergic receptor agonists could support a model for intracortical circuitry or thalamocortical circuitry or a model combining both types of afferents onto cortical L5 pyramidal cells.

Similar to the effect of 5-HT via 5-HT_{2A} receptors, activation of a number of $G_{q/11}$-coupled GPCRs appears to induce feedforward Glu release onto L5 pyramidal cells in the PFC and/or neocortex (Table 2 and Fig. 1). Where tested, common features for spontaneous neurotransmitter-induced EPSCs/EPSPCs included dependence on fast sodium channels (i.e., suppression by TTX) and AMPA receptors (i.e., suppression by LY293558), and suppression by a μ-opioid receptor agonist (DAMGO). Thus, norepinephrine (NE), (*S*)-3,5-dihydroxyphenylglycine (DHPG), hypocretin-2, and senktide all induce EPSCs/EPSPs via activation of α_1 adrenergic, $mGlu_5$, $orexin_2$ (OX_2), and NK_3 receptors, respectively (Lambe et al. 2007; Lambe and Aghajanian 2003; Marek and Aghajanian 1999; Marek and Zhang 2008; Rekling 2004).

The NE-induced spontaneous EPSCs are especially interesting given the existence of significant similarities between α_1-adrenoceptors (especially the α_{1B} subtype) and 5-HT_{2A} receptors with respect to regional brain distribution, laminar cortical protein distribution, and laminar cortical mRNA distribution, in addition to a range of physiological and behavioral effects with salient medial PFC (mPFC) involvement as discussed elsewhere (Marek and Aghajanian 1999; Santana et al. 2012) and later in this chapter. Expression of α_1-adrenoceptors in glutamatergic axons and terminals in the mPFC, including colocalization with vesicular glutamate transporter 1 and 2 (VGluT1 and VGluT2), are consistent with cells of origin in the PFC and midline and intralaminar thalamic nuclei, respectively (Mitrano et al. 2012). When compared to the quite modest and more controversial localization of

Table 2 Neurotransmitters that induce EPSCs/EPSPs via a feedforward mechanism

Neurotransmitter	Receptor	TTX	LY293558	DAMGO
norepinephrine	α_1-noradrenergic	Suppress	Suppress	Suppress
Glutamate	$mGlu_5$	Suppress	Suppress	Suppress
Hypocretin-2	Orexin OX_2	Suppress	*Not tested*	Suppress
Neurokinin	NK_3	*Not tested*	*Not tested*	*Not tested*
acetylcholine	$\alpha_4\beta_2$	Suppress	Suppress	Suppress

5-HT$_{2A}$ receptors to presynaptic neuronal compartments (primarily monoaminergic axons and varicosities) determined using well-known antibodies (Jakab and Goldman-Rakic 1998; Miner et al. 2003), the fact that the majority of α_1-adrenoceptors in the PFC, ventral tegmental area, and nucleus accumbens are expressed by unmyelinated axons and axon terminals rather than having a postsynaptic localization raises interesting questions about the reliability of monoamine antibody epitopes required to detect labeling in presynaptic terminals and axons.

In recordings from L5 pyramidal cells in rat brain slices, the type I mGlu receptor agonist DHPG induces an increase in spontaneous EPSCs that can be blocked by a selective negative allosteric modulator (NAM) of the mGlu$_5$ receptor (Marek and Zhang 2008). The pharmacology of DHPG-induced EPSCs is similar to 5-HT-induced EPSCs with respect to their suppression and enhancement by mGlu$_{2/3}$ receptor agonists and antagonists, respectively. In addition, a μ-opioid receptor agonist, as well as a AMPA/GluK5 receptor antagonist (LY293558), suppressed DHPG-induced EPSCs. DHPG was not observed to have consistent effects on membrane depolarization after blockade of sodium channels and impulse flow with TTX. Although mGlu$_5$ receptors are predominantly expressed at postsynaptic sites in the neocortex, according to light and electron microscopic studies, a minority of mGlu$_5$ receptors are localized in presynaptic axon terminals (Romano et al. 1995). The localization of mGlu$_5$ receptor mRNA and protein is consistent with a presynaptic role where the cells of origin might arise from either the thalamus, the PFC and neocortex, or both (Romano et al. 1995; Neto et al. 2000; Simonyi et al. 2005). Strong evidence for a presynaptic effect of orexin-B involving OX$_2$ receptors will be discussed shortly, including data indicating that the midline and intralaminar thalamic nuclei are a potential source for these afferents (Lambe et al. 2007; Lambe and Aghajanian 2003). The pharmacology of the senktide-induced increase in EPSC frequency has not been defined beyond blockade by NK$_3$ receptor antagonists. Immunohistochemical, receptor autoradiography, and mRNA studies in rats appear to suggest that NK$_3$ is largely a cortical receptor expressed in mid-cortical layers (layers IV and V) throughout the PFC and neocortex (Ding et al. 1996; Langlois et al. 2001; Saffroy et al. 2003; Shughrue et al. 1996). However, one study suggests that NK$_3$ receptors are present in the midline thalamic nuclei in primates (Rigby et al. 2005). Thus, agonists for a number of G$_{q/11}$-coupled receptors known to be expressed in the midline and intralaminar thalamic nuclei appear to induce an increase in the frequency of spontaneous EPSCs similar to the effect of 5-HT on L5 pyramidal neurons in the PFC and neocortex (Kolaj et al. 2014).

In addition to the neurotransmitters discussed above, acetylcholine (ACh) and nicotine were also found to induce EPSCs/EPSPs through activation of $\alpha_4\beta_2$ nicotinic receptors on thalamic afferents to L5 pyramidal cells (Lambe et al. 2003, 2005). Evidence also exists that ACh may increase the frequency of spontaneous EPSCs, although the pharmacology of this effect has not been defined beyond the involvement of muscarinic M$_1$ receptor activation. The potential for a G$_{i/o}$-coupled muscarinic receptor to suppress 5-HT-, NE-, orexin-B-, and DHPG-induced EPSCs should be explored as well.

Neurocircuitry: Thalamocortical versus cortical/claustrum projections to mPFC layer V neurons

At least a majority of the 5-HT-induced EPSCs, if not all, appear to originate from subcortical sources. The frequency of 5-HT-induced EPSCs in rat PFC slices was reduced 61–67% by fiber-sparing chemical or radiofrequency lesions of the midline and intralaminar thalamic nuclei that were made 12–18 or 6–14 days, respectively, before the intracellular recordings from L5 pyramidal cells (Marek et al. 2001). As described above, these thalamic lesions also reduced the expression of $mGlu_2$ and μ-opioid receptors, suggesting that about 40% of the $mGlu_2$ and μ-opioid receptors in the PFC are located on thalamic afferents presynaptic to the L5 pyramidal cells. The fact that these lesions greatly reduce the frequency of 5-HT-induced EPSCs suggests an important functional link to thalamic terminals. Postsynaptic $5\text{-}HT_{2A}$ receptors present in layers I and Va of the mPFC were significantly upregulated ($\sim$114% of control levels) in response to the chemical fiber-sparing thalamic lesions. Because $5\text{-}HT_{2A}$ receptors are primarily localized postsynaptically on pyramidal cells and interneurons in the PFC and neocortex, an upregulation of postsynaptic $5\text{-}HT_{2A}$ receptors would likely conceal the fact that a relatively minor pool of presynaptic $5\text{-}HT_{2A}$ receptors was eliminated by the thalamic lesions. Less than 25% of the prefrontal cortical $5\text{-}HT_{2A}$ receptors are present on presynaptic sites on axons, and most of those $5\text{-}HT_{2A}$ receptors are thought to be present on the axons of monoaminergic neurons (Miner et al. 2003). This assessment, which was made using currently available antibodies, suggests that glutamatergic terminals and axons expressing $5\text{-}HT_{2A}$ receptors are relatively rare in the mPFC (Jakab and Goldman-Rakic 1998). A subsequent report confirmed that most of the 5-HT-induced EPSPs recorded in PFC slices appear to originate from axons and terminals of thalamic projections (Lambe and Aghajanian2001). Additional experiments suggested that 5-HT induces EPSCs by closing Kv1.2-containing potassium channels (Lambe and Aghajanian2001). Confirming previous reports that presynaptic $\alpha_4\beta_2$ nicotinic acetylcholine (nACh) receptors in mid-cortical layers originate from the thalamus, Lambe and colleagues reported that thalamic lesions reduced the frequency of nicotine- and ACh-induced spontaneous EPSCs by $\sim$80% (Lambe et al. 2003). Thus, agonists for both the $5\text{-}HT_{2A}$ receptor and the $\alpha_4\beta_2$ nAChR induced an increase in spontaneous currents in the PFC via activation of afferents from the midline thalamus.

Around this time, compelling converging evidence emerged suggesting that activation of $orexin_2$ (OX_2) receptors induced an increase in spontaneous EPSCs via activation of afferents from the midline and intralaminar thalamic nuclei (Lambe and Aghajanian 2003). First, in recordings from layer V pyramidal cells, the hypocretin-2 peptide was found to induce spontaneous EPSCs without directly depolarizing the pyramidal cells. Hypocretin-2, an OX_2 receptor agonist, induced calcium transients in a small minority of spines (<10%) in the layer I and Va apical dendritic fields of the pyramidal cells. Similar calcium transients were not induced in basilar dendritic spines. Furthermore, these calcium transients were induced in spines generally in apposition to anterogradely labeled terminals originating from neurons in the midline and intralaminar thalamic nuclei. Midline thalamic lesions suppressed the hypocretin-induced EPSCs without affecting baseline EPSCs. TTX, μ-opioid agonists, and an AMPA receptor antagonist all suppressed the spontaneous EPSCs and the calcium transients, similar to the effects observed previously for 5-HT_{2A} receptor activation. These findings with hypocretin-2 are also consistent with the hypothesis that 5-HT_{2A} receptor activation induces Glu release onto the apical dendritic field of L5 pyramidal neurons via thalamocortical afferents.

The partial suppression (60–65%) of 5-HT-induced EPSCs in the mPFC by large thalamic lesions suggests that there is at least one additional source of afferents to the L5 pyramidal cells. The basolateral nucleus of the amygdala (BLA) projects to the deep portion of layer I, layer II, and the entire width of layer V, making this amygdaloid nucleus another potential subcortical source for 5-HT-induced EPSCs. However, complete destruction of the BLA bilaterally did not alter the frequency of 5-HT-induced EPSCs, ruling out the amygdala as a feedforward afferent to the L5 pyramidal cells (Marek et al. 2001).

Several cortical sources that could potentially mediate the EPSCs induced by 5-HT and serotonergic hallucinogens have also been proposed. Rodrigo Andrade and colleagues identified a subpopulation of neurons deep within the mPFC that are depolarized sufficiently by 5-HT_{2A} receptor activation to fire action potentials; these cells are a potential source of the 5-HT-induced EPSCs (Beique et al. 2007). This subpopulation of mPFC neurons needs to be defined in greater detail, especially with regard to the $G_{I/o}$ coupled GPCRs that suppress 5-HT-induced EPSCs and the other $G_{q/11}$-coupled GPCRs that produce similar increases of spontaneous EPSC frequency. To date, the μ-opioid receptor is the only GCPR known to modulate the activity of this population of deep cortical neurons. A more complete pharmacological characterization of these cells with respect to inhibitory effects of $mGlu_2$, $mGlu_4$, $mGlu_8$, adenosine A_1, dopamine $D_{1/5}$, β_2-adrenergic and 5-HT_1 receptors and the excitatory effects of OX_2, α_1-adrenergic, $mGlu_5$, $\alpha_4\beta_2$-nicotinic, and NK_3 receptors is necessary before concluding that these cells represent a cortical source of the afferents activated by 5-HT_{2A} receptors. Another cortical source for 5-HT-induced EPSCs may be L5 pyramidal neurons that send callosal and commissural projections to the contralateral cortex, which are excited by 5-HT (Avesar and Gulledge 2012). Unfortunately, the pharmacology of these cells with respect to inhibition and excitation by the range of GPCRs described above has not been explored.

One report, which analyzed 5-HT-induced EPSCs recorded from transgenic mice that were rescued from a constitutive knockout of 5-HT_{2A} receptors throughout the CNS, has suggested that 5-HT-induced EPSCs arise primarily from neocortical afferents intrinsic to the cerebral cortical mantle (Weisstaub et al. 2006). However, interpretation of the *htr2a* gene rescue experiment for the thalamus appears to have been compromised by the use of a promoter that is weakly expressed in the midline and intralaminar thalamic nuclei compared to a robust expression in the primary sensory thalamic nuclei (Lebrand et al. 1996; Narboux-Neme et al. 2008). Further, the *htr2a* gene rescue directed at the mPFC also included the claustrum as an area being rescued. Because the claustrum, which sends afferents to layers I and V of the mPFC and neocortex, is also a potential source of the afferents mediating the 5-HT-induced EPSCs, it is premature to conclude that the spontaneous 5-HT-induced EPSCs are mediated by cortical afferents based solely on this experiment. Clearly, further work is necessary to understand whether circuitry between the mPFC and claustrum plays a role in mediating the effects of hallucinogenic drugs and 5-HT_{2A} receptor activation. Additional studies directed at understanding whether hippocampal afferents to the mPFC could be involved are also warranted. Further studies using different technologies to confirm or refute the presence of 5-HT_{2A} receptors in midline and intralaminar thalamic nuclei, such as optogenetic recordings with gene-based targeting strategies, could clarify models of interactions between hallucinogenic drugs and Glu in the PFC (Barre et al. 2012). Nevertheless, as indicated by both thalamic lesions and the combined two-photon imaging and anterograde labeling of thalamic projections arising from the intralaminar and thalamic nuclei, these "non-specific" thalamic nuclei clearly seem to serve as an important substrate for the spontaneous EPSCs induced by 5-HT and hypocretin-2. Additional experiments are required to understand the potential role of claustral and/or neocortical afferents to L5 pyramidal cells.

1 From 5-HT_{2A} Receptor-Induced EPSCs and Recurrent Neuronal Activity to Salient Behavioral Effects Induced by 5-HT_{2A} Receptor Activation

Head-Twitch Response

Given that a wide range of neurotransmitter systems appear to have modulatory effects on Glu released from subcortical terminals and/or intrinsic cortical afferents onto the apical dendrites of L5 pyramidal cells, salient behavioral effects would be expected to arise from these interactions since L5 pyramidal cells are the principal source of output from the neocortex to subcortical regions. A range of behaviors induced by systemic administration of hallucinogenic drugs, including a relatively simple behavioral response known as head twitches or head shakes, appear to be modulated by the same neurotransmitter relationships outlined previously for subcortical glutamatergic afferents to L5 pyramidal cells (Canal and Morgan 2012; Handley and Singh 1986).

According to a report published almost 20 years ago, local infusion of DOI into the prelimbic region of the mPFC induces a head-twitch response (HTR) that is blocked by systemic administration of 5-HT_{2A} receptor antagonists but not by a selective 5-HT_{2C} receptor antagonist, suggesting that hallucinogen-induced head twitches are mediated by activation of 5-HT_{2A} receptors in the mPFC (Willins and Meltzer 1997). Given that a range of antidepressant drugs, as well as typical and atypical antipsychotic drugs, suppress the HTR induced by DOI, 5-HTP, and quipazine (Czyrak et al. 1993; Eison et al. 1990; Friedman et al. 1983; Goodwin et al. 1984; Moore et al. 1992; Nacca et al. 1998; Peroutka et al. 1981; Rojoz 2012; Sanchez and Arnt 2000; Wettstein et al. 1999), drugs acting on $mGlu_2$ receptors were tested for effects on the DOI-induced HTR. Several $mGlu_{2/3}$ receptor agonists (e.g., LY354740 and LY379268) and $mGlu_2$ receptor PAMs (e.g., BINA and CBiPES) were found to suppress DOI-induced head twitches, whereas the $mGlu_{2/3}$ receptor antagonist LY341495 potently enhances the frequency of DOI-induced head twitches (Fig. 2), consistent with an interaction between 5-HT_{2A} receptors and $mGlu_2$ autoreceptors on afferents to the mPFC (Benneyworth et al. 2007; Klodzinska et al. 2002; Benvenga et al. 2006; Gewirtz and Marek 2000; Gonzalez-Maeso et al. 2008; Moreno et al. 2011; Wieronska et al. 2013). These observations prompted efforts to determine whether a range of neurotransmitter receptor agonists, PAMs, or antagonists that modulate 5-HT-induced EPSCs can also alter the HTR induced by serotonergic hallucinogens. As expected, μ-opioid agonists (Marek 2003; Rojas-Corrales et al. 2007), adenosine A_1 receptor agonists (Marek 2009), AMPA receptor antagonists (Egashira et al. 2011; Zhang and Marek 2008), $mGlu_4$ receptor PAMs (Slawinska et al. 2013; Wieronska et al. 2012), $mGlu_5$ receptor negative allosteric modulators (NAMs; GJ Marek, unpublished observations), NK_3 receptor antagonists (GJ Marek, unpublished observations), α_1-adrenergic receptor antagonists (Dursun and Handley 1996; Schreiber et al. 1995), OX_2 receptor antagonists (GJ Marek, unpublished observations), and β_2-adrenergic receptor agonists (GJ Marek, unpublished observations) have been found to suppress head twitches induced by serotonergic hallucinogens (Fig. 2). Thus, impressive coherence exists between a range of targets modulating Glu release due to 5-HT_{2A} receptor activation and the ability of ligands acting at those targets to modulate the hallucinogen-induced HTR in rodents (Table 3).

In contrast to expectations from the molecular targets described above, dopamine $D_{1/5}$ receptor antagonists were found to suppress, rather than enhance, DOI-induced head twitches (Dursun and Handley 1996; Schreiber et al. 1995). However, the potency and efficacy of the dopamine D_1 receptor antagonists in suppressing DOI-induced head twitches was equally correlated with affinity for 5-HT_{2A} receptors and dopamine D_1 receptors, creating uncertainty regarding which mechanism was responsible for the behavioral action. Nicotine is another ligand that has been tested and found to have effects opposite to those expected based on its ability to increase spontaneous EPSCs. Nicotine suppresses DOI-induced head twitches (Gaynor and Handley 2001; Hayslett and Tizabi 2005; Tizabi et al. 2001). However, since hallucinogen-induced head twitches in rodents have been reported to have an inverted-U-shaped dose response relationship, it is not clear whether the apparent suppressant effect of nicotine on DOI-induced head twitches reflects a confounding effect of the position on the dose–response curve being studied.

Further pharmacological characterization with respect to Glu suggests that low doses of DOI (0.315–0.63 mg/kg) and the NMDA receptor antagonist MK-801 (0.2 mg) have synergistic effects on horizontal locomotor activity in rats when administered in combination (Zhang and Marek 2008). Taken together, these overall findings are consistent with the hypothesis that modulation of Glu spillover (arising from subcortical afferents and/or intrinsic cortical afferents and directed

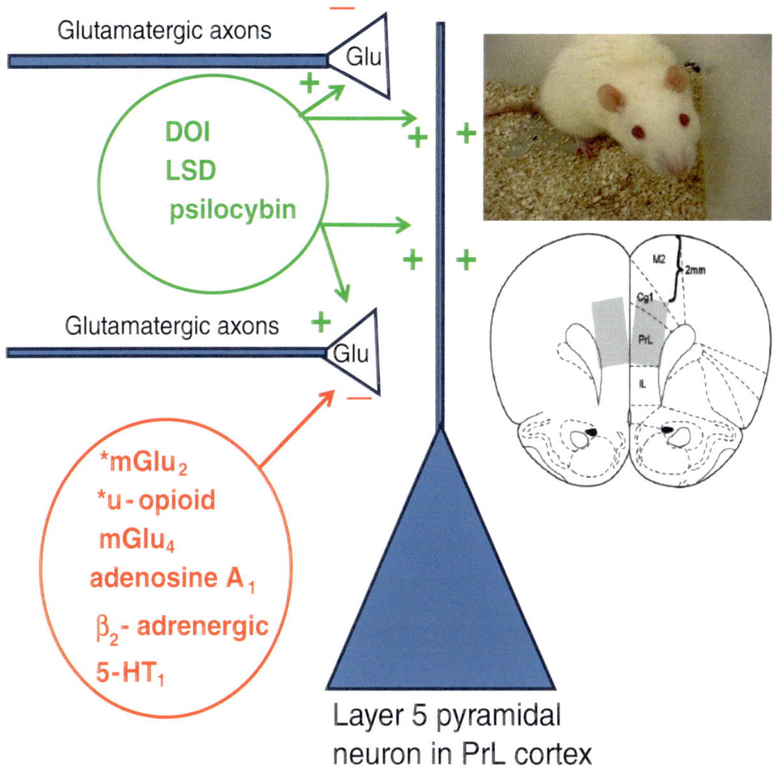

Fig. 2 Model of glutamatergic afferents to the apical dendrites of layer 5 pyramidal cells that release glutamate (Glu) in response to serotonergic hallucinogens such as LSD, DOI, and psilocybin (shown in *green*), which also induce head twitches in rats and mice. A variety of presynaptic receptors can suppress the hallucinogen-induced head twitches, including $mGlu_2$, $mGlu_4$, μ-opioid, adenosine A_1, β_2-adrenergic, and $5\text{-}HT_{1A}$ receptors. All of these receptors suppress recurrent network activity (late EPSPs) induced by stimulation of $5\text{-}HT_{2A}$ receptors on presumed thalamocortical afferents. Most of these agonists only partially suppress DOI-induced head shakes. The Glu released in response to $5\text{-}HT_{2A}$ receptor activation then stimulates AMPA receptors on the layer 5 pyramidal neurons. The inset picture on the top right shows a Sprague-Dawley rat. The inset illustration on the bottom right shows a coronal section through the brain with shading indicating the location of the prelimbic (PrL) region of the medial prefrontal cortex. As shown by David Willins and Herb Meltzer in 1997, infusion of DOI into the PrL region induces head-twitch behavior. *Abbreviations:* Cg1 cingulate area 1; IL infralimbic cortex; M2 premotor cortex

onto the apical dendritic field of L5 pyramidal cells in the mPFC) is a key substrate for modulating the HTR in mice and rats, a well-known behavioral effect of serotonergic hallucinogens.

Table 3 Correlation between PFC in vitro electrophysiology and in vivo screening for psychiatric drugs

Drug Type	Effect on feed forward neurotransmitter-induced EPSCs/EPSPs	Effect on feed forward and feedback recurrent network activity (late EPSPs)	Effect on fast EPSP/EPSC	Effect in the DRL 72-s behavioral paradigm	Effect on DOI-induced head shakes or NMDA receptor antagonist-induced hyperactivity
5-HT_{2A} receptor antagonist	Suppress	Suppress	No effect	Antidepressant-like	Suppress
Dextran	No effect	Suppress	No effect	Not tested	Not tested
$mGlu_{2/3}$ receptor agonist	Suppress	Suppress	Modest, less potent	No Effect	Suppress
$mGlu_2$ receptor PAM	Suppress	Not tested	No effect	Antidepressant-like	Suppress
Adenosine A_1 receptor agonist	Suppress	Suppress	No effect	Antidepressant-like	Suppress
$mGlu_{4/8}$ receptor agonist	Suppress	Not tested	No effect	Not tested	Suppress
μ-Opioid receptor agonist	Suppress	Not tested	No effect	No effect	Suppress
β_2-receptor agonist	Not tested	Suppress	No effect	Antidepressant-like	Suppress
Dopamine D_1 receptor agonist	Not tested	Suppress	No effect	Not tested	No effect
Orexin OX_2 receptor antagonist	Suppress	Not tested	No effect	AD-like	Suppress
$mGlu_5$ receptor NAM	Suppress	Not tested	No effect	No effect	Suppress
α_1-Adrenergic receptor antagonist	Suppress	Not tested	No effect	No effect	Suppress
NK_3 receptor antagonist	Suppress	Not tested	No effect	No effect	Suppress
Ketamine	Increase	Not tested	decrease	Antidepressant-like	Increased
Lithium (acute)	Increase	Not tested	Not tested	Not tested	Increased

Beyond their utility as an in vivo model for modulation of mPFC 5-HT_{2A} receptor activity, DOI-induced head twitches in rodents have been used as a relatively promiscuous screening model for potential antipsychotic and/or antidepressant drugs. The DOI-induced HTR has also been suggested to be a model for tic disorders, such as Tourette's syndrome. However, the actual physiological role for the HTR, observed to occur at a relatively low spontaneous rate in all vertebrate species outside of most primates and humans, remains to be delineated.

Motoric impulsivity studied using the 5-CSRTT

The five-choice serial reaction time test (5-CSRTT) is another intriguing behavioral paradigm that has been studied with respect to activation of 5-HT_{2A} receptors in the PFC. A particular type of impulsivity, motoric impulsivity or action impulsivity, is assessed using the 5-CSRTT. Important for our discussion, systemic administration of 5-HT_{2A} receptor antagonists such as ketanserin reduces motoric impulsivity (especially premature responses) in 5-CSRTT studies, but does not alter another form of impulsivity related to delayed reward presentation (Talpos et al. 2006). Anatomical validation of this behavior was demonstrated by the finding that infusion of selective 5-HT_{2A} receptor antagonists such as M100907 into the mPFC suppressed anticipatory responding induced by administration of a NMDA receptor antagonist (Carli et al. 2006). As expected based on in vitro electrophysiological recordings from L5 pyramidal cells and studies of DOI-induced head twitches, pretreatment with the $mGlu_{2/3}$ receptor agonist LY379268 blocks impulsivity or over responding induced by DOI in rats performing the 5-CSRTT (Wischhof and Koch 2012). Furthermore, intra-mPFC infusions of DOI were shown to increase impulsive responding in the 5-CSRTT whereas the $mGlu_{2/3}$ receptor agonist LY379268 suppressed this effect of DOI (Wischhof et al. 2011). Also, as expected from electrophysiological and HTR studies, the α_1-adrenergic receptor antagonist prazosin suppressed DOI-induced premature responses on the 5-CSRTT, in contrast to the lack of effect for an α_2-adrenergic receptor antagonist (Koskinen et al. 2003).

Both consistent and contradictory behavioral evidence for a modulatory effect of dopamine $D_{1/5}$ receptors has been found. For example, the dopamine $D_{1/5}$ receptor agonist SKF38393 enhanced the accuracy of attentional performance under low baseline conditions; a dopamine D_1 receptor antagonist blocked the effect of SKF38393 (Granon et al. 2000). Furthermore, systemic administration of the dopamine $D_{1/5}$ receptor agonist SKF38393 improved choice accuracy in the 5-CSRTT, whereas the 5-HT_{2A} antagonist ketanserin suppressed premature responding when administered systemically or directly into the mPFC (Passetti et al. 2003). These findings were extended to the striatum where infusions of a dopamine D_1 receptor agonist improved accuracy in the 5-CSRTT; these effects were also blocked by a dopamine D_1 receptor antagonist. In contrast to the previous reports, other experiments suggested that the dopamine D_1 receptor antagonist SCH23390 suppressed the effect of DOI on premature responding (Koskinen and Sirvio 2001). Hence, although further pharmacological characterization with a range of other GPCR agonists and antagonists is necessary, initial experiments have shown that the 5-CSRTT has intriguing consistencies with the electrophysiology of L5 pyramidal cells and the DOI-induced HTR.

DRL 72-s operant behavior: A behavioral screen for antidepressant-like effects

Operant lever pressing by rats under a differential-reinforcement-of-low rate 72-sec (DRL-72 s) schedule of reinforcement is another behavioral paradigm involving PFC-striatal-thalamic-amygdala loops with a critical relationship to motoric impulsive behavior (Marek et al. 2016). Antidepressant drugs as a class appear to bias the responding of rats operating under DRL 72-s schedules away from impulsive responding (e.g., failing to wait 72-s from the last response to lever press in order to obtain water or food) and increase the probability that responding will occur at time durations matching reinforcement contingencies (Marek et al. 2016; O'Donnell et al. 2005). Not surprisingly, selective blockade of 5-HT_{2A} receptors induces cohesive antidepressant-like rightward shifts in the inter-response time (IRT) distribution (along with increases in the reinforcement rate and decreases in the response rate) alone or when added to the SSRI fluoxetine, the tricyclic antidepressant desipramine, and the monoamine oxidase inhibitor tranylcypromine (Ardayfio et al. 2008; Marek et al. 2005). Similar rightward shifts in IRT distributions were also induced by the selective 5-HT_{2A} receptor antagonist M100907 following administration of the channel blocking NMDA receptor antagonist MK-801 (Ardayfio et al. 2008; Higgins et al. 2003). As expected from the relationships between studies of L5 pyramidal cell electrophysiological responses and the DOI-induced HTR, suppressing Glu release with a mGlu$_2$ receptor PAM or an adenosine A$_1$ receptor agonist also resulted in antidepressant-like effects in rats responding under DRL 72-s schedules (Fell et al. 2011; Marek 2012; Nikiforuk et al. 2010). More recently, the blockade of OX$_2$ receptors was reported to result in antidepressant effects on DRL 72-s behavior (Fitch et al. 2014). Interestingly, several decades ago, β_2-adrenergic receptor agonists such as clenbuterol were demonstrated to produce antidepressant-like activity on DRL 72-s behavior in rats (Dunn et al. 1993; O'Donnell 1987, 1988, 1990, 1993; O'Donnell et al. 1994), effects that appear consistent with preliminary clinical observations and also match their profile of effects on rat PFC slice electrophysiology and the HTR induced by DOI. While some receptor systems discussed above have not yet been tested (mGlu$_4$ receptor agonists, dopamine D$_{1/5}$ receptor agonists, and NK$_3$ receptor antagonists), μ-opioid receptor agonists and α_1-adrenergic receptor antagonists do not appear to induce antidepressant-like effects on DRL 72-s behavior.

The lack of antidepressant-like effects on DRL behavior for some agents known to suppress Glu release onto L5 pyramidal cells is not surprising given the likelihood that actions on μ-opioid receptors or α_1-adrenergic receptors elsewhere in prefrontal cortical–striatal–thalamic–amygdaloid circuits may counter their modulatory effects on the thalamic and non thalamic glutamatergic afferents to cortical L5 pyramidal cells. What is remarkable is that so many of the mechanisms that suppress 5-HT$_{2A}$ receptor-induced Glu release onto L5 pyramidal cells possess either preclinical antidepressant-like (e.g., DRL 72-s behavior) or antipsychotic-like behavioral effects (e.g., blockade of DOI-induced HTR and/or NMDA receptor antagonist-induced hyperactivity; Table 3).

Heterocomplexes or functional interactions between 5-HT$_{2A}$ and mGlu$_2$ receptors?

In vitro colocalization demonstrated using BRET and FRET technology and a wealth of both in vitro and in vivo functional interactions has led to the suggestion that a heterocomplex is formed between 5-HT$_{2A}$ and mGlu$_2$ receptors in L5 pyramidal cells in the PFC and neocortex (Gonzalez-Maeso et al. 2008; Moreno et al. 2011; Fribourg et al. 2011). The in vitro localization of apparent complexes between 5-HT$_{2A}$ and mGlu$_2$ receptors in transfected cell lines has been confirmed by others (Delille et al. 2012). However, functional evidence for heterocomplexes between 5-HT$_{2A}$ and mGlu$_2$ receptors could not be confirmed in in vitro experiments (Delille et al. 2012). Thus, the presence of 5-HT$_{2A}$ receptors did not appreciably appear to affect the binding of mGlu$_2$ radioligands and vice versa. Furthermore, it is not clear that heteromeric receptors are necessary for the occurrence of functional interactions such as those that exist between 5-HT$_{2A}$ receptors and mGlu$_2$ receptors. The only other G$_q$-coupled GPCR that may be closely localized with mGlu$_2$ receptors in heteromeric receptor complexes is the mGlu$_5$ receptor (Delille et al. 2012), but such relationships are not known to exist for other receptors that are capable of inducing EPSCs, such as α_1-adrenergic, OX$_2$, or NK_3 receptors (Delille et al. 2013). Similarly, heteromeric relationships have not been identified between $5\text{-}HT_{2A}$ receptors and a range of receptors that can suppress 5-HT-induced EPSCs (Table 1) (Delille et al. 2013). In contrast, Doumazane and colleagues did produce evidence for the existence of heterodimers between $mGlu_2$ and $mGlu_4$ receptors in overexpressing cell lines (Doumazane et al. 2011). The model for heterocomplexes between $5\text{-}HT_{2A}$ and $mGlu_2$ receptors places them both on L5 pyramidal cells and ignores the data suggesting that only a very minor subpopulation of $5\text{-}HT_{2A}$ receptors are expressed presynaptically, as well as the overwhelming evidence for a predominantly presynaptic function of the $mGlu_2$ receptor (Delille et al. 2013). Although it has been suggested that clozapine downregulates both $mGlu_2$ and $5\text{-}HT_{2A}$ receptors (Gonzalez-Maeso et al. 2008), this finding was not observed by others (Wright and Schoepp 2003). Furthermore, dissociations between $mGlu_2$ receptor distribution and $5\text{-}HT_{2A}$ receptor distribution appear to occur following sub-chronic antidepressant treatment in rodents or in depressed patients (Muguruza et al. 2014). Previous evidence for a dissociation between $mGlu_2$ and $5\text{-}HT_{2A}$ receptor distribution was derived from studies of midline thalamic lesions, where a $\sim$40% reduction of the density of presumed presynaptic $mGlu_2$ receptors occurred simultaneously with an increase in the density of postsynaptic $5\text{-}HT_{2A}$ receptors (Marek et al. 2001). In contrast to these previous lines of evidence, the in vitro and in vivo evidence that mutating three residues from the intracellular portion of the $mGlu_2$ receptor suppresses $5\text{-}HT_{2A}$ receptor activity strongly supports the existence of $5\text{-}HT_{2A}$- and $mGlu_2$-containing heterocomplexes (Moreno et al. 2012). Thus, it remains to be completely understood which of the following three hypotheses explains the in vivo interactions between $5\text{-}HT_{2A}$ and $mGlu_2$ ligands: (1) only functional relationships occur between $5\text{-}HT_{2A}$ and $mGlu_2$ receptors; (2) only heterocomplexes containing $5\text{-}HT_{2A}$ and $mGlu_2$ receptors mediate the interactions; or (3) both independently acting $5\text{-}HT_{2A}$ and $mGlu_2$ receptors and $5\text{-}HT_{2A}$–$mGlu_2$ heterocomplexes are involved.

As an aside with regard to receptor heteromers involving $5\text{-}HT_{2A}$ receptors, heterocomplexes between $5\text{-}HT_{2A}$ and cannabinoid CB_1 receptors appear to be involved in Δ^9-tetrahydrocannabinol (THC)-induced amnesia but not the locomotor, hypothermic, anxiogenic, or antinociceptive effects (Vinals et al. 2015). These $5\text{-}HT_{2A}$ and CB_1 heteromers appear to share some similarity to $5\text{-}HT_{2A}$–$mGlu_2$ heterocomplexes since the $5\text{-}HT_{2A}$ receptor is primarily a postsynaptic receptor while $mGlu_2$ and CB_1 receptors are primarily presynaptic receptors.

2 Clinical Relevance of the Interactions Between Glutamate and $5\text{-}HT_{2A}$ Receptors in the mPFC

Based initially on insights regarding the interaction of hallucinogenic drugs with the $5\text{-}HT_{2A}$ receptor, several $5\text{-}HT_{2A}$ receptor antagonists were discovered, characterized, and tested in patients with schizophrenia. Clinical results with drugs lacking selectivity at 5-HT_{2A} versus 5-HT_{2C} receptors (e.g., ritanserin) or with drugs having at least 30-fold selectivity for 5-HT_{2A} versus 5-HT_{2C} receptors (e.g., M100907, SR46349B, pimavanserin, and CYR-101) have been disappointing because none of the drugs exerted antipsychotic effects approaching those of either first- or second-generation antipsychotic drugs when tested as monotherapy for schizophrenia. At best, these drugs exert statistically significant but clinically disappointing intermediate efficacy between haloperidol and placebo (Marder 1999; Meltzer et al. 2004). Even when added to the atypical antipsychotic risperidone, the positive therapeutic effects of the 5-HT_{2A} inverse agonist pimavanserin were limited to those subjects on a very low (2 mg) daily dose of risperidone (Meltzer et al. 2012).

More recent results over the last 7 years with pimavanserin raise some hope that selective 5-HT_{2A} receptor antagonists or inverse agonists may provide clinically relevant therapeutic effects for psychosis in patients with neurodegenerative disease (Hacksell et al. 2014). Pimavanserin is only the second potent 5-HT_{2A} receptors antagonist or inverse agonist that has been shown to improve Parkinson's disease (PD) psychosis (Cummings et al. 2014; Meltzer et al. 2010). The other antipsychotic exerting both demonstrable antipsychotic efficacy and lacking clinically significant extrapyramidal side effects is clozapine (French Clozapine Parkinson Study Group 1999; Parkinson Study Group 1999). In April 2016, the FDA approved pimavanserin for psychosis associated with PD. With respect to the disease itself, multiple studies involving both postmortem data and in vivo PET imaging have reported increased levels of 5-HT_{2A} receptor expression in patients with Parkinson's disease (Ballanger et al. 2010; Chen et al. 1998; Huot et al. 2010; Rasmussen et al. 2016). Changes in brain $mGlu_2$ receptor expression in Parkinson's disease have yet to be examined in relationship to questions of functional interactions with 5-HT_{2A} receptors or the existence of an actual heterocomplex between these two receptors. Nevertheless, the effect of pimavanserin on Parkinson's disease does suggest that relatively selective 5-HT_{2A} blockade can suppress the delusions and hallucinations characteristic of this neurodegenerative disorder.

A phase 2 randomized, placebo-controlled trial addressing the effects of pimavanserin on psychosis associated with Alzheimer's disease (AD) was recently completed and reportedly showed positive results (www.ClinicalTrial.gov identifier NCT02035553). Thus, studies in psychosis associated with either PD or AD have brought to realization predictions made years ago by the late psychiatrist Dr. Daniel X. Freedman that the study of LSD would result in advances for the treatment of major mental illnesses.

The pimavanserin AD study is intriguing given that (1) down-regulation of 5-HT_{2A} receptors occurs in AD before other neurotransmitter receptors are altered (Blin et al. 1993; Cross et al. 1984, 1986; Hasselbalch et al. 2008; Lai et al. 2005; Lorke et al. 2006; Marner et al. 2011; Meltzer et al. 1999; Perry et al. 1984; Reynolds et al. 1984; Santhosh et al. 2009); (2) pharmacogenetic studies have revealed a potential association between certain 5-HT_{2A} receptor gene single nucleotide polymorphisms (SNPs) and psychosis in AD patients (Ramanathan and Glatt 2009); and (3) shared pathophysiology exists between AD and PD such as early degeneration of cholinergic and serotonergic neurons, amyloidosis, and the presence of Lewy bodies (Bohnen and Albin 2011; Halliday et al. 2011; Petrou et al. 2012; Trillo et al. 2013).

The recent studies showing that a 5-HT_{2A} receptor inverse agonist can be used to treat psychosis associated with either PD or AD raises the provocative question of whether $mGlu_{2/3}$ receptor agonists might have a larger effect size in PD or AD psychosis compared with the apparent efficacy of the $mGlu_{2/3}$ agonist LY2140023 (pomaglumetad methionil) in only a modest subset of patients with schizophrenia. Given that clinical lore and experience in methadone maintenance clinics indicates that μ-opioid receptor agonists may possess clinical antipsychotic activity (Marek and Aghajanian 1998), the work with 5-HT_{2A} receptor antagonists in psychosis associated with neurodegenerative diseases suggests that testing compounds acting at novel molecular targets for antipsychotic activity should not be solely limited to patients with schizophrenia.

As described earlier, a number of drugs targeting receptor sites that modulate 5-HT-induced EPSCs in rat mPFC slice preparations also exert antidepressant-like effects in the DRL 72-s schedule. This includes selective 5-HT_{2A} receptor antagonists such as M100907, ketanserin, and pipamperone (Ardayfio et al. 2008; Marek et al. 1989, 2005; Marek and Seiden 1988). 5-HT_{2A} receptor antagonists also enhance the therapeutic effects of SSRIs, tricyclic antidepressants, and monoamine oxidase inhibitors (Ardayfio et al. 2008; Marek et al. 2005). These findings are consistent with the results of double-blind, placebo-controlled studies suggesting antidepressant efficacy for drugs such as trazodone, nefazodone, mirtazapine, and mianserin (Marek et al. 2003), which block 5-HT_{2A} receptors. Some, but not all, studies examining augmentation with mirtazapine or mianserin have demonstrated improved antidepressant efficacy (Carpenter et al. 2002; Ferreri et al. 2001; Maes et al. 1999). Furthermore, these preclinical data are also consistent with a meta-analysis confirming that adding atypical antipsychotic drugs, which block 5-HT_{2A} receptors among a range of pharmacological effects, onto ongoing treatment with SSRIs or SNRIs improves the clinical antidepressant action (Nelson and Papakostas 2009). Previous open-label and controlled studies suggested that the β_2-adrenergic receptor agonist salbutamol might have antidepressant effects in depressed patients (Lecrubier et al. 1980; Simon et al. 1984). A $mGlu_2$ receptor PAM is the only novel mechanism suppressing 5-HT-induced EPSCs that has also been tested in a clinical study. The results of a single trial investigating the effect of adjunctive treatment with a $mGlu_2$ receptor PAM in combination with SSRIs/SNRIs largely appeared negative with a numerical but not a significant advantage for the $mGlu_2$ receptor PAM over placebo (Kent et al. 2016). The primary limitation of this $mGlu_2$ PAM trial in major depressive disorder (MDD), which only showed a weak trend toward antidepressant efficacy, was uncertainty whether optimal doses and/or dosing schedules had been used. A number of open-label trials with buprenorphine, a μ-opioid agonist, and κ-opioid antagonist, in patients with MDD, including several involving patients with treatment-resistant depression, have suggested that this compound has antidepressant activity (Bodkin et al. 1995; Emrich 1982; Karp et al. 2014; Kosten et al. 1990). These effects of buprenorphine are consistent with the fact that opiates had a long history of use in depressed patients prior to the development of other effective treatments (Carlson and Simpson 1963). A selective OX_2 receptor antagonist (Min-202 or JNJ-42847922) is currently being developed by Minerva Neurosciences and Janssen as a treatment for insomnia and MDD. Preliminary reports from phase 1b studies with this OX_2 receptor antagonist in patients with MDD indicate that there were improvements in both sleep and mood, but these results await confirmation in phase 2 proof-of-concept trials. Thus, a number of compounds that block 5-HT_{2A} receptors or suppress the effects of hallucinogens in either in vitro brain slice preparations or in vivo HTR studies in rodents appear to possess at least some potential antidepressant efficacy in patients.

A series of studies with the channel blocking NMDA receptor antagonist ketamine have raised the possibility that a unique pattern of results may have predictive utility for discovering drugs having positive effects in treatment refractory MDD or bipolar depression. In addition to having well-known pro-psychotic effect in subjects with schizophrenia, administration of subanesthetic doses of ketamine produces nearly immediate (90 min) but short-lived (up to approximately 2 weeks) antidepressant effects in subjects with treatment refractory MDD or bipolar depression (Berman et al. 2000; Preskorn et al. 2008; Zarate et al. 2006). When brain slices are prepared from rats treated 24 h earlier with ketamine, an increase in spontaneous EPSCs is observed in recordings from L5 pyramidal cells (Li et al. 2010). More remarkably, a single subanesthetic dose of ketamine (10 mg IP) induces the formation of new synaptic spines in the apical dendritic field of L5 pyramidal cells within 24 h. At the behavioral level, 24 h after ketamine administration antidepressant-like effects were observed in the forced swim test, the learned helplessness paradigm, and the novelty suppressed feeding test. At a mechanistic level, blockade of the mTOR pathway blocked these effects, which had been observed 24 h following an acute dose of ketamine. Deficits in synaptic electrophysiology, spine number, synaptic protein expression, and behavior induced by 21 days of chronic stress exposure are reversed by administration of a single dose of ketamine (Li et al. 2011). These effects of ketamine are also consistent with its antidepressant-like activity in a number of depression models and screens for antidepressant drugs (Table 3). Ketamine has even been found to exert antidepressant-like effects on DRL 72-s behavior (Hillhouse and Porter 2014), although this effect appears to be limited to a period ranging from minutes to 2 h after dosing and is not apparent 24 h after a single acute administration (GJ Marek, unpublished observations).

The evidence with ketamine is further strengthened by recent findings with lithium (Table 3). Lithium potentiates the electophysiological, synaptogenic, and antidepressant-like behavioral effects of ketamine in rats (Liu et al. 2013). These findings are of interest because lithium augmentation has long been a gold-standard treatment for treatment refractory MDD and bipolar disorder (Burgess et al. 2001). The fact that the muscarinic receptor antagonist scopolamine produces electrophysiological, synaptogenic, and antidepressant-like behavioral effects in preclinical rodent studies (Voleti et al. 2013) that parallel its clinical antidepressant effects observed for MDD and bipolar depression, including treatment refractory cases (Drevets and Furey 2010; Ellis et al. 2014; Furey and Drevets 2006;

Khajavi et al. 2012), provides additional support for the use of this paradigm to discover novel antidepressants for treatment refractory patients. From a theoretical standpoint, the effects of rapidly acting putative antidepressants such as ketamine and scopolamine on synaptogenesis may point to a fundamental link between depressive symptoms and the loss of the neuropil observed in MDD patients in postmortem histopathology studies and inferred from the decrease in cortical thickness seen in neuroimaging and postmortem neuropathology studies.

3 Conclusions

After the discovery of LSD's potent psychotropic effects by Albert Hofmann in Basel, Switzerland in April 1943, serotonergic hallucinogens have inspired research goals ranging from curing major neuropsychiatric diseases to understanding consciousness. Understanding the common shared effects of hallucinogens such as LSD, mescaline, and psilocybin has directed modern neuroscience toward the PFC and neocortex, where some of the highest densities of 5-HT_{2A} receptors are found. At the cellular level, a significant literature has described the effects of hallucinogens and 5-HT_{2A} receptor activation on L5 pyramidal cells, which are the principal output cells of the PFC and neocortex. A range of monoaminergic, glutamatergic, purinergic, and peptide neurotransmitter receptor ligands suppress Glu release from putative thalamic afferents and from other sources (potentially including the PFC, neocortex, and claustrum) where the release is induced by activation of 5-HT_{2A} receptors, as inferred from intracellular and whole cell recordings of L5 pyramidal cells in the PFC. Most of these transmitter systems also suppress the HTR induced by serotonergic hallucinogens. Administration of agonists or antagonists for most of these transmitter systems also results in antidepressant-like and/or antipsychotic-like effects for salient behaviors mediated by the PFC. Positive results have been observed for a number, though not all, of these targets when evidence for antidepressant or antipsychotic effects has been evaluated clinically in patients. One important recent lesson in CNS drug development regarding testing for clinical antipsychotic effects is that the recent positive results for pimavanserin in PD psychosis demonstrate that negative or disappointing results in schizophrenia do not always rule out a potential antipsychotic action in patients with neurodegenerative disease. Finally, recent studies with ketamine, lithium, and scopolamine raise the possibility that studying unique electrophysiological, synaptogenic, and behavioral profiles may help to uncover new medications for treatment refractory MDD and bipolar depression. However, there still remain a plethora of outstanding questions at multiple levels of inquiry ranging from receptor–receptor interactions, post-receptor transduction pathways, and both micro- and macro-circuit involvements in a range of behaviors, emotions, thoughts, and in consciousness itself.

References

Aghajanian GK (2009) Modeling "psychosis" in vitro by inducing disordered neuronal network activity in cortical brain slices. Psychopharmacology 206:575–585

Aghajanian GK, Marek GJ (1997) Serotonin induces excitatory postsynaptic potentials in apical dendrites of neocortical pyramidal cells. Neuropharmacology 36(3/4):589–599

Ardayfio PA et al (2008) The 5-hydroxytryptamine2A receptor antagonist R-(+)-a-(2,3-dimethoxyphenyl)-1-[2-(4-fluorophenyl)ethyl-4-piperidinemethanol (M100907) attenuates impulsivity after both drug-induced disruption (dizocilpine) and enhancement (antidepressant drugs) of differential-reinforcement-of-low-rate 72-s behavior in the rat. J Pharmacol Exp Ther 327:891–897

Avesar D, Gulledge AT (2012) Selective serotonergic excitation of callosal projection neurons. Front Neural circuits 6

Ballanger B et al (2010) Serotonin 2A receptors and visual hallucinations in Parkinson disease. Arch Neurol 67:416–421

Barre A et al (2012) Presynaptic serotonin 2A receptors modulate thalamocortical plasticity and associative learning. Proc Natl Acad Sci USA 113:E1382–E1391

Beique J-C et al (2007) Mechanism of the 5-hydroxytryptamine$_{2A}$ receptor-mediated facilitation of synaptic activity in prefrontal cortex. Proc Natl Acad Sci USA 104:9870–9875

Benneyworth MA et al (2007) A selective positive allosteric modulator of metabotropic glutamate receptor subtype 2 blocks a hallucinogenic drug model of psychosis. Mol Pharmacol 72:477–484

Benvenga MJ et al (2006) Attenuation of DOI-induced head twitches in mGluR2 KO mice. FASEB J 20:A246

Berman RM et al (2000) Antidepressant effects of ketamine in depressed patients. Biol Psychiatry 47:351–354

Blin J et al (1993) Loss of brain 5-HT$_2$ receptors in Alzheimer's disease. Brain 116:497–510

Bodkin JA et al (1995) Buprenorphine treatment of refractory depression. J Clin Psychopharmacol 15:49–57

Bohnen NI, Albin RL (2011) The cholinergic system and Parkinson disease. Behav Brain Res 221:564–573

Burgess S et al (2001) Lithium for maintenance treatment of mood disorders. Cochrane Database Syst. Rev., (2):CD003013

Canal CE, Morgan D (2012) Head-twitch response in rodents induced by the hallucinogen 2,5-dimethoxy-4-iodoamphetamine: a comprehensive history, a re-evaluation of mechanisms, and its utility as a model. Drug Test Anal 4:556–576

Carli M et al (2006) Dissociable contribution of 5-HT$_{1A}$A and 5-HT$_{2A}$ receptors in the medial prefrontal cortex to different aspects of executive control such as impulsivity and compulsive perseveration in rats. Neuropsychopharmacology 31:757–767

Carlson ET, Simpson MM (1963) Opium as a tranquilizor. Am J Psychiatry 120:112–117

Carpenter LL, Yasmin S, Price LH (2002) A double-blind placebo-controlled study of antidepressant augmentation with mirtazapine. Biol Psychiatry 51:183–188

Chen, CPL-H et al (1998) Post-synaptic 5-HT$_{1A}$ and 5-HT$_{2A}$ receptors are increased in Parkinson's disease neocortex. Ann NY Acad Sci 861: 288–289

Ciliax BJ et al (2000) Dopamine D5 receptor immunolocalization in rat and monkey brain. Synapse 37:125–145

Cross AJ et al (1984) Serotonin receptor changes in dementia of the Alzheimer type. J Neurochem 43:1574–1581

Cross AJ et al (1986) The selectivity of the reduction of serotonin S2 receptors in Alzheimer-type dementia. Neurobiol Aging 7:3–7

Cummings J et al (2014) Pimavanserin for patients with Parkinson's disease psychosis: a randomized, placebo-controlled phase 3 trial. Lancet 383:533–540

Czyrak A et al (1993) Pharmacological effects of zotipine and other antipsychotics on the central 5-HT_2 receptors. Pharmacopsychiatry 26:53–58
Delille HK et al (2012) Heterocomplex formation of $5\text{-HT}_{(2A)}$-$\text{mGlu}_{(2)}$ and its relevance for cellular signaling cascades. Neuropharmacology 62:2183–2190
Delille HK, Mezler M, Marek GJ (2013) The two faces of the pharmacological interaction of mGlu2 and 5-HT_{2A}—relevance of receptor heterocomplexes and interaction through functional brain pathways. Neuropharmacology 70:296–305
Ding YQ et al (1996) Localization of the neuromedin K receptor (NK3) in the central nervous system of the rat. J Comp Neurol 364:290–310
Doumazane E et al (2011) A new approach to analyze cell surface protein complexes reveals specific heterodimeric metabotropic glutamate receptors. FASEB J. 25:66–77
Drevets WC, Furey ML (2010) Replication of scopolamine's antidepressant efficacy in major depressive disorder: a randomized, placebo-controlled clinical trial. Biol Psychiatry 67:432–438
Dunn RT, Richards JB, Seiden LS (1993) Effects of salbutamol upon performance on an operant screen for antidepressant drugs. Psychopharmacology 113:1–10
Dursun SM, Handley SL (1996) Similarities in the pharmacology of spontaneous and DOI-induced head-shakes suggest 5-HT_{2A} receptors are active under physiological conditions. Psychopharmacology 128:198–205
Egashira N et al (2011) Role of endocannabinoid and glutamatergic systems in DOI-induced head-twitch response in mice. Pharmacol Biochem Behav 99:52–58
Eison AS et al (1990) Nefazodone: preclinical pharmacology of a new antidepressant. Psychopharmacol Bull 26:311–315
Ellis JS et al (2014) Antidepressant treatment history as a predictor of response to scopolamine: clinical implications. J Affect Disord 162:39–42
Emrich HM, Voght P, Herz (1982) Possible antidepressive effects of opioids: action of buprenorphine. Ann NY Acad Sci 398:108–112
Fell MJ et al (2011) N-(4-((2-trifluoromethyl)-3-hydroxy-4-(isobutyryl)phenoxy)methyl)benzyl)-1-methyl-1H-imidazole-4-carboxamide (THIIC), a novel metabotropic glutamate 2 potentiator with potential anxiolytic/antidepressant properties: in vivo profiling suggests a link between behavioral and central nervous system neurochemical changes. J Pharmacol Exp Ther 336:165–177
Ferreri M et al (2001) Benefits from mianserin augmentation of fluoxetine in patients with major depression non-responders to fluoxetine alone. Acta Psychiatr Scand 103:66–72
Fitch TE, Benvenga MJ, Jesudason CD, Zink C, Vandergriff AB, Menezes MM, Schober DA, Rorick-Kehn LM, (2014) LSN2424100: a novel, potent orexin-2 receptor antagonist with selectivity over orexin-1 receptors and activity in an animal model predictive of antidepressant-like efficacy. Front Neurosci 8(Article 5): 1–11
French Clozapine Parkinson Study Group (1999) Clozapine in drug-induced psychosis in Parkinson's disease. Lancet 353: 2041–2042
Fribourg M et al (2011) Decoding the signaling of a GPCR heteromeric complex reveals a unifying mechanism of action of antipsychotic drugs. Cell 147:1011–1023
Friedman E, Cooper TB, Dallob A (1983) Effects of chronic antidepressant treatment on serotonin receptor activity in mice. Eur J Pharmacol 89:69–76
Furey ML, Drevets WC (2006) Antidepressant efficacy of the antimuscarinic drug scopolamine. Arch Gen Psychiatry 63:1121–1129
Gao W-J, Krimer LS, Goldman-Rakic PS (2001) Presynaptic regulation of reccurent excitation by D1 receptors in prefrontal circuits. Proc Natl Acad Sci U S A 98:295–300
Gaynor CM, Handley SL (2001) Effects of nicotine on head-shakes and tryptophan metabolites. Psychopharmacology 153:327–333
Gewirtz JC, Marek GJ (2000) Behavioral evidence for interactions between a hallucinogenic drug and group II metabotropic glutamate receptors. Neuropsychopharmacology 23:569–576
Gonzalez-Maeso J et al (2008) Identification of a serotonin/glutamate receptor complex implicated in psychosis. Nature 452:93–99

Goodwin GM, Green AR, Johnson P (1984) $5\text{-}HT_2$ receptor characteristics in frontal cortex and $5\text{-}HT_2$ receptor-mediated head-twitch behavior following antidepressant treatment to mice. Br J Pharmacol 83:235–242

Gouzoulis-Mayfrank E et al (2005) Psychological effects of (S)-ketamine and N, N-dimethyltryptamine: a double-blind, cross-over study in healthy volunteers. Pharmacopsychiatry 38:301–311

Granon S et al (2000) Enhanced and impaired attentional performance after infusion of D1 dopaminergic receptor agents into rat prefrontal cortex. J Neurosci 20:1208–1215

Hacksell U et al (2014) On the discovery and development of pimavanserin: a novel drug candidate for Parkinson's psychosis. Neurochem Res 39:2008–2017

Halliday G, Lees A, Stern M (2011) Milestones in Parkinson's disease—Clinical and pathologic features. Mov Disord 26:1015–1021

Handley SL, Singh L (1986) Neurotransmitters and shaking behaviour—more than a 'gut-bath' for the brain. Trends Pharmacol Sci 7: 324–328

Hasselbalch SG et al (2008) Reduced $5\text{-}HT_{2A}$ receptor binding in patients with mild cognitive impairment. Neurobiol Aging 29:1380–1388

Hayslett RL, Tizabi Y (2005) Effects of donepezil, nicotine and haloperidol on the central serotonergic system in mice: Implications for Tourette's syndrome. Pharmacol Biochem Behav 81:879–886

Higgins GA et al (2003) The $5\text{-}HT_{2A}$ receptor antagonist M100,907 attenuates motor and 'impulsive-type' behaviors produced by NMDA receptor antagonism. Psychopharmacology 170:309–319

Hillhouse TM, Porter JH (2014) Ketamine, but not MK-801, produces antidepressant-like effects in rats responding on a differential-reinforcement-of-low-rate operant schedule. Behav Pharmacol 25:80–91

Huang Q et al (1992) Immunohistochemical localization of the D1 dopamine receptor in rat brain reveals its axonal transport, pre- and postsynaptic localization, and prevalence in the basal ganglia, limbic system, and thalamic reticular nucleus. Proc Natl Acad Sci U S A 89:11988–11992

Huot P et al (2010) Increased $5\text{-}HT_{2A}$ receptors in the temporal cortex of parkinsonian patients with visual hallucinations. Mov Disord 25:1399–1408

Jakab RL, Goldman-Rakic PS (1998) $5\text{-}HT_{2A}$ serotonin receptors in the primate cerebral cortex: possible site of action of hallucinogens in pyramidal cell apical dendrites. Proc Natl Acad Sci USA 95:735–740

Karp JF et al (2014) Safety, tolerability, and clinical effect of low-dose buprenorphine for treatment-resistant depression in midlife and older adults. J Clin Psychiatry 75:e785–e793

Kent JM et al (2016) Efficacy and safety of an adjunctive mGlu2 receptor positive allosteric modulator to a SSRI/SNRI in anxious depression. Prog Neuro-Psychopharmacol Biol Psychiatry 67:66–73

Khajavi D et al (2012) Oral scopolamine augmentation in moderate to severe major depressive disorder: A randomized, double-blind, placebo-controlled study. J Clin Psychiatry 73:1428–1433

Klodzinska A, et al (2002) Group II mGlu receptor agonists inhibit behavioral and electrophysiological effects of DOI in mice. Pharmacol. Biochem. Behav 73: 327–332

Kolaj M et al (2014) Intrinsic properties and neuropharmacology of midline paraventricular thalamic nucleus neurons. Front Behav Neurosci 8:132

Koskinen T, Sirvio J (2001) Studies on the involvement of the dopaminergic system in the $5\text{-}HT_2$ agonist (DOI)-induced premature responding in a five-choice serial reaction time task. Br Res Bull 54:65–75

Koskinen T, Haapalinna A, Sirvio J (2003) Alpha-adrenoceptor-mediated modulation of $5\text{-}HT_2$ receptor agonist induced impulsive responding in a 5-choice serial reaction time task. Pharmacol Toxicol 92:214–225

Kosten TR, Morgan C, Kosten TA (1990) Depressive symptoms during buprenorphine treatment of opioid abusers. J Substance Abuse Treatment 7:51–54

Lai MK et al (2005) Loss of serotonin 5-HT$_{2A}$ receptors in the postmortem temporal cortex correlates with rate of cognitive decline in Alzheimer's disease. Psychopharmacology 179:673–677

Lambe EK, Aghajanian GK (2001) The role of Kv1.2-containing potassium channels in serotonin-induced glutamate release from thalamocortical terminals in rat frontal cortex. J Neurosci 21: 9955–9963

Lambe EK, Aghajanian GK (2003) Hypocretin (orexin) induces calcium transients in single spines postsynaptic to identified thalamocortical boutons in prefrontal slice. Neuron 40:139–150

Lambe EK, Aghajanian GK (2007) Prefrontal cortical network activity: opposite effects of psychedelic hallucinogens and D1/D5 dopamine receptor activation. Neuroscience 145:900–910

Lambe EK, Picciotto MR, Aghajanian GK (2003) Nicotine induces glutamate release from thalamocortical terminals in prefrontal cortex. Neuropsychopharmacology 28:216–225

Lambe EK, Olausson P, Horst NK, Taylor JR, Aghajanian GK (2005) Hypocretin and nicotine excite the same thalamocortical synapses in prefrontal cortex: correlation with improved attention in rat. J Neurosci 25:5225–5229

Lambe EK, Liu RJ, Aghajanian GK (2007) Schizophrenia, hypocretin (orexin), and the thalamocortical activating system. Schizophr Bull 33:1284–1290

Langlois X et al (2001) Detailed distribution of neurokinin 3 receptors in the rat, guinea pig and gerbil brain: a comparative autoradiographic study. Neuropharmacology 40:242–253

Lebrand C et al (1996) Transient uptake and storage of serotonin in developing thalamic neurons. Neuron 17:991–1003

Lecrubier Y et al (1980) A beta adrenergic stimulant (salbutamol) versus clomipramine in depression: a controlled study. Br J Psychiatry 136:354–358

Li N et al (2010) mTOR-dependent synapse formation underlies the rapid antidepressant effects of NMDA antagonists. Science 329:959–964

Li N et al (2011) Glutamate N-methyl-D-aspartate receptor antagonists rapidly reverse behavioral and synaptic deficits caused by chronic stress exposure. Biol Psychiatry 69:754–761

Liu RJ, Fuchikami M, Dwyer JM, Lepack AE, Duman RS, Aghajanian GK (2013) GSK-3 inhibition potentiates the synaptogenic and antidepressant-like effects of subthreshold doses of ketamine. Neuropsychopharmacology 38:2268–2277

Lorke DE et al (2006) Serotonin 5-HT$_{2A}$ and 5-HT$_6$ receptors in the prefrontal cortex of Alzheimer and normal aging patients. BMC Neurosci 7:36

Maes M et al (1999) Pindolol and mianserin augment the antidepressant activity of fluoxetine in hospitalized major depressed patients, including those with treatment resistance. J Clin Psychopharmacol 19:177–182

Mansour A et al (1992) A comparison of D$_1$ receptor binding and mRNA in rat brain using receptor autoradiographic and in situ hybridization techniques. Neuroscience 46:959–971

Marder SR (1999) Limitations of dopamine-D2 antagonists and the search for novel antipsychotic strategies. Neuropsychopharmacology 21(S6):S117–S121

Marek GJ (2003) Behavioral evidence for μ-opioid and 5-HT$_{2A}$ receptor interactions. Eur J Pharmacol 474:77–83

Marek GJ (2009) Activation of adenosine$_1$ (A$_1$) receptors suppresses head shakes induced by a serotonergic hallucinogen in rats. Neuropharmacology 56:1082–1087

Marek GJ (2012) Activation of adenosine$_1$ receptors induces antidepressant-like, anti-impulsive effects on differential reinforcement of low-rate 72-s behavior in rats. J Pharmacol Exp Ther 341:564–570

Marek GJ, Aghajanian GK (1998a) 5-HT-induced EPSCs in neocortical layer V pyramidal cells: suppression by μ-opiate receptor activation. Neuroscience 86:485–497

Marek GJ, Aghajanian GK (1998b) The electrophysiology of prefrontal 5-HT systems: therapeutic implications for mood and psychosis. Biol Psychiat 44:1118–1127

Marek GJ, Aghajanian GK (1999) 5-HT$_{2A}$ or α_1-adrenoceptor activation induces excitatory postsynaptic currents in layer V pyramidal cells of the medial prefrontal cortex. Eur J Pharmacol 367:197–206

Marek GJ, Seiden LS (1988) Effects of selective 5-hydroxytryptamine-2 and nonselective 5-hydroxytryptamine antagonists on the differential-reinforcement-of-low-rate 72-second schedule. J Pharmacol Exp Ther 244(2):650–658

Marek GJ, Zhang C (2008) Activation of metabotropic glutamate 5 (mGlu5) receptors induces spontaneous excitatory synaptic currents in layer V pyramidal cells of the rat prefrontal cortex. Neurosci Lett 442(3):239–243

Marek GJ, Li AA, Seiden LS (1989) Selective 5-hydroxytryptamine2 antagonists have antidepressant-like effects on differential-reinforcement-of-low-rate 72-second schedule. J Pharmacol Exp Ther 250(1):52–59

Marek GJ et al (2000) Physiological antagonism between 5-hydroxytryptamine$_{2A}$ and group II metabotropic glutamate receptors in prefrontal cortex. J Pharmacol Exp Ther 292:76–87

Marek GJ et al (2001) A major role for thalamocortical afferents in serotonergic hallucinogen receptor function in the rat neocortex. Neuroscience 105:379–392

Marek GJ et al (2003) Synergistic action of 5-HT$_{2A}$ antagonists and selective serotonin reuptake inhibitors in neuropsychiatric disorders. Neuropsychopharmacology 28:402–412

Marek GJ et al (2005) The selective 5-HT$_{2A}$ receptor antagonist M100907 enhances antidepressant-like behavioral effects of the SSRI fluoxetine. Neuropsychopharmacology 30:2205–2215

Marek GJ, Day M, Hudzik TJ (2016) The utility of impulsive bias and altered decision making as predictors of drug efficacy and target selection: rethinking behavioral screening for antidepressant drugs. J Pharmacol Exp Ther 356:534–548

Marner L et al (2011) The reduction of baseline serotonin 2A receptors in mild cognitive impairment is stable at two-year follow-up. J Alzheimers Dis 23:453–459

Meltzer CC et al (1999) PET imaging of serotonin type 2A receptors in late-life neuropsychiatric disorders. Am J Psychiatry 156:1871–1878

Meltzer HY et al (2004) Placebo-controlled evaluation of four novel compounds for the treatment of schizophrenia and schizoaffective disorders. Am J Psychiatry 161:975–984

Meltzer HY et al (2010) Pimavanserin, a serotonin(2A) receptor inverse agonist, for the treatment of parkinson's disease psychosis. Neuropsychopharmacology 35:881–892

Meltzer HY et al (2012) Pimavanserin, a selective serotonin (5-HT)$_{2A}$-inverse agonist, enhances the efficacy and safety of risperidone, 2 mg/day, but does not enhance efficacy of haloperidol, 2 mg/day: comparison with reference dose risperidone, 6 mg/day. Schizophr Res 141:144–152

Miner LAH et al (2003) Ultrastructural localization of serotonin2A receptors in the middle layers of the rat prelimbic prefrontal cortex. Neuroscience 116:107–117

Mitrano DA et al (2012) Alpha-1 adrenergic receptors are localized on presynaptic elements in the nucleus accumbens and regulate mesolimbic dopamine transmission. Neuropsychopharmacology 37:2161–2172

Moore NA et al (1992) The pharmacology of olanzapine, a novel "atypical" antipsychotic agent. J Pharmacol Exp Ther 262:545–551

Moreno JL et al (2011) Metabotropic glutamate mGlu2 receptor is necessary for the pharmacological and behavioral effects induced by hallucinogenic 5-HT$_{2A}$ receptor agonists. Neurosci Lett 493:76–79

Moreno JL et al (2012) Identification of three residues essential for 5-hydroxtryptamine 2A-metabotropic glutamate 2 (5-HT$_{2A}$-mGlu$_2$) receptor heteromerizaiton and its psychoactive behavioral function. J Biol Chem 287:44301–44319

Muguruza C et al (2014) Evaluation of 5-HT$_{2A}$ and mGlu$_{2/3}$ receptors in postmortem prefrontal cortex of subjects with major depressive disorder: effect of antidepressant treatment. Neuropharmacology 86:311–318

Nacca A et al (1998) Brain-to-blood partition and in vivo inhibition of 5-hydroxytryptamine reuptake and quipazine-mediated behaviour of nefazodone and its main active metabolites in rodents. Br J Pharmacol 1998:1617–1623

Narboux-Neme N et al (2008) Serotonin transporter transgenic (SERT^{cre}) mouse line reveals developmental targets of serotonin specific reuptake inhbitors (SSRIs). Neuropharmacology 55:994–1005

Nelson JC, Papakostas GI (2009) Atypical antipsychotic augmentation in major depressive disorders: A meta-analysis of placebo-controlled randomized trials. Am J Psychiatry 166:980–991

Neto FL et al (2000) Differential distribution of metabotropic glutamate receptor subtype mRNAs in the thalamus of the rat. Brain Res 854:93–105

Nicholas AP, Pieribone VA, Hokfelt T (1993) Cellular localization of messenger RNA for beta-1 and beta-2 adrenergic receptors in rat brain: an in situ hybridization study. Neuroscience 56:1023–1039

Nikiforuk A, Popik P, Drescher KU, van Gaalen M, Relo A-L, Mezler M, Marek G, Schoemaker H, Gross G, Bespalov A (2010) Effects of a positive allosteric modulatory of group II metabotropic glutamate receptors, LY487379, on cognitive flexibility and impulsive-like responding in rats. J Pharmacol Exp Ther 335:665–673

O'Donnell JM (1990) Behavioral effects of beta adrenergic agonists and antidepressant drugs after down-regulation of beta-2 adrenergic receptors by clenbuterol. J Pharmacol Exp Ther 254 (1):147–157

O'Donnell JM (1993) Effects of the beta-2 adrenergic agonist zinterol on DRL behavior and locomotor activity. Psychopharmacology 113:89–94

O'Donnell JM (1987) Effects of clenbuterol and prenalterol on performance during differential reinforcement of low response rate in the rat. J Pharmacol Exp Ther 241:68–75

O'Donnell JM (1988) Behavioral consequences of activation of beta adrenergic receptors by clenbuterol: evidence for mediation by the central nervous system. Br Res 21(3):491–497

O'Donnell JM, Frith S, Wilkens J (1994) Involvement of beta-1 and beta-2 adrenergic receptors in the antidepressant-like effects of centrally administered isoproterenol. J Pharmacol Exp Ther 271:246–254

O'Donnell JM, Marek GJ, Seiden LS (2005) Antidepressant effects assessed using behavior maintained under a differential-reinforcement-of-low-rate (DRL) operant schedule. Neurosci Biobehav Rev 29:785–798

Palucha-Poniewierg A et al (2008) Peripheral administration of group III mGlu receptor agonist ACPT-I exerts potential antipsychotic effects in rodents. Neuropharmacology 55:517–524

Parkinson Study Group (1999) Low-dose clozapine for the treatment of drug-induced psychosis in Parkinson's disease. N Eng J Med 340:757–763

Passetti F, Dalley JW, Robbins TW (2003) Double dissociation of serotonergic and dopaminergic mechanisms on attentional performance using rodenet five-choice reaction time test. Psychopharmacology 2003:136–145

Peroutka SJ, Lebovitz RM, Snyder SH (1981) Two distinct central serotonin receptors with different physiological function. Science 212:827–829

Perry EK et al (1984) Cortical serotonin-S2 receptor binding abnormalities in patients with Alzheimer's disease: comparisons with Parkinson's disease. Neurosci Lett 51:353–357

Petrou M, Kotagel V, Bohnen NI (2012) An update on brain imaging in Parkinson's dementia. Imaging Med 4:201–213

Preskorn SH et al (2008) An innovative design to establish proof of concept of the antidepressant effects of the NR2B subunit selective N-methyl-D-aspartate antagonist, CP-101,606, in patients with treatment-refractory major depressive disorder. J Clin Psychopharmacol 28:631–637

Rainbow TC, Parsons B, Wolfe BB (1984) Quantitative autoradiography of β1- and β2-adrenergic receptors in rat brain. Proc Natl Acad Sci USA 81:1585–1589

Ramanathan S, Glatt SJ (2009) Serotonergic genes in psychosis of Alzheimer dementia: meta-analysis. Am J Geriatr Psychiatry 17:839–846

Rasmussen NB et al (2016) 5-HT_{2A} receptor binding in the frontal cortex of Parkinson's disease patients and alpha-synuclein overexpressing mice: a post-mortem study. Parkinson's Disease 2016

Rekling JC (2004) NK-3 receptor activation depolarizes and induces an after-depolarization in pyramidal neurons in gerbil cingulate cortex. Br Res Bull 63:85–90

Reynolds GP et al (1984) Reduced binding of [3H]ketanserin to cortical 5-HT_2 receptors in senile dementia of the Alzheimer type. Neurosci Lett 44:47–51

Rigby M, O'Donnell R, Rupniak NM (2005) Species differences in tachykinin receptor distribution: further evidence that the substance P (NK1) receptor predominates in human brain. J Comp Neurol 490:335–353

Rojas-Corrales, MO, Gilbert-Rahola J, Mico JA (2007) Role of atypical opiates in OCD. Experimental approach through the study of 5-HT$_{(2A/C)}$ receptor-mediated behavior. Psychopharmacology 190: 221–231

Rojoz Z (2012) Effect of co-treatment with mirtazapine and risperidone in animal models of the positive symptoms of schizophrrenia in mice. Pharmacol Rep 64:1567–1572

Romano C et al (1995) Distribution of metabotropic glutamate receptor mGluR5 immunoreactivity in rat brain. J. Comp. Neurol. 355:455–469

Rorick-Kehn LM et al (2007) Pharmacological and pharmacokinetic properties of a structurally novel, potent, and selective metabotropic glutamate 2/3 receptor agonist: In vitro characterization of agonist (-)-(1R,4S,5S,6S)-4-Amino-2-sulfonylbicyclo[3.1.0]-hexane-4,6-dicarboxylic acid (LY404039). J Pharmacol Exp Ther 321:308–317

Rotaru DC, Lewis DA, Gonzalez-Burgos G (2007) Dopamine D1 receptor activation regulates sodium-dependent EPSP amplication in rat prefrontal cortex pyramidal neurons. J Physiol 581 (3):981–1000

Saffroy M et al (2003) Autoradiographic distribution of tachykinin NK$_2$ binding sites in the rat brain: comparison with NK$_1$ and NK$_3$ binding sites. Neuroscience 116:761–773

Sanchez C, Arnt J (2000) In-vivo assessment of 5-HT$_{2A}$ and 5-HT$_{2C}$ antagonistic properties of newer antipsychotics. Behav Pharmacol 11:291–298

Santana N, Mengod G, Artigas F (2012) Expression of α_1-adrenergic receptors in rat prefrontal cortex: cellular co-localization with 5-HT$_{2A}$ receptors. Int J Neuropsychopharmacol 16:1139–1151

Santhosh L et al (2009) Regional distribution and behavioral correlates of 5-HT$_{(2A)}$ receptors in Alzheimer's disease with [(18)F]deuteroaltanserin and PET. Psychiatry Res 173:212–217

Schreiber R et al (1995) (1-(2,5-Dimethoxy-4 iodophenyl)-2-aminopropane)-induced head-twitches in the rat are mediated by 5-hydroxytryptamine(5-HT)$_{2A}$ receptors: Modulation by novel 5-HT$_{2A/2C}$ antagonists, D1 antagonists and 5-HT$_{1A}$ agonists. J Pharmacol Exp Ther 273:101–112

Shaw E, Woolley DW (1956) Some serotoninlike activities of lysergic acid diethylamide. Science 124:121–122

Shughrue PJ, Lane MV, Merchenthaler I (1996) In situ hybridization analysis of the distribution of neurokinin-3 mRNA in the rat central nervous system. J Comp Neurol 372:395–414

Simon P et al (1984) Beta-Receptor Stimulation in the Treatment of Depression. In: Usdin E (ed) Frontiers in Biochemical and Pharmacological Research in Depression, Raven PRess, New York, p 293

Simonyi A et al (2005) Expression of groups I and II metabotropic glutamate receptors in the rat brain during aging. Brain Res 1043(1–2):95–106

Slawinska A et al (2013) The antipsychotic-like effects of positive allosteric modulators of metabotropic glutamate mGlu4 receptors in rodents. Br J Pharmacol 169:1824–1839

Stutzman GE, Marek GJ, Aghajanian GK (2001) Adenosine preferentially suppresses serotonin2A receptor-enhanced excitatory postsynaptic currents in layer V neurons of the rat medial prefrontal cortex. Neuroscience 105:55–69

Talpos JC, Wilkinson LS, Robbins TW (2006) A comparison of multiple 5-HT receptors in two tasks measuring impulsivity. J. Psychopharmacol. 20:47–58

Tizabi Y et al (2001) Nicotine attenuates DOI-induced head-twitch response in mice: Implications for Tourette Syndrome. Prog. Neuro-Psychopharmacol. & Biol. Psychiat. 25:1445–1457

Trillo L et al (2013) Ascending monoaminergic systems alterations in Alzheimer's disease. Translating basic science into clinical care. Neurosci Biobehav Rev 37:1363–1379

Vinals X et al (2015) Cognitive impairment induced by delta9-tetrahydrocannabinol occurs through heteromers between cannobinoid CB1 and serotonin 5-HT$_{2A}$ receptors. PLoS Biol 13: e1002194

Voleti B, Navarria A, Liu RJ, Banasr M, Li M, Terwilliger R, Sanacora G, Eid T, Aghajanian G, Duman RS (2013) Scopolamine rapidly increases mammalian target of rapamycin complex 1 signaling, synaptogenesis, and antidepressant behavioral responses. Biol Psychiatry 74:742–749

Vollenweider FX et al (1998) Psilocybin induces schizophrenia-like psychosis in humans via a serotonin-2 agonist action. NeuroReport 9:3897–3902

Weisstaub NV et al (2006) Cortical 5-HT_{2A} receptor signalling modulates anxiety-like behaviors in mice. Science 313:536–540

Wettstein JG, Host M, Hitchcock JM (1999) Selectivity of action of typical and atypical anti-psychotic drugs as antagonists of the behavioral effects of 1-[2,5-dimethoxy-4-iodophenyl]-2-aminopropane (DOI). Prog Neuropsychopharmacol Biol Psychiatry 23:533–544

Wieronska JM et al (2012) Opposing efficacy of group III mGlu receptor activators, LSP1-2111 and AMN082, in animal models of positive symptoms of schizophrenia. Psychopharmacology 220:481–494

Wieronska JM et al (2013) The reversal of cognitive, but not negative or positive symptoms of schizophrenia by the $mGlu_{2/3}$ receptor agonist, LY379268, is 5-HT_{1A} dependent. Behav Brain Res 256:298–304

Willins DL, Meltzer HY (1997) Direct injection of 5-HT_{2A}receptor agonists into the medial prefrontal cortex produces a head-twitch response in rats. J Pharmacol Exp Ther 282:699–706

Wischhof L, Koch M (2012) Pretreatment with the mGlu2/3 receptor agonist LY379268 attenuates DOI-induced impulsive responding and regional c-Fos protein expression. Psychopharmacology 219:387–400

Wischhof, L., K.J. Hollensteiner, and M. Koch, *Impulsive behavior in rats induced by intracortical DOI infusions is antagonized by co-administration of an mGlu2/3 receptor agonist.* Behav. Pharmacol., 2011. **22**: p. 805–813

Wright RA, Schoepp DD (2003) Effect of chronic exposure to mGlu2/3 receptor ligands (LY354740, LY379268 or LY341495) or antipsychotic agents clozapine or haloperidol on binding to mGlu2/3 and 5-HT_{2A} receptors in rat prelimbic cortex. Neuroscience Meeting Planner, Society for Neuroscience

Wright RA et al (2013) CNS distribution of metabotropic glutamate 2 and 3 receptors: transgenic mice and [^{3}H]LY459477 autoradiography. Neuropharmacology 66:89–98

Zarate CA et al (2006) A randomized trial of an N-methyl-D-aspartate antagonist in treatment-resistant major depression. Arch Gen Psychiatry 63:856–864

Zhang C, Marek GJ (2007) Group III metabotropic glutamate receptor agonists selectively suppress excitatory synaptic currents in the rat prefrontal cortex induced by 5-hydroxytrypt-$amine_{2A}$ receptor stimulation. J Pharmacol Exp Ther 320:437–447

Zhang C, Marek GJ (2008a) AMPA receptor involvement in 5-hydroxytryptamine2A receptor-mediated pre-frontal cortical excitatory synaptic currents and DOI-induced head shakes. Prog Neuro-Psychopharmacol Biol Psychiatry 32:62–71

Zhang C, Marek GJ (2008b) AMPA receptors involvement in 5-$hydroxytryptamine_{2A}$ receptor-mediated prefrontal cortical excitatory synaptic currents and DOI-induced head shakes. Prog. Neuropsychopharmacol. & Biol. Psychiatry 32:62–71

Zhang L, Renaud LP, Kolaj M (2009) Properties of T-Type Ca^{2+} channel-activated slow afterhyperpolarization in thalamic paraventricular nucleus and other thalamic midline neurons. J Neurophysiol 101:2741–2750

The Effects of Hallucinogens on Gene Expression

David A. Martin and Charles D. Nichols

Abstract The classic serotonergic hallucinogens, or psychedelics, have the ability to profoundly alter perception and behavior. These can include visual distortions, hallucinations, detachment from reality, and mystical experiences. Some psychedelics, like LSD, are able to produce these effects with remarkably low doses of drug. Others, like psilocybin, have recently been demonstrated to have significant clinical efficacy in the treatment of depression, anxiety, and addiction that persist for at least several months after only a single therapeutic session. How does this occur? Much work has recently been published from imaging studies showing that psychedelics alter brain network connectivity. They facilitate a disintegration of the default mode network, producing a hyperconnectivity between brain regions that allow centers that do not normally communicate with each other to do so. The immediate and acute effects on both behaviors and network connectivity are likely mediated by effector pathways downstream of serotonin 5-HT2A receptor activation. These acute molecular processes also influence gene expression changes, which likely influence synaptic plasticity and facilitate more long-term changes in brain neurochemistry ultimately underlying the therapeutic efficacy of a single administration to achieve long-lasting effects. In this review, we summarize what is currently known about the molecular genetic responses to psychedelics within the brain and discuss how gene expression changes may contribute to altered cellular physiology and behaviors.

Keywords Psychedelics · 5-HT2A · IEG · Gene expression

D.A. Martin · C.D. Nichols (✉)
Department of Pharmacology and Experimental Therapeutics, LSU Health Sciences Center, New Orleans, LA 70112, USA
e-mail: cnich1@lsuhsc.edu

Curr Topics Behav Neurosci (2017) 36: 137–158
DOI 10.1007/7854_2017_479
Published Online: 05 July 2017

Contents

1 Introduction

The immediate behavioral effects that result from the administration of classic hallucinogens, or psychedelics, occur through activation of 5-HT_{2A} receptors and subsequent changes in synaptic transmission and action potential firing. Alterations in subjective experience as well as changes in neuronal activity are seen in a matter of seconds following the intravenous injection of psychedelics (Strassman et al. 1994; Carhart-Harris et al. 2011; Riga et al. 2014). Psychedelics and many other psychotropics also initiate neuronal signaling that occurs over longer intervals. This signaling includes alterations in patterns of mRNA expression in cells activated by the drug, along with corresponding alterations in protein translation. Lysergic acid diethylamide (LSD) and 2,5-dimethoxy-4-iodoamphetamine (DOI) are the most highly studied psychedelics with respect to transcriptional activation, and produce overlapping genetic changes in many parts of the brain, including cortical areas that are known to be important in mediating their effects.

Gene transcription within activated neurons is likely important for the long-term clinical phenomena that are observed following psychedelic administration. For example, the psychological changes that can last months after a single dose of LSD or psilocybin implicate processes that endure over long timescales, as does the existence of hallucinogen-persisting perceptual disorder (HPPD), which can persist for years following a single dose of LSD (Halpern and Pope 2003; Griffiths et al. 2006). Although the mechanisms by which these varied long-term phenomena occur remain obscure, they likely involve changes in long-term synaptic plasticity.

Synaptic events such as late long-term potentiation (late-LTP) require the transcription and translation of a number of genes and proteins, and are thought to be essential in forming memories and mediating learning (Frey et al. 1989; Alberini 2009). Therefore, an understanding of the transcriptional program initiated by psychedelics is necessary to gain insight into the potential long-term clinical benefits and risks of these drugs. Furthermore, examination of the patterns of mRNA transcription produced by psychedelic compounds can provide information about the signaling pathways these drugs modify acutely, and may offer insight into the mechanisms by which they produce sensory and cognitive changes. For example, the first microarray screen to examine LSD-induced gene expression changes in the prefrontal cortex identified the *ania3* transcript as induced (Nichols and Sanders-Bush 2002). This transcript is from the *homer* locus and encodes for a protein associated with post-synaptic metabotropic glutamate receptor signaling (Kammermeier 2008). Work by others at around the same time demonstrated functional interactions between the 5-HT_{2A} receptor and the mGluR2 receptor, and that metabotropic glutamate receptor signaling is involved in the behaviors produced by psychedelics (Marek et al. 2000).

2 Immediate Early Genes

By far, the most well-studied gene expression changes observed following psychedelics administration involve induction of a variety of immediate early genes (IEGs), whose transcription is begun within minutes following neuronal stimulation. IEGs are expressed as a result of a variety of signals that converge on the nucleus and effect transcriptional activators, allowing them to drive subsequent IEG expression. For example, elevations in Ca^{2+} signaling resulting from synaptic activity and membrane depolarization lead to phosphorylation of cAMP response element binding protein (CREB), which binds to specific upstream activating sequences of DNA to induce transcription of certain genes, including *c-Fos*, the prototypical member of a large family of IEGs (Sheng and Greenberg 1990). Fos proteins and other IEGs function primarily as short-lived transcription factors, which themselves initiate a complicated program of further transcription of late-response genes. The specific pattern of gene expression that occurs depends on many factors such as the stimulus, cell type, and second-messenger systems activated. Importantly, the activity-dependent gene expression that begins in the neuronal nucleus ultimately provides a mechanism by which long-term structural and connective changes can occur at the synapse (Kandel 2001; Cohen and Greenberg 2008).

Psychedelics were first shown to induce IEGs in rats through observation of c-Fos protein using immunohistochemistry following the administration of DOI, an agonist that is selective for 5-HT_2 receptors. Leslie et al. (1993) demonstrated clear c-Fos-labeling in the amygdala, mamillary nucleus, bed nucleus of the stria terminalis, nucleus accumbens, and cortex following 8 mg/kg DOI. Cortical c-Fos staining was concentrated in the cingulate cortex and in a laminar banding pattern along layer Va in the somatosensory cortex. c-Fos staining appeared nearly exclusively in a subset of neurons labeled with neuron-specific enolase, and never in GFAP-labled astrocytes. DOI-induced c-Fos-labeling was first detected 30 min after treatment and peaked at 3 h, subsequently declining to background levels by 6 h. Further, DOI-induced c-Fos immunostaining was largely eliminated by pretreatment with the 5-HT_{2A} receptor antagonist ritanserin (Leslie et al. 1993). These researchers later described a dose–response relationship concerning c-Fos expression using escalating doses of DOI ranging from 1 to 32 mg/kg. The density of c-Fos immunoreactivity in the parietal cortex increased over background at 2 mg/kg and reached a plateau at 12 mg/kg (Moorman and Leslie 1998). The authors also noted a high correlation between levels of c-Fos staining and 5-HT$_{2A}$ receptor–ligand binding sites, particularly in layer Va of the cortex (Mengod et al. 1990; Moorman and Leslie 1998). Those early studies clearly indicated that DOI produces a translational signature in neurons that progresses in a dose- and time-dependent manner. They also suggest that this induction follows from 5-HT$_{2A}$ activation in several cortical and sub-cortical regions.

An increase in mRNA levels of IEGs following DOI administration was first reported in the cortex, hippocampus, and cerebellum of rats using northern blot analysis (Tilakaratne and Friedman 1996). The transcription factor genes *c-Fos*, *ngf1c (egr4)*, and *tis1* (*nr4a1*), were increased significantly (219–327%) in the cortex 90 min following 4 mg/kg i.p. DOI. Similar though somewhat smaller responses were observed in the hippocampus and cerebellum in the case of *c-Fos* and *ngf1c*, but large increases in *tis1* expression were restricted to the cortex. All changes were blocked by the 5-HT$_2$ receptor antagonist, ketanserin (Tilakaratne and Friedman 1996). These data confirmed that, in addition to *c-Fos*, other transcription factors (*tis1* and *ngf1c*) are induced by DOI in the cortex, likely through modulation of 5-HT$_2$ receptors. Additionally, this work demonstrated that patterns of DOI-induced gene expression in the brain are regionally specific.

Other types of IEGs that are distinct in function from transcription factors are also induced by psychedelics. Brain-derived neurotrophic factor (BDNF) was the first of these to be observed increasing in response to DOI. BDNF is a widely expressed neurotrophin involved in neuron development, experience-dependent plasticity, and modification of dendritic morphology (Egan et al. 2003; Genoud et al. 2004; Chen et al. 2006). BDNF mRNA was shown with in situ hybridization to increase dose-dependently in the parietal cortex following DOI administration of 0.5 and 2 mg/kg. Interestingly, these same doses caused a decrease below baseline for BDNF expression in the dentate gyrus of the hippocampus. Modulation of BDNF expression by DOI in all regions was abolished using the 5-HT$_2$ receptor antagonist ketanserin and the selective 5-HT$_{2A}$ receptor antagonist MDL100907, indicating that activation of 5-HT$_{2A}$ receptors leads to the increase and decrease of BDNF expression in the cortex and dentate gyrus, respectively (Vaidya et al. 1997). The upregulation of BDNF provides one potential mechanism for psychedelics to modify synaptic strength and connectivity, especially considering the role of BDNF in potentiation of active synapses and ketamine-mediated synaptogenesis (Patterson et al. 1996; Liu et al. 2012).

Further evidence for the widespread initiation of transcriptional activity was supplied when DOI was found to induce *arc* mRNA expression in rats in a dose-dependent fashion (Pei et al. 2000). *Arc* (activity-regulated, cytoskeletal protein) is an effector IEG that displays an affinity for neuronal dendrites, is shuttled to and transcribed at active synapses, and modulates the development of late-LTP and long-term memory (Steward et al. 1998; Guzowski et al. 2000). The expression of *arc* mRNA, examined using in situ hybridization histochemistry, was found to increase above background after administration of 0.2 mg/kg DOI, with a further increase in magnitude through 2.0 mg/kg in several regions of the cortex, including parietal, frontal, cingulate, and orbital cortex. A slight increase in *arc* mRNA was also noted in the striatum at the highest dose. Consistent with reports for other IEGs, the 5-HT$_2$ receptor antagonist ketanserin completely prevented the increase in *arc* mRNA (Pei et al. 2000).

Arc protein was later shown with immunohistochemistry to follow a pattern of DOI induction similar to the mRNA. At a dose of 1 mg/kg DOI, Arc immunoreactivity was increased robustly throughout the frontal cortex, while slight increases were noted in the caudate putamen. The authors also noted in double-labeling studies that Arc and c-Fos staining were largely overlapping, indicating the same cells were producing these two IEGs (Pei et al. 2004). More recently it was shown that Arc induction following both DOI and stress is reduced upon inducible BDNF loss, suggesting a relationship between the expression of these two genes (Benekareddy et al. 2013). Although playing a role in LTP, Arc also reduces α-amino-3-hydroxy-5-methyl-4isoxazolepropionic acid (AMPA) glutamate receptor signaling (Chowdhury et al. 2006) and contributes to synaptic elimination (Wilkerson et al. 2014). Interestingly, levels of Arc in the cortex are increased after administration of ketamine, which also increases AMPA signaling (Li et al. 2015), demonstrating a complex role for Arc in mediating synaptic plasticity.

3 IEG and 5-HT$_{2A}$ Receptor Expression: Early Studies

Several early studies examining the upregulation of IEGs in response to psychedelics made attempts to characterize the cells that were responsive to the drugs. Interestingly, many studies failed to demonstrate c-Fos expression within cells that were labeled with antisera against the 5-HT$_{2A}$ receptor. The first study to report this absence of colocalization administered 8 mg/kg DOI to rats but could find no cells that were co-labeled with antisera against both c-Fos and 5-HT$_{2A}$ receptors (Mackowiak et al. 1999). Other investigators found similar patterns of non-overlapping c-Fos and 5-HT$_{2A}$ receptor staining, and showed that c-Fos$^+$ cells primarily band near the apical dendritic trunk of cells stained with 5-HT$_{2A}$ receptor antibody, but only very rarely found double-labeled pyramidal cells (Scruggs et al. 2000). Using the same receptor antibody as the above two studies, Pei et al. reported similar results showing that DOI-induced Arc protein did not coincide with 5-HT$_{2A}$ receptor antisera reactivity (2004). The dependence of DOI's effects on IEG expression through the 5-HT$_{2A}$ receptor is well documented (Leslie et al. 1993; Tilakaratne and Friedman 1996; Vaidya et al. 1997; Pei et al. 2000; Scruggs et al. 2000), so these observations were initially surprising. However, they should be interpreted with some caution given that more recent studies have indicated psychedelic-induced c-Fos expression occurs almost exclusively in cells that are positive for 5-HT$_{2A}$ mRNA expression (Gonzalez-Maeso et al. 2007). These experiments were performed using fluorescence in situ hybridization (FISH) in both neuronal cultures and mouse somatosensory cortex following LSD and DOI administration (Gonzalez-Maeso et al. 2007). The disparity between these findings might be explained by differences between mRNA and expression of the 5-HT$_{2A}$ receptor protein, or by the reliance of several of the previous immunohistochemistry studies on one particular antibody to the 5-HT$_{2A}$ receptor that was originally distributed by Pharmingen. Indeed, the pattern of somatodendritic staining produced by the Pharmingen 5-HT$_{2A}$ antibody does not match the staining pattern of other antibodies whose specificity was recently verified in mice lacking the 5-HT$_{2A}$ receptor (Weber and Andrade 2010).

The lack of 5-HT$_{2A}$ immunoreactivity in cells displaying DOI-induced IEG transcription implied that the effect of cellular activation by hallucinogens was indirect. Although the specificity of the antibodies used to make this conclusion is somewhat in doubt, the observations were originally interpreted in the context of early theories positing an important role for thalamocortical afferents in mediating the effects of hallucinogens. This idea was based on a variety of electrophysiological observations in cortical slices, and also explained a reduction of DOI-induced c-Fos expression seen following thalamic lesions (Scruggs et al. 2000; Marek et al. 2001). However, in the light of more recent evidence, cortical–cortical interactions appear critical for hallucinogenic activity. For example, electrolytic thalamic lesions using whole animals rather than slices did not alter cortical responses to DOI, implying that a local cortical mechanism controls neuronal excitability (Puig et al. 2003). Further, a subpopulation of excitatory pyramidal and inhibitory neurons in the cortex can be directly depolarized by 5-HT$_{2A}$ agonists (Beique et al. 2007; Weber and Andrade 2010), and re-introduction of 5-HT$_{2A}$ receptor into primarily cortical areas but not thalamic areas in 5-HT$_{2A}^{-/-}$ knockout mice restores behavioral and transcriptional effects of hallucinogens (Gonzalez-Maeso et al. 2007). These data suggested that IEG induction in vivo may result directly from activation of 5-HT$_{2A}$ receptors on transcriptionally activated cells, through indirect mechanisms (i.e., 5-HT$_{2A}$-mediated glutamate release), or a combination of these events.

4 Identification of Tissues that Transcriptionally Respond to Psychedelics: Early Studies

The identification of activated cell populations and the manner by which they initiate gene transcription may provide important information about the mechanisms through which psychedelics cause their effects. Co-labeling studies demonstrated that c-Fos is induced in a variety of cell types in response to psychedelics. For example, 12% of inhibitory GABAergic neurons identified through GAD$_{67}$ labeling in the prefrontal cortex (PFC) also co-stain for c-Fos following 5.0 mg/kg DOI, compared to only 1.4% under control conditions (Abi-Saab et al. 1999). Another unspecified population of cells that did not label for GAD$_{67}$, presumably pyramidal neurons, were also c-Fos^{+} (Abi-Saab et al. 1999). Increases in extracellular GABA following local PFC infusion of DOI also occurs (Abi-Saab et al.

1999). These results are consistent with the idea that, in addition to excitatory neurons, DOI activates inhibitory cells that release GABA in response to stimulation. This activation may be direct because 5-HT$_{2A}$ receptor expression is detected in a population of cortical interneurons, a large fraction of which are parvalbumin positive (de Almeida and Mengod 2007; Weber and Andrade 2010). Furthermore, 5-HT and hallucinogens can depolarize GABAergic interneurons in vitro and in vivo (Foehring et al. 2002; Weber and Andrade 2010; Zhang et al. 2010).

Within the somatosensory cortex, increases in extracellular GABA and glutamate have been observed following systemic DOI, but local infusion of DOI produced increases only in glutamate (Scruggs et al. 2003). Additionally, no GABAergic, c-Fos^{+} cells were located in the barrel cortex, indicating that a degree of variation exists between the effects of DOI on interneurons across brain regions (Scruggs et al. 2003). Other researchers have noted that DOI-induced c-Fos within the orbital cortex and dorsal medial PFC primarily occurred within GAD$_{67}^{+}$ cells (Wischhof and Koch 2012). However, little c-Fos and parvalbumin overlap was found in these areas, indicating other interneuron subtypes are activated by DOI (Wischhof and Koch 2012). Although substantial evidence for the activation of neurons has been published, glial cell populations have been largely unstudied with respect to hallucinogens. A single report, however, details co-expression of c-Fos and Olig1, an oligodendrocyte marker, in the PFC following LSD administration (Reissig et al. 2008).

In addition to DOI, other psychedelics produce a variety of genetic responses in the cortex. The first data indicating an IEG response to LSD was published in 1999, and mapped the induction of *c-Fos* mRNA using in situ hybridization. The *c-Fos* staining following 1.0 mg/kg LSD was quite widespread, labeling the frontal and parietal cortex, the striatum, nucleus accumbens, paraventricular nucleus, and a population of cells in the ventral central gray (Erdtmann-Vourliotis et al. 1999). Another report showed similar *c-Fos* mRNA induction, unaltered by morphine pretreatment, which mimicked the laminar pattern seen with DOI-induced *c-Fos* in the cortex (Erdtmann-Vourliotis et al. 2000). Induction of c-Fos protein in the rat brain following 0.5 mg/kg LSD was demonstrated in the mPFC, anterior cingulate cortex, parietal cortex, and in the amygdala (Gresch et al. 2002). The increase in c-Fos immunoreactivity in this study was attenuated by the selective 5-HT$_{2A}$ receptor antagonist M100,907, and was not found in the nucleus accumbens or the ventral striatum (Gresch et al. 2002). It was later demonstrated that a lower, but behaviorally active, dose of LSD (0.16 mg/kg) was able to induce c-Fos in a time-dependent manner in the anterior cingulate cortex, nucleus accumbens shell, paraventricular nucleus, and the bed nucleus of stria terminalis (Frankel and Cunningham 2002). Interestingly, no increases were seen in the prelimbic, frontal, or parietal cortical areas at this dose, nor did LSD increase c-Fos in the nucleus accumbens core (Frankel and Cunningham 2002).

These studies clearly demonstrate the ability of LSD to induce c-Fos expression in many of the same areas where DOI induces expression, with the exception of a less robust response following lower doses of LSD (0.16 mg/kg) in areas of the frontal and parietal cortex (Frankel and Cunningham 2002). Also, expression of IEGs in the nucleus accumbens has been demonstrated with both LSD and DOI

(Leslie et al. 1993; Erdtmann-Vourliotis et al. 2000; Frankel and Cunningham 2002), but is not universally reported with either of these two drugs (Gresch et al. 2002; Wischhof and Koch 2012).

5 IEG and 5-HT$_{2A}$ Receptor Expression in the Post-Genomic Era

Widespread surveys of the transcriptional response to psychedelics were made possible with the advent of DNA microarray technology. One of us performed the first unbiased microarray screen on the effects of LSD within the brain and assessed the effects of 1.0 mg/kg LSD on rat PFC 90 min after drug administration. In the first screen, a collection of five genes upregulated by LSD in the PFC were identified: *serum glucocorticoid kinase* (*sgk*), *Iκβ-α*, *neuron derived orphan receptor 1* (*nor1*; *nr4a3*), *ania3*, and *krox-20* (*egr-2*) (Nichols and Sanders-Bush 2002). These genes, along with *Arc* and *c-Fos*, were validated by RNase protection as differentially expressed in the PFC (Nichols and Sanders-Bush 2002).

We subsequently examined time course, dose–response, and sensitivity of the LSD response to 5-HT$_{2A}$ and 5-HT$_{1A}$ receptor antagonists (Nichols et al. 2003). The expression of most of these genes peaked at 90 min and returned to baseline 3–5 h following LSD treatment. The *nor1* gene, however, remained at maximum elevated levels through the final 5 h time point tested (Nichols et al. 2003). Two genes were significantly upregulated at the low dose of 0.20 mg/kg LSD (*krox-20*, *Iκβ-α*), and most expression levels increased with successively higher doses of drug from 0.5 to 1.0 mg/kg. Consistent with reports utilizing DOI, the transcriptional effects of LSD were unaffected by selective antagonism of the 5-HT$_{1A}$ receptor with WAY-100,635 but were significantly attenuated by the 5-HT$_{2A}$ receptor selective antagonist M100,907, with the exception of *sgk* and *Iκβ-α*, which were also unaffected by M100,907. These results indicate that the majority of LSD-related gene expression alterations were induced through activation of the 5-HT$_{2A}$ receptor, but that other receptors contribute to its effects (Nichols et al. 2003).

Extending this work, we performed a second microarray screen using a different Affymetrix gene chip version, and identified and validated three additional transcripts increased by LSD (1.0 mg/kg) in the rat PFC: *map kinase phosphatase 1* (*mkp1*), *core/enhancer binding protein β* (*C/EBP-β*), and the novel gene, *induced by lysergic acid diethylamide 1* (*ilad1*; subsequently renamed *arrestin domain containing* 2, *arrdc2*) (Nichols and Sanders-Bush 2004). Along with the other LSD-induced differentially expressed genes, these also followed a dose- and time-dependent expression pattern. At the highest dose of 1.0 mg/kg LSD, the expression of *mkp1*, *C/EBP-β*, and *ilad* was only partially blocked by M100,907, indicating that activation of multiple receptors is contributing to the effects of LSD on gene expression at this dose (Nichols and Sanders-Bush 2004). Indeed, LSD is a relatively non-selective serotonin (5-HT) and dopamine receptor ligand, with high to moderate affinity for a number of receptors that may contribute to its effects

(Nichols 2004). Interestingly, neither *c-Fos* nor *arc* changes were identified in these two microarray screens. Examination of the raw data suggested equivalent levels between the LSD and control groups. Furthermore, several additional genes were initially called as significantly increased in expression. Each of the genes identified in the primary screen was subject to validation by RNAse protection, where it was found that only 1 in 4–5 genes could be confirmed as differentially expressed. Further, there was no real correlation between expression levels changed between microarray and RNase protection. These results together demonstrate limitations of microarray screens, which are not quantitative, and can result in high false positive and negative rates.

The general functions of the genes induced by LSD are varied, and in the case of some genes mentioned above, little is known. However, a common theme linking the transcriptional changes is an effect on synaptic plasticity. For example, *nor1* is a member of the *Nr4a* family of activity-dependent transcription factors, which has been demonstrated to be important for transcription-dependent LTP in the hippocampus (Bridi and Abel 2013). Similarly, *sgk* has been shown to play a role in long-term memory and the expression of LTP in hippocampal neurons (Ma et al. 2006). *Ania3* is a splice variant within the *Homer1* gene family that encodes synaptic proteins, and has been implicated in mGluR-mediated plasticity (de Bartolomeis et al. 2014; O'Riordan et al. 2014). *C/EBP-β* is known to affect memory consolidation and synaptic strength (Alberini et al. 1994; Taubenfeld et al. 2001), and *Iκβ-α* inhibits NFκB, which is primarily known for its role in inflammatory pathways (Hinz et al. 2012), but is also important in synapse regulation (Salles et al. 2014). However, the manner in which these IEGs contribute to the downstream transcriptional, structural, and functional sequelae of neuronal activation remains poorly understood. Such long-term changes may be important in mediating a variety of behavioral effects of chronic LSD (Marona-Lewicka et al. 2011; Martin et al. 2014).

Transcriptional profiling of cells in culture following the administration of psychotropic agents is one strategy to reveal clues as to the mechanisms by which ligands produces their effects. Gonzalez-Maeso and colleagues initially studied the transcriptional effects of a series of ligands (5-HT, tryptamine, 5-methoxytryptamine, and DOI) in a cell culture system using 5-HT_{2A}-expressing HEK293 cells. Although these ligands produced no changes in HEK293 cells without 5-HT_{2A} receptor, they each produced a distinct profile of concentration-dependent gene induction when applied to cells expressing the 5-HT_{2A} receptor. DOI induced the transcription of several genes in this system, including *egr-3*, *egr-2 (krox-20)*, *cox2*, and *cyr61*, whereas 5-HT and tryptamine produced responses of larger magnitude. Both LSD and lisuride (a non-hallucinogenic structural analog of LSD with significantly higher affinity for dopamine D_2 receptors), produced very little gene induction in this in vitro system (Gonzalez-Maeso et al. 2003). These researchers further analyzed the effects of DOI on mouse somatosensory cortex using microarray analysis, and validated by RT-QPCR that a subset of genes was differentially regulated. These genes (*c-Fos*, *egr-2*, *N-10*, *Iκβ-α*, *sty kinase*) followed dose- and time-dependent responses, with *sty kinase* being the only downregulated gene (Gonzalez-Maeso et al. 2003). They also compared the transcriptional effects of LSD, lisuride, and DOI on the expression of 20 genes in mouse somatosensory cortex, which revealed that LSD and DOI upregulated expression of three genes (*egr-1*, *egr-2*, and *period-1*) that were not upregulated by lisuride (Gonzalez-Maeso et al. 2003). A further collection of three genes, including *c-Fos*, was also induced by each ligand in somatosensory cortex. Both LSD and lisuride upregulated $I\kappa\beta$-α in mice lacking the 5-HT_{2A} receptor, however, DOI produced no gene expression changes in these knockout mice (Gonzalez-Maeso et al. 2003), confirming our earlier results of 5-HT_{2A} receptor independent expression of LSD-induced $I\kappa\beta$-α.

These data provide evidence that functional selectivity, a phenomenon whereby different ligands acting through the same receptor can lead to different signaling patterns within the cell (Urban et al. 2007), is occurring at the 5-HT_{2A} receptor. That lisuride induces *c-Fos* expression through the 5-HT_{2A} receptor in primary cultures of mouse cortical neurons (Gonzalez-Maeso et al. 2007), yet does not elicit psychedelics behaviors in humans or produce head-twitch responses in mice, suggests that production of c-Fos is not always correlated with overt behaviors known to be mediated through 5-HT_{2A} receptor activation (Gerber et al. 1985; Gonzalez-Maeso et al. 2003; Halberstadt and Geyer 2013) The two genes, *egr-1* and *egr-2*, which were identified as robustly upregulated by DOI and LSD but not lisuride in somatosensory cortex, represent members of the zinc family of transcription factors whose expression is correlated with LTP and implicated in modulation of synaptic plasticity and memory formation (Richardson et al. 1992; Veyrac et al. 2014). These studies also highlight the difference between in vitro and in vivo models for the study of gene expression, because LSD and lisuride robustly induce gene expression in vivo, but not in vitro in HEK293 cells expressing 5-HT_{2A} receptors (Gonzalez-Maeso et al. 2003).

A subsequent report extended this work to provide more evidence that hallucinogenic and non-hallucinogenic 5-HT_{2A} receptor agonists can produce dissociable effects on transcription and behavior. Hallucinogens (DOI, DOM, DOB, psilocin, mescaline, and LSD) and non-hallucinogens (ergotamine, (*R*)-lisuride, (*S*)-lisuride) were tested in mice expressing the 5-HT_{2A} receptor ($htr2A^{+/+}$) and in the receptor knockout mice ($htr2A^{-/-}$). Each ligand produced a different transcriptional pattern among the 19 genes tested across several brain regions analyzed including prefrontal and somatosensory cortex, with the overall transcriptional response largely eliminated in the $htr2A^{-/-}$ mice. Whereas all ligands tested induced *c-Fos*, only the psychedelic-induced *egr-1* and *egr-2*. *Period-1*, which was specific to psychedelics in the earlier, smaller study, was induced by ergotamine, but not DOM in this larger study, indicating that the transcriptional response to this gene is not psychedelic-specific (Gonzalez-Maeso et al. 2007).

Studies performed in primary neuronal cultures found that whereas both c-Fos and *egr-2* expression were induced by LSD, only c-Fos was increased by R-lisuride (Gonzalez-Maeso et al. 2007). We have examined *egr-2* expression in response to lisuride in several cell types in culture and found that *egr-2* expression by lisuride is dependent on the type of cell used (unpublished data). Therefore, its proposed use as a biomarker for hallucinogenic properties of a drug must be considered with some degree of caution. Nevertheless, the finding that different 5-HT$_{2A}$ receptor agonists that produce different behaviors can induce differential expression responses in vivo suggests that differential recruitment of signal effector pathways may be important for the behavioral effects of psychedelics.

Interestingly, the application of tetrodotoxin to neuronal cultures, which blocks Na^+ channels and action potential firing, had no effect on the ability of LSD to induce transcription of *egr-2* or *c-Fos* (Gonzalez-Maeso et al. 2007). Additionally, these transcripts were present exclusively in cells positive for 5-HT$_{2A}$ mRNA, akin to results seen in layer 5 of somatosensory cortex (Gonzalez-Maeso et al. 2007). These data demonstrate that, at least in cell culture, activation of 5-HT$_{2A}$ receptors can directly alter gene expression without a requirement for neuronal depolarization. Taken together, the cell culture and brain tissue data indicate that the transcriptional response mediated by 5-HT$_{2A}$ receptor activation is dependent on the ligand, the cell type, and the environment in which the receptor resides, factors that vary considerably between neuronal cell culture and the brain.

6 Psychedelics, Gene Expression, and Cellular Signaling

The wide range of effector pathways recruited by the 5-HT$_{2A}$ receptor generates a complex signaling network. Although many effectors are presently known, there are likely many more that remain to be discovered. The canonical signaling pathway involves positive coupling to Gα_q, which activates PLC-β, leading to phosphatidylinositol (PI) hydrolysis, the release of intracellular calcium, and the activation of protein kinase C (PKC). The 5-HT$_{2A}$ receptor can also couple to other G proteins and downstream effector pathways. For example, receptor stimulation can activate phospholipase A$_2$ (PLA$_2$) and the production of arachidonic acid (AA) independently of Gα_q and PLC-β (Felder et al. 1990; Berg et al. 1998; Kurrasch-Orbaugh et al. 2003b). The activation of PLA$_2$ through the 5-HT$_{2A}$ receptor is complex, and can involve several pathways that include a G$\alpha_{i/o}$-associated Gβγ pathway through Src, and G$\alpha_{12/13}$ activation of Rho (Kurrasch-Orbaugh et al. 2003a). Interestingly, the interoceptive behavioral cues elicited by psychedelics in drug discrimination assays in rodents correlate with activation of the PLA$_2$ pathway through G$\alpha_{i/o}$ rather than the PLC-β pathway through Gα_q (Kurrasch-Orbaugh et al. 2003a). Additional pathways linked to 5-HT$_{2A}$ receptor stimulation include pERK activation through β-arrestin (Schmid and Bohn 2010), and phospholipase D (PLD) activation through the small G-protein ADP-ribosylation factor (ARF) (Barclay et al. 2011). Each of these pathways could conceivably recruit expression of different sets of genes such that, depending on the ligand used to activate the receptor and the nature of the cell it is expressed in, gene responses could be vastly different.

Further exploration of gene expression differences between LSD and lisuride in neuronal cell culture has found that inhibition of PLC-β with U73122 eliminates the transcriptional response to both LSD and (*R*)-lisuride, but that pertusis toxin

(PTX) only lowers the magnitude of transcriptional response to LSD (*c-Fos*, *egr-1*, *egr-2*) and not (*R*)-lisuride (*c-Fos*) (Gonzalez-Maeso et al. 2007). The Src inhibitor PP2 prevents LSD from inducing *egr-1* and *egr-2*, but allows for equivalent *c-Fos* induction between LSD and (*R*)-lisuride (Gonzalez-Maeso et al. 2007). These results suggest that both $G\alpha_q$ activation of PLC-β and activation of $G\alpha_{i/o}/G\beta\gamma$/Src is necessary for psychedelic relevant gene expression patterns, consistent with previous pharmacological data (Kurrasch-Orbaugh et al. 2003a; Gonzalez-Maeso et al. 2007). These findings do not, however, preclude the involvement or necessity of additional signaling pathways in the genetic response to psychedelics.

Several studies have attempted to modify the behavioral response to psychedelics by disrupting signaling pathways downstream of 5-HT_{2A} receptor stimulation. For example, dexamethasone, a glucocorticoid that inhibits PLA_2, and indomethacin, which prevents the conversion of AA into other signaling molecules through the inhibition of COX enzymes, were used in combination with DOI. Both indomethacin and dexamethasone each reduced c-Fos expression in the cortex by $\sim 50\%$, but did not eliminate it (Mackowiak et al. 2002). In mice lacking the $G\alpha_q$ protein, c-Fos induction following DOI is abolished, and the head-twitch response is markedly reduced (Garcia et al. 2007). Further, PLC-β activation is necessary for the head-bob response produced by intra-cortical DOI in rabbits, although LSD-induced head bobs were unaffected by PLC-β inhibition (Schindler et al. 2013). These studies provide further evidence that multiple pathways downstream of receptor activation are important for transcriptional and behavioral effects of psychedelics in vivo.

IEG expression has frequently been used as an output to measure perturbations of psychedelic drug mediated signaling. For example, no increase in c-Fos immunoreactivity was observed in somatosensory cortex when an AMPA receptor antagonist, GYKI 52466, preceded DOI administration (Scruggs et al. 2000). In a separate study, GYKI 52466 (25 mg/kg, i.p.) attenuated the increase in Arc protein seen following DOI in all cortical areas examined (Pei et al. 2004). The NMDA antagonist MK-801 blocked the increase of Arc in frontal, orbital, and cingulate cortex, but not in parietal cortex where MK-801 given alone induces Arc expression (Pei et al. 2004). These results suggest the importance of AMPA and NMDA receptor activation for the induction of Arc expression, and by extension implicate glutamate transmission as an essential element in DOI's transcriptional effects. Further supporting this idea, double immunofluorescence for mGluR2/3 receptors and NMDAR1 receptors revealed the great majority of c-Fos^+ cells are positive for these AMPA and NMDA subunits. These data indicate that ionotropic glutamate receptor activation is necessary for neurons to be transcriptionally activated by DOI (Pei et al. 2004). These results are consistent with a preponderance of evidence implicating glutamate release as critical for the electrophysiological and behavioral responses to hallucinogens (Aghajanian and Marek 2000; Scruggs et al. 2003; Muschamp et al. 2004).

The characterization of metabotropic glutamate receptor 2 (mGluR2) influences on 5-HT_{2A} signaling has also relied partly on measurement of IEG expression. For example, pretreatment of rats with the mGluR2/3 agonist LY35470 dose-dependently and completely (at 10 mg/kg) prevented the upregulation of *BDNF* mRNA in the medial prefrontal cortex (mPFC) caused by 5 mg/kg DOI (Gewirtz et al. 2002). This blockade extended to other frontoparietal regions of the cortex and to the claustrum, however, LY35470 did not prevent the DOI-mediated upregulation of *BDNF* in the intralaminar and midline thalamic nuclei. Consistent with these data, the mGlurR2/3 antagonist LY341495 significantly potentiated the upregulation of *BDNF* by 5 mg/kg DOI (Gewirtz et al. 2002). These results are consistent with electrophysiological and behavioral data showing that mGlurR2/3 receptor activity attenuates the effects of psychedelics acting at the 5-HT_{2A} receptor, and that 5-HT_{2A} and mGluR2/3 receptors are localized to similar structures in the mPFC (Gewirtz and Marek 2000; Marek et al. 2000). With respect to IEG expression, activation of mGlurR2/3 with LY379268 attenuates DOI-induced *c-Fos* expression in the mPFC, but not the frontoparietal or somatosensory cortex (Zhai et al. 2003). Because the positive allosteric modulator of the mGluR2 receptor, biphenyl-indanone A (BINA), reduces (*R*)-DOB-induced c-Fos expression in the mPFC, but not somatosensory cortex, it is likely that mGluR2 and not mGluR3 receptors are involved in psychedelic-induced c-Fos expression changes (Benneyworth et al. 2007). These data together indicate that there is a strong functional relationship between 5-HT_{2A} and mGluR2 receptors with respect to IEG expression, at least within the mPFC. This relationship extends to other aspects of psychedelic-induced effects because mGlurR2 activation also reduces psychedelic-induced EPSCs and behavioral head-twitch responses. The mechanism for the functional interaction between 5-HT_{2A} and mGluR2 signaling is not completely understood. Although there are reports of heterodimerization between these two receptors (Gonzalez-Maeso et al. 2008; Moreno et al. 2011), this conclusion has been controversial (Delille et al. 2013). Interestingly, in support of functional heterodimerization between these two receptors in the mechanism of action of psychedelics, presynayptic 5-HT_{2A} receptors have recently been identified on thalamic inputs in the cortex (Barre et al. 2016).

7 Identification and Characterization of the Cortical Cellular Population Responsive to Psychedelics

Although the importance of 5-HT_{2A} receptor signaling in the cortex for the transcriptional and behavioral effects of psychedelics has long been appreciated, the precise population of cells responsive to psychedelics that initiates the signaling that leads to psychedelic transcriptional and behavioral effects remains has only recently been studied. Experiments using Cre recombinase under control of the *Emx1* promoter to restore 5-HT_{2A} receptor expression in cortical pyramidal cells of *htr2A*$^{-/-}$ mice (Gorski et al. 2002) revealed that 5-HT_{2A} receptor signaling in these neurons is sufficient to recapitulate the transcriptional (*c-Fos*, *egr-1*, *egr-2*) response to LSD, along with the behavioral head-twitch responses to LSD and DOI

(Gonzalez-Maeso et al. 2007). Therefore, 5-HT_{2A} receptor signaling within the *Emx1* lineage, including glutamatergic neurons and glia in the cortex, but not GABAergic interneurons or non-cortical neurons, is necessary and sufficient for at least some of the effects of psychedelics.

Recently, we optimized neurocytometry methodology to isolate and analyze populations of cells within the brain that transcriptionally respond to psychedelics (Martin and Nichols 2016; Martin et al. 2017). Somewhat surprisingly we found that only ~5% of cortical neurons directly respond to psychedelics in vivo by increasing transcription of immediate early and other genes. These genes include those for *c-Fos*, *ΔfosB*, *krox20/erg2, and per1* (Martin and Nichols 2016). So far, the only feature found to distinguish these neurons from those that do not respond transcriptionally has been that the responding cells have a significantly higher level of *HTR2A* mRNA expression, which may result in higher 5-HT_{2A} receptor levels rendering the neurons more sensitive to agonists for this receptor. We speculate that activation of this 5% of cortical neurons, which we have termed the 'trigger population', is necessary to initiate the cascade of events leading to changes in the default mode network and behavioral alterations, and may be the same small population that was earlier identified by electrophysiological experiments to depolarize in the presence of 5-HT_{2A} receptor agonists (Beique et al. 2007). In addition to excitatory cortical neurons, we found that ~5 to 10% of inhibitory GABA neurons are transcriptionally activated by psychedelics. Because these activated interneurons, which are comprised of only the somatostatin and parvalbumin subclasses, do not express higher levels of the 5-HT_{2A} receptor than non-transcriptionally activated interneurons, we believe that their activation is primarily indirect (Martin and Nichols 2016). Interestingly, small populations of non-neuronal cells like astrocytes also become transcriptionally active for genes like *c-Fos* following administration of psychedelic drugs (Martin and Nichols 2016). We also found that transcriptional responses differed between brain regions analyzed. For example, somatostatin interneurons are transcriptionally activated in somatosensory cortex, but not medial prefrontal cortex, and *mGluR2* expression in general was higher in responding populations of neurons compared to non-responding neurons in somatosensory cortex compared to medial prefrontal cortex (Martin and Nichols 2016).

8 Chronic Effects of Psychedelics

In addition to producing acute molecular and behavioral effects, LSD also produces long-lasting changes in gene expression and behavior when given chronically. We initially reported that rats given 0.16 mg/kg LSD every other day for 90 days exhibit a variety of behavioral alterations, including hyperactivity in an open-field, reduced sucrose preference, and changes of social behaviors (Marona-Lewicka et al. 2011). Interestingly, some of these behaviors, such as increased locomotion, are persistent at full strength long after the drug is discontinued (Marona-Lewicka et al. 2011), indicating that long-term LSD administration in rats may permanently shift brain neurochemistry and gene expression from a normal to a pathological state. These altered phenotypes represent several domains of the Research Domain Criteria matrix (Morris and Cuthbert 2012), including negative and positive valence systems, and social processes.

To investigate how long-term LSD administration affects gene expression in the brain, we performed RNA sequencing on RNA isolated from the mPFC of rats four weeks after cessation of a 90-day treatment protocol with LSD or saline (Martin et al. 2014). We found several hundred relatively low-magnitude (two-fold) yet significant transcriptional changes in the mPFC of LSD-treated animals long after drug administration stopped. Functional clustering analysis indicated that the altered genes were significantly concentrated in pathways related to neurotransmission, synaptic plasticity, and metabolism (Martin et al. 2014). Several unanticipated clusters of genes were identified that included those involved in RNA processing and endocrine function (Martin et al. 2014). We also found a significant enrichment for altered transcripts whose homologs in humans have been implicated in schizophrenia by others. These include genes for the dopamine D_1 and D_2 receptors, BDNF, ERBB4, and various NMDA and GABA receptor subunits (Martin et al. 2014).

Persistent connectivity modifications produced by long-term LSD are likely mediated through general plasticity mechanisms that begin with the sustained activation of neuronal ensembles and resultant changes in transcription and translation that alter synaptic function (Leslie and Nedivi 2011). The wave of genes induced by psychedelic drug administration functions partially to initiate a stereotyped cascade of complex late-response transcription that can nevertheless alter neuronal function and connectivity in a highly coordinated fashion (Lyons and West 2011). Our early microarray studies demonstrated that that acute administration of LSD induces a small collection of immediate early genes and transcription factors. Although most of these return to baseline expression within several hours, some do not. We hypothesize that with repeated LSD administration, the genes that remain differentially expressed serve to both subtly alter cellular function and recruit additional genes to a dysregulated state. After a certain window of time, between 6 and 12 weeks of treatment, the cellular changes reach a critical and self-sustaining point such that when drug administration ceases the brain has shifted to an abnormal state. Because of the nature of the abnormal behaviors produced and the genes that are affected, we have proposed that rats treated with LSD for three months may serve as a useful platform to study mechanisms underlying behaviors relevant to certain psychiatric diseases (Martin et al. 2014).

9 Effects of Psychedelics Outside of the CNS

Outside of the CNS, there has been little study of the effects of psychedelics on gene expression. That may be because psychedelics are primarily thought of as CNS active agents devoid of effects in the periphery. The 5-HT_{2A} receptor is, however, the most widely expressed serotonin receptor in the mammalian body and found to be expressed in nearly every tissue and cell type. Psychedelics would therefore be predicted to have effects on these peripheral tissues, including effects on gene expression. We have investigated the role of psychedelics in the periphery, and have discovered them to be powerful anti-inflammatory agents. At extremely low doses, drugs like (*R*)-DOI and LSD can inhibit inflammation mediated by the proinflammatory agent tumor necrosis factor alpha (TNF-α) in both cell culture and in whole animal (Yu et al. 2008; Nau et al. 2013). When administered directly to the lung through nebulization, (*R*)-DOI potently prevents the development of allergic asthma and associated inflammation in a mouse model (Nau et al. 2014). The effects of 5-HT_{2A} receptor activation by (*R*)-DOI on gene expression in peripheral tissues (e.g., vascular, gut, lung) are consistent, and include a reduction in mRNA levels of several inflammatory related cytokines and chemokines such as *Il-6*, *Il-5*, *Il-1b*, *Il-13*, *GMCSF*, and *Mcp1* (Nau et al. 2013, 2014). Although the precise mechanism for inhibition of transcription of proinflammatory genes remains to be elucidated, we believe that stimulation of 5-HT_{2A} receptors with psychedelics acts through specific isoforms of PKC to inhibit signaling from the TNF-α receptor, and inhibits activation of NF-κB. Interestingly, our earlier microarray studies found that LSD increases expression of the gene encoding for IκB, the main inhibitory protein of NF-κB, in the brain (Nichols et al. 2003). There are no reports in the literature, however, examining the effects of psychedelics on neuroinflammation and associated gene expression and any potential effects remain to be fully elucidated.

10 Conclusion

Clinical studies on psychedelic compounds conducted through the early 1970s explored a variety of potential uses for psychedelics, including the treatment of various mental disorders and addictions (Baker 1964; Savage and McCabe 1973; Krebs and Johansen 2012). Renewed clinical interest in these drugs has followed along this path in recent years, and a small group of studies has been performed using psilocybin as an anxiolytic/antidepressant in terminal cancer patients, as a treatment for obsessive compulsive disorder, and as a treatment for nicotine addiction (Moreno et al. 2006; Grob et al. 2011; Johnson et al. 2014). LSD also has recently been tested as an adjunct to psychotherapy in terminal illness (Gasser et al. 2014). Our recent work with (*R*)-DOI may lead to clinical therapies for inflammatory disorders like asthma (Nau et al. 2014).

Generally, psychedelics have been recognized for their ability to occasion mystical-type experiences. In one study, a single administration of psilocybin to healthy humans had a positive effect on mood and well-being that persisted for at least 14 months (Griffiths et al. 2006, 2011), and there have been several recent publications describing the efficacy of one or two treatments with psilocybin to long-lasting antidepressant effects and treat addiction (Griffiths et al. 2016; Johnson et al. 2017). In patients, it is reasonable to speculate that a single administration of a psychedelic may be producing long-lasting positive behavioral and/or physiological changes through long-term alterations in gene expression. In the event that larger clinical studies can further establish therapeutic value for psychedelics, it will be very exciting to elucidate which changes in gene expression are ultimately responsible for their clinical efficacy.

References

Abi-Saab WM, Bubser M, Roth RH, Deutch AY (1999) 5-HT2 receptor regulation of extracellular GABA levels in the prefrontal cortex. Neuropsychopharmacology 20:92–96

Aghajanian GK, Marek GJ (2000) Serotonin model of schizophrenia: emerging role of glutamate mechanisms. Brain Res Brain Res Rev 31:302–312

Alberini CM (2009) Transcription factors in long-term memory and synaptic plasticity. Physiol Rev 89:121–145

Alberini CM, Ghirardi M, Metz R, Kandel ER (1994) C/EBP is an immediate-early gene required for the consolidation of long-term facilitation in Aplysia. Cell 76:1099–1114

Baker EF (1964) The use of lysergic acid diethylamide (LSD) in psychotherapy. Can Med Assoc J 91:1200–1202

Barclay Z, Dickson L, Robertson DN, Johnson MS, Holland PJ, Rosie R, Sun L, Fleetwood-Walker S, Lutz EM, Mitchell R (2011) 5-HT2A receptor signalling through phospholipase D1 associated with its C-terminal tail. Biochem J 436:651–660

Barre A, Berthoux C, De Bundel D, Valjent E, Bockaert J, Marin P, Becamel C (2016) Presynaptic serotonin 2A receptors modulate thalamocortical plasticity and associative learning. Proc Natl Acad Sci USA 113:E1382–E1391

Beique JC, Imad M, Mladenovic L, Gingrich JA, Andrade R (2007) Mechanism of the 5-hydroxytryptamine 2A receptor-mediated facilitation of synaptic activity in prefrontal cortex. Proc Natl Acad Sci USA 104:9870–9875

Benekareddy M, Nair AR, Dias BG, Suri D, Autry AE, Monteggia LM, Vaidya VA (2013) Induction of the plasticity-associated immediate early gene Arc by stress and hallucinogens: role of brain-derived neurotrophic factor. Int J Neuropsychopharmacol 16:405–415

Benneyworth MA, Xiang Z, Smith RL, Garcia EE, Conn PJ, Sanders-Bush E (2007) A selective positive allosteric modulator of metabotropic glutamate receptor subtype 2 blocks a hallucinogenic drug model of psychosis. Mol Pharmacol 72:477–484

Berg KA, Maayani S, Goldfarb J, Scaramellini C, Leff P, Clarke WP (1998) Effector pathway-dependent relative efficacy at serotonin type 2A and 2C receptors: evidence for agonist-directed trafficking of receptor stimulus. Mol Pharmacol 54:94–104

Bridi MS, Abel T (2013) The NR4A orphan nuclear receptors mediate transcription-dependent hippocampal synaptic plasticity. Neurobiol Learn Mem 105:151–158

Carhart-Harris RL, Williams TM, Sessa B, Tyacke RJ, Rich AS, Feilding A, Nutt DJ (2011) The administration of psilocybin to healthy, hallucinogen-experienced volunteers in a mock-functional magnetic resonance imaging environment: a preliminary investigation of tolerability. J Psychopharmacol 25:1562–1567

Chen ZY, Jing D, Bath KG, Ieraci A, Khan T, Siao CJ, Herrera DG, Toth M, Yang C, McEwen BS, Hempstead BL, Lee FS (2006) Genetic variant BDNF (Val66Met) polymorphism alters anxiety-related behavior. Science 314:140–143

Chowdhury S, Shepherd JD, Okuno H, Lyford G, Petralia RS, Plath N, Kuhl D, Huganir RL, Worley PF (2006) Arc/Arg3.1 interacts with the endocytic machinery to regulate AMPA receptor trafficking. Neuron 52:445–459

Cohen S, Greenberg ME (2008) Communication between the synapse and the nucleus in neuronal development, plasticity, and disease. Ann Rev Cell Dev Biol 24:183–209 de Almeida J, Mengod G (2007) Quantitative analysis of glutamatergic and GABAergic neurons expressing 5-HT(2A) receptors in human and monkey prefrontal cortex. J Neurochem 103:475–486

de Bartolomeis A, Latte G, Tomasetti C, Iasevoli F (2014) Glutamatergic postsynaptic density protein dysfunctions in synaptic plasticity and dendritic spines morphology: relevance to schizophrenia and other behavioral disorders pathophysiology, and implications for novel therapeutic approaches. Mol Neurobiol 49:484–511

Delille HK, Mezler M, Marek GJ (2013) The two faces of the pharmacological interaction of mGlu2 and 5-HT(2)A—relevance of receptor heterocomplexes and interaction through functional brain pathways. Neuropharmacology 70:296–305

Egan MF, Kojima M, Callicott JH, Goldberg TE, Kolachana BS, Bertolino A, Zaitsev E, Gold B, Goldman D, Dean M, Lu B, Weinberger DR (2003) The BDNF val66met polymorphism affects activity-dependent secretion of BDNF and human memory and hippocampal function. Cell 112:257–269

Erdtmann-Vourliotis M, Mayer P, Riechert U, Hollt V (1999) Acute injection of drugs with low addictive potential (delta(9)-tetrahydrocannabinol, 3,4-methylenedioxymethamphetamine, lysergic acid diamide) causes a much higher c-fos expression in limbic brain areas than highly addicting drugs (cocaine and morphine). Brain Res Mol Brain Res 71:313–324

Erdtmann-Vourliotis M, Mayer P, Riechert U, Hollt V (2000) Prior experience of morphine application alters the c-fos response to MDMA ('ecstasy') and cocaine in the rat striatum. Brain Res Mol Brain Res 77:55–64

Felder CC, Kanterman RY, Ma AL, Axelrod J (1990) Serotonin stimulates phospholipase A2 and the release of arachidonic acid in hippocampal neurons by a type 2 serotonin receptor that is independent of inositolphospholipid hydrolysis. Proc Natl Acad Sci USA 87:2187–2191

Foehring RC, van Brederode JF, Kinney GA, Spain WJ (2002) Serotonergic modulation of supragranular neurons in rat sensorimotor cortex. J Neurosci 22:8238–8250

Frankel PS, Cunningham KA (2002) The hallucinogen d-lysergic acid diethylamide (d-LSD) induces the immediate-early gene c-Fos in rat forebrain. Brain Res 958:251–260

Frey U, Krug M, Brodemann R, Reymann K, Matthies H (1989) Long-term potentiation induced in dendrites separated from rat's CA1 pyramidal somata does not establish a late phase. Neurosci Lett 97:135–139

Garcia EE, Smith RL, Sanders-Bush E (2007) Role of G(q) protein in behavioral effects of the hallucinogenic drug 1-(2,5-dimethoxy-4-iodophenyl)-2-aminopropane. Neuropharmacology 52:1671–1677

Gasser P, Kirchner K, Passie T (2014) LSD-assisted psychotherapy for anxiety associated with a life-threatening disease: a qualitative study of acute and sustained subjective effects. J Psychopharmacol

Genoud C, Knott GW, Sakata K, Lu B, Welker E (2004) Altered synapse formation in the adult somatosensory cortex of brain-derived neurotrophic factor heterozygote mice. J Neurosci 24:2394–2400

Gerber R, Barbaz BJ, Martin LL, Neale R, Williams M, Liebman JM (1985) Antagonism of L-5-hydroxytryptophan-induced head twitching in rats by lisuride: a mixed 5-hydroxytryptamine agonist-antagonist? Neurosci Lett 60:207–213

Gewirtz JC, Marek GJ (2000) Behavioral evidence for interactions between a hallucinogenic drug and group II metabotropic glutamate receptors. Neuropsychopharmacology 23:569–576

Gewirtz JC, Chen AC, Terwilliger R, Duman RC, Marek GJ (2002) Modulation of DOI-induced increases in cortical BDNF expression by group II mGlu receptors. Pharmacol Biochem Behav 73:317–326

Gonzalez-Maeso J, Yuen T, Ebersole BJ, Wurmbach E, Lira A, Zhou M, Weisstaub N, Hen R, Gingrich JA, Sealfon SC (2003) Transcriptome fingerprints distinguish hallucinogenic and nonhallucinogenic 5-hydroxytryptamine 2A receptor agonist effects in mouse somatosensory cortex. J Neurosci 23:8836–8843

Gonzalez-Maeso J, Weisstaub NV, Zhou M, Chan P, Ivic L, Ang R, Lira A, Bradley-Moore M, Ge Y, Zhou Q, Sealfon SC, Gingrich JA (2007) Hallucinogens recruit specific cortical 5-HT (2A) receptor-mediated signaling pathways to affect behavior. Neuron 53:439–452

Gonzalez-Maeso J, Ang RL, Yuen T, Chan P, Weisstaub NV, Lopez-Gimenez JF, Zhou M, Okawa Y, Callado LF, Milligan G, Gingrich JA, Filizola M, Meana JJ, Sealfon SC (2008) Identification of a serotonin/glutamate receptor complex implicated in psychosis. Nature 452:93–97

Gorski JA, Talley T, Qiu M, Puelles L, Rubenstein JL, Jones KR (2002) Cortical excitatory neurons and glia, but not GABAergic neurons, are produced in the Emx1-expressing lineage. J Neurosci 22:6309–6314

Gresch PJ, Strickland LV, Sanders-Bush E (2002) Lysergic acid diethylamide-induced Fos expression in rat brain: role of serotonin-2A receptors. Neuroscience 114:707–713

Griffiths RR, Richards WA, McCann U, Jesse R (2006) Psilocybin can occasion mystical-type experiences having substantial and sustained personal meaning and spiritual significance. Psychopharmacology (Berl) 187:268–283; discussion 284–292

Griffiths RR, Johnson MW, Richards WA, Richards BD, McCann U, Jesse R (2011) Psilocybin occasioned mystical-type experiences: immediate and persisting dose-related effects. Psychopharmacology 218:649–665

Griffiths RR, Johnson MW, Carducci MA, Umbricht A, Richards WA, Richards BD, Cosimano MP, Klinedinst MA (2016) Psilocybin produces substantial and sustained decreases in depression and anxiety in patients with life-threatening cancer: a randomized double-blind trial. J Psychopharmacol 30:1181–1197

Grob CS, Danforth AL, Chopra GS, Hagerty M, McKay CR, Halberstadt AL, Greer GR (2011) Pilot study of psilocybin treatment for anxiety in patients with advanced-stage cancer. Arch Gen Psychiatry 68:71–78

Guzowski JF, Lyford GL, Stevenson GD, Houston FP, McGaugh JL, Worley PF, Barnes CA (2000) Inhibition of activity-dependent arc protein expression in the rat hippocampus impairs the maintenance of long-term potentiation and the consolidation of long-term memory. J Neurosci 20:3993–4001

Halberstadt AL, Geyer MA (2013) Characterization of the head-twitch response induced by hallucinogens in mice: detection of the behavior based on the dynamics of head movement. Psychopharmacology 227:727–739

Halpern JH, Pope HG Jr (2003) Hallucinogen persisting perception disorder: what do we know after 50 years? Drug Alcohol Depend 69:109–119

Hinz M, Arslan SC, Scheidereit C (2012) It takes two to tango: IkappaBs, the multifunctional partners of NF-kappaB. Immunol Rev 246:59–76

Johnson MW, Garcia-Romeu A, Cosimano MP, Griffiths RR (2014) Pilot study of the 5-HT2AR agonist psilocybin in the treatment of tobacco addiction. J Psychopharmacol 28:983–992

Johnson MW, Garcia-Romeu A, Griffiths RR (2017) Long-term follow-up of psilocybin-facilitated smoking cessation. Am J Drug Alcohol Abuse 43:55–60

Kammermeier PJ (2008) Endogenous homer proteins regulate metabotropic glutamate receptor signaling in neurons. J Neurosci 28:8560–8567

Kandel ER (2001) The molecular biology of memory storage: a dialogue between genes and synapses. Science 294:1030–1038

Krebs TS, Johansen PO (2012) Lysergic acid diethylamide (LSD) for alcoholism: meta-analysis of randomized controlled trials. J Psychopharmacol 26:994–1002

Kurrasch-Orbaugh DM, Parrish JC, Watts VJ, Nichols DE (2003a) A complex signaling cascade links the serotonin2A receptor to phospholipase A2 activation: the involvement of MAP kinases. J Neurochem 86:980–991

Kurrasch-Orbaugh DM, Watts VJ, Barker EL, Nichols DE (2003b) Serotonin 5-hydroxytryptamine 2A receptor-coupled phospholipase C and phospholipase A2 signaling pathways have different receptor reserves. J Pharmacol Exp Ther 304:229–237

Leslie JH, Nedivi E (2011) Activity-regulated genes as mediators of neural circuit plasticity. Prog Neurobiol 94:223–237

Leslie RA, Moorman JM, Coulson A, Grahame-Smith DG (1993) Serotonin 2/1 C receptor activation causes a localized expression of the immediate-early gene c-fos in rat brain: evidence for involvement of dorsal raphe nucleus projection fibres. Neuroscience 53:457–463

Li Y, Pehrson AL, Waller JA, Dale E, Sanchez C, Gulinello M (2015) A critical evaluation of the activity-regulated cytoskeleton-associated protein (Arc/Arg3.1)'s putative role in regulating dendritic plasticity, cognitive processes, and mood in animal models of depression. Front Neurosci 9:279

Liu RJ, Lee FS, Li XY, Bambico F, Duman RS, Aghajanian GK (2012) Brain-derived neurotrophic factor Val66Met allele impairs basal and ketamine-stimulated synaptogenesis in prefrontal cortex. Biol Psychiatry 71:996–1005

Lyons MR, West AE (2011) Mechanisms of specificity in neuronal activity-regulated gene transcription. Prog Neurobiol 94:259–295

Ma YL, Tsai MC, Hsu WL, Lee EH (2006) SGK protein kinase facilitates the expression of long-term potentiation in hippocampal neurons. Learn Mem 13:114–118

Mackowiak M, Chocyk A, Fijal K, Czyrak A, Wedzony K (1999) c-Fos proteins, induced by the serotonin receptor agonist DOI, are not expressed in 5-HT2A positive cortical neurons. Brain Res Mol Brain Res 71:358–363

Mackowiak M, Czyrak A, Wedzony K (2002) Inhibition of arachidonic acid cascade attenuates the induction of c-Fos proteins by DOI, 5-HT2A/2C receptor agonist, in the rat cortex. Pol J Pharmacol 54:73–76

Marek GJ, Wright RA, Schoepp DD, Monn JA, Aghajanian GK (2000) Physiological antagonism between 5-hydroxytryptamine(2A) and group II metabotropic glutamate receptors in prefrontal cortex. J Pharmacol Exp Ther 292:76–87

Marek GJ, Wright RA, Gewirtz JC, Schoepp DD (2001) A major role for thalamocortical afferents in serotonergic hallucinogen receptor function in the rat neocortex. Neuroscience 105:379–392

Marona-Lewicka D, Nichols CD, Nichols DE (2011) An animal model of schizophrenia based on chronic LSD administration: old idea, new results. Neuropharmacology 61:503–512

Martin DA, Nichols CD (2016) Psychedelics recruit multiple cellular types and produce complex transcriptional responses within the brain. EBioMedicine 11:262–277

Martin DA, Marona-Lewicka D, Nichols DE, Nichols CD (2014) Chronic LSD alters gene expression profiles in the mPFC relevant to schizophrenia. Neuropharmacology 83:1–8

Martin D, Xu J, Porretta C, Nichols CD (2017) Neurocytometry: flow cytometric sorting of specific neuronal populations from human and rodent brain. ACS Chem Neurosci 8:356–367

Mengod G, Pompeiano M, Martinez-Mir MI, Palacios JM (1990) Localization of the mRNA for the 5-HT2 receptor by in situ hybridization histochemistry. Correlation with the distribution of receptor sites. Brain Res 524:139–143

Moorman JM, Leslie RA (1998) Paradoxical effects of lithium on serotonergic receptor function: an immunocytochemical, behavioural and autoradiographic study. Neuropharmacology 37:357–374

Moreno FA, Wiegand CB, Taitano EK, Delgado PL (2006) Safety, tolerability, and efficacy of psilocybin in 9 patients with obsessive-compulsive disorder. J Clin Psychiatry 67:1735–1740

Moreno JL, Holloway T, Albizu L, Sealfon SC, Gonzalez-Maeso J (2011) Metabotropic glutamate mGlu2 receptor is necessary for the pharmacological and behavioral effects induced by hallucinogenic 5-HT2A receptor agonists. Neurosci Lett 493:76–79

Morris SE, Cuthbert BN (2012) Research domain criteria: cognitive systems, neural circuits, and dimensions of behavior. Dialogues Clin Neurosci 14:29–37

Muschamp JW, Regina MJ, Hull EM, Winter JC, Rabin RA (2004) Lysergic acid diethylamide and [-]-2,5-dimethoxy-4-methylamphetamine increase extracellular glutamate in rat prefrontal cortex. Brain Res 1023:134–140

Nau F Jr, Yu B, Martin D, Nichols CD (2013) Serotonin 5-HT2A receptor activation blocks TNF-alpha mediated inflammation in vivo. PLoS ONE 8:e75426

Nau F Jr, Miller J, Saravia J, Ahlert T, Yu B, Happel KI, Cormier SA, Nichols CD (2014) Serotonin 5-HT2 receptor activation prevents allergic asthma in a mouse model. Am J Physiol Lung Cell Mol Physiol ajplung 00138:02013

Nichols DE (2004) Hallucinogens. Pharmacol Ther 101:131–181

Nichols CD, Sanders-Bush E (2002) A single dose of lysergic acid diethylamide influences gene expression patterns within the mammalian brain. Neuropsychopharmacology 26:634–642

Nichols CD, Sanders-Bush E (2004) Molecular genetic responses to lysergic acid diethylamide include transcriptional activation of MAP kinase phosphatase-1, C/EBP-beta and ILAD-1, a novel gene with homology to arrestins. J Neurochem 90:576–584

Nichols CD, Garcia EE, Sanders-Bush E (2003) Dynamic changes in prefrontal cortex gene expression following lysergic acid diethylamide administration. Brain Res Mol Brain Res 111:182–188

O'Riordan K, Gerstein H, Hullinger R, Burger C (2014) The role of Homer1c in metabotropic glutamate receptor-dependent long-term potentiation. Hippocampus 24:1–6

Patterson SL, Abel T, Deuel TA, Martin KC, Rose JC, Kandel ER (1996) Recombinant BDNF rescues deficits in basal synaptic transmission and hippocampal LTP in BDNF knockout mice. Neuron 16:1137–1145

Pei Q, Lewis L, Sprakes ME, Jones EJ, Grahame-Smith DG, Zetterstrom TS (2000) Serotonergic regulation of mRNA expression of Arc, an immediate early gene selectively localized at neuronal dendrites. Neuropharmacology 39:463–470

Pei Q, Tordera R, Sprakes M, Sharp T (2004) Glutamate receptor activation is involved in 5-HT2 agonist-induced Arc gene expression in the rat cortex. Neuropharmacology 46:331–339

Puig MV, Celada P, Diaz-Mataix L, Artigas F (2003) In vivo modulation of the activity of pyramidal neurons in the rat medial prefrontal cortex by 5-HT2A receptors: relationship to thalamocortical afferents. Cereb Cortex 13:870–882

Reissig CJ, Rabin RA, Winter JC, Dlugos CA (2008) d-LSD-induced c-Fos expression occurs in a population of oligodendrocytes in rat prefrontal cortex. Eur J Pharmacol 583:40–47

Richardson CL, Tate WP, Mason SE, Lawlor PA, Dragunow M, Abraham WC (1992) Correlation between the induction of an immediate early gene, zif/268, and long-term potentiation in the dentate gyrus. Brain Res 580:147–154

Riga MS, Soria G, Tudela R, Artigas F, Celada P (2014) The natural hallucinogen 5-MeO-DMT, component of Ayahuasca, disrupts cortical function in rats: reversal by antipsychotic drugs. Int J Neuropsychopharmacol 17:1269–1282

Salles A, Romano A, Freudenthal R (2014) Synaptic NF-kappa B pathway in neuronal plasticity and memory. J Physiol Paris 108:256–262

Savage C, McCabe OL (1973) Residential psychedelic (LSD) therapy for the narcotic addict. A controlled study. Arch Gen Psychiatry 28:808–814

Schindler EA, Harvey JA, Aloyo VJ (2013) Phospholipase C mediates (±)-1-(2,5-dimethoxy-4-iodophenyl)-2-aminopropane (DOI)-, but not lysergic acid diethylamide (LSD)-elicited head bobs in rabbit medial prefrontal cortex. Brain Res 1491:98–108

Schmid CL, Bohn LM (2010) Serotonin, but not N-methyltryptamines, activates the serotonin 2A receptor via a ss-arrestin2/Src/Akt signaling complex in vivo. J Neurosci 30:13513–13524

Scruggs JL, Patel S, Bubser M, Deutch AY (2000) DOI-Induced activation of the cortex: dependence on 5-HT2A heteroceptors on thalamocortical glutamatergic neurons. J Neurosci 20:8846–8852

Scruggs JL, Schmidt D, Deutch AY (2003) The hallucinogen 1-[2,5-dimethoxy-4-iodophenyl]-2-aminopropane (DOI) increases cortical extracellular glutamate levels in rats. Neurosci Lett 346:137–140

Sheng M, Greenberg ME (1990) The regulation and function of c-fos and other immediate early genes in the nervous system. Neuron 4:477–485

Steward O, Wallace CS, Lyford GL, Worley PF (1998) Synaptic activation causes the mRNA for the IEG Arc to localize selectively near activated postsynaptic sites on dendrites. Neuron 21:741–751

Strassman RJ, Qualls CR, Uhlenhuth EH, Kellner R (1994) Dose-response study of N, N-dimethyltryptamine in humans. II. Subjective effects and preliminary results of a new rating scale. Arch Gen Psychiatry 51:98–108

Taubenfeld SM, Milekic MH, Monti B, Alberini CM (2001) The consolidation of new but not reactivated memory requires hippocampal C/EBPbeta. Nat Neurosci 4:813–818

Tilakaratne N, Friedman E (1996) Genomic responses to 5-HT1A or 5-HT2A/2C receptor activation is differentially regulated in four regions of rat brain. Eur J Pharmacol 307:211–217

Urban JD, Clarke WP, von Zastrow M, Nichols DE, Kobilka B, Weinstein H, Javitch JA, Roth BL, Christopoulos A, Sexton PM, Miller KJ, Spedding M, Mailman RB (2007) Functional selectivity and classical concepts of quantitative pharmacology. J Pharmacol Exp Ther 320:1–13

Vaidya VA, Marek GJ, Aghajanian GK, Duman RS (1997) 5-HT2A receptor-mediated regulation of brain-derived neurotrophic factor mRNA in the hippocampus and the neocortex. J Neurosci 17:2785–2795

Veyrac A, Besnard A, Caboche J, Davis S, Laroche S (2014) The transcription factor Zif268/Egr1, brain plasticity, and memory. Prog Mol Biol Transl Sci 122:89–129

Weber ET, Andrade R (2010) Htr2a Gene and 5-HT(2A) Receptor Expression in the Cerebral Cortex Studied Using Genetically Modified Mice. Front Neurosci 4

Wilkerson JR, Tsai NP, Maksimova MA, Wu H, Cabalo NP, Loerwald KW, Dictenberg JB, Gibson JR, Huber KM (2014) A role for dendritic mGluR5-mediated local translation of Arc/Arg3.1 in MEF2-dependent synapse elimination. Cell Rep 7:1589–1600

Wischhof L, Koch M (2012) Pre-treatment with the mGlu2/3 receptor agonist LY379268 attenuates DOI-induced impulsive responding and regional c-Fos protein expression. Psychopharmacology 219:387–400

Yu B, Becnel J, Zerfaoui M, Rohatgi R, Boulares AH, Nichols CD (2008) Serotonin 5-hydroxytryptamine(2A) receptor activation suppresses tumor necrosis factor-alpha-induced inflammation with extraordinary potency. J Pharmacol Exp Ther 327:316–323

Zhai Y, George CA, Zhai J, Nisenbaum ES, Johnson MP, Nisenbaum LK (2003) Group II metabotropic glutamate receptor modulation of DOI-induced c-fos mRNA and excitatory responses in the cerebral cortex. Neuropsychopharmacology 28:45–52

Zhang QJ, Wang S, Liu J, Ali U, Gui ZH, Wu ZH, Hui YP, Wang Y, Chen L (2010) Unilateral lesion of the nigrostriatal pathway decreases the response of interneurons in medial prefrontal cortex to 5-HT 2A/2C receptor stimulation in the rat. Brain Res 1312:127–137

Effect of Hallucinogens on Unconditioned Behavior

Adam L. Halberstadt and Mark A. Geyer

Abstract Because of the ethical and regulatory hurdles associated with human studies, much of what is known about the psychopharmacology of hallucinogens has been derived from animal models. However, developing reliable animal models has proven to be a challenging task due to the complexity and variability of hallucinogen effects in humans. This chapter focuses on three animal models that are frequently used to test the effects of hallucinogens on unconditioned behavior: head twitch response (HTR), prepulse inhibition of startle (PPI), and exploratory behavior. The HTR has demonstrated considerable utility in the neurochemical actions of hallucinogens. However, the latter two models have clearer conceptual bridges to human phenomenology. Consistent with the known mechanism of action of hallucinogens in humans, the behavioral effects of hallucinogens in rodents are mediated primarily by activation of 5-HT_{2A} receptors. There is evidence, however, that other receptors may play secondary roles. The structure–activity relationships (SAR) of hallucinogens are reviewed in relation to each model, with a focus on the HTR in rats and mice.

Keywords Psychedelic · Wet dog shake · 5-HT_{2C} receptor · Lisuride · Quipazine · Lysergic acid diethylamide · LSD · DOI · SB-242,084 · M100907 · Mescaline · Psilocybin · Locomotor activity · 25I-NBOMe · 25CN-NBOH · 1-methylpsilocin · Rat · Mouse

---

A.L. Halberstadt (✉) · M.A. Geyer
Department of Psychiatry, University of California San Diego, La Jolla, CA 92093-0804, USA
e-mail: ahalberstadt@ucsd.edu

A.L. Halberstadt · M.A. Geyer
Research Service, VA San Diego Healthcare System, San Diego, CA, USA

Curr Topics Behav Neurosci (2017) 36: 159–200
DOI 10.1007/7854_2016_466
Published Online: 22 February 2017

Contents

1 Introduction

Because of the difficulties associated with hallucinogen studies in human subjects, investigations of the mechanism of action of serotonergic hallucinogens have relied heavily on animal models. Most clinical work with hallucinogens ceased in the early 1970s; although human testing resumed in 1994 and has accelerated in recent years (e.g., Griffiths et al. 2006; Grob et al. 2011; Carhart-Harris et al. 2012; Schmid et al. 2015), animal behavioral models remain the primary methodology used to characterize the pharmacology of hallucinogens in vivo. In fact, much of what is known about the actions of this class has been derived from animal behavioral models. Drug discrimination studies in rats provided some of the first evidence that 5-HT_{2A} activation is responsible for mediating the behavioral effects of hallucinogens (Glennon et al. 1983, 1984), a finding later confirmed in human volunteers (Vollenweider et al. 1998; Kometer et al. 2013). Animal models have also provided important insights into the structure–activity relationships of hallucinogens and the downstream signaling events responsible for mediating their effects. They have also helped to elucidate how the expression of hallucinogen effects is regulated by a variety of receptors and transporters (González-Maeso et al. 2007, 2008; Canal and Morgan 2012; Nichols 2012). Importantly, there is a highly significant and robust correlation between the potencies of hallucinogens in humans and in animal behavioral models (Glennon et al. 1984).

Given the complexity, variety, and variability of the effects of hallucinogens in humans, it has been difficult to define animal behavioral models of hallucinogenic activity. Few methods have satisfied the criteria for an animal model of acute hallucinogenic activity, as we have discussed elsewhere (Adams and Geyer 1985b; Segal and Geyer 1985; Geyer and Krebs 1994). Behavioral models of hallucinogen effects can be divided into two classes: (1) models based on behaviors that are analogous to the effects of hallucinogens in humans and (2) models with no human counterpart. The first type of model is often preferred because they have a conceptual link to human phenomenology, and therefore can be used to investigate the mechanisms responsible for the human response to hallucinogens. Such models may exhibit construct validity to the degree that they assess the same behavioral and physiological effects observed in humans administered hallucinogens. On the other hand, the second type of model often has considerable predictive validity but not construct validity since it is unclear how the behaviors being studied relate to the effects of hallucinogens in humans. Numerous animal models of hallucinogen effects have been proposed used over the last four decades, but most were found to lack specificity or to suffer from other drawbacks. In this chapter, we review the effects of hallucinogens on three unconditioned behaviors: HTR, prepulse inhibition (PPI) of startle, and exploratory behavior. Drug discrimination, another popular model of hallucinogen effects, will be addressed in a separate chapter.

2 Head Twitch Response

Serotonergic hallucinogens induce a variety of spontaneous unconditioned movements in laboratory animals, including ear scratching (mice), limb flicks (cats), and head bobs (rabbits). Hallucinogens also elicit the head twitch response (HTR), a paroxysmal rotational shaking of the head, in mice. The HTR is similar to the head shaking reflex induced by mechanical or chemical irritation of the pinna. The kinematics of the HTR induced by the hallucinogen 2,5-dimethoxy-4-iodoamphetamine (DOI) in mice have been extensively characterized in C57BL/6J mice (Halberstadt and Geyer 2013). When mice make a head twitch, the head repeatedly twists from side-to-side, with each individual head movement lasting approximately 10 ms. Rats also display head twitches after administration of hallucinogens, but in that species the response is a mixture of head shakes and whole-body shakes similar to the behavior of dogs emerging from the water. Hence, in rats the behavior is sometimes referred to as a wet dog shake (WDS).

Although the hallucinogen-induced HTR is most commonly associated with rats and mice, hallucinogen-induced shaking has been observed in many other species. According to observational studies, Stumptail macaque monkeys (*Macaca arctotdes*) exhibit a WDS-like behavior after administration of serotonergic hallucinogens (Schlemmer and Davis 1986). DOI reportedly produces a robust HTR in the least shrew (Darmani et al. 1994). In addition to limb flicks, cats also make head and body shakes after administration of LSD (Jacobs et al. 1976, 1977).

In recent years, the HTR has been widely adopted as a behavioral proxy in rodents for human hallucinogen effects. There are several reasons why use of the HTR has become so common in preclinical behavioral studies of hallucinogens. Most importantly, the HTR is one of only a few behaviors that can reliably distinguish hallucinogenic and non-hallucinogenic 5-HT_{2A} agonists (González-Maeso et al. 2007). The HTR has also proven useful for characterizing the behavioral and neurochemical interactions between the serotonergic system and a variety of other neurotransmitters and neuromodulators, including dopamine, norepinephrine, glutamate, opioids, and cannabinoids (Canal and Morgan 2012). Another reason for the widespread use of this behavior is that it can be assessed by direct observation, and therefore studies do not require specialized training or equipment. Nevertheless, although the observational methods used in HTR studies are accurate, they are also extremely labor intensive and do not allow the behavior to be analyzed in a qualitative fashion or with great temporal precision. As we have recently shown, however, a head-mounted magnet and a magnetometer coil can be used to detect head twitches with high sensitivity and specificity, providing a semi-automated assessment of the response (Halberstadt and Geyer 2013). Our studies to date have confirmed that this method markedly increases the throughput of HTR studies.

2.1 Ergolines

LSD is the first hallucinogen that was shown to induce shaking behavior in laboratory animals. In the course of a behavioral study with LSD in rats, Winter and Flataker (1956) made the following observation:

> When LSD is injected intraperitoneally…the first sign can be observed within 1–2 min, and consists of hyperactivity. The animal explores its cage more busily than usual. After 2 or 3 min, this activity is interrupted momentarily while the animal violently shakes his head; this shaking may be so pronounced that it involves not only the head but the entire body.

Later that year, it was reported (Keller and Umbreit 1956) that LSD induces head shaking in mice. However, it does not appear that the behavior observed by Keller and Umbreit is identical to the HTR because the response was only expressed in response to light tactile stimulation of the head and did not occur spontaneously in mice treated with LSD. Furthermore, mescaline did not elicit the same response, even though it is well known that mescaline induces the HTR in mice. Follow-up studies by other groups failed to replicate the finding that LSD facilitates shaking induced by tactile stimuli (Corne et al. 1963). Nevertheless, it was subsequently confirmed that LSD induces the HTR in mice, rats, and cats (Corne and Pickering

1967; Silva and Calil 1975; Jacobs et al. 1976; Bedard and Pycock 1977). LSD is highly potent, inducing the HTR with an ED_{50} of 53 μg/kg (133 nmol/kg) in mice (Halberstadt and Geyer 2013).

Very few lysergamides other than LSD have been investigated in HTR studies. Mice display head twitches in response to 1-methyl-LSD, which is reportedly about half as potent as LSD (Corne and Pickering 1967). The potency in mice is consistent with human studies indicating that 100 μg 1-methyl-LSD produces hallucinogenic effects roughly comparable to 35 μg LSD (Abramson et al. 1958). 1-Propionyl-LSD (1P-LSD) has recently been sold online as a "research chemical" and reportedly produces LSD-like hallucinogenic effects. We have confirmed that 1P-LSD induces the HTR in mice with about one-third the potency of LSD (1P-LSD: ED_{50} = 159 μg/kg or 350 nmol/kg; Brandt et al. 2016). Two other lysergamides sold as "research chemicals," N^6-allyl-6-*nor*-LSD (AL-LAD) and (2′*S*,4′*S*)-lysergic acid 2,4-dimethylazetidide (LSZ), have also been shown to induce the HTR (Brandt et al. 2017). The potency of LSZ (ED_{50} = 114 nmol/kg) is approximately the same as LSD, whereas AL-LAD is slightly less potent (ED_{50} = 175 nmol/kg). The ergot alkaloid ergometrine (also known as ergonovine or (+)-lysergic acid β-propanolamide) is a uterotonic and is used to reduce postpartum bleeding. According to anecdotal reports, ergometrine produces minor LSD-like effects at p.o. doses of 2–10 mg (Wasson et al. 1978; Bigwood et al. 1979). Consistent with the weak hallucinogenic activity of ergometrine, two groups have observed the HTR in mice treated with high doses of ergometrine (Corne and Pickering 1967; Balsara et al. 1986). Finally, as discussed below, non-hallucinogenic lysergamides such as lisuride and ergotamine fail to induce head twitches in mice and rats.

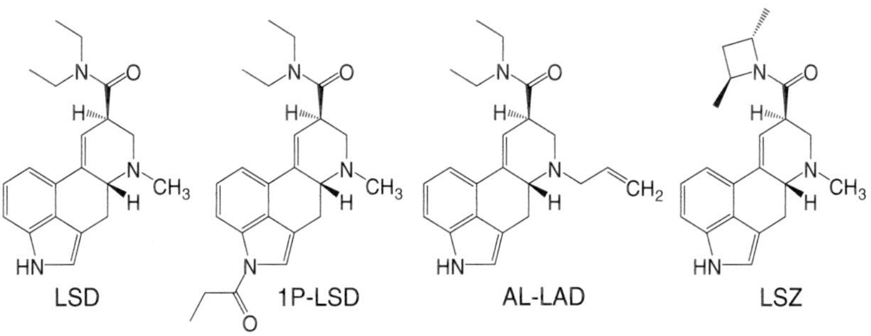

The activity of LSD is very sensitive to halogenation at the 2-position of the indole ring. 2-Bromo-LSD, also known as BOL-148, is a non-hallucinogenic derivative of LSD that acts as a 5-HT_{2A} antagonist (Burris et al. 1991). As expected, 2-bromo-LSD does not produce the HTR in mice or cats (Corne and Pickering 1967; Jacobs et al. 1976).

2.2 Phenylisopropylamines

Many 4-substituted derivatives of 2,5-dimethoxyamphetamine (2,5-DMA) are potent hallucinogens and several are known to induce the HTR. The most important examples are the 4-methyl (DOM), 4-bromo (DOB), and 4-iodo (DOI) derivatives. DOI is widely used in HTR studies due to the fact that it produces a robust response, is commercially available, and is uncontrolled throughout most of the world (Canal and Morgan 2012). According to Schreiber et al. (1995), the ED_{50} for DOI in rats is 0.78 mg/kg (2.43 μmol/kg). To our knowledge, the ED_{50} of DOI in mice has not been reported. Most studies have shown that the response to DOI peaks at doses between 1 and 5 mg/kg (Darmani et al. 1990; Schreiber et al. 1995; Canal et al. 2010; Fantegrossi et al. 2010). DOM and DOB are essentially equipotent with DOI (Yamamoto and Ueki 1975; Wieland et al. 1990; Fantegrossi et al. 2005a). Reports also indicate that DOM, DOB, and *trans*-2-(2,5-dimethoxy-4-methylphenyl)-cyclopropylamine (DMCPA) produce head shakes and body shakes in cats (Nichols et al. 1978; Rusterholz et al. 1978).

Phenylisopropylamine hallucinogens are chiral molecules with one stereocenter. According to several reports, DOM, DOB, DOI, and DOET have stereoselective effects in humans, and the *R*-(–)-isomers are the eutomers (Shulgin et al. 1971; Shulgin 1973; Snyder et al. 1974; Shulgin and Shulgin 1991). Drug discrimination studies have confirmed that that the *R*-(–)-isomers of phenylisopropylamine hallucinogens are more potent than their *S*-(+) enantiomers (Glennon et al. 1982a, b, 1986, 1987). Several head twitch studies have compared the enantiomeric potency of the stereoisomers of phenylisopropylamine hallucinogens. For example, Glennon and colleagues reported that 2.5 mg/kg *R*-(–)-DOI produces almost twice as many head twitches as 2.5 mg/kg *S*-(+)-DOI in mice (Darmani et al. 1990). Another study compared the effects of the stereoisomers of DOM and DMCPA in cats (Nichols et al. 1978). When administered at 0.125 mg/kg, the *R*-(–)-isomers provoked head and body shaking, whereas their *S*-(+)-enantiomers failed to elicit a response. Likewise, tests of *R*-(–)- and *S*-(+)-DOB over a wide range of doses confirmed that *R*-(–)-DOB is more potent and effective than *S*-(+)-DOB (Rusterholz et al. 1978). These results clearly show that phenylisopropylamine hallucinogens produce the HTR in a stereospecific manner.

Methylenedioxy-substituted phenylisopropylamines have been assessed in head twitch studies. Racemic 3,4-methylenedioxyamphetamine (MDA) and *S*-(+)-MDA reportedly induce WDS in monkeys and rats, respectively (Schlemmer and Davis 1986; Hiramatsu et al. 1989). Although (±)-3,4-methylenedioxymethamphetamine (MDMA, Ecstasy) does not induce the HTR in mice, both of the stereoisomers of MDMA have been shown to elicit the response (Fantegrossi et al. 2004, 2005b). 5-HT depletion inhibits the response to *S*-(+)-MDMA but does not alter the response to *R*-(–)-MDMA, suggesting the isomers act through different mechanisms (Fantegrossi et al. 2005b). This suggestion is consistent with the fact that *S*-(+)- and *R*-(–)-MDMA exhibit qualitatively distinct pharmacological profiles, with the *S*-(+)-isomer working primarily as a monoamine releaser (Johnson et al. 1986; Baumann et al. 2008; Murnane et al. 2010) and the *R*-(–)-enantiomer acting directly through 5-HT_{2A} receptors (Lyon et al. 1986; Nash et al. 1994). In contrast to their effects in mice, Hiramatsu reported that *S*-(+)- and *R*-(–)-MDMA fail to produce WDS in rats (Hiramatsu et al. 1989). The discrepant findings with MDMA in mice and rats may reflect species differences in sensitivity to the HTR (see below for further discussion).

2.3 Phenethylamines

There is some disagreement in the literature regarding whether phenethylamine hallucinogens induce the HTR. Mescaline produces the HTR in rats and mice (Silva and Calil 1975; Yamamoto and Ueki 1975; González-Maeso et al. 2007). Likewise, 2,5-dimethoxy-4-*n*-propylthiophenethylamine (2C-T-7) has been shown to induce the HTR in mice (Fantegrossi et al. 2005a; Smith et al. 2014). Mice treated with theconformationally restricted phenethylamine (4-bromo-3,6-dimethoxybenzo-cyclobuten-1-yl)methylamine (TCB-2) exhibit head twitches (Fox et al. 2010a). In contrast to those findings, Moya and colleagues reported that 2,5-dimethoxy-4-iodophenethylamine (2C-I), 2,5-dimethoxy-4-bromophenethylamine (2C-B), and 2,5-dimethoxy-4-methylphenethylamine (2C-D) do not produce the HTR in rats, even after administration of relatively high doses (Moya et al. 2007). Our experiments, however, have shown that 2C-I induces the HTR in mice with an ED_{50} of 0.83 mg/kg (2.42 μmol/kg; Halberstadt and Geyer 2014). It appears that mice are more sensitive than rats to the HTR induced by 5-HT_{2A} partial agonists, which may explain why mouse and rat studies sometimes yield disparate results. For example, *m*-trifluoromethylphenylpiperazine (TFMPP), a weak 5-HT_{2A} partial agonist (Grotewiel et al. 1994), produces head twitches in mice (Yarosh et al. 2007) but not in rats (Arnt and Hyttel 1989; Schreiber et al. 1995). 2C-I has relatively low intrinsic activity at 5-HT_{2A} (Acuña-Castillo et al. 2002; Parrish et al. 2005; Moya et al. 2007), so may not have sufficient efficacy to provoke the HTR in rats.

Although *N*-alkyl substitution attenuates the potency and 5-HT receptor affinity of phenethylamine hallucinogens (Shulgin and Shulgin 1991; Glennon et al. 1994), the addition of an *N*-benzyl group results in a dramatic increase in potency (Braden et al. 2006). Potency is increased even further if a polar oxygen substituent is present at the 2-position of the *N*-benzyl group. For example, 2C-I is active in humans at doses ranging from 14 to 22 mg (Shulgin and Shulgin 1991), whereas the *N*-(2-methoxybenzyl)-substituted derivative 25I-NBOMe is active at doses of 500–800 μg. We found that 25I-NBOMe is highly potent in mice, inducing the HTR with an ED_{50} of 78 μg/kg (0.17 μmol/kg), making it 14-fold more potent than 2C-I (Halberstadt and Geyer 2014) and only slightly less potent than LSD. The 4-bromo analog 25B-NBOMe reportedly induces the HTR in mice at 0.5 mg/kg but not at 0.05 mg/kg (Ettrup et al. 2013). Fantegrossi et al. reported that the HTR to *N*-(2-hydroxybenzyl)-2,5-dimethoxy-4-cyanophenethylamine (25CN-NBOH), another *N*-benzyl-substituted phenethylamine, peaks at 1 mg/kg (Fantegrossi et al.

2014). Studies in our laboratory have confirmed that 25CN-NBOH produces the HTR with an ED_{50} = 0.36 mg/kg (1.03 μmol/kg; Halberstadt et al. 2016a).

We have used the HTR to characterize a series of 25I-NBOMe analogs with varied substituents on the *N*-benzyl ring (Nichols et al. 2015). As shown in Table 1, the ability of 25I-NBOMe to induce the HTR is dramatically altered if the position of the methoxy group on the *N*-benzyl ring is altered or if the group is replaced by bromine. Compared with 25I-NBOMe, the analog with a meta-methoxy group (25I-NB3OMe) is markedly less potent, whereas the para-methoxy analog (25I-NB4OMe) is completely inactive. Replacing the ortho-methoxy moiety of 25I-NBOMe with bromine resulted in a dramatic reduction in potency. Although the ortho-bromo-substituted compound (25I-NB2B) is active, rearrangement of the bromine to either the meta- or para-position was not tolerated, and no HTR was observed with 25I-NB3B or 25I-NB4B at doses up to 30 mg/kg. The *N*-(2,3-methylenedioxybenzyl) analog 25I-NBMD is significantly less potent than 25I-NBOMe and shows the same potency as the non-benzylated parent compound 2C-I. Therefore, it appears that for *N*-benzylphenethylamines the highest potency in the HTR assay is associated with an ortho-substituent on the benzyl ring, especially if the substituent contains an oxygen atom.

Table 1 Head twitch response induced by *N*-benzylphenethylamines

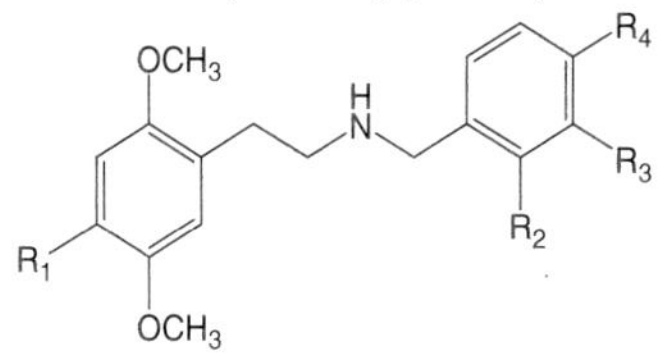

Compound	R_1	R_2	R_3	R_4	ED_{50} (mg/kg)	ED_{50} (μmol/kg)	Reference[1]
25I-NBOMe	–I	$–OCH_3$	–H	–H	0.078	0.17	A
25I-NB3OMe	–I	–H	$–OCH_3$	–H	4.34	9.36	B
25I-NB4OMe	–I	–H	–H	$–OCH_3$	nd (>30)		B
25I-NB2B	–I	–Br	–H	–H	2.31	4.50	B
25I-NB3B	–I	–H	–Br	–H	nd (>30)		B
25I-NB4B	–I	–H	–H	–Br	nd (>30)		B
25I-NBMD	–I	$–OCH_2O–$		–H	1.13	2.36	A
25CN-NBOH	–CN	–OH	–H	–H	0.36	1.03	C

nd not determined

[1] *A* Halberstadt and Geyer (2014); *B* Nichols et al. (2015); *C* Halberstadt et al. (2016a)

2.4 Tryptamines

A variety of tryptamine hallucinogens have been tested for HTR activity. There have been several reports that DMT elicits head twitches in mice (Corne and Pickering 1967; González-Maeso et al. 2007; Carbonaro et al. 2015). One study found that DMT did not produce a significant HTR (Fantegrossi et al. 2006), but that may have been a consequence of using a strain of mice (NIH-Swiss) that is hypo-responsive to hallucinogen-induced HTR (Canal and Morgan 2012). The HTR has also been observed in rodents treated with *N*-methyl-*N*-ethyltryptamine (MET), *N,N*-diethyltryptamine (DET), *N,N*-dipropyltryptamine (DPT), *N,N*-diisopropyltryptamine (DIPT), and *N,N*-diallyltryptamine (DALT) (Fantegrossi et al. 2008; Smith et al. 2014; Carbonaro et al. 2015; Halberstadt and Klein, unpublished observations). Likewise, the HTR is induced by ring-substituted tryptamines, including psilocin, psilocybin, 5-MeO-DMT, and 5-MeO-DIPT (Corne and Pickering 1967; Bedard and Pycock 1977; Fantegrossi et al. 2006; González-Maeso et al. 2007; Halberstadt et al. 2011). The HTR also occurs in mice after administration of 1-methylpsilocin (Halberstadt et al. 2011). As shown in Fig. 1, 1-methylpsilocin acts with an ED_{50} of 0.70 mg/kg (3.22 μmol/kg).

Several side-chain-substituted tryptamine hallucinogens have been assessed. α-Methyltryptamine (AMT) induces the HTR in mice but is not very potent (Corne

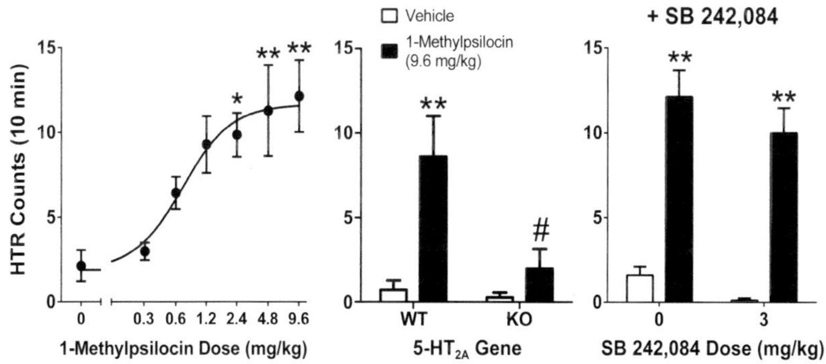

Fig. 1 Effect of 1-methylpsilocin on the head twitch response (HTR) in mice. *Left panel* Dose-response of 1-methylpsilocin in male C57BL/6J mice. *Middle panel* Response to 1-methylpsilocin (9.6 mg/kg) in wild-type (WT) and 5-HT_{2A} receptor knockout (KO) mice. *Right panel* Effect of pretreatment with the 5-HT_{2C} antagonist SB-242,084 (3 mg/kg) on the response to 1-methylpsilocin (9.6 mg/kg) in male C57BL/6J mice (n = 8/group, 32 total). SB-242,084 was injected SC 20 min prior to testing; 1-methylpsilocin was injected SC immediately prior to testing. Data are shown as mean ± SEM. The data in the first two panels are redrawn from: Halberstadt et al. (2011). $^{*}p < 0.05$, $^{**}p < 0.01$ versus the respective vehicle control group (Tukey's test). $^{\#}p < 0.01$ versus 1-methylpsilocin in WT mice (Tukey's test)

Table 2 Head twitch response induced by *N*-benzyl-5-methoxytryptamines

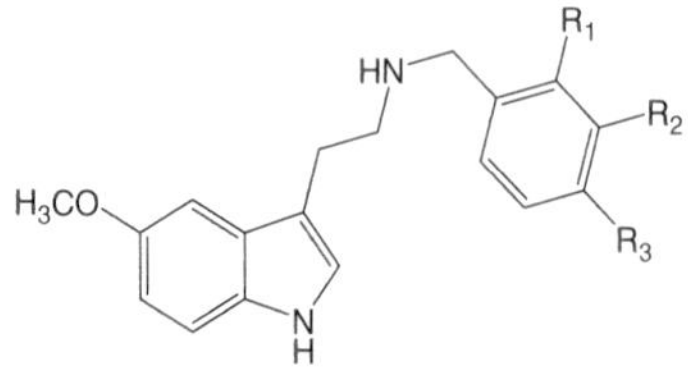

Compound	R_1	R_2	R_3	ED_{50} (mg/kg)	ED_{50} (μmol/kg)
5MT-NB2OMe	$-OCH_3$	–H	–H	3.15	9.08
5MT-NB3OMe	–H	$-OCH_3$	–H	3.28	7.69
5MT-NB4OMe	–H	–H	$-OCH_3$	nd (>30)	
5MT-NB2B	–Br	–H	–H	nd (>30)	
5MT-NB3B	–H	–Br	–H	5.18	10.89
5MT-NB4B	–H	–H	–Br	nd (>30)	
5MT-NB3F	–H	–F	–H	3.33	8.04
5MT-NB3Cl	–H	–Cl	–H	4.43	10.29
5MT-NB3I	–H	–I	–H	7.77	14.88
5MT-NB3Me	–H	$-CH_3$	–H	2.31	5.63
5MT-NB3TFM	–H	$-CF_3$	–H	nd (>30)	
5MT-NB3SMe	–H	$-SCH_3$	–H	nd (>30)	

nd not determined. Data from: Nichols et al. (2015)

and Pickering 1967). The 5-fluoro- and 6-fluoro- derivatives of AMT also induce head twitches (Tadano et al. 1995). The HTR has also been observed in mice treated with 5-MeO-AMT (May et al. 2006), which is known to be a potent hallucinogen with long-lasting effects (Shulgin and Nichols 1978; Kantor et al. 1980).

A series of *N*-benzyl derivatives of 5-methoxytryptamine have been characterized (Table 2). Although several *N*-benzyl-5-methoxytryptamines produce the HTR, none of the compounds are particularly potent (Nichols et al. 2015). In contrast to the *N*-benzylphenethylamines, where compounds with an ortho-substituted benzyl group were the most active, activity in the *N*-benzyl-5-methoxytryptamines was linked to the presence of a meta-substituent. Almost all the meta-substituted *N*-benzylated-5-methoxytryptamines were active, including compounds with 3-methyl, 3-methoxy, 3-fluoro, 3-chloro, 3-bromo, and 3-iodo groups (see Table 2). With the exception of the 2-methoxy-substitued compound (5MT-NB2OMe), which elicited the HTR with an ED_{50} of 3.15 mg/kg (9.08 μmol/kg), none of the compounds with ortho- or para-substituted benzyl groups produced a response at doses up to 30 mg/kg.

2.5 Receptor Mechanisms for the Hallucinogen Head Twitch Response

The 5-HT$_{2A}$ receptor is believed to be responsible for the hallucinogen HTR. It was reported as early as 1967 that the nonselective 5-HT antagonists methysergide and cyproheptadine block the ability of LSD and other hallucinogens to induce the HTR (Corne and Pickering 1967); subsequent studies confirmed that many other 5-HT antagonists ameliorate the response. The first evidence that the 5-HT$_{2A}$ receptor is specifically involved appeared in 1982 when Leysen reported that the potency of 19 5-HT antagonists to block the HTR to mescaline is highly correlated ($r = 0.88$) with their 5-HT$_{2A}$ binding affinity (Leysen et al. 1982). Subsequently, a similar correlation ($r = 0.83$) was shown to exist for blockade of the HTR to DOI (Schreiber et al. 1995; Dursun and Handley 1996). The ability of hallucinogens to elicit the HTR is blocked by the highly selective 5-HT$_{2A}$ antagonists M100907 and MDL 11,939 (Vickers et al. 2001; Fantegrossi et al. 2006, 2008, 2010; Fox et al. 2010a; Halberstadt and Geyer 2014; Carbonaro et al. 2015). Several studies have also reported that hallucinogens do not produce head twitches in knockout mice lacking the 5-HT$_{2A}$ receptor gene (González-Maeso et al. 2007; Halberstadt et al. 2011; Fig. 1). Importantly, however, the ability of hallucinogens to elicit the HTR can be rescued in 5-HT$_{2A}$ knockout mice by selective restoration of 5-HT$_{2A}$ receptors in forebrain regions (González-Maeso et al. 2007).

5-HT$_{2C}$ activation is not required for hallucinogens to induce head twitches. DOI produces head twitches in 5-HT$_{2C}$ knockout mice, although the response is somewhat blunted compared to wild-type mice (Canal et al. 2010). Furthermore, extensive testing has confirmed that the response to DOI is not blocked by the 5-HT$_{2C/2B}$ antagonist SB 200,646A (Kennett et al. 1994; Schreiber et al. 1995; Wettstein et al. 1999) or by the selective 5-HT$_{2C}$ antagonists SB-242,084 and RS102221 (Vickers et al. 2001; Fantegrossi et al. 2010). It is also important to note that 25CN-NBOH, a 5-HT$_{2A}$-selective agonist (Hansen et al. 2014), induces the HTR in mice (Fantegrossi et al. 2014; Halberstadt et al. 2016a). The fact that a selective 5-HT$_{2A}$ agonist produces head twitches indicates that 5-HT$_{2C}$ activation is not required to elicit shaking behavior. Indeed, the response to 25CN-NBOH is completely ameliorated by pretreatment with M100907 but not by RS102221 (Fantegrossi et al. 2014).

Evidence demonstrates that the 5-HT$_{2C}$ receptor modulates expression of head twitch behavior. There is disagreement in the literature, however, regarding whether the response is inhibited or augmented by 5-HT$_{2C}$ activation. According to Vickers and colleagues (Vickers et al. 2001), the nonselective 5-HT$_{2C}$ agonists Ro 60-0175, MK-212, and mCPP do not induce the HTR when administered alone, but they do provoke the behavior if they are administered in combination with a 5-HT$_{2C}$ antagonist. Several studies have also reported that 5-HT$_{2C}$ agonists attenuate the HTR produced by treatment with DOI (Berendsen and Broekkamp 1990; Schreiber et al. 1995; Siuciak et al. 2007; Fantegrossi et al. 2010; Canal et al. 2013). Taken together, these findings indicate that 5-HT$_{2C}$ activation inhibits the HTR.

Conversely, one group has claimed that pretreatment with a 5-HT$_{2C}$ antagonist reduces the magnitude of the HTR produced by DOI by ~50% in C57BL/6J and DBA/2J mice (Canal et al. 2010, 2013). Although the latter findings indicate that 5-HT$_{2C}$ activation augments the HTR, it is not clear why 5-HT$_{2C}$ blockade attenuates the HTR in some studies but has no effect in others (see above). Although it has been proposed that strain differences may underlie these differences (Fantegrossi et al. 2010), unpublished studies in our laboratory have confirmed that SB-242,084 does not significantly reduce the intensity of the HTR induced by 1-methylpsilocin in C57BL/6J mice (Fig. 1). Therefore, 5-HT$_{2C}$ antagonists do not appear to consistently attenuate the HTR in C57BL6J mice. Because 1-methylpsilocin has higher affinity for 5-HT$_{2C}$ versus 5-HT$_{2A}$ sites (Sard et al. 2005), we hypothesized that it was an excellent candidate to test whether crosstalk occurs between 5-HT$_2$ subtypes.

Detailed pharmacological analysis of the DOI dose-response curve has provided additional evidence that the 5-HT$_{2C}$ receptor can block or usurp 5-HT$_{2A}$-mediated responses. DOI produces biphasic dose-responses in some strains of mice and rats, with a marked response decrement occurring at higher doses. This pattern of response occurs in Sprague-Dawley rats (Pranzatelli 1990; Wettstein et al. 1999), outbred Swiss mouse strains (Fantegrossi et al. 2010), and in DBA/2J mice (Canal et al. 2010). Although SB-242,084 has no effect on the ascending arm of the DOI response function in NIH-Swiss and Swiss Webster mice, it shifts the descending arm of the DOI response to the right in those strains, indicating that the response elicited by high doses of DOI is attenuated by recruitment of 5-HT$_{2C}$ (Fantegrossi et al. 2010).

Like DOI, the HTR induced by 25CN-NBOH shows a similar biphasic dose-response function (Fantegrossi et al. 2014). In contrast to DOI, however, 5-HT$_{2C}$ blockade does not alter the descending arm of the response to 25CN-NBOH. The fact that a selective 5-HT$_{2A}$ agonist produces a biphasic response demonstrates that the inhibition of the HTR at high doses is not always a consequence of competing activity at 5-HT$_{2C}$. Alternatively, high levels of 5-HT$_{2A}$ activation may provoke competing behaviors that interfere with expression of shaking behaviors. Along these lines, high doses of quipazine, 5-MeO-DMT, and LSD are known to produce stereotypic behaviors that can mask head shakes and wet dog shakes in rats (Bedard and Pycock 1977; Matthews and Smith 1980; Heal et al. 1986).

2.6 *Effect of Lisuride*

The ergoline derivative lisuride is a structural analog of LSD that shows a similar binding profile at monoamine receptors. Although lisuride acts as a 5-HT$_{2A}$ agonist (Egan et al. 1998; Kurrasch-Orbaugh et al. 2003; Cussac et al. 2008), it does not produce hallucinogenic effects (Herrmann et al. 1977; Verde et al. 1980; Raffaelli et al. 1983; Benes et al. 2006). Nevertheless, lisuride produces false-positive results in some animal models used to study hallucinogens. For example, some drug

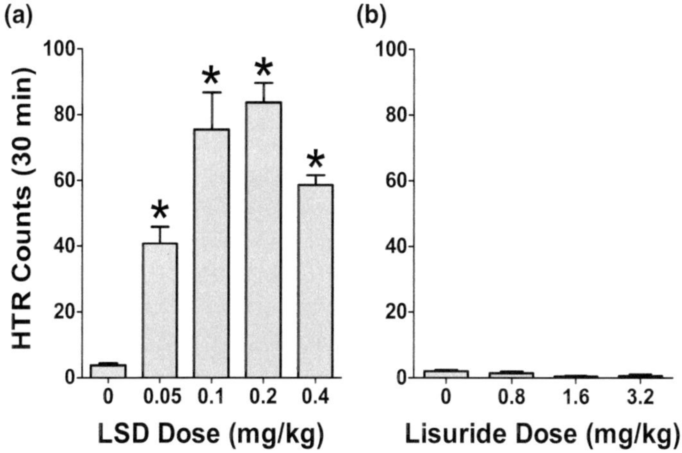

Fig. 2 Effects of LSD (**a**) and lisuride (**b**) on the head twitch response in male C57BL/6J mice. Data are shown as mean ± SEM. $^{**}p < 0.01$ versus vehicle control (Tukey's test). The data are redrawn from: Halberstadt and Geyer (2013)

discrimination studies have shown that rats trained with LSD, DOI, and DOM will generalize to lisuride (White and Appel 1982; Glennon and Hauck 1985; Fiorella et al. 1995; Appel et al. 1999). Likewise, both lisuride and LSD produce limb flicks and abortive grooming in cats (Marini et al. 1981). One of the reasons why the HTR is widely used as a behavioral model for human hallucinogen effects is that it can reliably distinguish hallucinogenic and non-hallucinogenic 5-HT_{2A} agonists (González-Maeso et al. 2007). Even high doses of lisuride (e.g., 3.2 mg/kg) fail to elicit HTR in mice; in comparison, LSD produces head twitches at doses as low as 0.05 mg/kg (Halberstadt and Geyer 2013). The effects of LSD and lisuride on HTR are illustrated in Fig. 2. It is not clear why lisuride is not hallucinogenic but the behavioral differences between LSD and lisuride may be a consequence of 5-HT_{2A} functional selectivity (González-Maeso et al. 2007). Specifically, LSD and other hallucinogens may activate signaling mechanisms that are not recruited by lisuride.

An alternative explanation for the lack of a HTR to lisuride is that the drug does not activate the 5-HT_{2A} receptor with sufficient efficacy to induce the behavior. Lisuride acts as a weak 5-HT_{2A} partial agonist (Rabin et al. 2002; Cussac et al. 2008). Results obtained with lisuride in discrimination studies are consistent with the action of a partial agonist. Specifically, lisuride produces full substitution in rats trained to discriminate DOM (1.0 mg/kg IP), but the response to the training drug is attenuated by 50% when it is administered in combination with 0.01 mg/kg lisuride (Glennon 1991). Lisuride reportedly produces head twitches in the least shrew (Darmani et al. 1994), a non-rodent species that is highly sensitive to 5-HT_{2A} activation. Hence, it cannot be excluded that lisuride fails to induce the HTR in rats and mice because it is a weak partial agonist.

2.7 *Effect of Quipazine*

The arylpiperazine quipazine is another example of a drug that is commonly classified as a hallucinogen false-positive. Quipazine acts as a nonselective 5-HT receptor agonist. In addition to inducing the HTR in rats and mice (Malick et al. 1977; Schreiber et al. 1995), quipazine also produces cross-generalization with serotonergic hallucinogens in drug discrimination studies (White et al. 1977; Winter 1979; Colpaert et al. 1982). Quipazine is considered to be a false-positive result because it did not produce hallucinogenic effects when tested in humans at a dose of 25 mg p.o. (Winter 1994). However, subjects treated with quipazine experienced nausea and gastrointestinal discomfort—potentially due to 5-HT$_3$ receptor activation—which severely limits the dose that can be administered to humans. According to Shulgin, quipazine does produce a psychedelic response when administered in combination with a 5-HT$_3$ antagonist (Shulgin et al. 2011). In summary, quipazine may not actually be a hallucinogen false-positive.

2.8 *Neuroanatomical Locus for the Head Twitch Response Induced by 5-HT$_{2A}$ Receptor Activation*

The HTR induced by the 5-HT precursor 5-hydroxytryptophan (5-HTP) is not blocked by xylamidine, an antagonist of peripheral 5-HT$_2$ receptors (Matthews and Smith 1980). Furthermore, DOI induces the HTR when administered by the intracerebroventricular (ICV) route (Hawkins et al. 2002; Nakagawasai et al. 2007). These findings are consistent with a central site of action, but there is some disagreement regarding the neuroanatomical locus for the HTR. Injection of DOI directly into the medial prefrontal cortex (PFC) induces the HTR in rats (Willins and Meltzer 1997; Ciccocioppo et al. 1999). Likewise, the loss of the HTR in 5-HT$_{2A}^{-/-}$ knockout mice can be rescued by restoration of 5-HT$_{2A}$ receptors to glutamatergic forebrain neurons (González-Maeso et al. 2007). Other evidence, however, indicates that the PFC is not required for the HTR induced by 5-HT agonists. The ability of quipazine and 5-HTP to induce the HTR in rats is unaffected by ablation of the frontal cortex (Lucki and Minugh-Purvis 1987). The HTR still occurs if the brain is sliced transversely at the level of the anterior commissure, but is abolished by transection at the level of the posterior commissure, indicating that the HTR is mediated by structures in the caudal diencephalon and medial brainstem (Bedard and Pycock 1977). According to studies in rats, WDS can be induced by infusion of DOI into the ventromedial brainstem (Watson and Gorzalka 1992) or by infusion of 5-MeO-DMT into the caudal part of the periaqueductal gray (Webster et al. 1982). In summary, there appear to be multiple populations of 5-HT$_{2A}$ receptors in different brain regions that can elicit the HTR, although the region(s) that are responsible for the HTR induced by systemic administration of 5-HT receptor agonists have not been conclusively identified.

2.9 Signal Transduction Mechanisms Underlying the Head Twitch Response

The 5-HT_{2A} receptor can couple to multiple signaling pathways but the specific effector mechanisms responsible for the HTR have not been identified conclusively. The $G_{q/11}$-PLCβ cascade is the canonical signaling pathway recruited by 5-HT_{2A} activation, resulting in phosphoinositide (PI) hydrolysis and mobilization of intracellular Ca^{2+}. Although the G_q-PLC-PI cascade is usually assumed to mediate the behavioral response to hallucinogens, that does not actually appear to be the case. There is no correlation between 5–HT_{2A}-induced PI hydrolysis and LSD-like stimulus effects (Rabin et al. 2002). Despite having profound hallucinogenic effects, LSD stimulates PI hydrolysis with low efficacy (Kurrasch-Orbaugh et al. 2003; Knauer et al. 2009). Conversely, lisuride stimulates PI hydrolysis via 5-HT_{2A} receptors (Burris et al. 1991; Rabin et al. 2002; Kurrasch-Orbaugh et al. 2003; González-Maeso et al. 2007; Cussac et al. 2008), but is not hallucinogenic. Indeed, the ability of DOI to induce the HTR is only blunted 35–40% in $G_q^{-/-}$ mice (Garcia et al. 2007), indicating that other signaling pathways are likely involved in the behavior.

We recently found that for several *N*-benzyl-substituted tryptamines and phenethylamines there is a significant correlation ($r = 0.69$) between ED_{50} values for head twitch and EC_{50} values for increasing levels of intracellular Ca^{2+} via 5-HT_{2A} (Nichols et al. 2015). Conversely, HTR potency is not correlated with 5-HT_{2C} activation ($r = 0.17$). Although Ca^{2+} mobilization is usually thought to be a direct downstream consequence of G_q-PLC activation, other contributing factors may exist. Lisuride and other non-hallucinogenic 5-HT_{2A} agonists exhibit functional selectivity with regard to G_q-PLC activation and Ca^{2+} release. Compared with its effects on Ca^{2+}, lisuride activates G_q signaling with >1000-fold higher potency (Cussac et al. 2008). According to another study, lisuride is 300-fold selective for PI hydrolysis versus Ca^{2+} mobilization (Strachan et al. 2010). By contrast, LSD stimulates G_q signaling and Ca^{2+} mobilization with only an 18-fold difference in potency, and DOI activates both pathways non-selectively (Strachan et al. 2010). Further studies are necessary to identify whether non-G_q-dependent pathways contribute to 5-HT_{2A} receptor-induced Ca^{2+} mobilization.

2.10 Role of Glutamatergic Signaling in the Head Twitch Response

The glutamatergic system may represent a common final pathway for hallucinogenesis (Vollenweider and Geyer 2001). Serotonergic hallucinogens increase recurrent glutamatergic network activity in the PFC and other cortical regions (Beique et al. 2007), resulting in increased glutamate release (Scruggs et al. 2003; Muschamp et al. 2004) and subsequent activation of AMPA receptors (Aghajanian and Marek 1997; Zhang and Marek 2008). The enhancement of glutamatergic activity may play an important role in the HTR induced by hallucinogens. Most importantly, as we have reviewed elsewhere (Halberstadt 2015), manipulations that suppress recurrent glutamatergic network activity also block the HTR. The HTR is attenuated by activation of $mGlu_{2/3}$ receptors (Gewirtz and Marek 2000; Klodzinska et al. 2002), which function as terminal autoreceptors and inhibit glutamate release. Additionally, expression of the HTR is dependent on AMPA receptor activation (Gorzalka et al. 2005; Zhang and Marek 2008; Egashira et al. 2011). Based on these findings, increases of glutamatergic signaling and subsequent AMPA receptor activation appear to play a key role in the HTR induced by $5\text{-}HT_{2A}$ receptor activation.

There is evidence that $mGlu_2$ and $5\text{-}HT_{2A}$ receptors can form heterodimers (González-Maeso et al. 2008), potentially explaining the crosstalk between the receptors. Alternatively, the crosstalk may be purely functional and occur at the circuit level (Delille et al. 2012, 2013). The HTR induced by LSD and DOI is absent in $mGlu_2^{-/-}$ mice (Moreno et al. 2011), which indicates that $mGlu_2$ is required for expression of the behavior. Nevertheless, $mGlu_2$ receptors are known to play an important role in the regulation of glutamatergic transmission (Schoepp 2001) and $5\text{-}HT_{2A}$ signaling (Molinaro et al. 2009), so constitutive adaptations that restrain $5\text{-}HT_{2A}$ responses may occur in $mGlu_2^{-/-}$ mice. Although the level of $5\text{-}HT_{2A}$ expression is not altered in $mGlu_2$ knockout mice (Moreno et al. 2011), $5\text{-}HT_{2A}$ receptor desensitization or alterations of glutamatergic signaling may have occurred, potentially explaining why the HTR is absent in the knockout mice.

2.11 *Effect of Serotonin Releasers and Precursors*

Amphetamine and methamphetamine, which act primarily by increasing carrier-mediated release of dopamine and norepinephrine, do not provoke head twitches (Corne and Pickering 1967; Silva and Calil 1975; Yamamoto and Ueki 1975; Jacobs et al. 1976; Bedard and Pycock 1977; Halberstadt and Geyer 2013). By contrast, the 5-HT releasing drugs fenfluramine and *p*-chloroamphetamine (PCA) do produce a robust HTR (Singleton and Marsden 1981; Darmani 1998a). Fenfluramine and PCA are thought to act indirectly, by increasing carrier-mediated release of 5-HT, because the response can be blocked by inhibition of the 5-HT transporter (Balsara et al. 1986; Darmani 1998a) or by depletion of 5-HT (Singleton and Marsden 1981; Balsara et al. 1986).

In addition to fenfluramine and PCA, other manipulations that increase the level or functional activity of 5-HT in the central nervous system (CNS) produce head twitches. Examples include the 5-HT precursors L-tryptophan and L-5-hydroxytryptophan (5-HTP), $5\text{-}HT_{1A}$ antagonists (Darmani 1998b; Fox et al. 2010b), and the cannabinoid CB_1 receptor antagonist SR 141716A (Darmani and Pandya 2000; Darmani et al. 2003). Head twitches also occur when 5-HT is infused directly into the brain (Drust and Connor 1983). There is evidence that the HTR induced by high central levels of 5-HT may be mediated—at least in part—by hallucinogenic *N*-methylated metabolites. 5-HT is a substrate for *N*-methyltranferases, potentially resulting in the formation of *N*-methylserotonin and *N,N*-dimethylserotonin (Axelrod 1962; Kärkkäinen et al. 2005). One example is indolethylamine *N*-methyltransferase (INMT), which is expressed in the brain and spinal cord (Thompson et al. 1999; Mavlyutov et al. 2012) and can be inhibited by *N*,*N*'bis-(3-methyl-2-thiazolidinylidene)succinamide (MTZ). According to Schmid and Bohn (2010) the response to 5-HTP is significantly attenuated in mice pretreated with MTZ, whereas the response to *N*-methylserotonin and 5-MeO-DMT is not affected. Although MTZ does not completely block the ability of 5-HTP to produce head twitches—indicating that the response reflects the combined effects of *N*-methyltryptamines and 5-HT—the shaking induced by 5-HT appears to be mediated by a different signaling cascade than the shaking produced by hallucinogens. Specifically, the response to 5-HT is dependent on activation of a β-arrestin2/Src/Akt signaling cascade whereas the response to DOI, 5-MeO-DMT, and *N*-methylserotonin is independent of β-arrestin2 (Schmid et al. 2008; Schmid and Bohn 2010).

Because indirect 5-HT agonists such as fenfluramine, PCA, and 5-HTP are not hallucinogenic (Van Praag et al. 1971; Brauer et al. 1996; Turner et al. 2006), their effects on HTR can potentially be classified as false-positive responses. There is some disagreement regarding whether the HTR is a valid model of hallucinogenesis or merely serves as a convenient readout of 5-HT_{2A} activation in rodents (Canal and Morgan 2012). Given the possibility that endogenous *N*-methyltryptamines contribute to the response to 5-HTP and other indirect agonists, and because 5-HT_{2A} functional selectivity may also play a role, further studies are necessary to determine whether they are really false-positive responses.

3 Sensorimotor Gating

The startle reflex is a transient defensive motor response elicited by sudden intense stimuli such as loud and unexpected sounds (Dodge and Louttit 1926; Fleshler 1965). It is well established that the amplitude of the startle response is attenuated if a weak prestimulus is presented immediately prior to the startle pulse (Hoffman and Searle 1965, 1968; Hoffman and Ison 1980). The ability of prepulses to inhibit the startle response—a phenomenon known as PPI—is often used as a laboratory measure of sensorimotor gating. PPI can be assessed in humans and in laboratory animals using similar procedures. PPI deficits have been observed in patients suffering from a variety of illnesses associated with gating or filtering impairments (see Braff et al. 2001 for review), including schizophrenia (Braff et al. 1978; Grillon et al. 1992; Mackeprang et al. 2002), Tourette's syndrome (Swerdlow et al. 1994), and obsessive compulsive disorder (Swerdlow et al. 1993). Hallucinogens have been postulated to work by disrupting sensory filtering mechanisms, resulting in sensory overload and cognitive dysfunction (Vollenweider 1994; Vollenweider and

Geyer 2001). Studies of hallucinogen effects on PPI have been used to assess their influence on gating mechanisms as well as to model the filtering deficits found in schizophrenia patients.

3.1 Hallucinogen Effects on Prepulse Inhibition in Rats

Classical hallucinogens disrupt PPI in rats, indicating that they impair sensorimotor gating. Examples of hallucinogens that have been shown to reduce PPI include DOI (Sipes and Geyer 1994; Varty and Higgins 1995; Padich et al. 1996), DOB (Johansson et al. 1995), LSD (Ouagazzal et al. 2001; Leng et al. 2003; Halberstadt and Geyer 2010; Palenicek et al. 2010), mescaline (Páleníсek et al. 2008), and 2C-B (Páleníсek et al. 2013). The putative 5-HT_{2A}-selective agonist 25CN-NBOH also disrupts PPI in rats (Fig. 3). The effect of DOI on PPI is blocked by M100907, whereas SB 242,084 and the mixed 5-$HT_{2C/2B}$ antagonist SER-082 are ineffective (Sipes and Geyer 1995b; Padich et al. 1996). Likewise, the effect of LSD appears to be solely attributable to 5-HT_{2A} activation because it can be blocked by M100907 and MDL 11,939 but not by the 5-HT_{2C} antagonist SB 242,084, the 5-HT_{1A}

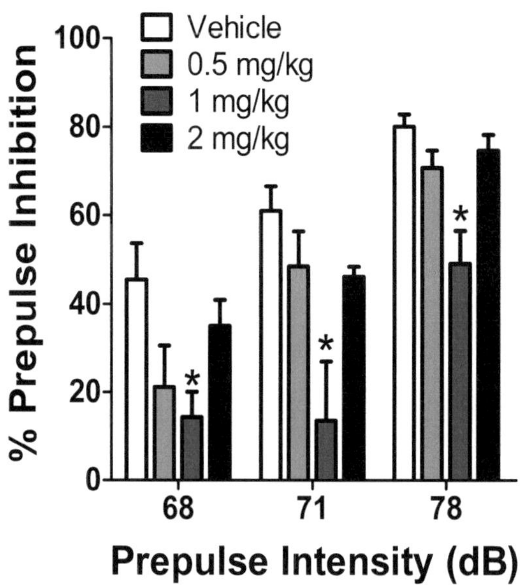

Fig. 3 The selective 5-HT_{2A} agonist 25CN-NBOH reduces prepulse inhibition of startle (PPI) in rats. Male Sprague-Dawley rats (n = 8/group, 32 total) were injected SC with vehicle (water containing 5% Tween-80) or 25CN-NBOH hydrochloride 10 min before placement in the startle test chambers. Values represent mean ± SEM for each group. *$p < 0.05$, significantly different from vehicle control (Tukey's test)

antagonist *S*-(+)-WAY-100,135, the 5-HT$_6$ antagonists Ro 04-6790 and Ro 65-7199, or the dopamine D$_2$ antagonist haloperidol (Ouagazzal et al. 2001; Leng et al. 2003; Halberstadt and Geyer 2010).

In contrast to the PPI disruption produced by hallucinogens, 5-HT$_{2C}$ receptor activation tends to *increase* PPI in rats. For example, the PPI disruption induced by the dopamine D$_{2/3/4}$ receptor agonist apomorphine is ameliorated by the 5-HT$_{2C}$ agonist CP-809,101 (Siuciak et al. 2007). It has also been shown that the 5-HT$_{2C}$-selective agonist WAY-163909 antagonizes the PPI disruption produced by DOI and the NMDA receptor antagonist MK-801 (Marquis et al. 2007). The fact that 5-HT$_{2C}$ agonists are capable of reversing PPI deficits produced by a variety of pharmacological manipulations is consistent with evidence that they have an antipsychotic-like profile of effects. Furthermore, given the effects of CP-809,101 and WAY-163909 on PPI, it is unlikely that the 5-HT$_{2C}$ receptor is responsible for mediating the PPI disruption produced by DOI and other hallucinogens.

The ventral pallidum (VP) is a critical neural substrate for the PPI-disruptive effects of DOI. The VP is part of a descending corticostriatal circuit that regulates PPI (Swerdlow et al. 2001). DOI produces a significant disruption of PPI when infused directly into the VP, but has no effect when infused into the nucleus accumbens (Sipes and Geyer 1997). Likewise, infusion of M100907 into the VP blocked the PPI disruption produced by subcutaneous administration of DOI. Although these findings indicate that 5-HT$_{2A}$ receptors in the VP mediate the PPI disruption induced by DOI, the results of the M100907 infusion experiment are confounded by the fact that the drug was infused in a vehicle containing dimethyl sulfoxide (DMSO). Because DMSO is a highly lipophilic solvent that exhibits poor tissue retention, some of the M100907 infused into the VP may have diffused into surrounding regions. Indeed, the effect of DOI was partially blocked by infusing M100907 into the caudate nucleus, indicating the VP may not be the only site where the 5-HT$_{2A}$ receptors mediating the PPI-disruptive effects of DOI are located.

There is some disagreement in the literature regarding whether dopamine receptor antagonists block the PPI disruption produced by DOI. Although it has been reported that haloperidol can ameliorate the effect of DOI on PPI (Sipes and Geyer 1994; Brea et al. 2006), most studies have found that haloperidol and the selective dopamine D$_{2/3}$ antagonist raclopride are ineffective at blocking the response to DOI (Varty and Higgins 1995; Padich et al. 1996; Marquis et al. 2007). It is unlikely that DOI acts directly through dopamine receptors to disrupt PPI because it has low affinity for D$_2$ receptors (K_i > 10,000 nM; Mos et al. 1992). However, it is now recognized that 5-HT$_{2A}$ and D$_2$ receptors can form heterodimers (Borroto-Escuela et al. 2010; Lukasiewicz et al. 2010), so D$_2$ ligands could potentially modulate the response to DOI by binding to D$_2$-5-HT$_{2A}$ oligomers.

5-MeO-DMT has also been found to disrupt PPI in rats (Rigdon and Weatherspoon 1992; Krebs-Thomson and Geyer 1996; Halberstadt 2016). However, in contrast to other hallucinogens, the effect of 5-MeO-DMT on PPI is blocked by the 5-HT$_{1A}$ antagonist WAY-100635 but not by M100907 (Krebs-Thomson and Geyer 1996). These findings demonstrate that the PPI disruption produced by 5-MeO-DMT is mediated by 5-HT$_{1A}$ but not by 5-HT$_{2A}$

receptors. 5-MeO-DMT binds to 5-HT_{1A} sites with high nanomolar affinity (Spencer et al. 1987; Glennon et al. 1988; McKenna et al. 1990) and acts as a potent agonist at 5-HT_{1A} receptors negatively coupled to adenylyl cyclase (Blair et al. 2000). Like 5-MeO-DMT, other potent 5-HT_{1A} agonists, including 8-OH-DPAT, buspirone, and gepirone, disrupt PPI in rats (Rigdon and Weatherspoon 1992; Sipes and Geyer 1995a; Johansson et al. 1995). Many of the behavioral effects of 5-MeO-DMT in rodents are mediated by 5-HT_{1A}, with the 5-HT_{2A} receptor playing only a minor role (Lucki et al. 1984; Spencer et al. 1987; Berendsen et al. 1989; Sanchez et al. 1996; Winter et al. 2000; van den Buuse et al. 2011).

We recently demonstrated that it is possible to enhance the interaction of 5-MeO-DMT with 5-HT_{2A} receptors by increasing the accumulation of the drug in the central nervous system (Halberstadt 2016). The primary route of metabolism for 5-MeO-DMT is oxidative deamination by monoamine oxidase-A (MAO_A) (Agurell et al. 1969; Squires 1975; Suzuki et al. 1981; Shen et al. 2010). Studies in rats have shown that inhibition of MAO_A increases the concentration of 5-MeO-DMT in the brain by more than an order of magnitude and reduces the clearance rate (Sitaram et al. 1987; Halberstadt 2016). We found that pretreatment with a behaviorally inactive dose of the MAO_A inhibitor clorgyline or the $MAO_{A/B}$ inhibitor pargyline markedly prolongs the effect of 5-MeO-DMT on PPI in rats (Halberstadt 2016). Furthermore, the combined effect of 5-MeO-DMT and pargyline on PPI can be antagonized by pretreatment with either WAY-100635 or MDL 11,939, indicating the effect is mediated by both 5-HT_{1A} *and* 5-HT_{2A} receptors. These results confirm that 5-HT_{2A} receptors can play a significant role in the behavioral response to 5-MeO-DMT, at least under certain conditions.

Studies have also compared the effects of LSD and lisuride on PPI in rats (Halberstadt and Geyer 2010). Lisuride produces a marked disruption of PPI and is even more potent than LSD. However, in contrast to LSD, the effect of lisuride was not blocked by MDL 11,939 but was antagonized by the dopamine $D_{2/3}$ antagonist raclopride. Therefore, the effects of LSD and lisuride on PPI are mediated by different receptor mechanisms. The fact that lisuride disrupts PPI by activating $D_{2/3}$ receptors is consistent with extensive evidence that lisuride is a potent dopaminergic drug in vitro (Piercey et al. 1996; Newman-Tancredi et al. 2002) and in vivo (Holohean et al. 1982; Kimura et al. 1991; Baladi et al. 2010).

3.2 *Hallucinogen Effects on Prepulse Inhibition in Mice*

In contrast to rats, 5-HT_{2A} receptor activation does not consistently alter PPI in mice. Dulawa and Geyer reported that administration of DOM at doses ranging from 0.5 to 4 mg/kg had no effect on PPI in C57BL/6, 129 Sv, or ICR mice (Dulawa and Geyer 2000). Another study found that 1 mg/kg DOI had no effect on PPI in wild-type (WT) mice on the C57BL/6J $\times$ 129S6/SvEvTac hybrid background, although it did disrupt PPI in pituitary adenylate cyclase-activating polypeptide (PACAP) $^{+/-}$ mice on the same background (Hazama et al. 2014). It was reported, however, that 1 and 5 mg/kg DOI significantly disrupted PPI in WT mice on a mixed 129/Sv and C57BL/6 background but had no effect on PPI in caveolin-1$^{-/-}$ mice (Allen et al. 2011).

Another difference between rats and mice is that tryptamine hallucinogens *increase* PPI in mice. The ability of tryptamine hallucinogens to increase PPI in mice was first observed with DMT (Freedland and Mansbach 1999) and was later found to occur with 5-MeO-DMT and psilocin (Halberstadt and Geyer 2011). The 5-HT$_{1A}$ receptor appears to play a role in the PPI increase produced by tryptamine hallucinogens because the effect of 5-MeO-DMT is partially attenuated by WAY-100635. Likewise, although selective 5-HT$_{1A}$ agonists disrupt PPI in rats, 8-OH-DPAT has been shown to increase PPI in mice (Dulawa et al. 2000; Gogos et al. 2008).

3.3 Hallucinogen Effects on Prepulse Inhibition in Humans

Hallucinogens alter PPI in humans but their effects depend on the interstimulus interval (ISI). Psilocybin reduces PPI at a short ISI (30 ms), but increases PPI at longer ISIs (100–2000 ms)(Gouzoulis-Mayfrank et al. 1998; Vollenweider et al. 2007). The PPI reduction produced by psilocybin at 30 ms ISI is completely blocked by pretreatment with 40 mg p.o. ketanserin (Quednow et al. 2012), indicating that the effect is mediated by 5-HT$_{2A}$ receptors. Administration of LSD at a moderately high dose (200 μg p.o.) reduces PPI at 30–120 ms ISIs (Schmid et al. 2015). In contrast to psilocybin and LSD, which alter PPI in humans, the hallucinogen DMT appears to be an exception. DMT had no effect on PPI when administered in an *ayahuasca* preparation (Riba et al. 2002) or by continuous i.v. infusion (Heekeren et al. 2007). Interestingly, the 5-HT$_{2A}$ A-1438G and T102C polymorphisms influence PPI levels in schizophrenia patients (Quednow et al. 2008) and in healthy control subjects (Quednow et al. 2009; Bräuer et al. 2009). These findings provide additional confirmation that the 5-HT$_{2A}$ receptor plays a role in the regulation of sensorimotor gating.

4 Exploratory and Investigatory Behavior

Animals are motivated to explore their surroundings in order to reduce perceptual uncertainty about the environment (Berlyne 1950, 1955, 1966; Fowler 1965). Measures of exploratory behavior, such as the amount of locomotor activity exhibited by rats in an open field, are often used to characterize the effects of psychoactive drugs. Studies investigating the effects of hallucinogens on the behavior of rats in an open field have produced inconsistent findings and fail to distinguish hallucinogens from other drug classes (Brimblecombe 1963; Dandiya et al. 1969; Gupta et al. 1971; Silva and Calil 1975). The inconclusive results are not surprising given the complexity of hallucinogen effects; univariate measures of spontaneous activity can detect arousing or sedating drug effects but reveal nothing about the qualitative nature of behavior. Locomotion alone is not necessarily the most reliable measure of exploration because it does not distinguish between nonspecific motor activity and specific exploratory responses to environmental stimuli (Hughes 1972).

The Behavioral Pattern Monitor (BPM) was developed to address some of the weaknesses associated with activity measurements in rodents. The BPM combines the features of activity and holeboard chambers and assesses both the quantity and several aspects of the quality of activity by monitoring response frequencies and spatial and temporal sequences of activity (Geyer et al. 1986). Investigatory hole-pokes and rearings are used as specific measures of inspective and diversive exploration, respectively (Berlyne 1960, 1966). By comparing measures of locomotor activity and investigatory responding, it is possible to discriminate changes in the responsivity of animals to environmental stimuli from more general stimulant or depressant effects. Statistical assessment of the geometrical and dynamical structure of motor behavior in the BPM has proven very useful in characterizing drug effects in rodents (Geyer et al. 1986; Gold et al. 1988; Paulus and Geyer 1991; Geyer and Paulus 1992). This approach has proven to be especially useful for studies of different classes of psychostimulants, which at certain doses produce comparable increases in locomotion but marked qualitative differences in behavior.

4.1 Effects in Rats

Phenylalkylamine hallucinogens (mescaline, DOI, DOM, and DOET) and indoleamine hallucinogens (LSD, DMT, 5-MeO-DMT, and psilocin) produce a characteristic pattern of effects when rats are tested in novel BPM chambers: (1) locomotor activity is reduced; (2) investigatory behaviors (rearings and hole-pokes) are suppressed; and (3) the animals spend less time in the center of the chamber (Adams and Geyer 1985a, b; Wing et al. 1990; Krebs-Thomson et al. 1998a, 2006). The effects of hallucinogens in the BPM are most apparent during the period immediately after the animals are placed in the chambers. When hallucinogens are tested in a familiar environment, the effects on locomotor activity and investigatory behavior are markedly attenuated (see Fig. 4), and therefore likely reflect potentiation of the neophobia exhibited by rats in novel settings (Adams and Geyer 1985a, b; Wing et al. 1990). In other words, hallucinogens enhance the threatening nature of the unfamiliar test environment, but hallucinogen-treated animals are more likely to explore the BPM chambers once the stimuli associated with the apparatus become less threatening due to habituation. Other groups have reported that hallucinogens reduce locomotor activity in a novel environment (Tilson et al. 1975; Hillegaart et al. 1996; Reyes-Parada et al. 1996; Scorza et al. 1996; Palenicek et al. 2010) but have no effect or increase activity in a familiar environment (Tilson et al. 1975; Ouagazzal et al. 2001; Filip et al. 2004; Zaniewska

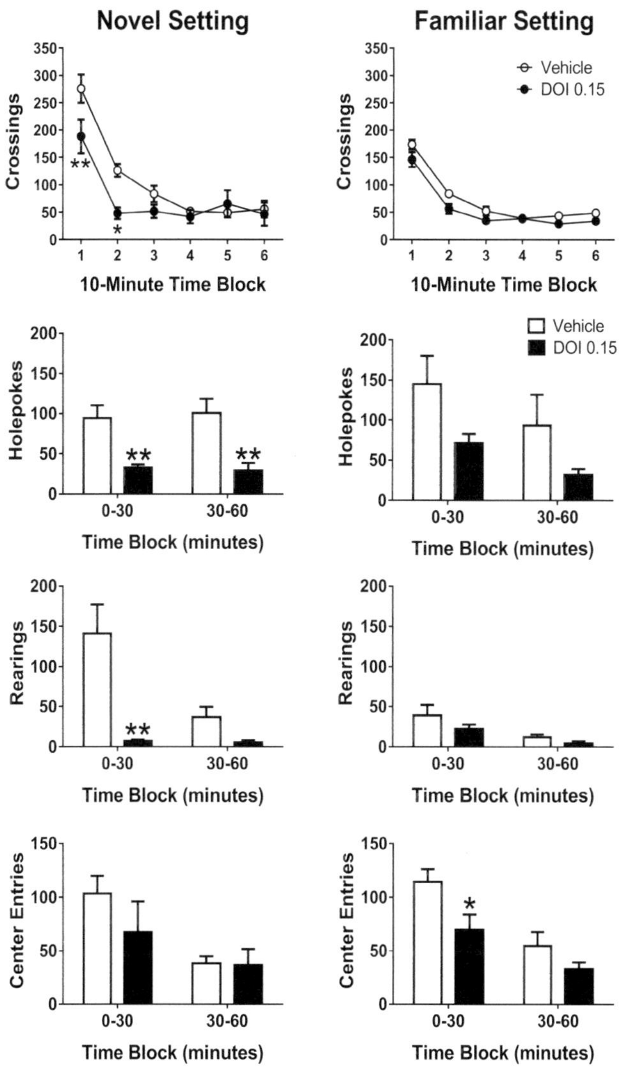

Fig. 4 Effect of habituation on the response to DOI in the rat Behavioral Pattern Monitor (BPM). Male Sprague-Dawley rats were tested in novel BPM chambers (*Left panels*) or familiar BPM chambers (*Right panels*). In the novel environment paradigm, rats (n = 5/group, 10 total) were injected SC with vehicle (saline) or DOI (0.15 mg/kg), 10 min prior to being placed in the BPM for a 60-min test session. In the familiar environment paradigm, rats (n = 8/group, 16 total) were tested in the BPM on three occasions at 48-h intervals, but only injected with the drug prior to their third exposure to the BPM chamber; rats were injected SC with vehicle (saline) or DOI (0.15 mg/kg), 10 min prior to being placed in the BPM for a 60-min test session. Data from the test sessions are presented in 10-min time blocks (crossings, a measure of locomotor activity) or 30-min time blocks (holepokes, rearings, center entries). Data are shown as mean ± SEM. *$p < 0.05$, **$p < 0.01$ versus vehicle control (Bonferroni's multiple comparisons test)

et al. 2009; Baisley et al. 2012). Hallucinogens increase center avoidance in novel *and* familiar settings, suggesting they probably also act by enhancing the fear of open spaces (agoraphobia), which is normally displayed by rodents even after habituation. In humans, it is well-known that the effects of LSD and other hallucinogens are dependent on environmental setting, with reactivity to the environment, and especially to unpleasant or threatening stimuli, being markedly enhanced (Salvatore and Hyde 1956; Cohen 1960). Hence, the increased avoidance of novel and open areas observed in rats after administration of hallucinogens may have construct validity as being analogous to the enhanced reactivity to environmental stimuli observed in humans.

In addition to the characteristic effects described above, certain hallucinogens produce behavioral profiles with greater complexity. Like other hallucinogens, LSD reduces investigatory behavior and center duration (Adams and Geyer 1985b), but it has biphasic effects on locomotor activity, initially suppressing activity but then increasing it as time progresses (Mittman and Geyer 1991). 5-MeO-DMT alone produces a brief reduction in locomotion (Wing et al. 1990; Krebs-Thomson et al. 2006), but administration of 5-MeO-DMT in combination with a behaviorally inactive dose of an MAO inhibitor (MAOI) such as harmaline, clorgyline, or pargyline produces biphasic effects on locomotor activity that are identical to those of LSD (Halberstadt et al. 2008, 2012). As with other hallucinogen effects, the delayed hyperactivity is not observed in rats habituated to the BPM chambers, indicating that the animals may be exhibiting an increased drive to explore the novel environment. Further work demonstrated that α,α,β,β-tetradeutero-5-MeO-DMT, a 5-MeO-DMT isotopologue that is resistant to metabolism by MAO, does not require an MAOI to produce a biphasic locomotor profile (Halberstadt et al. 2012). The latter finding indicates that MAOIs alter the behavioral profile of 5-MeO-DMT by slowing its biotransformation, which allows high levels of the drug to accumulate in the brain. Indeed, we found that the concentration of 5-MeO-DMT in the CNS is increased ~20-fold in animals pretreated with an MAOI (Halberstadt 2016). Reports have also appeared in the literature demonstrating that very high doses of mescaline (100 mg/kg), 2C-B (25 mg/kg), and DOM (≥ 5 mg/kg) can produce biphasic locomotor effects in rats (Yamamoto and Ueki 1975; Pálenícek et al. 2008, 2013).

The effects of hallucinogens on exploratory and investigatory behavior in the BPM are unique and are not reproduced by other pharmacological classes and thus have considerable predictive validity. Hallucinogen effects are distinct from those of 5-HT releasers (e.g., MDA and MDMA), phencyclidine (PCP) and other NMDA receptor antagonists, psychostimulants such as cocaine and (+)-amphetamine, DA receptor agonists, cholinergic agonists and antagonists, and antidepressants (Geyer et al. 1986, 1987; Gold et al. 1988; Callaway et al. 1990; Lehmann-Masten and Geyer 1991; Rempel et al. 1993; Halberstadt et al. 2016b). Selective 5-HT_{1A} agonists such as 8-OH-DPAT, ipsapirone, and gepirone produce hallucinogen-like on effects on exploratory and investigatory behavior, but these effects occur in both novel and familiar settings and therefore are thought to reflect a generalized sedative influence rather than a change in responsiveness to environmental stimuli

(Mittman and Geyer 1989). It is also possible to distinguish the effects of lisuride from those of LSD and other hallucinogens. The effects of lisuride are similar to those of apomorphine and other dopamine receptor agonists, with low doses (5–15 μg/kg SC) producing sedative effects and higher doses (60–80 μg/kg SC) producing highly stereotyped preservative patterns of hyperactivity (Adams and Geyer 1985c; Paulus and Geyer 1991).

Most of the effects of hallucinogens in the BPM are mediated by activation of the 5-HT_{2A} receptor. The effects of DOI, DOM, and mescaline on locomotor activity, investigatory behavior, and center activity are blocked by pretreatment with the 5-HT_2 antagonists ritanserin and ketanserin (Wing et al. 1990; Mittman and Geyer 1991; Hillegaart et al. 1996; Krebs-Thomson et al. 1998b). Additional studies demonstrated that the effects of DOI are blocked by M100907 but not by SER-082 (Krebs-Thomson et al. 1998a), demonstrating that they are mediated by 5-HT_{2A} but not by 5-HT_{2C} receptors. Compared with phenylalkylamine hallucinogens, the mechanism for the effects of indoleamines is more complex. For example, the mixed 5-HT_1/β-adrenergic antagonist propranolol (Mittman and Geyer 1991) and the selective 5-HT_{1A} antagonist WAY-100635 (Krebs-Thomson and Geyer 1996) block the initial suppression of locomotor activity induced by LSD, whereas ritanserin (Mittman and Geyer 1991) and M100907 (Ouagazzal et al. 2001) block the hyperactivity induced by LSD. Additionally, chronic pretreatment with either 8-OH-DPAT or DOI produces cross-tolerance to the effects of LSD on exploratory behavior (Krebs and Geyer 1994). For 5-MeO-DMT, the effects of low doses are blocked by WAY-100635 but not by M100907 (Krebs-Thomson et al. 2006). However, the delayed hyperactivity induced by administration of 5-MeO-DMT in combination with the MAOI clorgyline is blocked by the highly selective 5-HT_{2A} antagonist MDL 11,939 but not by WAY-100635 (Halberstadt et al. 2008). Thus, it appears that the effects of LSD and 5-MeO-DMT in the BPM are mediated by 5-HT_{1A} and 5-HT_{2A} receptors.

4.2 *Effects in Mice*

We have also tested hallucinogens in a mouse version of the BPM system (Tanaka et al. 2012). Administration of phenylalkylamine hallucinogens to C57BL/6J mice by the IP route produces a reduction in investigatory behavior, as well as effects on locomotor activity that follow a bell-shaped dose-response function, increasing activity at low to moderate doses and reducing activity at high doses. This specific profile of effects occurs with DOI, DOM, DOET, and mescaline, as well as with TCB-2 (Halberstadt et al. 2009, 2013). Consistent with our findings, other groups have reported that low doses of phenylalkylamine hallucinogens increase locomotor activity and high doses reduce activity in mice (Huang and Ho 1973; Yamamoto and Ueki 1975; Darmani et al. 1996; Brookshire and Jones 2009; Carlsson et al. 2011). The increase in locomotor activity produced by a low dose of DOI (1 mg/kg), mescaline (25 mg/kg), or TCB-2 (3 mg/kg) is blocked by pretreatment with M100907, and does not occur in 5-HT_{2A} $^{-/-}$ knockout mice (Halberstadt et al. 2009, 2013; Fig. 5). By contrast, the reduction in locomotor activity produced by a high dose of DOI (10 mg/kg) is potentiated in 5-HT_{2A} $^{-/-}$ knockout mice and attenuated by SER-082. Taken together, these findings indicate that low doses of phenylalkylamine hallucinogens increase locomotor activity by activating the 5-HT_{2A} receptor, and high doses reduce activity by activating the 5-HT_{2C} receptor. This conclusion is supported by the fact that selective 5-HT_{2C} agonists such as WAY-161,503 and CP-801,909 reduce locomotor activity in mice (Halberstadt et al. 2009; Fletcher et al. 2009). It is also interesting to note that the hypoactivity induced by atypical antipsychotics is reportedly mediated by blockade of the 5-HT_{2A} receptor, and that 5-HT_{2A} $^{-/-}$ but not 5-HT_{2C} $^{-/-}$ knockout mice are resistant to the suppression of locomotor activity by clozapine and olanzapine (McOmish et al. 2012). By contrast, 5-HT_{2C} blockade with the selective antagonist SB-242,084 increases locomotor activity in mice. Hence, it appears that 5-HT_{2A} and 5-HT_{2C} receptors have a countervailing influence on locomotor activity.

Compared with phenylalkylamines, tryptamine hallucinogens produce a disparate profile of effects in the mouse BPM. Administration of psilocin or 5-MeO-DMT produces a profound suppression of locomotor activity, investigatory holepokes and rearings, and center duration in C57BL/6J mice (Halberstadt et al. 2011). Most of these effects are blocked by pretreatment with the 5-HT_{1A}

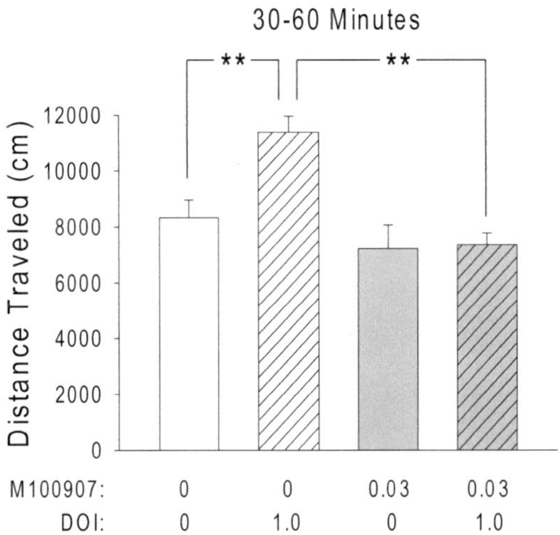

Fig. 5 Pretreatment with M100907 blocks the hyperactivity induced by DOI in mice. Male C57BL/6J mice (n = 10/group, 40 total) were treated SC with vehicle (water containing 5% Tween-80) or M100907 (0.03 mg/kg) 15 min prior to IP vehicle (saline) or DOI hydrochloride (1 mg/kg). Animals were placed in the mouse Behavioral Pattern Monitor (BPM) chambers 15 min after treatment with DOI and tested for 60 min. Administration of 1 mg/kg DOI increased locomotor activity (measured as distance traveled, in cm) during the second half of the test session; the effect of DOI was completely blocked by 0.03 mg/kg M100907. Data are shown as mean ± SEM for the second half of the test session (30–60 min). $^{**}p < 0.01$ (Tukey's test)

antagonist WAY-100635, whereas the 5-HT_{2C} antagonist SB242084 is ineffective. Likewise, the reduction of locomotor activity induced by 5-MeO-DMT is largely absent in $5\text{-HT}_{1A}{}^{-/-}$ knockout mice (van den Buuse et al. 2011). The 5-HT_{1A} agonist 8-OH-DPAT was found to produce a similar profile of effects that were not observed in mice lacking the 5-HT_{1A} receptor. It appears that tryptamines alter exploratory and investigatory activity in mice by activating 5-HT_{1A} receptors, whereas phenylalkylamines act through a mechanism involving both 5-HT_{2A} and 5-HT_{2C} receptors. The fact that phenylalkylamine and tryptamine hallucinogens produce different effects in the mouse BPM indicates this paradigm may have utility in the detection of subtle behavioral differences between these two classes of hallucinogens.

5 Conclusions

The HTR, PPI, and BPM paradigms are three of the most popular behavioral assays used to study the pharmacology of classical hallucinogens in laboratory animals. As is the case in humans, the behavioral responses induced by hallucinogens in animals are primarily mediated by 5-HT_{2A} receptor activation, although 5-HT_{1A} and 5-HT_{2C} receptors also contribute to or modulate the responses induced by certain compounds. The HTR has been widely adopted due to its insensitivity to false-positive responses and because it can be used to probe interactions between the 5-HT_{2A} receptor and other receptors and signaling pathways. In contrast to HTR, which has limited construct validity with regard to the effects of hallucinogens in humans, PPI is a cross-species readout that can be assessed in humans and animals using similar methodologies and represents clearly analogous and potentially homologous behaviors across species. Finally, the BPM paradigm has been used to assess hallucinogen effects in rodents and has clear conceptual relevance to human hallucinogen phenomenology and hence substantial construct validity. In rats, it appears that the BPM model also has predictive validity for identifying the class of hallucinogenic drugs. Importantly, however, a human version of the BPM has been developed (Young et al. 2007; Perry et al. 2009), potentially enabling human hallucinogen studies to be conducted.

This chapter has focused on the acute effects of hallucinogens, as assessed using measures of unconditioned behavior. Operant tasks such as drug discrimination have demonstrated considerable utility in studies of hallucinogen structure-activity relationships and have helped to elucidate some of the neurochemical effects of hallucinogens. However, the drug discrimination paradigm requires extended training and repeated administration of the hallucinogens. By contrast, unconditioned behavioral responses are more suited to the study of the acute effects of hallucinogens, which exhibit rapid tolerance and are more likely to be related to the subjective effects reported by humans. Though behavioral paradigms that necessitate repeated drug treatment have taught us much about the actions of hallucinogens, it is important that our field also employ behavioral tests in which acute effects that do exhibit tolerance are examined. Further, because the category of hallucinogens is defined by subjective reports of distortions of perception and affect, it is useful for animal studies of hallucinogens to incorporate specific measures of the responsivity to exteroceptive stimuli and their sensitivity to novelty (Mittman and Geyer 1989, 1991; Wing et al. 1990; Geyer 1998). Use of such measures increases the likelihood of obtaining results in laboratory animals that have construct validity and are translatable to man.

Acknowledgements This work was supported by grants from NIDA (R01 DA002925 and R01 DA041336), NIMH (K01 MH100644), and the Veteran's Affairs VISN 22 MIRECC.

References

Abramson HA, Sklarofsky B, Baron MO, Fremont-Smith N (1958) Lysergic acid diethylamide (LSD-25) antagonists. II. Development of tolerance in man to LSD-25 by prior administration of MLD-41 (1-methyl-d-lysergic acid diethylamide). Arch Neurol Psychiatry 79:201–207

Acuña-Castillo C, Villalobos C, Moya PR, Sáez P, Cassels BK, Huidobro-Toro JP (2002) Differences in potency and efficacy of a series of phenylisopropylamine/phenylethylamine pairs at 5-HT_{2A} and 5-HT_{2C} receptors. Br J Pharmacol 136:510–519

Adams LM, Geyer MA (1985a) Effects of DOM and DMT in a proposed animal model of hallucinogenic activity. Prog Neuropsychopharmacol Biol Psychiatry 9:121–132

Adams LM, Geyer MA (1985b) A proposed animal model for hallucinogens based on LSD's effects on patterns of exploration in rats. Behav Neurosci 99:881–900

Adams LM, Geyer MA (1985c) Patterns of exploration in rats distinguish lisuride from lysergic acid diethylamide. Pharmacol Biochem Behav 23:461–468

Aghajanian GK, Marek GJ (1997) Serotonin induces excitatory postsynaptic potentials in apical dendrites of neocortical pyramidal cells. Neuropharmacology 36:589–599

Agurell S, Holmstedt B, Lindgren JE (1969) Metabolism of 5-methoxy-*N*,-*N* dimethyltryptamine-^{14}C in the rat. Biochem Pharmacol 18:2771–2781

Allen JA, Yadav PN, Setola V, Farrell M, Roth BL (2011) Schizophrenia risk gene CAV1 is both pro-psychotic and required for atypical antipsychotic drug actions in vivo. Transl Psychiatry 1:e33

Appel JB, West WB, Rolandi WG, Alici T, Pechersky K (1999) Increasing the selectivity of drug discrimination procedures. Pharmacol Biochem Behav 64:353–358

Arnt J, Hyttel J (1989) Facilitation of 8-OHDPAT-induced forepaw treading of rats by the 5-HT2 agonist DOI. Eur J Pharmacol 161:45–51

Axelrod J (1962) The enzymatic N-methylation of serotonin and other amines. J Pharmacol Exp Ther 138:28–33

Baisley SK, Fallace KL, Rajbhandari AK, Bakshi VP (2012) Mutual independence of 5-HT_2 and α1 noradrenergic receptors in mediating deficits in sensorimotor gating. Psychopharmacology 220:465–479

Baladi MG, Newman AH, France CP (2010) Dopamine D3 receptors mediate the discriminative stimulus effects of quinpirole in free-feeding rats. J Pharmacol Exp Ther 332:308–315

Balsara JJ, Bapat TR, Nandal NV, Gada VP, Chandorkar AG (1986) Head-twitch response induced by ergometrine in mice: behavioural evidence for direct stimulation of central 5-hydroxytryptamine receptors by ergometrine. Psychopharmacology 88:275–278

Baumann MH, Clark RD, Rothman RB (2008) Locomotor stimulation produced by 3,4-methylenedioxymethamphetamine (MDMA) is correlated with dialysate levels of serotonin and dopamine in rat brain. Pharmacol Biochem Behav 90:208–217

Bedard P, Pycock CJ (1977) "Wet-dog" shake behaviour in the rat: a possible quantitative model of central 5-hydroxytryptamine activity. Neuropharmacology 16:663–670

Beique JC, Imad M, Mladenovic L, Gingrich JA, Andrade R (2007) Mechanism of the 5-hydroxytryptamine 2A receptor-mediated facilitation of synaptic activity in prefrontal cortex. Proc Natl Acad Sci USA 104:9870–9875

Benes H, Deissler A, Rodenbeck A, Engfer A, Kohnen R (2006) Lisuride treatment of Restless Legs Syndrome: first studies with monotherapy in de novo patients and in combination with levodopa in advanced disease. J Neural Transm 113:87–92

Berendsen HH, Broekkamp CL (1990) Behavioural evidence for functional interactions between 5-HT-receptor subtypes in rats and mice. Br J Pharmacol 101:667–673

Berendsen HH, Jenck F, Broekkamp CL (1989) Selective activation of 5HT1A receptors induces lower lip retraction in the rat. Pharmacol Biochem Behav 33:821–827

Berlyne DE (1950) Novelty and curiosity as determinants of exploratory behavior. Br J Psychol 41:68–80

Berlyne DE (1955) The arousal and satiation of perceptual curiosity in the rat. J Comp Physiol Psychol 48:238–246

Berlyne DE (1960) Conflict, arousal and curiosity. Macgraw-Hill, New-York

Berlyne DE (1966) Curiosity and exploration. Science 153:25–33

Bigwood J, Ott J, Thompson C, Neely P (1979) Entheogenic effects of ergonovine. J Psychedelic Drugs 11:147–149

Blair JB, Kurrasch-Orbaugh D, Marona-Lewicka D, Cumbay MG, Watts VJ, Barker EL, Nichols DE (2000) Effect of ring fluorination on the pharmacology of hallucinogenic tryptamines. J Med Chem 43:4701–4710

Borroto-Escuela DO, Romero-Fernandez W, Tarakanov AO, Marcellino D, Ciruela F, Agnati LF, Fuxe K (2010) Dopamine D2 and 5-hydroxytryptamine 5-HT_{2A} receptors assemble into functionally interacting heteromers. Biochem Biophys Res Commun 401:605–610

Braden MR, Parrish JC, Naylor JC, Nichols DE (2006) Molecular interaction of serotonin 5-HT_{2A} receptor residues Phe339(6.51) and Phe340(6.52) with superpotent *N*-benzyl phenethylamine agonists. Mol Pharmacol 70:1956–1964

Braff DL, Stone C, Callaway E, Geyer M, Glick I, Bali L (1978) Prestimulus effects on human startle reflex in normals and schizophrenics. Psychophysiology 15:339–343

Braff DL, Geyer MA, Swerdlow NR (2001) Human studies of prepulse inhibition of startle: normal subjects, patient groups, and pharmacological studies. Psychopharmacology 156:234–258

Brandt SD, Kavanagh PV, Westphal F, Elliott SP, Wallach J, Colestock T, Burrow TE, Chapman SJ, Stratford A, Nichols DE, Halberstadt AL (2017) Return of the lysergamides. Part II: analytical and behavioural characterization of *N*6-allyl-6-norlysergic acid diethylamide (AL-LAD) and (2'*S*,4'*S*)-lysergic acid 2,4-dimethylazetidide (LSZ). Drug Test Anal 9:38–50

Brandt SD, Kavanagh PV, Westphal F, Stratford A, Elliott SP, Hoang K, Wallach J, Halberstadt AL (2016) Return of the lysergamides. Part I: Analytical and behavioral characterization of 1-propionyl-d-lysergic acid diethylamide (1P-LSD). Drug Test Anal 8:891–902

Brauer LH, Johanson CE, Schuster CR, Rothman RB, de Wit H (1996) Evaluation of phentermine and fenfluramine, alone and in combination, in normal, healthy volunteers. Neuropsychopharmacology 14:233–241

Bräuer D, Strobel A, Hensch T, Diers K, Lesch KP, Brocke B (2009) Genetic variation of serotonin receptor function affects prepulse inhibition of the startle. J Neural Transm 116:607–613

Brea J, Castro M, Loza MI, Masaguer CF, Raviña E, Dezi C, Pastor M, Sanz F, Cabrero-Castel A, Galán-Rodríguez B, Fernández-Espejo E, Maldonado R, Robledo P (2006) QF2004B, a potential antipsychotic butyrophenone derivative with similar pharmacological properties to clozapine. Neuropharmacology 51:251–262

Brimblecombe RE (1963) Effects of psychotropic drugs on open-field behaviour in rats. Psychopharmacologia 4:139–147

Brookshire BR, Jones SR (2009) Direct and indirect 5-HT receptor agonists produce gender-specific effects on locomotor and vertical activities in C57 BL/6J mice. Pharmacol Biochem Behav 94:194–203

Burris KD, Breeding M, Sanders-Bush E (1991) (+)Lysergic acid diethylamide, but not its nonhallucinogenic congeners, is a potent serotonin 5HT1C receptor agonist. J Pharmacol Exp Ther 258:891–896

Callaway CW, Wing LL, Geyer MA (1990) Serotonin release contributes to the locomotor stimulant effects of 3,4-methylenedioxymethamphetamine in rats. J Pharmacol Exp Ther 254:456–464

Canal CE, Morgan D (2012) Head-twitch response in rodents induced by the hallucinogen 2,5-dimethoxy-4-iodoamphetamine: a comprehensive history, a re-evaluation of mechanisms, and its utility as a model. Drug Test Anal 4:556–576

Canal CE, Olaghere da Silva UB, Gresch PJ, Watt EE, Sanders-Bush E, Airey DC (2010) The serotonin 2C receptor potently modulates the head-twitch response in mice induced by a phenethylamine hallucinogen. Psychopharmacology 209:163–174

Canal CE, Booth RG, Morgan D (2013) Support for 5-HT2C receptor functional selectivity in vivo utilizing structurally diverse, selective 5-HT2C receptor ligands and the 2,5-dimethoxy-4-iodoamphetamine elicited head-twitch response model. Neuropharmacology 70:112–121

Carbonaro TM, Eshleman AJ, Forster MJ, Cheng K, Rice KC, Gatch MB (2015) The role of 5-HT2A, 5-HT 2C and mGlu2 receptors in the behavioral effects of tryptamine hallucinogens N, N-dimethyltryptamine and N, N-diisopropyltryptamine in rats and mice. Psychopharmacology 232:275–284

Carhart-Harris RL, Erritzoe D, Williams T, Stone JM, Reed LJ, Colasanti A, Tyacke RJ, Leech R, Malizia AL, Murphy K, Hobden P, Evans J, Feilding A, Wise RG, Nutt DJ (2012) Neural correlates of the psychedelic state as determined by fMRI studies with psilocybin. Proc Natl Acad Sci USA 109:2138–2143

Carlsson ML, Burstein ES, Kloberg A, Hansson S, Schedwin A, Nilsson M, Rung JP, Carlsson A (2011) I. In vivo evidence for partial agonist effects of (−)-OSU6162 and (+)-OSU6162 on 5-HT2A serotonin receptors. J Neural Transm 118:1511–1522

Ciccocioppo R, Angeletti S, Colombo G, Gessa G, Massi M (1999) Autoradiographic analysis of 5-HT2A binding sites in the brain of Sardinian alcohol-preferring and nonpreferring rats. Eur J Pharmacol 373:13–19

Cohen M (1960) LSD: Side effects and complications. J Nerv Ment Dis 130:39–45

Colpaert FC, Niemegeers CJ, Janssen PA (1982) A drug discrimination analysis of lysergic acid diethylamide (LSD): in vivo agonist and antagonist effects of purported 5-hydroxytryptamine antagonists and of pirenperone, a LSD-antagonist. J Pharmacol Exp Ther 221:206–214

Corne SJ, Pickering RW (1967) A possible correlation between drug-induced hallucinations in man and a behavioural response in mice. Psychopharmacologia 11:65–78

Corne SJ, Pickering RW, Warner BT (1963) A method for assessing the effects of drugs on the central actions of 5-hydroxytryptamine. Br J Pharmacol Chemother 20:106–120

Cussac D, Boutet-Robinet E, Ailhaud MC, Newman-Tancredi A, Martel JC, Danty N, Rauly-Lestienne I (2008) Agonist-directed trafficking of signalling at serotonin 5-HT_{2A}, 5-HT_{2B} and 5-$HT_{2C\text{-}VSV}$ receptors mediated Gq/11 activation and calcium mobilisation in CHO cells. Eur J Pharmacol 594:32–38

Dandiya PC, Gupta BD, Gupta ML, Patni SK (1969) Effects of LSD on open field performance in rats. Psychopharmacologia 15:333–340

Darmani NA (1998a) Cocaine and selective monoamine uptake blockers (sertraline, nisoxetine, and GBR 12935) prevent the *d*-fenfluramine-induced head-twitch response in mice. Pharmacol Biochem Behav 60:83–90

Darmani NA (1998b) The silent and selective 5-HT1A antagonist, WAY 100635, produces via an indirect mechanism, a 5-HT2A receptor-mediated behaviour in mice during the day but not at night. Short communication. J Neural Transm 105:635–643

Darmani NA, Pandya DK (2000) Involvement of other neurotransmitters in behaviors induced by the cannabinoid CB1 receptor antagonist SR 141716A in naive mice. J Neural Transm 107:931–945

Darmani NA, Martin BR, Pandey U, Glennon RA (1990) Do functional relationships exist between 5-HT1A and 5-HT2 receptors? Pharmacol Biochem Behav 36:901–906

Darmani NA, Mock OB, Towns LC, Gerdes CF (1994) The head-twitch response in the least shrew (*Cryptotis parva*) is a 5-HT_2- and not a 5-HT_{1C}-mediated phenomenon. Pharmacol Biochem Behav 48:383–396

Darmani NA, Shaddy J, Gerdes CF (1996) Differential ontogenesis of three DOI-induced behaviors in mice. Physiol Behav 60:1495–1500

Darmani NA, Janoyan JJ, Kumar N, Crim JL (2003) Behaviorally active doses of the CB1 receptor antagonist SR 141716A increase brain serotonin and dopamine levels and turnover. Pharmacol Biochem Behav 75:777–787

Delille HK, Becker JM, Burkhardt S, Bleher B, Terstappen GC, Schmidt M, Meyer AH, Unger L, Marek GJ, Mezler M (2012) Heterocomplex formation of 5-HT_{2A}-$mGlu_2$ and its relevance for cellular signaling cascades. Neuropharmacology 62:2184–2191

Delille HK, Mezler M, Marek GJ (2013) The two faces of the pharmacological interaction of $mGlu_2$ and 5-HT_{2A}—relevance of receptor heterocomplexes and interaction through functional brain pathways. Neuropharmacology 70:296–305

Dodge R, Louttit CM (1926) Modification of the pattern of the guinea pig's reflex response to noise. J Com Psychol 60:267–285

Drust EG, Connor JD (1983) Pharmacological analysis of shaking behavior induced by enkephalins, thyrotropin-releasing hormone or serotonin in rats: evidence for different mechanisms. J Pharmacol Exp Ther 224:148–154

Dulawa SC, Geyer MA (2000) Effects of strain and serotonergic agents on prepulse inhibition and habituation in mice. Neuropharmacology 39:2170–2179

Dulawa SC, Gross C, Stark KL, Hen R, Geyer MA (2000) Knockout mice reveal opposite roles for serotonin 1A and 1B receptors in prepulse inhibition. Neuropsychopharmacology 22:650–659

Dursun SM, Handley SL (1996) Similarities in the pharmacology of spontaneous and DOI-induced head-shakes suggest 5HT2A receptors are active under physiological conditions. Psychopharmacology 128:198–205

Egan CT, Herrick-Davis K, Miller K, Glennon RA, Teitler M (1998) Agonist activity of LSD and lisuride at cloned 5HT2Aand 5HT2Creceptors. Psychopharmacology 136:409–414

Egashira N, Shirakawa A, Okuno R, Mishima K, Iwasaki K, Oishi R, Fujiwara M (2011) Role of endocannabinoid and glutamatergic systems in DOI-induced head-twitch response in mice. Pharmacol Biochem Behav 99:52–58

Ettrup A, Holm S, Hansen M, Wasim M, Santini MA, Palner M, Madsen J, Svarer C, Kristensen JL, Knudsen GM (2013) Preclinical safety assessment of the 5-HT_{2A} receptor agonist PET radioligand [^{11}C]Cimbi-36. Mol Imaging Biol 15:376–383

Fantegrossi WE, Kiessel CL, Leach PT, Van Martin C, Karabenick RL, Chen X, Ohizumi Y, Ullrich T, Rice KC, Woods JH (2004) Nantenine: an antagonist of the behavioral and physiological effects of MDMA in mice. Psychopharmacology 173:270–277

Fantegrossi WE, Harrington AW, Eckler JR, Arshad S, Rabin RA, Winter JC, Coop A, Rice KC, Woods JH (2005a) Hallucinogen-like actions of 2,5-dimethoxy-4-(n)-propylthiophenethylamine (2C-T-7) in mice and rats. Psychopharmacology 181:496–503

Fantegrossi WE, Kiessel CL, De la Garza R, Woods JH (2005b) Serotonin synthesis inhibition reveals distinct mechanisms of action for MDMA and its enantiomers in the mouse. Psychopharmacology 181:529–536

Fantegrossi WE, Harrington AW, Kiessel CL, Eckler JR, Rabin RA, Winter JC, Coop A, Rice KC, Woods JH (2006) Hallucinogen-like actions of 5-methoxy-N, N-diisopropyltryptamine in mice and rats. Pharmacol Biochem Behav 83:122–129

Fantegrossi WE, Reissig CJ, Katz EB, Yarosh HL, Rice KC, Winter JC (2008) Hallucinogen-like effects of N, N-dipropyltryptamine (DPT): possible mediation by serotonin 5-HT1A and 5-HT2A receptors in rodents. Pharmacol Biochem Behav 88:358–365

Fantegrossi WE, Simoneau J, Cohen MS, Zimmerman SM, Henson CM, Rice KC, Woods JH (2010) Interaction of 5-HT2A and 5-HT2C receptors in R(−)-2,5-dimethoxy-4-iodoamphetamine-elicited head twitch behavior in mice. J Pharmacol Exp Ther 335:728–734
Fantegrossi WE, Gray BW, Bailey JM, Smith DA, Hansen M, Kristensen JL (2014) Hallucinogen-like effects of 2-([2-(4-cyano-2,5-dimethoxyphenyl) ethylamino]methyl)phenol (25CN-NBOH), a novel N-benzylphenethylamine with 100-fold selectivity for 5-HT_{2A} receptors, in mice. Psychopharmacology 232:1039–1047
Filip M, Bubar MJ, Cunningham KA (2004) Contribution of serotonin (5-hydroxytryptamine; 5-HT) 5-HT2 receptor subtypes to the hyperlocomotor effects of cocaine: acute and chronic pharmacological analyses. J Pharmacol Exp Ther 310:1246–1254
Fiorella D, Rabin RA, Winter JC (1995) Role of 5-HT2A and 5-HT2C receptors in the stimulus effects of hallucinogenic drugs. II: reassessment of LSD false positives. Psychopharmacology 121:357–363
Fleshler M (1965) Adequate acoustic stimulus for the startle reflex in the rat. J Com Physiol Psychol 60:200–207
Fletcher PJ, Tampakeras M, Sinyard J, Slassi A, Isaac M, Higgins GA (2009) Characterizing the effects of 5-HT_{2C} receptor ligands on motor activity and feeding behaviour in 5-HT_{2C} receptor knockout mice. Neuropharmacology 57:259–267
Fowler H (1965) Curiosity and exploratory behavior. Macmillan, New York
Fox MA, French HT, LaPorte JL, Blackler AR, Murphy DL (2010a) The serotonin 5-HT_{2A} receptor agonist TCB-2: a behavioral and neurophysiological analysis. Psychopharmacology 212:13–23
Fox MA, Stein AR, French HT, Murphy DL (2010b) Functional interactions between 5-HT2A and presynaptic 5-HT1A receptor-based responses in mice genetically deficient in the serotonin 5-HT transporter (SERT). Br J Pharmacol 159:879–887
Freedland CS, Mansbach RS (1999) Behavioral profile of constituents in ayahuasca, an Amazonian psychoactive plant mixture. Drug Alcohol Depend 54:183–194
Garcia EE, Smith RL, Sanders-Bush E (2007) Role of G_q protein in behavioral effects of the hallucinogenic drug 1-(2,5-dimethoxy-4-iodophenyl)-2-aminopropane. Neuropharmacology 52:1671–1677
Gewirtz JC, Marek GJ (2000) Behavioral evidence for interactions between a hallucinogenic drug and group II metabotropic glutamate receptors. Neuropsychopharmacology 23:569–576
Geyer MA (1998) Behavioral studies of hallucinogenic drugs in animals: implications for schizophrenia research. Pharmacopsychiatry 31:73–79
Geyer MA, Krebs KM (1994) Serotonin receptor involvement in an animal model of the acute effects of hallucinogens. NIDA Res Monograph Ser 146:124–156
Geyer MA, Paulus MP (1992) Multivariate and nonlinear approaches to characterizing drug effects on the locomotor and investigatory behavior of rats. NIDA Res Monogr 124:203–235
Geyer MA, Russo PV, Masten VL (1986) Multivariate assessment of locomotor behavior: pharmacological and behavioral analyses. Pharmacol Biochem Behav 25:277–288
Geyer MA, Russo PV, Segal DS, Kuczenski R (1987) Effects of apomorphine and amphetamine on patterns of locomotor and investigatory behavior in rats. Pharmacol Biochem Behav 28:393–399
Glennon RA (1991) Discriminative stimulus properties of hallucinogens and related designer drugs. NIDA Res Monogr 116:25–44
Glennon RA, Hauck AE (1985) Mechanistic studies on DOM as a discriminative stimulus. Pharmacol Biochem Behav 23:937–941
Glennon RA, Young R, Rosecrans JA (1982a) Discriminative stimulus properties of DOM and several molecular modifications. Pharmacol Biochem Behav 16:553–556
Glennon RA, Young R, Rosecrans JA (1982b) A comparison of the behavioral effects of DOM homologs. Pharmacol Biochem Behav 16:557–559
Glennon RA, Young R, Rosecrans JA (1983) Antagonism of the effects of the hallucinogen DOM and the purported 5-HT agonist quipazine by 5-HT2 antagonists. Eur J Pharmacol 91:189–196

Glennon RA, Titeler M, McKenney JD (1984) Evidence for 5-HT2 involvement in the mechanism of action of hallucinogenic agents. Life Sci 35:2505–2511

Glennon RA, McKenney JD, Lyon RA, Titeler M (1986) 5-HT_1 and 5-HT_2 binding characteristics of 1-(2,5-dimethoxy-4-bromophenyl)-2-aminopropane analogues. J Med Chem 29:194–199

Glennon RA, Titeler M, Seggel MR, Lyon RA (1987) N-Methyl derivatives of the 5-HT2 agonist 1-(4-bromo-2,5-dimethoxyphenyl)-2-aminopropane. J Med Chem 30:930–932

Glennon RA, Titeler M, Lyon RA, Slusher RM (1988) N, N-di-n-propylserotonin: binding at serotonin binding sites and a comparison with 8-hydroxy-2-(di-n-propylamino)tetralin. J Med Chem 31:867–870

Glennon RA, Dukat M, el-Bermawy M, Law H, De los Angeles J, Teitler M, King A, Herrick-Davis K (1994) Influence of amine substituents on 5-HT2A versus 5-HT2C binding of phenylalkyl- and indolylalkylamines. J Med Chem 37:1929–1935

Gogos A, Bogeski M, van den Buuse M (2008) Role of serotonin-1A receptors in the action of antipsychotic drugs: comparison of prepulse inhibition studies in mice and rats and relevance for human pharmacology. Behav Pharmacol 19:548–561

Gold LH, Koob GF, Geyer MA (1988) Stimulant and hallucinogenic behavioral profiles of 3,4-methylenedioxymethamphetamine and N-ethyl-3,4-methylenedioxyamphetamine in rats. J Pharmacol Exp Ther 247:547–555

González-Maeso J, Weisstaub NV, Zhou M, Chan P, Ivic L, Ang R, Lira A, Bradley-Moore M, Ge Y, Zhou Q, Sealfon SC, Gingrich JA (2007) Hallucinogens recruit specific cortical 5-HT_{2A} receptor-mediated signaling pathways to affect behavior. Neuron 53:439–452

González-Maeso J, Ang RL, Yuen T, Chan P, Weisstaub NV, López-Giménez JF, Zhou M, Okawa Y, Callado LF, Milligan G, Gingrich JA, Filizola M, Meana JJ, Sealfon SC (2008) Identification of a serotonin/glutamate receptor complex implicated in psychosis. Nature 452:93–97

Gorzalka BB, Hill MN, Sun JC (2005) Functional role of the endocannabinoid system and AMPA/kainate receptors in 5-HT2A receptor-mediated wet dog shakes. Eur J Pharmacol 516:28–33

Gouzoulis-Mayfrank E, Heekeren K, Thelen B, Lindenblatt H, Kovar KA, Sass H, Geyer MA (1998) Effects of the hallucinogen psilocybin on habituation and prepulse inhibition of the startle reflex in humans. Behav Pharmacol 9:561–566

Griffiths RR, Richards WA, McCann U, Jesse R (2006) Psilocybin can occasion mystical-type experiences having substantial and sustained personal meaning and spiritual significance. Psychopharmacology 187:268–283

Grillon C, Ameli R, Charney DS, Krystal J, Braff DL (1992) Startle gating deficits occur across prepulse intensities in schizophrenic patients. Biol Psychiatry 32:939–943

Grob CS, Danforth AL, Chopra GS, Hagerty M, McKay CR, Halberstadt AL, Greer GR (2011) Pilot study of psilocybin treatment for anxiety in patients with advanced-stage cancer. Arch Gen Psychiatry 68:71–78

Grotewiel MS, Chu H, Sanders-Bush E (1994) m-Chlorophenylpiperazine and m-trifluoromethylphenylpiperazine are partial agonists at cloned 5-HT2A receptors expressed in fibroblasts. J Pharmacol Exp Ther 271:1122–1126

Gupta BD, Dandiya PC, Gupta ML, Gabba AK (1971) An examination of the effect of central nervous system stimulant and anti-depressant drugs on open field performance in rats. Eur J Pharmacol 13:341–346

Halberstadt AL (2015) Recent advances in the neuropsychopharmacology of hallucinogens. Behav Brain Res 277:99–120

Halberstadt AL (2016) Behavioral and pharmacokinetic interactions between monoamine oxidase inhibitors and the hallucinogen 5-methoxy-N, N-dimethyltryptamine. Pharmacol Biochem Behav 143:1–10

Halberstadt AL, Geyer MA (2010) LSD but not lisuride disrupts prepulse inhibition in rats by activating the 5-HT_{2A} receptor. Psychopharmacology 208:179–189

Halberstadt AL, Geyer MA (2011) Multiple receptors contribute to the behavioral effects of indoleamine hallucinogens. Neuropharmacology 61:364–381

Halberstadt AL, Geyer MA (2013) Characterization of the head-twitch response induced by hallucinogens in mice: detection of the behavior based on the dynamics of head movement. Psychopharmacology 227:727–739

Halberstadt AL, Geyer MA (2014) Effect of the hallucinogen 2C-I and two superpotent *N*-benzyl derivatives on the head twitch response. Neuropharmacology 77:200–207

Halberstadt AL, Buell MR, Masten V, Risbrough VB, Geyer MA (2008) Modification of the effects of 5-methoxy-N,N-dimethyltryptamine on exploratory behavior in rats by monoamine oxidase inhibitors. Psychopharmacology 201:55–66

Halberstadt AL, van der Heijden I, Ruderman M, Risbrough VB, Gingrich JA, Geyer MA, Powell SB (2009) Opposing effects of 5-HT_{2A} and 5-HT_{2C} receptors on locomotor activity in mice. Neuropsychopharmacology 34:1958–1967

Halberstadt AL, Koedood L, Powell SB, Geyer MA (2011) Differential contributions of serotonin receptors to the behavioral effects of indoleamine hallucinogens in mice. J Psychopharmacol 25:1548–1561

Halberstadt AL, Nichols DE, Geyer MA (2012) Behavioral effects of α, α, β, β-tetradeutereo-5-MeO-DMT in rats: comparison with 5-MeO-DMT administered in combination with a monoamine oxidase inhibitor. Psychopharmacology 221:709–718

Halberstadt AL, Powell SB, Geyer MA (2013) Role of the 5-HT_{2A} receptor in the locomotor hyperactivity produced by phenylalkylamine hallucinogens in mice. Neuropharmacology 70:218–227

Halberstadt AL, Sindhunata I, Scheffers K, Flynn AD, Sharp RF, Geyer MA, Young JW (2016a) Effect of 5-HT_{2A} and 5-HT_{2C} receptors on temporal discrimination by mice. Neuropharmacology 107:364–375

Halberstadt AL, Slepak N, Hyun J, Buell MR, Powell SB (2016b) The novel ketamine analog methoxetamine produces dissociative-like behavioral effects in rodents. Psychopharmacology 233:1215–1225

Hansen M, Phonekeo K, Paine JS, Leth-Petersen S, Begtrup M, Bräuner-Osborne H, Kristensen JL (2014) Synthesis and structure-activity relationships of N-benzyl phenethylamines as 5-HT2A/2C agonists. ACS Chem Neurosci 5:243–249

Hawkins MF, Uzelac SM, Baumeister AA, Hearn JK, Broussard JI, Guillot TS (2002) Behavioral responses to stress following central and peripheral injection of the 5-HT_2 agonist DOI. Pharmacol Biochem Behav 73:537–544

Hazama K, Hayata-Takano A, Uetsuki K, Kasai A, Encho N, Shintani N, Nagayasu K, Hashimoto R, Reglodi D, Miyakawa T, Nakazawa T, Baba A, Hashimoto H (2014) Increased behavioral and neuronal responses to a hallucinogenic drug in PACAP heterozygous mutant mice. PLoS ONE 9:e89153

Heal DJ, Philpot J, O'Shaughnessy KM, Davies CL (1986) The influence of central noradrenergic function on 5-HT2-mediated head-twitch responses in mice: possible implications for the actions of antidepressant drugs. Psychopharmacology 89:414–420

Heekeren K, Neukirch A, Daumann J, Stoll M, Obradovic M, Kovar KA, Geyer MA, Gouzoulis-Mayfrank E (2007) Prepulse inhibition of the startle reflex and its attentional modulation in the human S-ketamine and N, N-dimethyltryptamine (DMT) models of psychosis. J Psychopharmacol 21:312–320

Herrmann WM, Horowski R, Dannehl K, Kramer U, Lurati K (1977) Clinical effectiveness of lisuride hydrogen maleate: a double-blind trial versus methysergide. Headache 17:54–60

Hillegaart V, Estival A, Ahlenius S (1996) Evidence for specific involvement of 5-HT1A and 5-HT2A/C receptors in the expression of patterns of spontaneous motor activity of the rat. Eur J Pharmacol 295:155–161

Hiramatsu M, Nabeshima T, Kameyama T, Maeda Y, Cho AK (1989) The effect of optical isomers of 3,4-methylenedioxymethamphetamine (MDMA) on stereotyped behavior in rats. Pharmacol Biochem Behav 33:343–347

Hoffman HS, Ison JR (1980) Reflex modification in the domain of startle: I. Some empirical findings and their implications for how the nervous system processes sensory input. Psychol Rev 87:175–189

Hoffman HS, Searle JL (1965) Acoustic variables in the modification of startle reaction in the rat. J Comp Physiol Psychol 60:53–58
Hoffman HS, Searle JL (1968) Acoustic and temporal factors in the evocation of startle. J Acoustic Soc Am 43:269–282
Holohean AM, White FJ, Appel JB (1982) Dopaminergic and serotonergic mediation of the discriminable effects of ergot alkaloids. Eur J Pharmacol 81:595–602
Huang JT, Ho BT (1973) Differentiation of pressor and behavioral effects of d-amphetamine and 2,5-dimethoxy-4-methylamphetamine (STP) by cinanserin. Can J Physiol Pharmacol 51:976–980
Hughes RN (1972) Chlordiazepoxide modified exploration in rats. Psychopharmacologia 24:462–469
Jacobs BL, Trulson ME, Stern WC (1976) An animal behavior model for studying the actions of LSD and related hallucinogens. Science 194:741–743
Jacobs BL, Trulson ME, Stern WC (1977) Behavioral effects of LSD in the cat: proposal of an animal behavior model for studying the actions of hallucinogenic drugs. Brain Res 132:301–314
Johansson C, Jackson DM, Zhang J, Svensson L (1995) Prepulse inhibition of acoustic startle, a measure of sensorimotor gating: effects of antipsychotics and other agents in rats. Pharmacol Biochem Behav 52:649–654
Johnson MP, Hoffman AJ, Nichols DE (1986) Effects of the enantiomers of MDA, MDMA and related analogues on [^{3}H]serotonin and [^{3}H]dopamine release from superfused rat brain slices. Eur J Pharmacol 132:269–276
Kantor RE, Dudlettes SD, Shulgin AT (1980) 5-Methoxy-α-methyltryptamine (α, O-dimethylserotonin) a hallucinogenic homolog of serotonin. Biol Psychiatry 15:349–352
Kärkkäinen J, Forsström T, Tornaeus J, Wähälä K, Kiuru P, Honkanen A, Stenman UH, Turpeinen U, Hesso A (2005) Potentially hallucinogenic 5-hydroxytryptamine receptor ligands bufotenine and dimethyltryptamine in blood and tissues. Scand J Clin Lab Invest 65:189–199
Keller DL, Umbreit WW (1956) Permanent alteration of behavior in mice by chemical and psychological means. Science 124:723–724
Kennett GA, Wood MD, Glen A, Grewal S, Forbes I, Gadre A, Blackburn TP (1994) In vivo properties of SB 200646A, a 5-HT2C/2B receptor antagonist. Br J Pharmacol 111:797–802
Kimura KK, Akai TT, Nakamura KK, Yamaguchi MM, Nakagawa H, Oshino NN (1991) Dual activation by lisuride of central serotonin 5-HT$_{1A}$ and dopamine D$_2$ receptor sites: drug discrimination and receptor binding studies. Behav Pharmacol 2:105–112
Klodzinska A, Bijak M, Tokarski K, Pilc A (2002) Group II mGlu receptor agonists inhibit behavioural and electrophysiological effects of in mice. Pharmacol Biochem Behav 73:327–332
Knauer CS, Campbell JE, Chio CL, Fitzgerald LW (2009) Pharmacological characterization of mitogen-activated protein kinase activation by recombinant human 5-HT2C, 5-HT2A, and 5-HT2B receptors. Naunyn Schmiedebergs Arch Pharmacol 379:461–471
Kometer M, Schmidt A, Jäncke L, Vollenweider FX (2013) Activation of serotonin 2A receptors underlies the psilocybin-induced effects on α oscillations, N170 visual-evoked potentials, and visual hallucinations. J Neurosci 33:10544–10551
Krebs KM, Geyer MA (1994) Cross-tolerance studies of serotonin receptors involved in behavioral effects of LSD in rats. Psychopharmacology 113:429–437
Krebs-Thomson K, Geyer MA (1996) The role of 5-HT1A receptors in the locomotor-suppressant effects of LSD: WAY-100635 studies of 8-OH-DPAT, DOI and LSD in rats. Behav Pharmacol 7:551–559
Krebs-Thomson K, Lehmann-Masten V, Naiem S, Paulus MP, Geyer MA (1998a) Modulation of phencyclidine-induced changes in locomotor activity and patterns in rats by serotonin. Eur J Pharmacol 343:135–143
Krebs-Thomson K, Paulus MP, Geyer MA (1998b) Effects of hallucinogens on locomotor and investigatory activity and patterns: influence of 5-HT2A and 5-HT2C receptors. Neuropsychopharmacology 18:339–351

Krebs-Thomson K, Ruiz EM, Masten V, Buell M, Geyer MA (2006) The roles of 5-HT1A and 5-HT2 receptors in the effects of 5-MeO-DMT on locomotor activity and prepulse inhibition in rats. Psychopharmacology 189:319–329

Kurrasch-Orbaugh DM, Watts VJ, Barker EL, Nichols DE (2003) Serotonin 5-hydroxytryptamine$_{2A}$ receptor-coupled phospholipase C and phospholipase A_2 signaling pathways have different receptor reserves. J Pharmacol Exp Ther 304:229–237

Lehmann-Masten VD, Geyer MA (1991) Spatial and temporal patterning distinguishes the locomotor activating effects of dizocilpine and phencyclidine in rats. Neuropharmacology 30:629–636

Leng A, Ouagazzal A, Feldon J, Higgins GA (2003) Effect of the 5-HT6 receptor antagonists Ro04-6790 and Ro65-7199 on latent inhibition and prepulse inhibition in the rat: comparison to clozapine. Pharmacol Biochem Behav 75:281–288

Leysen JE, Niemegeers CJ, Van Nueten JM, Laduron PM (1982) [^{3}H]Ketanserin (R 41 468), a selective 3H-ligand for serotonin2 receptor binding sites. Binding properties, brain distribution, and functional role. Mol Pharmacol 21:301–314

Lucki I, Minugh-Purvis N (1987) Serotonin-induced head shaking behavior in rats does not involve receptors located in the frontal cortex. Brain Res 420:403–406

Lucki I, Nobler MS, Frazer A (1984) Differential actions of serotonin antagonists on two behavioral models of serotonin receptor activation in the rat. J Pharmacol Exp Ther 228:133–139

Lukasiewicz S, Polit A, Kędracka-Krok S, Wędzony K, Maćkowiak M, Dziedzicka-Wasylewska M (2010) Hetero-dimerization of serotonin 5-HT$_{2A}$ and dopamine D_2 receptors. Biochim Biophys Acta 1803:1347–1358

Lyon RA, Glennon RA, Titeler M (1986) 3,4-Methylenedioxymethamphetamine (MDMA): stereoselective interactions at brain 5-HT$_1$ and 5-HT$_2$ receptors. Psychopharmacology 88:525–526

Mackeprang T, Kristiansen KT, Glenthoj BY (2002) Effects of antipsychotics on prepulse inhibition of the startle response in drug-naïve schizophrenic patients. Biol Psychiatry 52:863–873

Malick JB, Doren E, Barnett A (1977) Quipazine-induced head-twitch in mice. Pharmacol Biochem Behav 6:325–329

Marini JL, Jacobs BL, Sheard MH, Trulson ME (1981) Activity of a non-hallucinogenic ergoline derivative, lisuride, in an animal behavior model for hallucinogens. Psychopharmacology 73:328–331

Marquis KL, Sabb AL, Logue SF, Brennan JA, Piesla MJ, Comery TA, Grauer SM, Ashby CR Jr, Nguyen HQ, Dawson LA, Barrett JE, Stack G, Meltzer HY, Harrison BL, Rosenzweig-Lipson S (2007) WAY-163909 [(7*bR*,10*aR*)-1,2,3,4,8,9,10,10*a*-octahydro-7*bH*-cyclopenta-[*b*][1,4]di-azepino[6,7,1hi]indole]: a novel 5-hydroxytryptamine 2C receptor-selective agonist with preclinical antipsychotic-like activity. J Pharmacol Exp Ther 320:486–496

Matthews WD, Smith CD (1980) Pharmacological profile of a model for central serotonin receptor activation. Life Sci 26:1397–1403

Mavlyutov TA, Epstein ML, Liu P, Verbny YI, Ziskind-Conhaim L, Ruoho AE (2012) Development of the sigma-1 receptor in C-terminals of motoneurons and colocalization with the N, N'-dimethyltryptamine forming enzyme, indole-N-methyl transferase. Neuroscience 206:60–68

May JA, Dantanarayana AP, Zinke PW, McLaughlin MA, Sharif NA (2006) 1-((S)-2-aminopropyl)-1H-indazol-6-ol: a potent peripherally acting 5-HT2 receptor agonist with ocular hypotensive activity. J Med Chem 49:318–328

McKenna DJ, Repke DB, Lo L, Peroutka SJ (1990) Differential interactions of indolealkylamines with 5-hydroxytryptamine receptor subtypes. Neuropharmacology 29(3):193–198

McOmish CE, Lira A, Hanks JB, Gingrich JA (2012) Clozapine-induced locomotor suppression is mediated by 5-HT2A receptors in the forebrain. Neuropsychopharmacology 37:2747–2755

Mittman SM, Geyer MA (1989) Effects of 5HT-1A agonists on locomotor and investigatory behaviors in rats differ from those of hallucinogens. Psychopharmacology 98:321–329

Mittman SM, Geyer MA (1991) Disassociation of multiple effects of acute LSD on exploratory behavior in rats by ritanserin and propranolol. Psychopharmacology 105:69–76

Molinaro G, Traficante A, Riozzi B, Di Menna L, Curto M, Pallottino S, Nicoletti F, Bruno V, Battaglia G (2009) Activation of mGlu2/3 metabotropic glutamate receptors negatively regulates the stimulation of inositol phospholipid hydrolysis mediated by 5-hydroxytryptamine2A serotonin receptors in the frontal cortex of living mice. Mol Pharmacol 76:379–387

Moreno JL, Holloway T, Albizu L, Sealfon SC, González-Maeso J (2011) Metabotropic glutamate mGlu2 receptor is necessary for the pharmacological and behavioral effects induced by hallucinogenic 5-HT2A receptor agonists. Neurosci Lett 493:76–79

Mos J, Olivier B, Tulp MThM (1992) Ethopharmacological studies differentiate the effects of various serotonergic compounds on agression in rats. Drug Dev Res 26:343–360

Moya PR, Berg KA, Gutiérrez-Hernandez MA, Sáez-Briones P, Reyes-Parada M, Cassels BK, Clarke WP (2007) Functional selectivity of hallucinogenic phenethylamine and phenylisopropylamine derivatives at human 5-hydroxytryptamine (5-HT)2A and 5-HT2C receptors. J Pharmacol Exp Ther 321:1054–1061

Murnane KS, Fantegrossi WE, Godfrey JR, Banks ML, Howell LL (2010) Endocrine and neurochemical effects of 3,4-methylenedioxymethamphetamine and its stereoisomers in rhesus monkeys. J Pharmacol Exp Ther 334:642–650

Muschamp JW, Regina MJ, Hull EM, Winter JC, Rabin RA (2004) Lysergic acid diethylamide and [−]-2,5-dimethoxy-4-methylamphetamine increase extracellular glutamate in rat prefrontal cortex. Brain Res 1023:134–140

Nakagawasai O, Murata A, Arai Y, Ohba A, Wakui K, Mitazaki S, Niijima F, Tan-No K, Tadano T (2007) Enhanced head-twitch response to 5-HT-related agonists in thiamine-deficient mice. J Neural Transm (Vienna) 114:1003–1010

Nash JF, Roth BL, Brodkin JD, Nichols DE, Gudelsky GA (1994) Effect of the R(−) and S(+) isomers of MDA and MDMA on phosphatidyl inositol turnover in cultured cells expressing 5-HT_{2A} or 5-HT_{2C} receptors. Neurosci Lett 177:111–115

Newman-Tancredi A, Cussac D, Audinot V, Nicolas JP, De Ceuninck F, Boutin JA, Millan MJ (2002) Differential actions of antiparkinson agents at multiple classes of monoaminergic receptor. II. Agonist and antagonist properties at subtypes of dopamine D_2-like receptor and $alpha_1$/$alpha_2$-adrenoceptor. J Pharmacol Exp Ther 303:805–814

Nichols DE (2012) Structure-activity relationships of serotonin 5-HT_{2A} agonists. Wiley Interdiscip Rev Membr Transp Signal 1:559–579

Nichols DE, Pfister WR, Yim GK (1978) LSD and phenethylamine hallucinogens: new structural analogy and implications for receptor geometry. Life Sci 22:2165–2170

Nichols DE, Sassano MF, Halberstadt AL, Klein LM, Brandt SD, Elliott SP, Fiedler WJ (2015) N-benzyl-5-methoxytryptamines as potent serotonin 5-HT2 receptor family agonists and comparison with a series of phenethylamine analogues. ACS Chem Neurosci 6:165–1175

Ouagazzal A, Grottick AJ, Moreau J, Higgins GA (2001) Effect of LSD on prepulse inhibition and spontaneous behavior in the rat. A pharmacological analysis and comparison between two rat strains. Neuropsychopharmacology 25:565–575

Padich RA, McCloskey TC, Kehne JH (1996) 5-HT modulation of auditory and visual sensorimotor gating: II. Effects of the 5-HT2A antagonist MDL 100,907 on disruption of sound and light prepulse inhibition produced by 5-HT agonists in Wistar rats. Psychopharmacology 124:107–116

Palenicek T, Hlinak Z, Bubenikova-Valesova V, Novak T, Horacek J (2010) Sex differences in the effects of N, N-diethyllysergamide (LSD) on behavioural activity and prepulse inhibition. Prog Neuropsychopharmacol Biol Psychiatry 34:588–596

Pálenícek T, Balíková M, Bubeníková-Valesová V, Horácek J (2008) Mescaline effects on rat behavior and its time profile in serum and brain tissue after a single subcutaneous dose. Psychopharmacology 196:51–62

Pálenícek T, Fujáková M, Brunovskỳ M, Horácek J, Gorman I, Balíková M, et al. (2013) Behavioral, neurochemical and pharmaco-EEG profiles of the psychedelic drug 4-bromo-2,5-dimethoxyphenethylamine (2C-B) in rats. Psychopharmacology (Berl) 225:75–93

Parrish JC, Braden MR, Gundy E, Nichols DE (2005) Differential phospholipase C activation by phenylalkylamine serotonin 5-HT$_{2A}$ receptor agonists. J Neurochem 95:1575–1584
Paulus MP, Geyer MA (1991) A temporal and spatial scaling hypothesis for the behavioral effects of psychostimulants. Psychopharmacology 104:6–16
Perry W, Minassian A, Paulus MP, Young JW, Kincaid MJ, Ferguson EJ, Henry BL, Zhuang X, Masten VL, Sharp RF, Geyer MA (2009) A reverse-translational study of dysfunctional exploration in psychiatric disorders: from mice to men. Arch Gen Psychiatry 66:1072–1080
Piercey MF, Hoffmann WE, Smith MW, Hyslop DK (1996) Inhibition of dopamine neuron firing by pramipexole, a dopamine D$_3$ receptor-preferring agonist: comparison to other dopamine receptor agonists. Eur J Pharmacol 312:35–44
Pranzatelli MR (1990) Evidence for involvement of 5-HT2 and 5-HT1C receptors in the behavioral effects of the 5-HT agonist 1-(2,5-dimethoxy-4-iodophenyl aminopropane)-2 (DOI). Neurosci Lett 115:74–80
Quednow BB, Kühn K-U, Mössner R, Schwab SG, Schuhmacher A, Maier W, Wagner M (2008) Sensorimotor gating of schizophrenia patients is influenced by 5-HT2A receptor polymorphisms. Biol Psychiatry 64:434–437
Quednow BB, Schmechtig A, Ettinger U, Petrovsky N, Collier DA, Vollenweider FX, Wagner M, Kumari V (2009) Sensorimotor gating depends on polymorphisms of the serotonin-2A receptor and catechol-O-methyltransferase, but not on neuregulin-1 Arg38Gln genotype: a replication study. Biol Psychiatry 66:614–620
Quednow BB, Kometer M, Geyer MA, Vollenweider FX (2012) Psilocybin-induced deficits in automatic and controlled inhibition are attenuated by ketanserin in healthy human volunteers. Neuropsychopharmacology 37:630–640
Rabin RA, Regina M, Doat M, Winter JC (2002) 5-HT$_{2A}$ receptor-stimulated phosphoinositide hydrolysis in the stimulus effects of hallucinogens. Pharmacol Biochem Behav 72:29–37
Raffaelli E Jr, Martins OJ, dos Santos P, Dãgua Filho A (1983) Lisuride in cluster headache. Headache 23:117–121
Rempel NL, Callaway CW, Geyer MA (1993) Serotonin1B receptor activation mimics behavioral effects of presynaptic serotonin release. Neuropsychopharmacology 8:201–211
Reyes-Parada M, Scorza C, Romero V, Silveira R, Medina JH, Andrus D, Nichols DE, Cassels BK (1996) (±)-1-(2,5-Dimethoxy-4-ethylthiophenyl)-2-aminopropane (ALEPH-2), a novel putative anxiolytic agent lacking affinity for benzodiazepine sites and serotonin-1A receptors. Naunyn Schmiedebergs Arch Pharmacol 354:579–585
Riba J, Rodríguez-Fornells A, Barbanoj MJ (2002) Effects of ayahuasca on sensory and sensorimotor gating in humans as measured by P50 suppression and prepulse inhibition of the startle reflex, respectively. Psychopharmacology 165:18–28
Rigdon GC, Weatherspoon JK (1992) 5-Hydroxytryptamine$_{1a}$ receptor agonists block prepulse inhibition of acoustic startle reflex. J Pharmacol Exp Ther 263:486–493
Rusterholz DB, Spratt JL, Long JP, Kelly TF (1978) Serotonergic and dopaminergic involvement in the mechanism of action of R-(-)-2,5-dimethoxy-4-bromoamphetamine (DOB) in cats. Life Sci 23:1499–1506
Salvatore S, Hyde RW (1956) Progression of effects of lysergic acid diethylamide (LSD). Arch Neurol Psychiat 2:50–59
Sanchez C, Arnt J, Moltzen E (1996) Assessment of relative efficacies of 5-HT$_{1A}$ receptor ligands by means of in vivo animal models. Eur J Pharmacol 315:245–254
Sard H, Kumaran G, Morency C, Roth BL, Toth BA, He P, Shuster L (2005) SAR of psilocybin analogs: discovery of a selective 5-HT$_{2C}$ agonist. Bioorg Med Chem Lett 15:4555–4559
Schlemmer RF Jr, Davis JM (1986) A primate model for the study of hallucinogens. Pharmacol Biochem Behav 24:381–392
Schmid CL, Bohn LM (2010) Serotonin, but not N-methyltryptamines, activates the serotonin 2A receptor via a ß-arrestin2/Src/Akt signaling complex in vivo. J Neurosci 30:13513–13524
Schmid CL, Raehal KM, Bohn LM (2008) Agonist-directed signaling of the serotonin 2A receptor depends on beta-arrestin-2 interactions in vivo. Proc Natl Acad Sci USA 105:1079–1084

Schmid Y, Enzler F, Gasser P, Grouzmann E, Preller KH, Vollenweider FX, Brenneisen R, Müller F, Borgwardt S, Liechti ME (2015) Acute effects of lysergic acid diethylamide in healthy subjects. Biol Psychiatry 78:544–553

Schoepp DD (2001) Unveiling the functions of presynaptic metabotropic glu-tamate receptors in the central nervous system. J Pharmacol Exp Ther 299:12–20

Schreiber R, Brocco M, Audinot V, Gobert A, Veiga S, Millan MJ (1995) (1-(2,5-Dimethoxy-4 iodophenyl)-2-aminopropane)-induced head-twitches in the rat are mediated by 5-hydroxytryptamine (5-HT)$_{2A}$ receptors: modulation by novel 5-HT$_{2A/2C}$ antagonists, D_1 antagonists and 5-HT$_{1A}$ agonists. J Pharmacol Exp Ther 273:101–112

Scorza MC, Reyes-Parada M, Silveira R, Viola H, Medina JH, Viana MB, Zangrossi H Jr, Graeff FG (1996) Behavioral effects of the putative anxiolytic (±)-1-(2,5-dimethoxy-4-ethylthiophenyl)-2-aminopropane (ALEPH-2) in rats and mice. Pharmacol Biochem Behav 54:355–361

Scruggs JL, Schmidt D, Deutch AY (2003) The hallucinogen 1-[2,5-dimethoxy-4-iodophenyl]-2-aminopropane (DOI) increases cortical extracellular glutamate levels in rats. Neurosci Lett 346:137–140

Segal DS, Geyer MA (1985) Animal models of psychopathology. In: Judd LL, Groves PM (eds) Psychobiolgical foundations of clinical psychiatry. J.B. Lippincott Co., Philadelphia, pp 1–14

Shen HW, Jiang XL, Winter JC, Yu AY (2010) Psychedelic 5-methoxy-N, N-dimethyltryptamine: metabolism, pharmacokinetics, drug interactions, and pharmacological actions. Curr Drug Metabol 11:659–666

Shulgin AT (1973) Stereospecific requirements for hallucinogenesis. J Pharm Pharmacol 25:271–272

Shulgin AT, Nichols DE (1978) Characterization of three new psychotomimetics In: Stillman RC, Willette RE (eds) The psychopharmacology of hallucinogens. Pergamon Press, New York, p 74

Shulgin AT, Shulgin A (1991) PIHKAL: a chemical love story. Transform Press, Berkeley, CA

Shulgin AT, Sargent T, Naranjo C (1971) 4-Bromo-2,5-dimethoxyphenylisopropylamine, a new centrally active amphetamine analog. Pharmacology 5:103–107

Shulgin AT, Manning T, Daley PF (2011) The shulgin index. vol 1. Psychedelic phenethylamines and related compounds. Transform Press, Berkeley, CA

Silva MT, Calil HM (1975) Screening hallucinogenic drugs: systematic study of three behavioral tests. Psychopharmacologia 42:163–171

Singleton C, Marsden CA (1981) Circadian variation in the head twitch response produced by 5-methoxy-N1, N1-dimethyltryptamine and p-chloroamphetamine in the mouse. Psychopharmacology 74:173–176

Sipes TA, Geyer MA (1994) Multiple serotonin receptor subtypes modulate prepulse inhibition of the startle response in rats. Neuropharmacology 33:441–448

Sipes TA, Geyer MA (1995a) 8-OH-DPAT disruption of prepulse inhibition in rats: reversal with (+)WAY 100,135 and localization of site of action. Psychopharmacology 117:41–48

Sipes TE, Geyer MA (1995b) DOI disruption of prepulse inhibition of startle in the rat is mediated by 5-HT$_{2A}$ and not by 5-HT$_{2C}$ receptors. Behav Pharmacol 6:839–842

Sipes TE, Geyer MA (1997) DOI disrupts prepulse inhibition of startle in rats via 5-HT2A receptors in the ventral pallidum. Brain Res 761:97–104

Sitaram BR, Lockett L, Talomsin R, Blackman GL, McLeod WR (1987) In vivo metabolism of 5-methoxy-N, N-dimethyltryptamine and N, N-dimethyltryptamine in the rat. Biochem Pharmacol 36:1509–1512

Siuciak JA, Chapin DS, McCarthy SA, Guanowsky V, Brown J, Chiang P, Marala R, Patterson T, Seymour PA, Swick A, Iredale PA (2007) CP-809,101, a selective 5-HT2C agonist, shows activity in animal models of antipsychotic activity. Neuropharmacology 52:279–290

Smith DA, Bailey JM, Williams D, Fantegrossi WE (2014) Tolerance and cross-tolerance to head twitch behavior elicited by phenethylamine- and tryptamine-derived hallucinogens in mice. J Pharmacol Exp Ther 351:485–491

Snyder SH, Unger S, Blatchley R, Barfknecht CF (1974) Stereospecific actions of DOET (2,5-dimethoxy-4-ethylamphetamine) in man. Arch Gen Psychiatry 31:103–106

Spencer DG Jr, Glaser T, Traber J (1987) Serotonin receptor subtype mediation of the interoceptive discriminative stimuli induced by 5-methoxy-N, N-dimethyltryptamine. Psychopharmacology 93:158–166

Squires RF (1975) Evidence that 5-methoxy-N, N-dimethyltryptamine is a specific substrate for MAO-A in the rat: implications for the indoleamine dependent behavioral syndrome. J Neurochem 24:47–50

Strachan RT, Sciaky N, Cronan MR, Kroeze WK, Roth BL (2010) Genetic deletion of p90 ribosomal S6 kinase 2 alters patterns of 5-hydroxytryptamine$_{2A}$ serotonin receptor functional selectivity. Mol Pharmacol 77:327–338

Suzuki O, Katsumata Y, Oya M (1981) Characterization of eight biogenic indoleamines as substrates for type A and type B monoamine oxidase. Biochem Pharmacol 30:1353–1358

Swerdlow NR, Benbow CH, Zisook S, Geyer MA, Braff DL (1993) A preliminary assessment of sensorimotor gating in patients with obsessive compulsive disorder (OCD). Biol Psychiatry 33:298–301

Swerdlow NR, Zinner S, Hartston H, Filion D, Magulac M (1994) Central inhibitory deficits in obsessive compulsive disorder and Tourette syndrome. Biol Psychiatry 35(664):1994

Swerdlow NR, Geyer MA, Braff DL (2001) Neural circuit regulation of prepulse inhibition of startle in the rat: current knowledge and future challenges. Psychopharmacology 156:194–215

Tadano T, Neda M, Hozumi M, Yonezawa A, Arai Y, Fujita T, Kinemuchi H, Kisara K (1995) alpha-Methylated tryptamine derivatives induce a 5-HT receptor-mediated head-twitch response in mice. Neuropharmacology 34:229–234

Tanaka S, Young JW, Halberstadt AL, Masten VL, Geyer MA (2012) Four factors underlying mouse behavior in an open field. Behav Brain Res 233:55–61

Thompson MA, Moon E, Kim UJ, Xu J, Siciliano MJ, Weinshilboum RM (1999) Human indolethylamine N-methyltransferase: cDNA cloning and expression, gene cloning, and chromosomal localization. Genomics 61:285–297

Tilson HA, Baker TG, Chamberlain JH (1975) Behavioral and neuropharmacological analysis of amphetamine and 2,5-dimethoxy-4-methylamphetamine in rats. Psychopharmacologia 44:229–239

Turner EH, Loftis JM, Blackwell AD (2006) Serotonin a la carte: supplementation with the serotonin precursor 5-hydroxytryptophan. Pharmacol Ther 109:325–338

van den Buuse M, Ruimschotel E, Martin S, Perrin K, Risbrough VB, Halberstadt AL (2011) Enhanced effects of amphetamine but reduced effects of the hallucinogen, 5-MeO-DMT, on locomotor activity in 5-HT1A receptor knockout mice: implications for schizophrenia. Neuropharmacology 61:209–216

van Praag HM, Schut T, Bosma E, van den Bergh R (1971) A comparative study of the therapeutic effects of some 4-chlorinated amphetamine derivatives in depressive patients. Psychopharmacologia 20:66–76

Varty GB, Higgins GA (1995) Examination of drug-induced and isolation-induced disruptions of prepulse inhibition as models to screen antipsychotic drugs. Psychopharmacology 122:15–26

Verde G, Chiodini PG, Liuzzi A, Cozzi R, Favales F, Botalla L et al (1980) Effectiveness of the dopamine agonist lisuride in the treatment of acromegaly andpathological hyperprolactinemic states. J Endocrinol Invest 3:405–414

Vickers SP, Easton N, Malcolm CS, Allen NH, Porter RH, Bickerdike MJ, Kennett GA (2001) Modulation of 5-HT2A receptor-mediated head-twitch behaviour in the rat by 5-HT2C receptor agonists. Pharmacol Biochem Behav 69:643–652

Vollenweider FX (1994) Evidence for a cortical–subcortical imbalance of sensory information processing during altered states of consciousness using positron emission tomography and [^{18}F]fluorodeoxyglucose In: Pletscher A, Ladewig D (eds) 50 years of LSD: current status and perspectives of hallucinogens. Parthenon, London, pp 67–86

Vollenweider FX, Geyer MA (2001) A systems model of altered consciousness: integrating natural and drug-induced psychoses Brain Res Bull 56:495–507

Vollenweider FX, Vollenweider-Scherpenhuyzen MF, Bäbler A, Vogel H, Hell D (1998) Psilocybin induces schizophrenia-like psychosis in humans via a serotonin-2 agonist action. NeuroReport 9:3897–3902

Vollenweider FX, Csomor PA, Knappe B, Geyer MA, Quednow BB (2007) The effects of the preferential 5-HT2A agonist psilocybin on prepulse inhibition of startle in healthy human volunteers depend on interstimulus interval. Neuropsychopharmacology 32:1876–1887

Wasson RG, Hofmann A, Ruck CAP (1978) The road to Eleusis: unveiling the secret of the mysteries. Harcourt, Brace, Jovanovich, New York

Watson NV, Gorzalka BB (1992) Concurrent wet dog shaking and inhibition of male rat copulation after ventromedial brainstem injection of the 5-HT2 agonist DOI. Neurosci Lett 141:25–29

Webster VA, Griffiths EC, Slater P (1982) Induction of wet-dog shaking in rats by analogues and metabolites of thyrotrophin-releasing hormone (TRH). Regul Pept 5:43–51

Wettstein JG, Host M, Hitchcock JM (1999) Selectivity of action of typical and atypical anti-psychotic drugs as antagonists of the behavioral effects of1-[2,5-dimethoxy-4-iodophenyl]-2-aminopropane (DOI). Prog Neuropsychopharmacol Biol Psychiatry 23:533–544

White FJ, Appel JB (1982) Lysergic acid diethylamide (LSD) and lisuride: differentiation of their neuropharmacological actions. Science 216:535–537

White FJ, Kuhn DM, Appel JB (1977) Discriminative stimulus properties of quipazine. Neuropharmacology 16:827–832

Wieland S, Kreider MS, McGonigle P, Lucki I (1990) Destruction of the nucleus raphe obscurus and potentiation of serotonin-mediated behaviors following administration of the neurotoxin 3-acetylpyridine. Brain Res 520:291–302

Willins DL, Meltzer HY (1997) Direct injection of 5-HT2A receptor agonists into the medial prefrontal cortex produces a head-twitch response in rats. J Pharmacol Exp Ther 282:699–706

Wing LL, Tapson GS, Geyer MA (1990) 5HT-2 mediation of acute behavioral effects of hallucinogens in rats. Psychopharmacology 100:417–425

Winter JC (1979) Quipazine-induced stimulus control in the rat. Psychopharmacology 60:265–269

Winter JC (1994) The stimulus effects of serotonergic hallucinogens in animals. NIDA Res Monograph Ser 146:157–182

Winter CA, Flataker L (1956) Effects of lysergic acid diethylamide upon performance of trained rats. Proc Soc Exp Biol Med 92:285–289

Winter JC, Filipink RA, Timineri D, Helsley SE, Rabin RA (2000) The paradox of 5-methoxy-N, N-dimethyltryptamine: an indoleamine hallucinogen that induces stimulus control via 5-HT1A receptors. Pharmacol Biochem Behav 65:75–82

Yamamoto T, Ueki S (1975) Behavioral effects of 2,5-dimethoxy-4-methylamphetamine (DOM) in rats and mice. Eur J Pharmacol 32:156–162

Yarosh HL, Katz EB, Coop A, Fantegrossi WE (2007) MDMA-like behavioral effects of N-substituted piperazines in the mouse. Pharmacol Biochem Behav 88:18–27

Young JW, Minassian A, Paulus MP, Geyer MA, Perry W (2007) A reverse-translational approach to bipolar disorder: rodent and human studies in the Behavioral Pattern Monitor. Neurosci Biobehav Rev 31:882–896

Zaniewska M, McCreary AC, Filip M (2009) Interactions of serotonin (5-HT)2 receptor-targeting ligands and nicotine: locomotor activity studies in rats. Synapse 63:653–661

Zhang C, Marek GJ (2008) AMPA receptor involvement in 5-hydroxytryptamine2A receptor-mediated pre-frontal cortical excitatory synaptic currents and DOI-induced head shakes. Prog Neuropsychopharmacol Biol Psychiatry 32:62–71

Hallucinogens in Drug Discrimination

Lisa E. Baker

Abstract Hallucinogens comprise a diverse collection of chemicals with multifarious receptor actions in the central nervous system. Preclinical drug screening methods have proven invaluable in the evaluation and characterization of hallucinogen psychopharmacology. Used in concert with structural chemistry and receptor pharmacology methods, preclinical drug discrimination research has informed our current understanding of hallucinogens and the neurochemical receptor mechanisms responsible for their interoceptive stimulus effects. This chapter summarizes the strengths and limitations of drug discrimination as an in vivo drug detection method and offers a brief review of historical and contemporary drug discrimination research with classical hallucinogens.

Keywords Hallucinogens · Preclinical drug screening · Drug discrimination methods · Psychopharmacology · Behavioral pharmacology

Contents

L.E. Baker (✉)
Department of Psychology, Western Michigan University, Kalamazoo, MI, USA
e-mail: lisa.baker@wmich.edu

Curr Topics Behav Neurosci (2017) 36: 201–220
DOI 10.1007/7854_2017_476
Published Online: 09 May 2017

1 Introduction and Overview

Hallucinogens, broadly defined, comprise a diverse collection of psychoactive drugs that have been categorized into a variety of subclasses based on chemical structures, pharmacological actions, and subject-reported psychological effects. Considerable overlap exists among these classification domains. As such, hallucinogens with similar pharmacological actions in the central nervous system (CNS) generally tend to produce similar subjective effects, although there are notable exceptions. The application of behavioral screening methods utilizing nonhuman models has proven invaluable in elucidating the links among chemical structures, receptor pharmacology, and the psychological effects of hallucinogens. Among the various preclinical drug screening tools available, the drug discrimination paradigm is a particularly robust, sensitive, and specific model (Appel et al. 1982). The primary strengths of this behavioral assay include pharmacological specificity and predictive validity.

While no single method is solely sufficient to characterize the behavioral or pharmacological mechanisms of drug action, drug discrimination can be a very powerful tool, especially when it is used in concert with other research methods, such as in vitro assays of receptor binding or intracellular signaling. This chapter describes the drug discrimination paradigm with an emphasis on methodological variables, and offers a brief review of historical and contemporary research on the evaluation of classical hallucinogens utilizing these methods. Examples of research in which drug discrimination methods are paired with methods of structural chemistry and pharmacology are emphasized. The chapter ends with suggestions for future avenues of research.

2 Hallucinogen Classifications

The classical hallucinogens consist of two broad categories, indolealkylamines (also called indoleamines) and phenylalkylamines, each with subclassifications based on structural differences (Glennon 1994). The subclasses of indolealkylamines include simple tryptamines [e.g., *N,N*-dimethyltryptamine (DMT)], methyltryptamines (e.g., 5-methoxy-α-methyltryptamine), ergolines [e.g., lysergic acid diethylamide (LSD)], and ß-carbolines (e.g., harmala alkaloids). The phenylalkylamines consist of phenethylamines (e.g., mescaline) and phenylisopropylamines [e.g., 1-(2,5-dimethoxy-4-methyl-phenyl)-2-aminopropane (DOM)]. A more recent review on the behavioral pharmacology of hallucinogens offers a simpler distinction between two main chemical classes of hallucinogens, based on the structural backbone of either phenethylamine or tryptamine (Fantegrossi et al. 2008), both subclasses of the broader categories previously delineated by Glennon (1994). According to this simpler classification, DOM, and other related compounds, DOB and DOI are all characterized as phenethylamines. Other phenethylamines are noted for their structural similarities to the CNS stimulant, amphetamine [3,4-methylenedioxymethamphetamine (MDMA), and 3,4-methylenedioxyamphetamine (MDA)]. Although these substances are sometimes broadly classified as hallucinogens, they are more appropriately designated as emphathogens or entactogens (Nichols 1986).

Other categories of hallucinogens, structurally and pharmacologically distinct from the classical hallucinogens, include the dissociative anesthetics [e.g., ketamine, phencyclidine (PCP)], anticholinergics (e.g., scopolamine, atropine), and kappa-opioid receptor agonists (e.g., salvinorin A). For the sake of brevity, this chapter emphasizes drug discrimination research with the classical (serotonergic) hallucinogens.

3 The Drug Discrimination Paradigm

Drug discrimination is a well-established experimental method, commonly employed to classify psychoactive drugs and to characterize their neuropharmacological actions. This method has been utilized with nonhuman subjects (rats, mice, nonhuman primates) and human research participants. With some exceptions, there is considerable overlap across species regarding the discriminative stimulus effects of hallucinogens. The drug discrimination paradigm takes advantage of the fact that psychoactive drugs produce physiological changes in the nervous system that can function as interoceptive stimuli. Through differential reinforcement of specific behaviors (e.g., a lever press) in the presence or absence of such stimuli, an organism can be trained to indicate when these stimuli are present. Thus, the drug discrimination paradigm offers a rigorous and robust method for in vivo drug detection.

Of particular importance to the validity of drug discrimination as an in vivo drug-detection method, psychoactive drugs with similar discriminative stimulus effects in nonhumans tend to also produce similar subject-reported psychological effects in humans with considerable reliability (Young 2009). For example, responses in animals trained to discriminate lysergic acid diethylamide (LSD) reliably generalize to other classical hallucinogens (e.g., psilocybin, mescaline, or DMT), but do not typically generalize to the dissociative anesthetics (e.g., phencyclidine or ketamine) or to other psychoactive drug classes, such as central nervous system (CNS) stimulants, sedatives, or opiate analgesics. The entactogens (e.g., MDMA, MDA) have been reported to produce similar discriminative stimulus effects to LSD, although research findings are not always consistent regarding these similarities (Oberlender and Nichols 1988; Schechter 1998; Callahan and Appel 1988; Goodwin et al. 2003).

Indeed, drug discrimination is a particularly attractive investigative paradigm due to the specificity of discriminative stimuli correlating with underlying cellular and molecular mechanisms of drug action (Holtzman and Locke 1988; Colpaert 1999). For example, while mescaline (a phenethylamine derivative) is structurally different from LSD (an ergoline), these drugs share similar pharmacological actions at 5-HT_{2A} receptors and also reliably substitute for one another in nonhuman drug discrimination experiments (Appel and Callahan 1989). Such findings are consistent with reports that mescaline and LSD produce similar subject-reported psychological effects in humans (Winter 2009). Specific examples of mechanistic studies utilizing drug discrimination methods are addressed later in this chapter.

4 Drug Discrimination Methodology

A variety of behavioral strategies have been employed to establish discriminative stimulus control with psychoactive drugs in nonhumans, some utilizing classical conditioning methods (e.g., conditioned taste aversion) and others using operant conditioning methods with either negative reinforcement (e.g., conditioned shock avoidance), or positive reinforcement (e.g., appetitive conditioning). In contemporary drug discrimination research, the most commonly employed drug discrimination methods utilize operant conditioning technology in which food-restricted animals (rats, mice, or nonhuman primates) are reinforced with food delivery for responding on one operandum (e.g., pressing a lever in an operant conditioning chamber) within a specified time period following drug injections and they are reinforced for emitting an alternate response (pressing a different lever) following vehicle (e.g., saline) injections. Drug or vehicle training sessions typically occur once per day, in an alternating, semi-random order.

The specific training methods and the mastery criteria for stimulus control vary among drug discrimination studies from different laboratories. One methodological variable of interest is the schedule of reinforcement used to establish the discrimination. Variable interval (VI) and fixed ratio (FR) schedules are two common reinforcement schedules employed in drug discrimination with nonhumans, so will be discussed briefly here. The key difference between interval and ratio schedules is what determines the delivery of reinforcement, which in turn influences the frequency of reinforcement. With interval schedules, the delivery of reinforcement is dependent on passage of time; with variable interval schedules, the passage of time is unpredictable. For example, in a VI 30-s schedule, the average time interval between opportunities to earn a reinforcer is 30 s, but it may be shorter or longer. Training for an extended period under a VI reinforcement schedule tends to produce a moderate but steady response rate, with minimal pausing after reinforcement delivery. A key advantage of VI schedules is that they are resistant to extinction. With ratio schedules, the delivery of reinforcement is dependent on the number of responses emitted by the organism. For example, an FR 20 schedule requires the organism to emit 20 responses to receive a reinforcer. This schedule engenders a high, steady response rate until the delivery reinforcement, with a brief response pause following each reinforcer delivery.

It has long been established that reinforcement schedules can influence the development of discriminative stimulus control by drugs (Overton 1979; Koek and Slangen 1982; Stolerman 1989; McMillan and Wenger 1984; McMillan et al. 2001).

While these studies are of historical significance to drug discrimination researchers, investigations on the influence of reinforcement schedule or other methodological variables are extremely scarce in the drug discrimination literature with hallucinogens. Kueh and Baker (2007) compared FR 20 and VI 15-sec reinforcement schedules on the acquisition of discriminative stimulus control by MDMA. While stimulus control was established more rapidly under the FR 20 schedule, only minor differences were observed in MDMA dose-response functions. However, LSD produced a higher percentage of MDMA-lever appropriate responding in rats trained to discriminate MDMA on the VI 15-sec schedule compared to rats trained on the FR 20 schedule. This finding suggests training methods can influence the outcome of stimulus generalization tests.

For good reason, the majority of drug discrimination research with hallucinogens has emphasized pharmacological questions, in an attempt to discern the neural systems and receptor-mediated activities contributing to the interoceptive stimulus effects of these drugs. However, the possibility that training methods can influence the outcome of these studies should not be ignored. Inasmuch as training methods can be conceptualized as a convenient manipulation of behavioral history, inquiry into the behavioral mechanisms involved in hallucinogen discrimination may warrant further investigation.

When comparing the results obtained from different laboratories, it is important to attend to certain methodological differences. In addition to differences in reinforcement schedules as noted above, the criteria required to establish stimulus control also vary among studies. Mastery criteria for stimulus control typically range between 75 and 83% for a specified number of training sessions (e.g., five consecutive sessions or eight out of 10 consecutive sessions). Most studies include the criteria that discrimination accuracy be attained prior to the delivery of the first reinforcer in each session as well as for the remainder of each training session before commencing stimulus generalization tests. Once stimulus control is established with a training drug, stimulus generalization tests are then conducted with a range of doses of that substance as well as several other substances. Some investigators conduct these tests under extinction, not allowing for reinforcement of responding under potentially different stimulus conditions. Other investigators allow for programmed reinforcement of responses on either lever during these assessments.

The results obtained from stimulus generalization tests are typically used to plot dose-response curves for quantitative comparisons. For example, dose-response curves may be compared with respect to the magnitude of stimulus generalization, or height of the dose-response curves, to determine whether a test drug substitutes for the training drug. The magnitude of stimulus generalization can be expressed as the average percentage of drug-lever responses among the animals tested or as the percentage of animals that selected the drug lever. The former is referred to as a quantitative measure, while the latter is referred to as a quantal measure. Both measures provide an index of similarity between the interoceptive stimulus effects of the training drug and the test drugs. Most researchers generally consider a minimum of 80% drug-lever responses or 80% of rats selecting the drug lever as evidence for full stimulus generalization (or substitution).

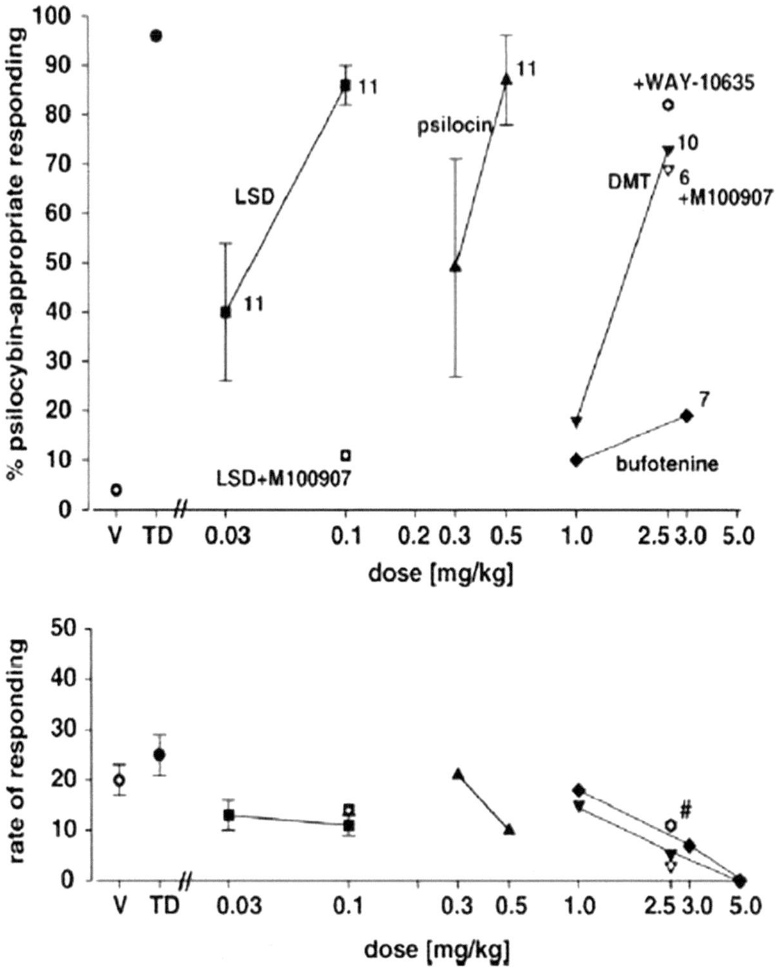

Fig. 1 Reprinted from Winter et al. (2007)

The term "generalization" is often used incorrectly and interchangeably with the term "substitution" in the drug discrimination literature. The following two sentences differentiate the proper use of these terms. Responses may *generalize* from one stimulus to other stimuli. A stimulus may *substitute* for other stimuli. In the case of drug discrimination, responses are the allocation of lever presses and the stimuli are drug-induced interoceptive stimuli. An example of stimulus generalization (i.e., drug substitution) is depicted in Fig. 1, reprinted from a study reported by Winter et al. (2007). This figure illustrates dose-response curves obtained with LSD, psilocin, DMT, and bufotenine in rats trained to discriminate 0.5 mg/kg psilocybin. LSD and psilocin produced full substitution for psilocybin, whereas DMT produced only 73% psilocybin-appropriate responding and bufotenine produced less than 20% psilocybin-appropriate responding at the doses tested.

Dose-response curves generated from drug substitution tests also allow for comparison of drug potency. The median-effective dose (ED_{50}) may be estimated from a linear regression of dose-response curves that are typically plotted on a logarithmic scale. The ED_{50} values are compared among the test drugs and the training drug. Drugs with a lower ED_{50} value are considered more potent than those with a higher ED_{50} value. In the aforementioned example, LSD is more potent than psilocin.

In addition to the assessment of substitution between various test drugs and a particular training drug, response rate is another quantitative measure of interest. As such, response rate can provide an index of drug-induced suppression of behavior. As a general rule, increasing doses of a test drug are assessed until either full substitution is observed or until response rate is significantly reduced. For example, a study by Killinger et al. (2010) assessed salvinorin A, a unique hallucinogen with selective kappa receptor affinity in animals trained to discriminate LSD. Figure 2 depicts the dose-response function determined for LSD and salvinorin A. LSD produced dose-dependent increases in responding on the LSD-associated lever. Salvinorin A failed to produce more than 40% LSD-lever responding and significantly reduced response rate.

A particularly attractive feature of drug discrimination methodology is that it allows for the assessment of neurochemical actions and/or receptor signaling mechanisms that contribute to the discriminative stimulus effects of a drug. For such assessments, pharmacological antagonists are given in combination with the training drug or in combination with another drug that substituted for the training drug to ascertain whether their stimulus effects can be attenuated or blocked. For example, several different drugs with varying receptor affinities and selectivity can be compared for antagonism of the drug stimulus. In one such study, Fiorella et al. (1995a) evaluated 12 different pharmacological antagonists with varying affinities for $5\text{-}HT_{2A}$ and $5\text{-}HT_{2C}$ receptors. Each antagonist was assessed in combination

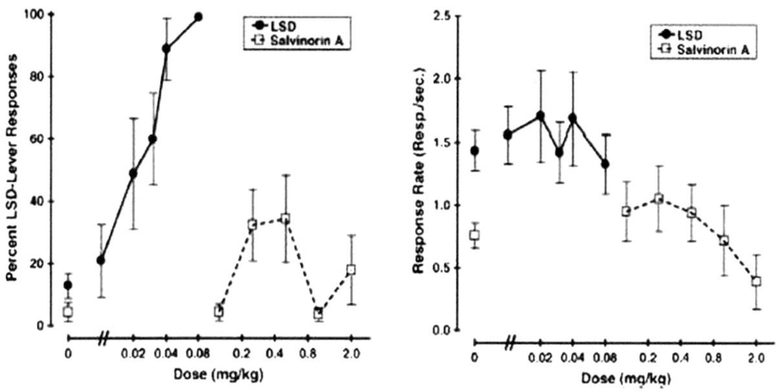

Fig. 2 Reprinted from Killinger et al. (2010)

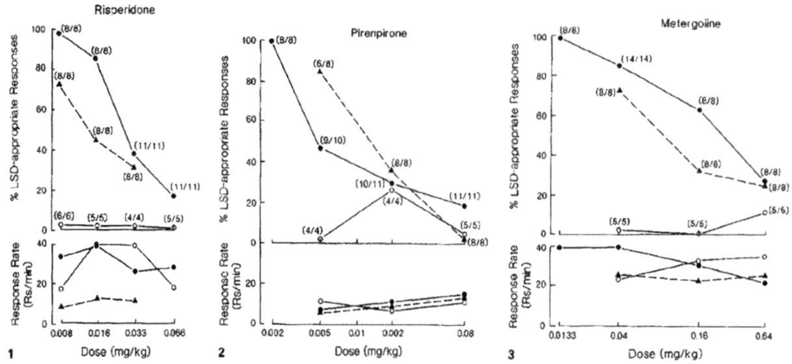

Fig. 3 Reprinted from Fiorella et al. (1995a)

with LSD or (−)-DOM in rats trained to discriminate LSD. For illustration, Fig. 3 depicts three of the 12 dose-response curves reported in that study. These graphs depict the dose-inhibition functions for risperidone, pirenpirone, and metergoline. The open circles represent the effects of each antagonist alone. The closed circles represent each drug in the presence of 0.1 mg/kg LSD, and the closed triangles represent each antagonist in the presence of 0.4 mg/kg (−)-DOM. The results of this study are discussed later in this chapter under the heading Mechanistic Studies.

Another common strategy in drug discrimination studies involves the assessment of pharmacological agents in combination with the training drug to determine whether they potentiate its effects. For example, although the noncompetitive NMDA glutamate antagonists do not fully substitute for serotonergic hallucinogens, these substances have been shown to potentiate the effects of DOM and LSD (Winter et al. 2000, 2004).

Despite the obvious limitations inherent in any attempt to model a complex and uniquely human subjective experience using infrahuman species, the strengths of drug discrimination research with hallucinogens are irrefutable. When paired with tests of pharmacological agents with varying receptor selectivity and affinity, drug discrimination studies have aided in elucidating multiple neurotransmitter receptor subtypes contributing to the complex interoceptive stimuli produced by hallucinogens. Considering the recent resurgence in the recreational use as well renewed interests in the therapeutic potential of some hallucinogens, preclinical drug discrimination studies continue to inform clinical investigations of these drugs.

5 Preclinical Drug Discrimination with Classical Hallucinogens

An exhaustive review of drug discrimination research on hallucinogens over the past four decades is beyond the scope of this chapter. However, some prominent examples of early research with classical hallucinogens (e.g., LSD, mescaline, psilocybin) and a few leading examples of contemporary research will be highlighted. Hirschhorn and Winter (1971) published the first known report that LSD and mescaline can establish discriminative stimulus control in rats. Subsequent research over the next decade revealed that the discriminative stimulus effects of mescaline and LSD are distinct from hallucinogens outside the indoleamine or phenethylamine subclasses (Shannon 1981; Silverman and Ho 1978; Swedberg and Jarbe 1986). Early studies also established that psilocybin induces stimulus control in rats (Harris and Balster 1971; Schechter and Rosecrans 1972; Koerner and Appel 1982).

Research findings accumulated over the past four decades indicate that drug-induced stimulus control by the classical hallucinogens (i.e., LSD, phenethylamines, tryptamines) is primarily dependent on their interactions with serotonin (5-HT) receptors (Appel et al. 1982; Colpaert et al. 1982; Glennon et al. 1982, 1984a, b; Kuhn et al. 1978; Winter 1978; Young et al. 1982). Despite numerous mechanistic studies on the interoceptive stimulus effects of these hallucinogens, the precise mechanism of action underlying these effects remains elusive. The 5-HT_{2A} receptor subtype is thought to be necessary, but not sufficient, and there is considerable evidence that both the 5-HT_{2C} and 5-HT_{1A} receptor subtypes also contribute to the interoceptive stimulus effects of these substances (Fiorella et al. 1995a; Winter 2009; Carbonaro et al. 2015; also see Nichols 2004 and Fantegrossi et al. 2008 for reviews).

Winter et al. (2007) were the first researchers to examine thoroughly the involvement of serotonin receptors in psilocybin discrimination in the rat. In tests of stimulus generalization, DOM, LSD, psilocin, and DMT all substituted fully for psilocybin, while only partial substitution was observed with 2C-T-7 (2,5-dimethoxy-4-propylthiophenethylamine) and mescaline. These authors also reported that MDL 100907, a 5-HT_{2A} inverse agonist, partially blocked psilocybin discrimination, whereas the 5-$HT_{1A/7}$ receptor antagonist, WAY-100635, and the dopamine D_2 receptor antagonist remoxipride failed to block psilocybin discrimination. Winter et al. (2007) concluded that 5-HT_{2A} receptor activities play a prominent but incomplete role in the compound stimulus induced by psilocybin. They further noted that unlike other closely related hallucinogens, 5-HT_{1A} receptors do not appear to contribute to psilocybin-induced stimulus control.

Despite an abundance of evidence for the involvement of serotonin receptors in the discriminative stimulus effects of the classical hallucinogens, a myriad of research data indicate that interactions among multiple receptor systems contribute to their psychopharmacological effects. For example, González-Maeso et al. (2007) demonstrated that the effects of the tryptamine hallucinogens are dependent on secondary signaling pathways and are not limited simply to 5-HT_{2A} receptor activation. Of particular interest is evidence indicating the involvement of group II glutamate receptors (mGluR2/3) in mediating hallucinogenic effects (González-Maeso et al. 2007, 2008; Delille et al. 2012; Moreno et al. 2011; Winter et al. 2004; Carbonaro et al. 2015).

In a recent study, Carbonaro et al. (2015) explored the involvement of serotonin and group II glutamate receptors in the discrimination of a naturally occurring tryptamine hallucinogen DMT and a synthetic analog of this substance, *N,N*-diisopropyltryptamine (DIPT). Two separate groups of 16 rats were trained to discriminate either 5 mg/kg DMT or 5 mg/kg DIPT from saline. Tests were conducted with the 5-HT_{2A} inverse agonist, MDL 100907, the 5-HT_{2C} antagonist, SB242084, the mGluR2/3 agonist, LY379268, and the mGluR2/3 antagonist, LY341495. MDL 100907 fully blocked the discriminative stimulus effects of DMT, whereas SB242084 produced minimal attenuation of the DMT cue. Both MDL 100907 and SB 242084 only partially attenuated DIPT discrimination. LY379268 only partially blocked the discriminative stimulus effects of DMT and had minimal effects on DIPT discrimination, whereas LY341495 potentiated the effects of both DMT and DIPT. The authors concluded that 5-HT_{2A} receptors predominantly mediate the discriminative stimulus effects of DMT and DIPT, while both 5-HT_{2C} and mGluR2 receptors may modulate their discriminative stimulus effects to some extent.

In a brief but very informative review of drug discrimination research on the classical hallucinogens, Winter (2009) highlighted two key factors that complicate the characterization of these drugs: (1) the complexity of the serotonin (5-HT) receptor family and (2) the lack of selective antagonists for specific 5-HT receptor subtypes. Winter's review addresses historical and contemporary research on the neurochemical mediation of stimulus control by hallucinogens, indicating a prominent role for 5-HT_{2A} receptor with significant modulation by 5-HT_{2C} and 5-HT_{1A} receptors. Also noted in this review are active investigations on the contribution of dopamine and glutamate to the complex stimulus effects of the classical hallucinogens. At least three crucial points may be gleaned from Winter's commentary on this research: (1) Hallucinogens represent a diverse collection of agents with compound stimulus properties; (2) selective ligands for various receptor subtypes are essential tools for examining the complex stimulus properties of hallucinogens; and (3) preclinical models of drug discrimination, involving mainly rats and some nonhuman primates, have assisted in the characterization of the uniquely complex stimulus properties of hallucinogens.

The challenges Winter (2009) highlighted still exist in behavioral pharmacology research with hallucinogens. Determining the precise mechanistic interactions among serotonergic, glutamatergic, and dopaminergic brain systems involved in drug-induced psychological phenomena is encumbered by the fact that there are multiple receptor subtypes within each of these systems and there are few highly selective agents for these receptors readily available to most researchers. The application of cellular and molecular biology techniques can aid in the evaluation of these mechanistic interactions. As such, evidence from in vitro electrophysiology studies with rodent cortical tissue slices suggests hallucinogens exert their effects through serotonergic–glutamatergic interactions. For example, 5-HT_{2A} receptor stimulation is correlated with an increase in glutamate-mediated synaptic activity in the rat prefrontal cortex (Lambe et al. 2000; Lambe and Aghajanian 2001). Additionally, dopaminergic D_1/D_5 receptors attenuate glutamatergic activity and oppose the effects of both phenethylamines and tryptamines (Lambe and Aghajanian 2007). Béïque et al. (2007) combined cellular and molecular approaches to examine the mechanisms of interaction between 5-HT_{2A} receptors and glutamatergic synaptic activity in the prefrontal cortex. They located a subpopulation of pyramidal cells that were strongly excited by 5-HT_{2A} receptor activation, and they suggested that 5-HT_{2A} receptors facilitate intrinsic networks within the PFC.

Translating the findings from in vitro electrophysiology studies into mechanistic explanations of drug-induced psychological experiences requires cautious interpretation. While the cellular and molecular actions of hallucinogens in the brain contribute to discriminative stimulus control by these drugs, contextual features of the training environment must also be considered. Just as we cannot ignore the influence of environmental setting on the subjective experience induced by hallucinogens in human users, we must also consider the influence of environmental context when interpreting mechanistic studies of drug discrimination in nonhuman models. Nevertheless, mechanistic approaches in drug discrimination have proven to be an invaluable tool in determining the receptor mechanisms involved in the complex stimulus properties of hallucinogens.

6 Mechanistic Studies

The power of drug discrimination as an in vivo assay of drug–receptor interactions is particularly evident when it is used in concert with other experimental methods, including in vitro receptor binding assays, intracellular signaling assays, and various methods designed to alter the activities or expression of receptors. As such, the majority of published drug discrimination studies involving hallucinogens have emphasized mechanistic approaches to delineate the neural systems involved in their discrimination.

In a series of cleverly designed mechanistic studies, Winter and colleagues conducted fundamental research to differentiate the respective roles of 5-HT_{2A} and 5-HT_{2C} receptors in the discriminative stimulus effects of LSD and (−)-DOM (Fiorella et al. 1995a, b). Two mechanistic approaches utilized in these studies are reviewed by Winter et al. (1999). The first approach utilizes an antagonist correlational analysis. Using this approach, a diverse group of antagonists are tested for blockade of drug-induced stimulus control. In separate in vitro radioligand competition experiments, receptor affinity values are determined with the same collection of antagonists. The IC_{50} values obtained from tests of stimulus antagonism are then analyzed for correlation with binding affinities determined from the receptor binding experiments. In one such study, 12 different antagonists were assessed with a range of doses for antagonism of LSD discrimination and of stimulus generalization to (−)-DOM in rats trained to discriminate LSD (Fiorella et al. 1995a). A few of the graphs from this study were previously displayed (see Fig. 3) to illustrate an example of antagonism tests in drug discrimination. Receptor affinity values for the same set of antagonists were determined in vitro from radioligand competition experiments. Results revealed that the in vivo potency of antagonists to block LSD discrimination and stimulus generalization to (−)-DOM was highly correlated with the in vitro binding affinity of these antagonists to 5-HT_{2A} receptors and poorly correlated with affinity for 5-HT_{2C} receptors. The authors concluded that the 5-HT_{2A} receptor subtype plays a predominant role in the stimulus effects of LSD and (−)-DOM.

A second mechanistic approach reviewed by Winter et al. (1999) involves pretreatment with serotonin-depleting agents to assess the consequences of such treatment on LSD discrimination. Citing several landmark studies, Winter et al. (1999) reviewed the evidence that pretreatment with the 5-HT-depleting agent 5,7-dihydroxytryptamine (5,7-DHT) potentiates LSD discrimination (White et al. 1980), increases 5-HT_{2C} receptor density, and upregulates 5-HT_{2C}-mediated phosphoinositide (PI) turnover (Conn et al. 1987; Pranzatelli 1990) but does not upregulate 5-HT_{2A} receptors (Conn and Sanders-Bush 1986). Collectively, these findings indicate a modulatory role of 5-HT_{2C} receptors in LSD's stimulus effects. A subsequent study by Fiorella et al. (1995b) tested the hypothesis that 5-HT_{2C} receptor upregulation mediates supersensitivity to LSD discrimination following 5-HT depletion. They compared the effects of two 5-HT-depleting agents, *p*-chlorophenylalanine (PCPA) and *p*-chloroamphetamine (PCA), on LSD discrimination and on 5-HT_{2A} receptor-mediated and 5-HT_{2C} receptor-mediated phosphoinositide (PI) hydrolysis. These chemical agents reduce brain serotonin by different mechanisms; PCPA blocks 5-HT biosynthesis, whereas PCA produces the loss of presynaptic 5-HT neurons through a toxic mechanism. Pretreatment with PCPA, but not PCA, shifted the LSD dose-response curve to the left, indicative of enhanced LSD discrimination. Further, PCPA, but not PCA, significantly increased 5-HT_{2C} receptor-mediated PI hydrolysis. However, 5-HT_{1A} receptor-mediated PI hydrolysis was not affected by either 5-HT-depleting agent. These results suggest the involvement of 5-HT_{2C} receptor-mediated PI hydrolysis in the discriminative stimulus effects of LSD.

In considering the multiple neural systems involved in hallucinogen discrimination, it is important to note that different neuronal systems may differentially contribute to the stimulus effects of a drug at different post-injection intervals. Drug discrimination methods can readily be applied to explore the temporal dynamics of the interoceptive stimulus effects of drugs. For example, Marona-Lewicka and Nichols (2007) utilized drug discrimination methods to explore the distinction between serotonergically mediated and dopaminergically mediated discriminative stimulus effects of LSD. Rats were trained to discriminate LSD from saline following either a 30 or 90-min preinjection interval, and a variety of agonists and antagonists from distinct pharmacological classes were assessed for substitution or antagonism. Other serotonergic hallucinogens, including psilocin and mescaline, substituted only in the rats trained to discriminate LSD with a 30-min preinjection interval. In contrast, several dopamine receptor agonists either fully or partially substituted for the LSD 90-min discriminative stimulus. The authors concluded that dopaminergically mediated effects play a more prominent role in the delayed temporal effects of LSD.

7 False Positives

Despite its pharmacological specificity, drug discrimination is not without limitations. One commonly cited limitation of this paradigm is the occasional occurrence of false positives, when a substance produces substitution for a training drug, but those two drugs do not typically produce similar subject-reported effects in humans. The antimigraine medication, lisuride, is a frequently cited example of a false positive for LSD discrimination. Lisuride is structurally similar to LSD and has a high affinity for 5-HT$_{2A}$ receptors, but lisuride does not produce hallucinations in humans. Nevertheless, lisuride has been reported to produce full substitution in rats trained to discriminate LSD (White and Appel 1982; Holohean et al. 1982), although others have found only partial substitution (Marona-Lewicka et al. 2002). As noted previously, methodological differences should be considered carefully when comparing results from different laboratories. As such, a variety of methodological differences between these studies, such as the type of reinforcer and reinforcement schedule, the presession injection interval, or the use of quantal versus quantitative measures of drug substitution, could contribute to the discrepant findings regarding LSD and lisuride.

The implementation of alternative drug discrimination methods, such as a three-lever drug discrimination (e.g., LSD–vehicle–lisuride) or a drug versus other (e.g., LSD versus multiple other drug stimuli) discrimination, has proven useful in reducing or eliminating the detection of lisuride as a false positive. For example, Callahan and Appel (1990) successfully trained rats to discriminate lisuride from LSD using a three-lever drug discrimination procedure. In another study, one group of rats was trained to discriminate LSD from saline, while another group was trained to discriminate LSD from either saline, cocaine, or pentobarbital (Appel et al. 1999). When lisuride was tested for substitution in both groups, there were fewer instances of false positives in the second group. These studies exemplify how the sensitivity of the drug discrimination assay is dependent on the discrimination training methods.

Lisuride's substitution for LSD's discriminative stimulus effects in nonhumans is likely due to its high affinity for 5-HT$_{2A}$ receptors. However, González-Maeso et al. (2007) differentiated the actions of LSD and lisuride with regard to their functional selectivity via 5-HT$_{2A}$ receptor-mediated intracellular signaling pathways. This study exemplifies how the methods of molecular biology and biochemistry are essential complementary tools to behavioral studies of hallucinogens. Only when multiple methodological approaches are used in concert can we begin to delineate the precise cellular and molecular mechanisms responsible for mediating the complex psychoactive properties of these drugs.

8 Structure–Activity Relationships

When used in the context of drug development and design, drug discrimination serves well for investigating structure–activity relationships (SAR). The specific aim of SAR studies is to assess the influence of chemical structure on pharmacological activity. When paired with drug design strategies, drug discrimination methods contribute to SAR investigations by providing both qualitative and quantitative information (Glennon and Young 2011). Besides determining whether a test agent produces qualitatively similar discriminative stimulus effects as the training drug, potency comparisons can be made by comparing the ED_{50} values of various test agents to the training stimulus. A recently published book devoted to drug discrimination by Glennon and Young (2011) includes a chapter on the implementation of this drug-detection method in SAR investigations with hallucinogens. In one example, they highlight a study in which rats were trained to discriminate 1.0 mg/kg DOM and subsequently tested with several methoxy-substituted analogs of the basic phenylisopropylamine structure. For a detailed description and graphic depiction of their findings, the interested reader may consult the chapter by Glennon and Young (2011) or the original research report (Glennon and Young 1982). In short, all of the monomethoxy compounds tested failed to substitute for DOM. Of the six positional isomers of dimethoxyamphetamine (DMA) tested, only 2,4-DMA and 2,5-DMA substituted for DOM, and all five trimethoxyamphetamine (TMA) analogs tested produced full substitution for the DOM training stimulus. The authors concluded that the presence of the 4-methyl group of DOM contributes to the stimulus characteristics and the potency of 2,5-DMA as a DOM-like substance.

9 Alternative Training Methods in Drug Discrimination

The most common application of drug discrimination methodology involves training an organism to distinguish a drug from the absence of that drug (Drug vs. Vehicle). Alternative approaches that are currently underutilized include training a discrimination between two drugs (Drug A vs. Drug B), training a discrimination between one drug versus a variety of other drugs (Drug vs. Other), or training a three-lever discrimination among two different drugs and the absence of either drug (Drug A vs. Vehicle vs. Drug B). As mentioned above, some of these methods were employed to differentiate the stimulus effects of LSD and lisuride. In particular, the three-lever drug discrimination procedure has been reported to be a more sensitive tool with which to investigate the stimulus properties of psychoactive drugs (Stolerman 1993). To briefly elaborate on the utility of the three-lever discrimination method, a series of studies are summarized below in which this method was used to characterize the unique discriminative stimulus effects of the entactogen, MDMA.

Traditional two-lever discrimination procedures were employed in the late 1980s and early 1990s to characterize the interoceptive stimulus effects of MDMA. Some of these studies suggested the optical isomers of MDMA may exert differential stimulus effects and that (+)-MDMA may produce more stimulant-like effects compared to (−)-MDMA. For example, Glennon et al. (1988) reported that (+)-MDMA substituted for *d*-amphetamine, whereas (−)–MDMA failed to do so.

However, Oberlender and Nichols (1988) found neither isomer to substitute for *d*-amphetamine. Baker et al. (1995) trained separate groups of rats to discriminate either (+)-MDMA or (−)-MDMA from saline and tested both stimulants (*d*-amphetamine, cocaine) and hallucinogens (DOM, LSD, mescaline) for substitution. None of the stimulants nor the hallucinogens substituted for (+)-MDMA, and only LSD produced full substitution for (–)–MDMA in that study.

In an effort to further differentiate the MDMA isomers, Baker and Taylor (1997) trained rats to discriminate LSD and *d*-amphetamine from saline using three-lever discrimination methods. Both MDMA isomers produced partial substitution for LSD and failed to substitute for d-amphetamine, indicating the isomers do not produce distinct discriminative stimulus effects as previously suggested. In a subsequent three-lever discrimination study, rats were trained to discriminate MDMA from *d*-amphetamine (Goodwin and Baker 2000). In this study, LSD produced dose-dependent increases in MDMA-lever responses but not quite full substitution (78%) for MDMA. However, full substitution for MDMA was obtained with the 5-HT releaser, fenfluramine, indicating that the serotonergic actions of MDMA were particularly salient in maintaining stimulus control in animals trained to discriminate MDMA from *d*-amphetamine, a dopamine releaser. In a follow-up study, it was determined that rats can also be trained to discriminate MDMA from LSD using a similar three-lever discrimination procedure (Goodwin et al. 2003). In that study, *d*-amphetamine produced only partial substitution for MDMA and fenfluramine still produced complete substitution for MDMA. MDL 100907 only partially blocked the stimulus effects of MDMA, but completely antagonized LSD discrimination in these rats. The dopamine D_2 antagonist haloperidol also failed to block MDMA discrimination. These results indicate that 5-HT release remains a salient feature in MDMA's discriminative stimulus effects, even when rats are trained to discriminate MDMA from another serotonin agonist, LSD. Furthermore, this series of three-lever discrimination studies provided conclusive evidence that MDMA produces complex stimulus effects, distinct from both stimulants and hallucinogens.

10 Future Directions

If this chapter has served its purpose, the reader should be convinced that the drug discrimination paradigm is a valuable investigative tool for assessing the psychopharmacology of hallucinogens. Since its inception, this sensitive and pharmacologically specific in vivo drug-detection method has garnered considerable evidence for the involvement of complex neural systems in the interoceptive stimulus effects of hallucinogens. As technology advances within the fields of structural chemistry, molecular biology, and genetic engineering, pairing these technologies with drug discrimination methods can prove fruitful in the continued quest to understand the complexities of hallucinogens and related psychedelic drugs. For example, recent developments in transgenic and genetic knockout rodent models can make use of drug discrimination methods to discern the importance of specific gene expression to the interoceptive stimulus effects of selected drugs.

As scientific research progresses with emphasis on rediscovering the medicinal values of hallucinogens, we must take into account individual differences in sensitivity to the putative therapeutic effects as well as possible adverse effects. As such, exploration of genetic and sex differences is a worthwhile endeavor in both preclinical and clinical investigations. To date, nonhuman drug discrimination investigations with hallucinogens have used male subjects exclusively. The evaluation of sex differences in the interoceptive stimulus effects of hallucinogens is a potentially lucrative future research direction. Moreover, greater inclusion of female subjects in preclinical behavioral pharmacology is long overdue. Research employing drug discrimination methods to evaluate genetic and sex differences in concert with structure–activity and mechanistic studies of hallucinogens can serve to inform further clinical investigations with hallucinogens.

References

Appel JB, Callahan PM (1989) Involvement of 5-HT receptor subtypes in the discriminative stimulus properties of mescaline. Eur J Pharmacol 159:41–46

Appel JB, White FJ, Holohean AM (1982) Analyzing mechanism(s) of hallucinogenic drug action with drug discrimination procedures. Neurosci Biobehav Rev 6:529–536

Appel JB, West WB, Rolandi WG, Alici T, Pechersky K (1999) Increasing the selectivity of drug discrimination procedures. Pharmacol Biochem Behav 64:353–358

Baker LE, Taylor MM (1997) Assessment of the MDA and MDMA optical isomers in a stimulant-hallucinogen discrimination. Pharmacol Biochem Behav 57:737–748

Baker LE, Broadbent J, Michael EK, Matthews CA, Metosh RB, West WB, Appel JB (1995) Assessment of the discriminative stimulus effects of the optical isomers of ecstasy (3,4-methylenedioxymethamphetamine; MDMA). Behav Pharmacol 6:263–275

Béïque JC, Imad M, Mladenovic L, Gingrich JA, Andrade R (2007) Mechanism of the 5-hydroxytryptamine 2A receptor-mediated facilitation of synaptic activity in prefrontal cortex. Proc Natl Acad Sci 104(23):9870–9875

Callahan PM, Appel JB (1988) Differences in the stimulus properties of 3,4-methylenedioxyamphetamine and 3,4-methylenedioxymethamphetamine in animals trained to discriminate hallucinogens from saline. J Pharmacol Exp Ther 246:866–870

Callahan PM, Appel JB (1990) Differentiation between the stimulus effects of LSD and lisuride using a three-choice drug discrimination procedure. Psychopharmacology 100:13–18

Carbonaro TM, Eshleman AJ, Forster MJ, Cheng K, Rice KC, Gatch MB (2015) The role of 5-HT_{2A}, 5-HT_{2C} and mGlu2 receptors in the behavioral effects of tryptamine hallucinogens N, N-dimethyltryptamine and N,N-diisopropyltryptamine in rats and mice. Psychopharmacology 232: 275–284

Colpaert FC (1999) Drug discrimination in neurobiology. Pharmacol Biochem Behav 64:337–345

Colpaert FC, Niemegeers CJ, Janssen PA (1982) A drug discrimination analysis of lysergic acid diethylamide (LSD): in vivo agonist and antagonist effects of purported 5-hydroxytryptamine antagonists and of pirenperone, a LSD-antagonist. J Pharmacol Exp Ther 221:206–214

Conn PJ, Sanders-Bush E (1986) Regulation of serotonin-stimulated phosphoinositide hydrolysis: relation to the 5-HT_2 binding site. J Neurosci 6:3669–3675

Conn PJ, Janowsky A, Sanders-Bush E (1987) Denervation supersensitivity of the 5-HT_{1C} receptors in rat choroid plexus. Brain Res 400:396–398

Delille HK, Becker JM, Burkhardt S, Bleher B, Terstappen GC, Schmidt M, Meyer AH, Unger L, Marek GJ, Mezler M (2012) Heterocomplex formation of 5-HT_{2A}-mGlu2 and its relevance for cellular signaling cascades. Neuropharmacology 62:2184–2191

Fantegrossi WE, Murnane KS, Reissig CJ (2008) The behavioral pharmacology of hallucinogens. Biochem Pharmacol 75:17–33

Fiorella D, Rabin RA, Winter JC (1995a) The role of the 5-H_{2A} and 5-HT_{2C} receptors in the stimulus effects of hallucinogenic drugs I: antagonist correlation analysis. Psychopharmacology 121:347–356

Fiorella D, Helsley SE, Lorraine DS, Palumbo PA, Rabin RA, Winter JC (1995b) The role of the 5-H_{2A} and 5-HT_{2C} receptors in the stimulus effects of hallucinogenic drugs III: The mechanistic basis for supersensitivity to the LSD stimulus following serotonin depletion. Psychopharmacology 121:364–372

Glennon RA (1994) Classical hallucinogens: an introductory overview. NIDA Res Monogr 146:4–32

Glennon RA, Young R (1982) Comparison of behavioral properties of di- and tri-methoxyphenylisopropylamines. Pharmacol Biochem Behav 17(4):603–607

Glennon RA, Young R (2011) Drug discrimination and in vivo structure-activity relationships. In: Glennon R, Young R (eds) Drug discrimination: applications to medicinal chemistry and drug studies, 1st edn. Wiley, Hoboken, pp 163–181

Glennon RA, Rosecrans JA, Young R (1982) The use of the drug discrimination paradigm for studying hallucinogenic agents. In: Colpaert FC, Slangen JL (eds) Drug discrimination: applications in CNS pharmacology. Elsevier Biomedical Press, Amsterdam, pp 69–96

Glennon RA, Titeler M, Mckenney JD (1984a) Evidence for 5-HT2 involvement in the mechanism of action of hallucinogenic agents. Life Sci 35:2505–2511

Glennon RA, Rosecrans JA, Young R (1984b) Drug-induced discrimination: a description of the paradigm and a review of its specific application to the study of hallucinogenic agents. Med Res Rev 3:289–376

Glennon RA, Yousif M, Patrick G (1988) Stimulus properties of 1-(3,4-methylenedioxyphenyl)-2-aminopropane (MDA) analogs. Pharmacol Biochem Behav 29:443–449

González-Maeso J, Weisstaub NV, Zhou M, Chan P, Ivic L, Ang R et al (2007) Hallucinogens recruit specific cortical 5-HT_{2A} receptor-mediated signaling pathways to affect behavior. Neuron 53:439–452

González-Maeso J, Ang RL, Yuen T, Chan P, Weisstaub NV, Lopez-Gimenez JF, Zhou M, Okawa Y, Callado LF, Milligan G, Gingrich JA, Filizola M, Meana JJ, Sealfon SC (2008) Identification of a serotonin/glutamate receptor complex implicated in psychosis. Nature 452:93–97

Goodwin AK, Baker LE (2000) A three-choice drug discrimination procedure dissociates the discriminative stimulus effects of MDMA and d-amphetamine. Exp Clin Psychopharmacol 8(3):415–423

Goodwin AK, Pynnonen DM, Baker LE (2003) Serotonergic-dopaminergic mediation of MDMA's discriminative stimulus effects in a three-choice discrimination. Pharmacol Biochem Behav 74:987–995

Harris RT, Balster RL (1971) An analysis of the function of drugs in the stimulus control of operant behavior. In: Thompson T, Pickens R (eds) Stimulus properties of drugs. Appleton-Century Crofts, New York, pp 111–132

Hirschhorn I, Winter JC (1971) Mescaline and LSD as discriminative stimuli. Psychopharmacologia 22:64–71

Holohean AM, White FJ, Appel JB (1982) Dopaminergic and serotonergic mediation of the discriminable effects of ergot alkaloids. Eur J Pharmacol 81(4):595–602

Holtzman SG, Locke KW (1988) Neural mechanisms of drug stimuli: experimental approaches. Psychopharmacology 4:138–153

Killinger BA, Peet MM, Baker LE (2010) Salvinorin A fails to substitute for the discriminative stimulus effects of LSD or ketamine in Sprague-Dawley rats. Pharmacol Biochem Behav 96(3):260–265

Koek W, Slangen JL (1982) Effects of reinforcement differences between drug and saline sessions on discriminative stimulus properties of fentanyl. In: Colpaert FC, Slangen (eds) Drug discrimination: applications in CNS pharmacology. Elsevier, Amsterdam, pp 343–354

Koerner J, Appel JB (1982) Psilocybin as a discriminative stimulus: lack of specificity in an animal behavior model for 'hallucinogens'. Psychopharmacology 76(2):130–135

Kueh D, Baker LE (2007) Reinforcement schedule effects in rats trained to discriminate 3,4-methylenedioxymethamphetamine (MDMA) or cocaine. Psychopharmacology 189:447–457

Kuhn DM, White FJ, Appel JB (1978) the discriminative stimulus properties of LSD: mechanism of action. Neuropharmacology 17:257–263

Lambe EK, Aghajanian GK (2007) Prefrontal cortical network activity: opposite effects of psychedelic hallucinogens and D1/D5 dopamine receptor activation. Neuroscience 145:900–910

Lambe EK, Goldman-Rakic PS, Aghajanian GK (2000) Serotonin induces EPSCs preferentially in layer V pyramidal neurons of the frontal cortex in the rat. Cereb Cortex 10:974–980

Lambe EK, Aghajanian GK (2001) The role of Kv1.2-containing potassium channels in serotonin-induced glutamate release from thalamocortical terminals in rat frontal cortex. J Neurosci 21:9955–9963

Marona-Lewicka D, Nichols DE (2007) Further evidence that the delayed temporal dopaminergic effects of LSD are mediated by a mechanism different than the first temporal phase of action. Pharmacol Biochem Behav 87:453–461

Marona-Lewicka D, Kurrasch-Orbaugh DM, Selken JR, Cumbay MG, Lisnicchia JG, Nichols DE (2002) Re-evaluation of lisuride pharmacology: 5-hydroxytryptamine$_{1A}$ receptor-mediated behavioral effects overlap its other properties in rats. Psychopharmacology 64(1):93–107

McMillan DE, Wenger GR (1984) Bias of phencyclidine discrimination by the schedule of reinforcement. J Exp Anal Behav 42:51–66

McMillan DE, Hardwick WC, Li M (2001) Discrimination of pentobarbital doses and drug mixtures under fixed-ratio and fixed-interval reinforcement schedules. Behav Pharmacol 12(3): 195–208

Moreno JL, Holloway T, Albizu L, Sealfon SC, González-Maeso J (2011) Metabotropic glutamate mGlu2 receptor is necessary for the pharmacological and behavioral effects induced by hallucinogenic 5-HT$_{2A}$ receptor agonists. Neurosci Lett 493:76–79

Nichols DE (1986) Differences between the mechanism of Action of MDMA, MBDB, and the classic hallucinogens. Identification of a new therapeutic class. Entactogens J Psychoactive Drugs 18(4):305–313

Nichols DE (2004) Hallucinogens. Pharmacol Ther 101:131–181

Oberlender R, Nichols DE (1988) Drug discrimination studies with MDMA and amphetamine. Psychopharmacology 95:71–76

Overton DA (1979) Influence of shaping procedure and schedules on performance in the two-bar drug discrimination task: a methodological report. Psychopharmacology 65:291–298

Pranzatelli MR (1990) Neonatal, 5,7-DHT lesions up-regulate (3H)mesulergine-labelled spinal 5-HT1C binding sites in the rat. Brain Res Bull 25:151–153

Schechter MD (1998) MDMA-like stimulus effects of hallucinogens in male fawn-hooded rats. Pharmacol Biochem Behav 59:265–270

Schechter MD, Rosecrans JA (1972) Lysergic acid diethylamide (LSD) as a discriminative cue: drugs with similar stimulus properties. Psychopharmacologia 26:313–316

Shannon HE (1981) Evaluation of phencyclidine analogs on the basis of their discriminative stimulus properties in the rat. J Pharmacol Exp Ther 216:543–551

Silverman PB & Ho BT (1978) Stimulus properties of DOM: commonality with other hallucinogens. In: Colpaert FC, Rosecrans JA (eds) Stimulus properties of drugs: ten years of progress. Elsevier/North Holland, Amsterdam, pp 189–198

Stolerman IP (1989) Discriminative stimulus effects of nicotine in rats trained under different schedules of reinforcement. Psychopharmacology 97:131–138

Stolerman IP (1993) Drug discrimination. In: Van Haaren F (ed) Methods in behavioral pharmacology. Elsevier Science, Amsterdam, pp 217–224

Swedberg MDB, Jarbe TUC (1986) Drug discrimination procedures: differential characteristics of the drug A vs drug B and the drug A vs drug B vs no drug cases. Psychopharmacology 90:341–346

White FJ, Appel JB (1982) Lysergic acid diethylamide (LSD) and lisuride: Differentiation of their neuropharmacological actions. Science 216:535–537

White FJ, Simmons MA, West KB, Holohean AM, Appel JB (1980) The effect of serotonin depletion on the discriminability of LSD. Pharmacol Biochem Behav 13:567–574

Winter JC (1978) Stimulus properties of phenethylamine hallucinogens and lysergic acid diethylamide: the role of 5-hydroxytryptamine. J Pharmacol Exp Ther 204:416–423

Winter JC, Fiorella DJ, Timineri DM, Filipink RA, Helsley SE, Rabin RA (1999) Serotonergic receptor subtypes and hallucinogen-induced stimulus control. Pharmacol Biochem Behav 64:283–293

Winter JC, Doat M, Rabin RA (2000) Potentiation of DOM-induced stimulus control by non-competitive NMDA antagonists. A link between glutamatergic and serotonergic hypotheses of schizophrenia. Life Sci 68:337–344

Winter JC, Eckler JR, Rabin RA (2004) Serotonergic/glutamatergic interactions: the effects of $mGlu_{2/3}$ receptor ligands in rats trained with LSD and PCP as discriminative stimuli. Psychopharmacology 172:233–240

Winter JC, Rice KC, Amorosi DH, Rabin RA (2007) Psilocybin-induced stimulus control in the rat. Pharmacol Biochem Behav 87:472–480

Winter JC (2009) Hallucinogens as discriminative stimuli in animals: LSD, phenethylamines, and tryptamines. Psychopharmacology 203:251–263

Young R (2009) Drug discrimination. In: Buccafusco JJ (ed) Methods of behavior analysis in neuroscience, 2nd edn. Retrieved from http://www.ncbi.nlm.nih.gov/books/NBK5225/

Young R, Rosecrans JA, Glennon RA (1982) Comparative discriminative stimulus effects of 5-methoxy-N, N-dimethoxytryptamine and LSD. Life Sci 30:2057–2062

Phenomenology, Structure, and Dynamic of Psychedelic States

Katrin H. Preller and Franz X. Vollenweider

Abstract Classic serotonergic hallucinogens or psychedelics produce an altered states of consciousness (ASC) that is characterized by profound alterations in sensory perception, mood, thought including the perception of reality, and the sense of self. Over the past years, there has been considerable progress in the search for invariant and common features of psychedelic states. In the first part of this review, we outline contemporary approaches to characterize the structure of ASCs by means of three primary etiology-independent dimensions including oceanic boundlessness, anxious ego-dissolution, and visionary restructuralization as well as by 11 lower-order factors, all of which can be reliably measured by the altered state of consciousness questionnaire (APZ-OAV). The second part sheds light on the dynamic nature of psychedelic experiences. Frequently, psychedelic subjects progress through different stages over time and levels of changes along a perception-hallucination continuum of increasing arousal and ego-dissolution. We then review in detail the acute effects of psychedelics on sensory perception, emotion, cognition, creativity, and time perception along with possible neural mechanisms underlying them. The next part of this review outlines the influence of non-pharmacological factors (predictors) on the acute psychedelic experience, such as demographics, genetics, personality, mood, and setting, and also discusses some long-term effects succeeding the acute experience. The last part presents some recent concepts and models attempting to understand different facets of psychedelic states of consciousness from a neuroscientific perspective.

The original version of this chapter was revised: Revised manuscript has been processed. The erratum to this chapter is available at 10.1007/7854_2017_477

K.H. Preller (✉) · F.X. Vollenweider (✉)
Neuropsychopharmacology and Brain Imaging Unit, Heffter Research Center Zurich
Department of Psychiatry, Psychotherapy and Psychosomatics, Psychiatric University Hospital, University of Zurich, Lenggstrasse 31, 8032 Zurich, Switzerland
e-mail: preller@bli.uzh.ch

F.X. Vollenweider
e-mail: vollen@bli.uzh.ch

Curr Topics Behav Neurosci (2016) 36: 221–256
DOI 10.1007/7854_2016_459
Published Online: 27 December 2016

Keywords Psilocybin • Lysergic acid diethylamide (LSD) • Hallucinogens • Psychedelics • Altered states of consciousness • Serotonin • Human

Contents

1 Introduction

Classic hallucinogens or psychedelics such as LSD, psilocybin, DMT, and mescaline comprise a group of compounds all of which produce a so-called altered state of consciousness (ASC) that is characterized by profound alterations of sensory perception, mood, thought, perception of reality, and the sense of self (Hofmann 1968). All of these classic hallucinogens are agonists or partial agonists at serotonin 5-HT2 receptors; additionally, the tryptamines activate 5-HT1, 5-HT6, and 5-HT7 receptors, whereas LSD also directly stimulates dopamine D1 and D2 receptors (Nichols 2004). Substantial evidence demonstrates that activation of 5-HT2A receptors by classic hallucinogens is a primary key mechanism to psychedelic symptom formation in humans (Vollenweider 1998; Vollenweider et al. 1998). Since the 1990s, there has been a growing interest in the investigation of the complex phenomenology and neuronal correlates of hallucinogen-induced ASCs in humans, facilitated by recent advances in concepts and techniques in cognitive neurosciences (Geyer and Vollenweider 2008; Vollenweider 1998; Vollenweider and Geyer 2001; Vollenweider and Kometer 2010).

The study of hallucinogens in humans is important for several reasons, as follows. First, hallucinogens affect a number of psychological domains and brain functions, including emotional and social behaviors, cognitive-thoughtful functions, and self-awareness, that typically characterize the human mind and cannot be reliably studied in behavioral animal models (Hanks and Gonzalez-Maeso 2013). Second, hallucinogens can elicit a clinical syndrome that in several aspects resembles the incipient stages of the spectrum of schizophrenic psychoses or the acute stages of mania-like disorders (Geyer and Vollenweider 2008; Gouzoulis-Mayfrank et al. 1998a; Vollenweider 1998; Vollenweider et al. 1998). Third, there is accumulating evidence that hallucinogens have clinically relevant effects as adjuncts in the treatment of depression, anxiety, obsessive–compulsive disorder, and addictions (Bogenschutz et al. 2015; Grob et al. 2011; Leuner et al. 1994; Moreno and Delgado 1997). Finally, hallucinogens represent a unique heuristic experimental tool that can provide novel insights into the dimensions of the human unconscious and the basis for creativity (Masters and Houston 2000; Sessa 2008).

Over the years, several attempts have been made to find a universally accepted taxonomy to describe the different facets of the phenomenology produced by classic hallucinogens (Leuner 1981; Schultes and Hofmann 1980). At the present time, the term *hallucinogen* is the most common designator in the scientific literature, although it is somewhat of a misnomer because true hallucinations rarely occur at low-to-medium doses. *Hallucinogen* is often used interchangeably with the term "psychotomimetic" (psychosis-mimicking), which emphasizes the psychosis-like symptoms induced by these agents. More recently, the more popular term "psychedelic," meaning mind-manifesting, has also been employed in scientific publications; *psychedelic* was coined to convey the fact that the psychological state induced by hallucinogens is not defined by pathological features but rather by a general activation and manifestation of the experiential and behavioral repertoire of the psyche (Osmond 1957).

By the late 1970s, more than 1000 papers had been published describing the psychological, behavioral, or clinical effects of classic hallucinogens in normal subjects or in psychiatric patients. Several attempts were made—using a plethora of approaches—to describe and categorize the wide range of phenomena occurring during psychedelic states. The methods and instruments used to describe the acute phenomenology of psychedelic states ranged from narrative reports and subjective first-person accounts to a variety of psychological, cognitive, behavioral, and psychopathological measures, but also included process-related constructs and terms borrowed from psychoanalytical theories, as well as Gestalt and transpersonal psychology. The various approaches used to probe the topography and dynamics of the psychedelic experience are well documented in three comprehensive monographs (Grof 1975; Leuner 1962; Masters and Houston 2000). For example, Master and Houston (2000) suggested that the psychological features and processes characterizing psychedelic states in healthy subjects fall into four broad categories: changes occurring at the sensory level, at the recollective–psychodynamic level, at the affective and symbolic level, and at the deep integral level of self-transcendence. Similar experiential categories and comparable processes were described by Grof (1975) in psychiatric patients undergoing psychedelic-assisted psychotherapy.

Although these early categorization attempts provided important insight into the content and dynamics of the psychedelic experience, an accurate detection of patterns in the features of the psychedelic experience remained problematic. The identification of structural features was hampered by the fact that the theoretical framework and methods used differed markedly across studies and that some of the proposed categories denote the method used to get to a particular state rather than the experiential state itself. Moreover, personal narratives, unvalidated self-report measures, and clinical descriptions may be well suitable to extract a lower-order classification of psychedelic states, but they are not sufficient to detect a higher-order structure, e.g., orthogonal dimensions of psychedelic experiences.

To overcome some of these methodological problems, Dittrich and his co-workers (Dittrich 1985, 1998; Dittrich et al. 1981, 1985) conducted a series of studies to empirically test the hypothesis that pharmacological and non-pharmacological-induced ASCs share certain core dimensions, independent of the method of induction or intensity (Ludwig 1966). This initial work and subsequent validation studies yielded a 3-factorial structure of ASCs and led to the development of the altered state of consciousness questionnaire (APZ) (Dittrich 1985), as well as its revised versions, the APZ-OAV (Dittrich 1998) and the 5D-ASC (for a review, see Studerus et al. 2010). The 5D-ASC comprises three primary etiology-independent dimensions: oceanic boundlessness (OBN), anxious ego-dissolution (AED), and visionary restructuralization (VIR), as well as two secondary dimensions: acoustic alterations (AA) and vigilance reduction (VR); these five dimensions measure characteristic patterns of changes ranging from sensations to self-awareness. Dittrich had theorized that detection of such invariant core dimensions of ASCs may lead to a more coherent taxonomy of ASCs and may facilitate the identification of neuronal correlates of the different facets of the subjective experiences in ASCs. More recently, using a large dataset, Studerus et al. (2010) demonstrated that ASCs can be described by 11 lower-order factors that are highly measurement invariant across various drug conditions and settings. Two other instruments that are used to measure the subjective experience of ASCs in current hallucinogen research are the Phenomenology of Consciousness Inventory (Pekala et al. 1986) and the Hallucinogen Rating Scale (HRS), which was developed by Strassman et al. (1994) and later validated by Riba et al. (2001).

Since the 1990s, the APZ questionnaires has been included in more than 80 studies to assess the psychological effects of different classes of psychoactive drugs including psilocybin, ketamine, DMT, and mescaline and to link first-person accounts to various cognitive, behavioral, and neurophysiological measures of ASCs including positron emission tomography (PET), electroencephalography (EEG-ERP), magnetoencephalography (MEG), and functional magnetic resonance imaging (fMRI). Several studies focused on the pharmacokinetics, metabolism, and receptor profile of psilocybin and its relationship to the various facets of the psychedelic experience, while other studies have explored the dose-response, tolerability, and side effects of psilocybin in healthy subjects. Moreover, some early studies using psychophysiological approaches explored the relationship between specific subjective symptoms (e.g., the sensation of a color) and objective physical measures (e.g., a change in brightness) induced by psilocybin and LSD in healthy subjects (for review see Fischer 1971). Many of the recent studies used psilocybin or DMT within the model psychosis framework as tools to identify the neuronal underpinnings of psychotic symptom formation, including hallucinations, thought disorder, and disturbances of the self (Geyer and Vollenweider 2008). For instance, some studies tested the hypothesis that deficits of basic information processing functions (such as sensory gating and perceptual learning) contribute to the generation of psychosis-like symptoms and cognitive alterations in psychedelic states. Other studies have specifically focused on the neural basis of perceptual alterations such as the sense of time, binocular rivalry, and hallucinations. More recent work has also re-investigated the potential for psilocybin to induce a positively experienced self-dissolution as a central feature of so-called mystical experiences and its ability to produce long-lasting positive changes in attitude, mood, and behavior in healthy subjects (Griffiths et al. 2006).

The present review is concerned with the phenomenology of psychedelic states and particular with characterization of the effects of classic hallucinogens on perception, emotion, cognition, and the sense of self in healthy subjects. Given that most of the recent studies were conducted with psilocybin, and because the psychological peak effects of psilocybin could not be distinguished from those of LSD or mescaline in controlled studies, we will focus on recent and early studies that attempted to identify neuronal correlates of psychedelic features induced by psilocybin. Furthermore, we attempt to characterize psilocybin-induced states by content, structure, and distinct stages of a psilocybin-induced dynamic process and will discuss some conceptual issues and difficulties encountered.

2 Structure and Basic Dimensions of Psychedelic Experiences

The term "altered state of consciousness" (ASC), coined by Ludwig in (1966), refers to a mental state that is induced by physiological, psychological, or pharmacological manipulations and significantly deviates from normal waking consciousness (Ludwig 1966). Following Moreau De Tours (1973), Ludwig formulated the hypothesis that deliberately and disease-induced ASCs share common core dimensions. Ludwig's hypothesis was empirically tested and validated by Dittrich (1985) in a series of studies using various ASC-inducing methods (e.g., hallucinogens, hypnosis, and sensory overload). This work revealed that the structure of ASCs comprises three primary etiology-independent dimensions: oceanic boundlessness (OBN), anxious ego-dissolution (AED), and visionary restructuralization (VIR), which can be reliably measured by the APZ-OAV altered state of consciousness questionnaire (Dittrich 1985, 1998).

The OBN dimension reflects a highly pleasurable state of self-dissolution and comprises items that measure alterations of ego-boundaries, an experience of unity and transcendence of space and time, spiritual experiences, a sense of intuitive understanding, and changes in emotions ranging from heightened mood to bliss and ecstasy. The items in the OBN dimension can be categorized into five item-groups: depersonalization and derealization, changes in the sense of time, positive mood, and mania-like symptoms (see Fig. 3). Given that many of the OBN items have been based on six of the nine core characteristics of so-called "mystical" experiences defined by Stace (1961), high OBN scores may be indicative of mystical experiences, as described in the scientific literature on the psychology of religion (Forman 1990).

The AED dimension denotes a state of anxious ego-dissolution that is characterized by disintegration of the coherent self/ego or separation of the self from the world, loss of self control over one's autonomy, feelings of estrangement from the environment, cognitive impairment, anxiety or panic, and motor disturbances or catatonia. The respective item-groups assess negatively experienced derealization, thought disorder, paranoia, loss of thought control, and loss of body control. High AED scores can reflect a highly unpleasant state that is often referred to colloquially and in the literature as "bad trip." The disruption of the ego as a fundamental unifying drive for integration, organization, and meaning, and impairment of differentiation between ego and non-ego spheres, are highly reminiscent of the acute stages of schizophrenic decompensations (for review see Chapter—Krähenmann, Hermle).

The VIR dimension comprises items assessing elementary and complex pseudohallucinations or true hallucinations, illusions, audio-visual synesthesias, changed meaning of percepts, vivid recollection of memories and imagery from memory, and facilitated imagination. Together, these perceptual phenomena may represent what has been described as hallucinatory visions or "visionary experiences" by religious and historical figures such as Theresa von Avila, Plotin, Eckhart or Shankara (Forman 1990).

However, Dittrich's original work and subsequent investigations suggested that two additional—although etiology-dependent—dimensions exist: **acoustic alterations** (AA), relating to distortions, illusions, and hallucinations in the acoustic sphere, and **vigilance reduction** (VR), which depicts dreaminess, reduced alertness, and drowsiness (Dittrich 1998; Studerus et al. 2010). These additional dimensions were included in the five dimensions of altered states of consciousness (5D-ASC), a revised version of the APZ-OAV questionnaire.

Although the three original OBN (O), AED (A), and VRS (V) dimensions were believed to be factorially invariant across ASC induction methods (Dittrich 1998), a recent re-examination of the APZ-OAV structure using pooled data from 43 controlled studies conducted with psilocybin, (*S*)-ketamine, and MDMA was only able to partially confirm this assumption. Using advanced statistical methods, Studerus and co-workers found that the primary APZ-OAV scales are multidimensional and the VRS dimension can be merged with the OBN dimension at the highest level of construct hierarchy (Studerus et al. 2010). Moreover, Studerus and co-workers were able to extract 11 new homogenous lower-order factors (subscales), which are highly measurement invariant across drug condition, setting, and gender (Studerus et al. 2010). The subscales were termed as follows: experience of unity, spiritual experiences, blissful state, insightfulness, disembodiment, impaired control and cognition, anxiety, elementary imagery, complex imagery, audio-visual synesthesiae, and changed meaning of percepts (Fig. 1).

These new subscales have recently been used to compare the subjective effects of psilocybin with those of (*S*)-ketamine and MDMA in healthy subjects. As shown in Fig. 2, despite some similarities, the profile of the subjective effects of classic hallucinogens can clearly be distinguished from those produced by a dissociative NMDA receptor antagonist ((*S*)-ketamine) and a mood activating entactogen (MDMA). These results suggest that the new subscales may be useful to identify more specific experiential features of ASCs and their respective neural correlates.

Specific aspects of ASCs can also be qualified using other psychometric scales. For example, Griffiths and co-workers have shown that psilocybin-induced mystical effects can be assessed with the Mystical Experience Questionnaire (MEQ) (Griffiths et al. 2008). The MEQ is a non-validated self-report measure (Pahnke 1963) covering 7 of the 9 core characteristics of mystical experiences defined by Stace (1961): feelings of unity, transcendence of time and space, noetic

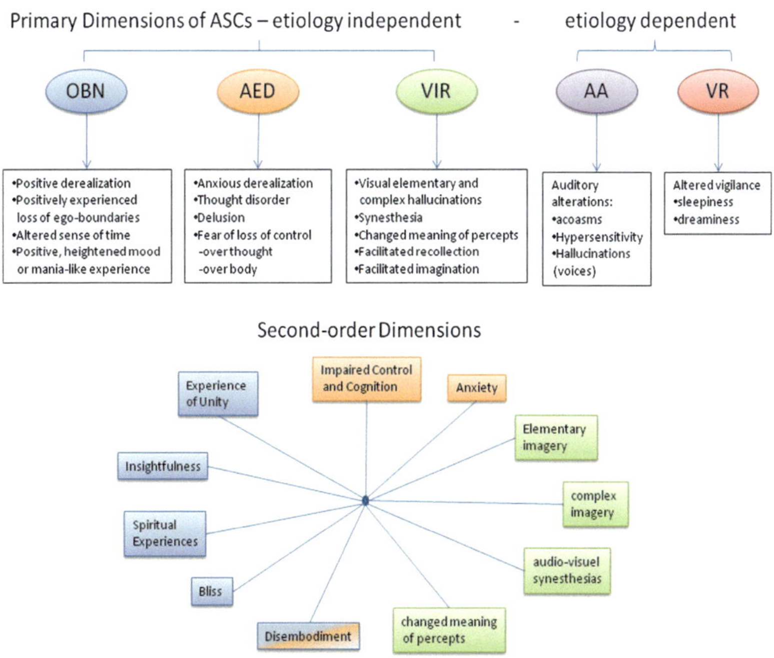

Fig. 1 Structure of altered states of consciousness (ASCs) comprises three primary etiology-independent dimensions/factors: oceanic boundlessness (*OBN*), anxious ego-dissolution (*AED*), and visionary restructuralization (*VIR*), and two etiology-dependent dimensions/factors: acoustic alterations (*AA*) and vigilance reduction (*VR*). The content of these 5 primary dimensions of ASCs can be described by their respective item clusters. According to Studerus et al. (2011), the structure of ASCs can also be described by 11 homogenous lower-order factors

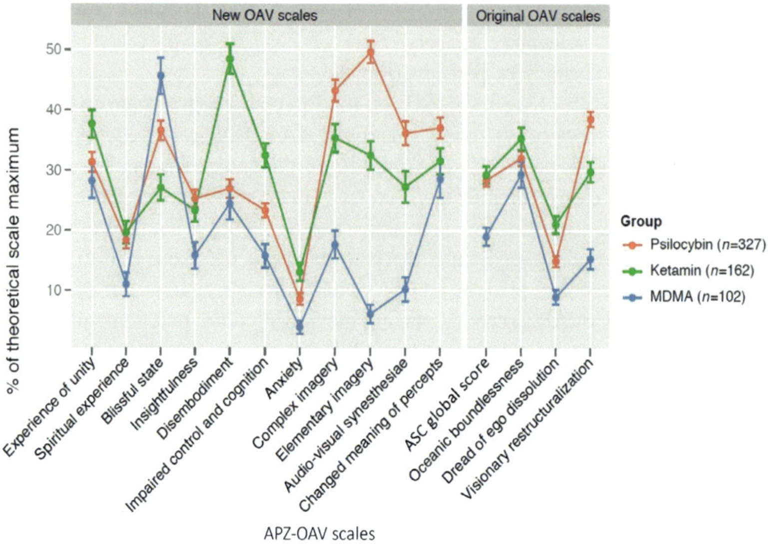

Fig. 2 Known-group validities of the 11 new lower-order and the original APZ-OAV scales (*ASC global score*), oceanic boundlessness (*OBN*), anxious ego-dissolution (*AED*), and visionary restructuralization (*VIR*). *Error bars* represent standard errors (from Studerus et al. 2010)

quality, sacredness, positive mood, ineffability, and paradoxicality. A recent exploratory factor analysis of MEQ data revealed that these characteristic features of mystical states can be reduced to 4 factors covering the MEQ dimensions: F1: unity, noetic quality, and sacredness; F2: positive mood; F3: transcendence of time/space; and F4: ineffability (MacLean et al. 2012). These results indicate that the content of the MEQ dimensions markedly overlaps with the content of OBN dimension and therefore is likely indicative of a specific facet of ASCs.

3 Ego-Dissolution as a Dynamic Process

Several authors have emphasized that the psychedelic experience is not a firmly defined state but rather a dynamic process (Leuner 1962; Masters and Houston 2000). In fact, many of the subjects under the influence of a psychedelic drug appear to progress through different stages over time and levels of changes along a perception-hallucination continuum of increasing arousal and ego-dissolution. Although the intensity, duration, and depth of the stages depend most critically on the dosage, the specific psychedelic drug, and the route of administration, other factors such as personality structure, the nature and dynamics of unconscious material activated, and the expectancy of the subject are also important. Furthermore, the setting—including the physical, cultural, and social environment and the investigator's theoretical concept, methods, and instructions used—also appears to influence the outcome of the experience.

The psychedelic experience can be conceptualized as having two interrelated components: the experiential content and the experiential process. At the core of this process is the loosening of self-boundaries and the diminishing of ordinary ego-functions, which unfolds along a perception-hallucination continuum with increasing arousal, culminating in ego-dissolution and a state of oneness with the external world. The phenomenological ego as used here is the internal image of a person-as-a-whole, the "I" or "self" as it appears in conscious experience. According to Metzinger (2009), the ego is the content of a self-model; this conscious self-model constructed by the brain allows us to interact with our internal world as well as with the external environment in a holistic manner. In a broad sense, the self encompasses features such as a first-person perspective, feelings of agency, ownership ("mineness") and immediacy ("nowness"), spatial perspective, autobiographical memory, emotions, perceptions, thoughts and acts of will, as well as the feeling of being embedded in our bodily sensations (Metzinger 2009; Northoff 2011). Another function of the ego serves is to help control and plan our behavior and to understand the behavior of others. By representing the process of representation itself, we can catch ourselves in the act of knowing. Ultimately, the subjective experience of the ego arises from dynamic self-related information processing, which is the result of a self-organizing brain system interacting with its environment, because no such things as selves exist in the world (Metzinger 2009).

According to Masters and Houston (2000), modified after (Austin 1998), the dropout of the phenomenological self/ego and ego-functions arises sequentially and unfolds to various degrees along with alterations at: (1) the perceptual level, (2) the recollective–psychodynamic level, (3) the symbolic existential level, and (4) the deep integral level of self-transcendence (Fig. 3).

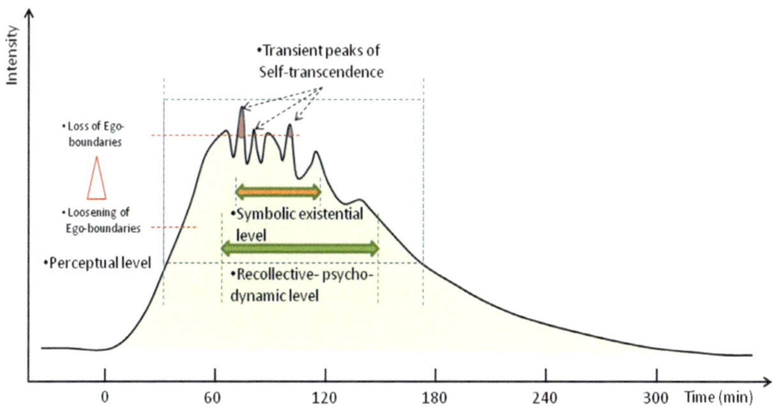

Fig. 3 Temporal dynamics and stages of a psilocybin-induced psychedelic experience. Adapted from Leuner (1962)

Perceptual changes appear with the onset of the reaction to psilocybin or LSD and are the most frequent and robust features of the psychedelic experience. Although perceptual changes can occur in all sensory modalities, the perceptual effects are dominated by visual phenomena, ranging from vividly colored, rapidly moving, and evolving elementary geometric figures to complex images and scenes involving persons, animals, architecture, or landscapes. Neither type has much meaning or function for subjects. In parallel, transformations of the environment and alterations of the body image are frequently reported.

Recollective–psychodynamic level: With increasing arousal toward and during the peak experience, visual images become more personalized, and boundaries between consciousness and unconsciousness dissolve, causing recall and re-enacting of past experiences and memories and releasing emotions into the process. Many of the phenomena and processes occurring at this level can be understood according to concepts of psychoanalytic theory.

Symbolic existential level: In the subsequent level, approaching the peak effects, ideas, eidetic images, or even the entire environment can become symbolized. Subjects become more personally involved and emotionally engaged as a participant in the ongoing psychedelic scenario, also referred to as a symbolic drama. The themes often have mythological and ritualistic overtones, and subjects may identify features of their own existence in legendary historical figures, fairy tales, and archetypal themes or other symbols and play out their personal drama on these allegoric terms. When the subjects encounter and struggle with these dramas, they can achieve a "solution," e.g., by imagination, ideation, sensations, and affective or kinesthetic involvement. This can result in a quiet but powerful emotional response and tension release that appears to be transformative and beneficial to the person. Although ego-boundaries are often transiently markedly reduced or may even disappear for seconds or minutes during these states, subjects are still aware of the situation and its ambiguity.

Deep integral level of self-transcendence: Along with the increasing dissolution of the ego, the psychedelic experience can peak in a state where subjects can become immersed for seconds or minutes in a profound awareness of oneness in which all boundaries disappear and objects are unified into a totality. When subjects have their eyes open, they retrospectively describe this state as an intense emotionally charged new and unfamiliar perspective, almost like a direct encounter with the "ultimate" reality, which can inspire feelings of awe, sacredness, and eternity. This novel experience is also characterized by a pervasive sense of deep insight into the nature and structure of the universe that is far beyond the person's usual mode of thinking. This intuitive awareness of unity appears to resemble the quality of "suchness" as described in states of awakening in advanced Zen mediation (Austin 1998) or as reported in the so-called extrovertive type of mystical states (Stace 1961; Forman 1990). However, when subjects have their eyes closed and turn their attention inward, a state of internal absorption may unfold, with subjects witnessing a vast internal space of objectless infinity that lacks not only the sense of the self, but also all sensory experiences and distracting thoughts. This very rare and transient state of extraordinary absorption is also an essential quality of advanced states in various forms of concentrative mediation (Austin 1998, 2006) and may reflect what has been termed as the "pure consciousness event" or as the introvertive type of mystical experience in the philosophy of religion (Forman 1990). Such deep integral levels of self-transcendence occur in mature and emotionally stable persons after repeated psychedelic experiences and after subjects have worked through their psychodynamic levels (Grof 1975; Leuner 1962, 1971; Leuner et al. 1994; Masters and Houston 2000).

In general, both the intensity of the psychedelic experience and the loosening of ego-boundaries are dose-dependent, so that ego-dissolution involving disorientation in person, place, and time is rarely induced by medium doses of psilocybin (15–25 mg po) or LSD (100–250 μg po) (Leuner 1962, 1971). However, the same dose may produce more psychotic ego-dissolution, including disorganization, fear, and paranoid ideation, in the same person on different occasions.

According to Leuner (1962, 1981), the type and course of ego-dissolution is driven by four key processes: increasing internal sensory stimulus production, activation of basic emotions and autobiographic memories, regression of ego-functions, and auto-production of symbolic imagery (and meaning). Leuner emphasized that low-to-medium doses of psilocybin or LSD primarily result in a reduction in ego-boundaries and in a "normal and scenic" type of ego-dissolution, which is characterized by vivid imagery and an intensification of affect that is accompanied by intact and adequate emotion processing and by a coherent though expanded "observing" self. In contrast, the "fragmentary or psychotic" type of ego-dissolution that mainly occurs after higher doses (psilocybin >25 mg po; LSD >250 μg po) is dominated by sensory overload including frightening symbolic imagery and excessive affective activation (i.e., "emotional overdrive"); such effects lead, in turn, to disruption of emotional and cognitive integrity, and subsequently to the experience of a fragmented and endangered self, often resulting in feelings of anxiety or panic.

The impact of various psilocybin doses (45–315 mg/kg po) on the two types of ego-dissolution was assessed in healthy volunteers using the "oceanic boundlessness" and "anxious ego-dissolution" 5D-ASC dimensions and is exemplified in Fig. 4.

The neural underpinnings of these two types of ego-dissolution (as indexed by "oceanic boundlessness" and "anxious ego-dissolution") induced by different hallucinogenic drugs have been investigated using PET neuroimaging with the radiotracer ^{18}F-fluorodeoxyglucose (^{18}FDG) (Vollenweider et al. 1997a, b, c). Correlational analysis between normalized metabolic activity and psychological scores on the APZ questionnaire revealed that the magnitude of OBN scores induced by both psilocybin and ketamine correlated positively with regional cerebral metabolic rate of glucose (CMRglu) bilaterally in frontomedial superior, frontolateral, and left inferolateral prefrontal cortex, anterior cingulate, as well as bilaterally in inferior parietal and occipitomedial cortex (Vollenweider and Geyer 2001). There were negative correlations between OBN and CMRglu bilaterally in the hippocampus and caudate nucleus, and left amygdala and ventral striatum. The severity of anxious ego-dissolution was positively correlated with CMRglu in the

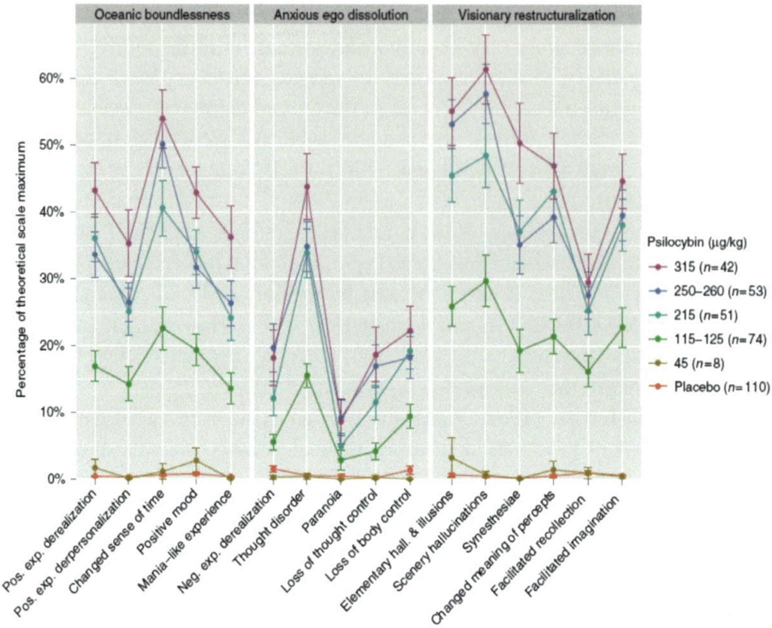

Fig. 4 Dose-dependent percentage scores of item clusters of the 3 etiology-independent primary dimensions: oceanic boundlessness (*OBN*), anxious ego-dissolution (*AED*), and visionary restructuralization (*VIR*) of the 5D-ASC rating scale. Error bars represent standard errors. Ratings were obtained during peak drug effects (60–270 min after drug administration). From Studerus et al. (2011)

thalamus and left temporomedial gyrus and negatively correlated with CMRglu bilaterally in orbitofrontal cortex and adjacent anterior cingulate. Thus, it appears that AED and the associated thought disorder depend mainly on thalamic overactivity and orbitofrontal underactivity. The hyperfrontality finding in ASC was further supported by further brain imaging studies in healthy subjects treated with the classic phenylethylamine hallucinogen mescaline, as well as with psilocybin and DMT (Gouzoulis et al. 1997; Hermle et al. 1993; Riba et al. 2006).

4 Sensory Perception

4.1 Visual Perception

Visual alterations, ranging from illusions to pseudohallucinations, and hallucinations, are a prominent feature of the psilocybin-induced ASC (Studerus et al. 2011; Díaz 2010). However, alterations of visual perception induced by psilocybin rarely represent true hallucinations, since they can usually be distinguished from real perceptions, at least at moderate doses. Phenomena that are often described include the perception of more intense colors and textures, geometric shapes, rhythmic movements of objects, micropsia and macropsia, after images of objects in movement, and objects, animals, or subjects which are not present (Díaz 2010).

Several experimental studies have described hallucinogen-induced illusions. Siegel and Jarvik (1975) reported that low doses of hallucinogens induced an increase in the subjective brightness of surrounding objects and colors, whereas at medium doses, objects also appeared to drift and pulsate (Fischer and Rockey 1967; Fischer et al. 1969a). Increased brightness in response to visual stimulation after psilocybin intake may be related to an increased P1 amplitude (Kometer et al. 2011).

Increased responsiveness to visual stimuli was related to a reduction in sensory thresholds (Hill et al. 1969; Fischer and Rockey 1967), which might contribute to subjects reporting that everything looks new, fascinating, and more intense (Díaz 2010). However, there are also reports of elevated visual thresholds after LSD (Silverman 1971). These apparent contradictory findings were thought to be due to differences in stimulus intensity: Serotonergic hallucinogens may induce a hypersensitivity to low-to-moderate intensity stimuli, but increased tolerance to high-intensity stimuli (Silverman 1971). However, these assumptions still have to be tested experimentally.

The elementary visual hallucinations induced by serotonergic hallucinogens are often characterized by geometric shapes with rhythmic kaleidoscopic and automatic movement (Díaz 2010). Additionally, the occurrence of simple phosphene-like visualizations such as blobs and light flashes has been described (Shanon 2002; Knoll et al. 1963). Furthermore, a recent study investigating imagination after the intake of moderate doses of LSD found that the vividness/realism of suggested scenarios was significantly higher for the LSD than the placebo condition (Carhart-Harris et al. 2015). These visual alterations were investigated in early work performed by Klüver (1966) and are classified as elementary geometrical patterns known as "form constants" (Siegel and Jarvik 1975; Siegel 1977; see Chapter Kometer and Vollenweider). Furthermore, alterations of motion perception and binocular rivalry are induced by psilocybin (Carter et al. 2005, 2007). Increased completion of partially deleted text as well as reduced object completion of Kanizsa figures was also reported to occur during psilocybin's peak effect (Fischer and Rockey 1967; Kometer et al. 2011). Furthermore, an enlargement of nearby visual space, as inferred from changes of apparent frontal plane (AFP) curve and an enlargement of handwriting area, was measured in psilocybin states (Fischer et al. 1970; Fischer and Mead 1966). This phenomenon was linked to an elevation of the so-called apparent horizon after the intake of LSD (Krus et al. 1963, 1966; Knoll et al. 1963). Under the influence of psilocybin, optical distortions induced by prism spectacles were interpreted as a transformation or unlearning of perceptual constancies that are important for maintaining the stability of the world (Fischer et al. 1970). Fischer et al. (1970) suggested that in psychedelic states the subject's perception is more and more dependent on the projection of central nervous system activity, while outside information is gradually reduced along the perception-hallucination continuum (Fischer et al. 1970). A similar shift in processing of external versus internal stimuli was suggested by Kometer et al. (2013).

Compared to elementary hallucinations, complex hallucinations are not as frequently reported to be experienced after psilocybin or LSD intake (Kometer et al. 2011; Studerus et al. 2011; Schmid et al. 2015). Complex hallucinations comprise objects, animals, or people and contain more semantic content in comparison with elementary hallucinations (Díaz 2010; Shanon 2002). Additionally, the perception of whole scenes and landscapes has been described (Shanon 2002; Leuner 1962). However, the content varies largely within and between subjects and may even differ between cultures (Díaz 2010). Complex hallucinatory content related to autobiographic memory or personal life situation has also been reported (Dittrich 1998; Studerus et al. 2011). Therefore, visual alterations induced by serotonergic hallucinogens are connected to emotional processing systems and contribute to changes in the meaning of percepts. Hence, they may represent significant experiences for participants. Alterations in the perception of the surroundings associated with feelings of unreality have also been termed "derealization phenomena."

4.2 Auditory Perception

Psilocybin and LSD have been shown to alter auditory perception to a lesser extent than visual perception (Leuner 1962; Studerus et al. 2011; Grof 1975). Studerus et al. (2011) reported that the auditory alterations scale of the 5D-ASC was only moderately affected by psilocybin, since true auditory hallucinations such as hearing voices rarely occurred at the moderate doses tested (Leuner 1962). The auditory alterations mostly comprised intensification of sounds or misperceptions of real auditory stimuli. Fischer et al. (1951) described the occurrence of LSD-induced auditory illusions, such as the perception of rustling noises or perceived alterations of the subject's own voice. Weber (1967) reported that musicians listening to music after ingesting psilocybin described an intensive, exhilarating sound experience. However, the dynamic structure of the music was reported to be difficult to comprehend, which possibly relates to disruptions of time perception (see below). Silverman (1971) measured lower thresholds to auditory stimuli, but also increased tolerance for strong auditory simulation, after the intake of 50 μg LSD. Furthermore, a psilocybin-induced reduction in the N1 sensory ERP in an auditory paradigm was reported, which suggests a reduced processing of the intensity of auditory stimuli (Umbricht et al. 2003). Additionally, psilocybin neither reduced mismatch negativity (MMN) amplitude nor implicit perceptual learning indexed by the frontal memory trace in classical and roving auditory MMN paradigms (Schmidt et al. 2012; Umbricht et al. 2003). These results indicate that psilocybin does not influence auditory processing and sensory learning at moderate doses. However, it is possible that higher doses are necessary to induce alterations in auditory processing. In line with these results, animal studies suggest that disruptive effects of serotonergic hallucinogens are observed on visual tasks at lower dosages than on auditory tasks (Silverman 1971; Caldwell and Domino 1967). However, a direct comparison in humans is still missing.

4.3 Tactile Perception

Participants reported of tickling sensations in hands and feet and sensations of warmth and cold after LSD intake (Leuner 1962). Furthermore, alterations of body experiences and body scheme, particularly in the perception of extremities, are common (Grof 1975). For example, participants reported muscular tension and feelings of heaviness in the limbs (Hermle et al. 1992). Furthermore, hands and arms were perceived as larger than normal (Leuner 1962; Fischer et al. 1951). Additionally, participants reported feelings of separation from body parts, e.g., the hand was perceived as belonging to someone else (Leuner 1962). Therefore, altered somatic experiences occurring after the intake of serotonergic hallucinogens are closely linked to alterations in self-experience (see above), as well as to detachment from reality and the observed self, which is often summarized under the term "depersonalization phenomena." In experimental studies, tolerance for painful stimulation was found to be elevated after the intake of LSD (Kast and Collins 1964).

4.4 Olfactory and Gustatory Perception

Moderate doses of psilocybin dose-dependently reduced the amount of sodium saccharinate necessary for the detection of taste (Fischer 1967; Fischer and Kaelbling 1966; Fischer et al. 1965a). However, there are also reports of blunted olfactory and gustatory perception after the intake of LSD (Fischer et al. 1951). Olfactory or gustatory hallucinations are more commonly described in terms of synesthesia experiences (see below).

4.5 Synesthesia

Synesthesia experiences are reported to be a common feature of psilocybin- and LSD-induced experiences, occurring even at low doses (Studerus et al. 2011; Schmid et al. 2015). Particularly after ingestion of LSD, audio-visual synesthesias are one of the most prominent experiences (Schmid et al. 2015). After ingestion of LSD, increased visual imagery with eyes closed has been reported to be associated with increased functional connectivity between the visual cortex and the parahippocampus (Kaelen et al. 2016). Music and sounds are most often reported to induce synesthesia, but tactile, gustatory, olfactory, or emotional stimuli can also induce synesthesia, with the stimuli primarily being translated to the visual domain (Sinke et al. 2012). An experience that is only found in drug-induced synesthesia is the perception of an altered body image (e.g., the body morphs in form and size) in response to visual, acoustical, or tactile stimuli (Sinke et al. 2012). Additionally, synesthesia comprising more than one modality has been described. Leuner (1962) described a visual–tactil–olfactory synesthesia experience: One subject reported that the blue color of a visual stimulus was associated with the perception of an ozone-like smell and the feeling of being electrically shocked. In contrast to acquired and genuine synesthesias, drug-induced synesthesias are highly dynamic and more flexible and show a high inter- and intrapersonal variance (Sinke et al. 2012). This may be one reason why experimental studies investigating drug-induced synesthesias and their underlying neural processes are very rare (Brogaard 2013).

5 Emotion

Serotonergic hallucinogens have been used in psychotherapy since the 1950s to facilitate emotional insight (Sandison and Whitelaw 1957). Various authors (Leuner 1971, 1962; Unger 1963; Johnson et al. 2008; Grof 1975) reported that psilocybin and LSD can intensify all forms of affective responses and may activate vivid memory traces with pronounced emotional undertones. In line with those findings, depressive patients made more personal statements when treated with low doses of LSD (Dahlberg and Jaffe 1979). Descriptions of LSD-induced experiences emphasized a state of euphoria that can take different forms, such as exhilarated elation with unmotivated laughter, deep feelings of peace, exuberant joy, and hedonistic pleasure (Grof 1975; Leuner 1962). Studerus et al. (2011) showed that psilocybin significantly increased heightened mood, emotional excitation, and sensitivity during the peak of the drug effect (60–95 min after administration) on the EWL affective mood rating scale in a large sample of 110 healthy subjects. Recently, Schmid et al. (2015) reported that LSD similarly increased emotional excitation and well-being in 16 healthy volunteers. Recent studies indicate that psilocybin selectively attenuates the encoding and processing of negative stimuli such as negative facial expressions and increases goal-directed behavior toward positive compared with negative cues (Kometer et al. 2012; Schmidt et al. 2013). Using EEG, it was shown that psilocybin valence-dependently decreases the P300 component (Kometer et al. 2012) and the N170 component of the ERP in response to fearful faces (Schmidt et al. 2013). These studies suggest that 5-HT2A receptor stimulation by psilocybin may exhibit antidepressive properties and may have the potential to normalize dysfunctional negative emotional biases (Vollenweider and Kometer 2010). A recent fMRI study corroborates this view by showing that psilocybin reduces amygdala reactivity to negative stimuli and that this attenuation is related to an increase of positive mood in healthy participants (Kraehenmann et al. 2014). Furthermore, it has been shown that psilocybin augments subjective and neural responses to positive autobiographical memory cues (Carhart-Harris et al. 2012b). Recently, the lasting antidepressive and anxiolytic potential of LSD and psilocybin has received support through the use of psychedelics as adjuncts in the psychotherapeutic treatment of depression and anxiety in patients with life-threatening diseases (Grob et al. 2011; Gasser et al. 2014, 2015).

Although psilocybin can induce very positive and sometimes even mystical-type experiences in healthy volunteers over a large dose range (0.115–0.315 mg/kg), it can also occasionally evoke frightening and unpleasant thoughts, memories, and emotions (Studerus et al. 2011). Within that dose range, unpleasant emotional responses were mostly transient and related to the fear of losing control; such effects can be attenuated by personal support (Johnson et al. 2008; Studerus et al. 2011; Kometer et al. 2012). However, the incidence for anxiety and panic increases with higher doses of psilocybin (Griffiths et al. 2006; Leuner 1971).

6 Cognition

There are numerous anecdotal reports of impaired attention and disturbed cognitive functioning after use of serotonergic hallucinogens (Leuner 1962, 1971). Leuner (1962) found that thinking was experienced as less target-orientated and the integration and linking of thoughts was reported to be difficult after ingestion of LSD. Self-ratings on the 5D-ASC questionnaire confirm that psilocybin can induce a subjective reduction in cognitive functioning (Studerus et al. 2011; Vollenweider et al. 2007; Hasler et al. 2004). Studies testing cognitive functions objectively revealed that psilocybin produces impairments of attention, working memory, and associative learning, while executive functions remain mostly unaffected (Carter et al. 2005; Quednow et al. 2012; Vollenweider et al. 2007; Geyer and Vollenweider 2008; Vollenweider et al. 1998). Therefore, psilocybin induces behavioral impairments in healthy subjects that mimic those seen in the earliest phases of schizophrenia, supporting the idea that serotonergic hallucinogens provide a model psychosis (Geyer and Vollenweider 2008). Several early studies showed that LSD prolonged the time participants required to solve arithmetic problems and impaired performance on the Raven Progressive Matrices Test and inhibitory performance on the Stroop Test (Krus et al. 1963; Wapner and Krus 1960). The later finding was replicated using psilocybin (Quednow et al. 2012). In addition to consistently slowing reaction times (Gouzoulis-Mayfrank et al. 2002; Gouzoulis-Mayfrank et al. 1998b; Quednow et al. 2012; Spitzer et al. 1996), psilocybin impairs sustained attentional performance on the Frankfurter Attention Inventory in a dose-dependent manner (Vollenweider et al. 2007). Psilocybin also impairs attentional tracking ability on a multiple object tracking tasks, an effect that may reflect a reduction in the ability to suppress or ignore distracting stimuli (Carter et al. 2005). Furthermore, psilocybin reduces binocular rivalry switch rate, which seems to be linked to reductions in arousal and attentional functions

(Carter et al. 2007). In line with those results, Gouzoulis-Mayfrank et al. (2002) also reported a psilocybin-induced deficit of response inhibition and difficulties in disengaging attention from previously attended locations. Furthermore, psilocybin triggered deficits in the AX-Continuous Performance Task that were characterized by a failure to use contextual information (Umbricht et al. 2003). Spatial working memory is reportedly impaired after high doses of psilocybin (250 μg/kg) (Wittmann et al. 2007; Vollenweider et al. 1998), but not after administration of lower doses (215 μg/kg) (Carter et al. 2005). Moreover, psilocybin has been shown to induce an increased indirect semantic priming effect, which may be due to a decreased capacity to use contextual information to focus semantic processing (Spitzer et al. 1996). Increased indirect semantic priming has also been observed in schizophrenia patients and has been linked with formal thought disorder (Spitzer et al. 1993). Vollenweider et al. (1998) showed that psilocybin-induced working memory deficits, but not attentional deficits (Carter et al. 2005, 2007), are due to excessive 5-HT2A receptor stimulation. Further studies also revealed that the ability of psilocybin to disrupt inhibition and attentional control measured with the Color Word Stroop Interference Test are attenuated by pretreatment with ketanserin or lamotrigine, indicating that those effects are attributable to 5-HT2A receptor stimulation and suggesting downstream effects on the glutamate system (Quednow et al. 2012; Vollenweider 2008).

A further aspect of cognition is creativity. Since the 1960s, there have been various reports of artists and musicians, as well as scientists and entrepreneurs, who either openly proclaim themselves to be "psychedelic artists," or acknowledge the influence of psychedelic drug-enhanced creativity on their work (Sessa 2008). However, if and how creativity is enhanced by the use of psychedelic drugs is still poorly understood (Sessa 2008). The lack of understanding may be caused by various factors; for example, it is not always possible to devise objective criteria that can be used to measure creativity, and therefore, subjective assessments are often made (Amabile 1982). Furthermore, criteria for determining whether something is viewed as being creative may vary over time and the ability of a drug to enhance creativity may not apply to every individual (Fischer and Scheib 1971; Krippner 1977). The interpretation of scientific data is further complicated since by contemporary standards many early studies had methodological flaws, for example, the omission of placebo controls (Berlin et al. 1955). Fischer et al. (Fischer and Rockey 1967; Fischer and Landon 1972) describe the creative state as an excited state that falls within the continuum of arousal ranging from normal waking consciousness to psychotic and ecstatic states. Those authors reported that elevated closure-making, i.e., an accelerated and enhanced transformation of information to meaning, under excitation induced by psilocybin may underlie the creative state. Sessa (2008) also identifies possible creativity-enhancing features of the hallucinogen-induced experience: (1) increase in complexity and openness, challenging the usual ego-bound restraints, and (2) in line with Fischer et al. (Fischer and Scheib 1971), the tendency to assign unique and novel meanings to the experience, together with the feeling of "oneness" with the world.

Fischer and Scheib (1971) tested these assumptions in a study by measuring creativity prior to, at the peak of, and also at the termination of a psilocybin-induced experience. The authors reported that an individual's ability to float in a dreamy state free of anxiety and to find symbolic meaning in a changing environment may be a prerequisite for psilocybin-induced creative experiences. More specifically, they proposed that sensitive and intuitive people with a field-independent cognition style are more likely to have creative experiences during hallucinogenic drug-induced states. Furthermore, the authors categorized subjects as being either creative performers or creative experiencers (creative performances can be recorded and evaluated, whereas creative experiences produce no record due to the fleeting nature of subjective experience). They further distinguished three types of creative experiencers and performers:

1. uncreative individuals.
2. sensitive, creative experiencers who:
   (a) are creative performers without drugs, but are unable to produce aesthetically pleasing art in drug-induced states or
   (b) score low on creativity without drugs but are creative performer in a drug-induced state
3. a minority who have creative hallucinatory experiences and are creative performers (Fischer and Scheib 1971).

Further studies explored the effect of LSD upon the work of graphic artists and found that art critics judged the LSD-influenced paintings as having "greater aesthetic value" than the artists' usual work (Berlin et al. 1955). In a placebo-controlled study, Zegans et al. (1967) investigated the effect of LSD upon creativity in graduate students using several different tests (the Mednick Association Test, the Modified Word Association Test, the Mosaic Design Test, and the Free Association Test). Although the LSD group performed significantly better than the control group on the test assessing the originality of word associations, no other results were statistically significant. The authors concluded that although administration of LSD cannot enhance creativity in a random group of people, it may increase the accessibility of remote ideas and associations (Zegans et al. 1967). These conclusions were later corroborated by a double-blind and placebo-controlled study where psilocybin was administered to healthy volunteers: Psilocybin induced a larger semantic and indirect semantic priming effect than placebo, suggesting an increased availability of remote associations under the influence of psilocybin (Spitzer et al. 1996). Reportedly, this effect is specific for hallucinogenic drugs (Gouzoulis-Mayfrank et al. 1998b). Additionally, an increase in novel, figurative language was reported in patients under the influence of LSD during psychoanalysis (Natale et al. 1978).

In summary, serotonergic hallucinogens can induce creativity-enhancing experiences related to reduced inhibition, increased fluency and flexibility of ideas, and increased visual imagery, empathy, and capacity to restructure problems

(Sessa 2008). Creative individuals may sometimes be able to transform these experiences into a creative work, but it is unlikely that creativity is enhanced by serotonergic hallucinogens in individuals who do not ordinarily show creative behavior (Krippner 1977; Zegans et al. 1967).

7 Time Perception

Even early phenomenological and psychological studies reported that serotonergic hallucinogens produce a strongly altered experience of time and caused misjudgments of elapsed time intervals (Leuner 1962; Heimann 1963; Grof 1975). Participants reported a feeling that the passage of time was speeding up or slowing down, or experienced feelings of timelessness, effects that were often associated with visual perceptual changes and alterations in self-experience (Kenna and Sedman 1964; Heimann 1963; Grof 1975). Heimann believed the psilocybin-induced feeling of time stopping was linked to an inner psychic state involving the loss of serial temporal order and an impairment of the ability to determine causal relationships and serial integration. Additionally, Heimann associated disturbances in time perception with impairments of the coordination and perception of movement (the motion of moving objects was not perceived as being continuous, but rather appeared as a series of discontinuous static images) (Heimann 1994). A reduction in temporal motion integration has also been described in a later placebo-controlled study with psilocybin (Carter et al. 2004). The fact that hallucinogens cause some musicians to feel that they have lost their musical talent may also be a consequence of altered temporal perception. While listening to music or playing music after ingesting psilocybin, musicians reported that sounds were experienced as stretched and that "music was standing still." Furthermore, psilocybin made it difficult for the musicians to distinguish rhythms; the dynamic structure of music was often disturbed, with songs experienced as an unrelated sequence of parallel tones, and temporal coherency was lost (Weber 1967). Using items in the 5D-ASC rating scale related to temporal perception, Studerus (Studerus et al. 2011) confirmed that psilocybin produced a dose-dependent alteration of the sense of time in a large sample of participants (n = 110). Recent studies using standardized measures of temporal processing were able to confirm that psilocybin dose-dependently alters time perception and temporal control of behavior (Wackermann et al. 2008; Wittmann et al. 2007). Wackermann et al. (2008) reported an increase in the loss rate of internal time representation even under a very low dose of psilocybin (12 μg/kg body weight). This study suggests that psilocybin might cause a fusion of the "inner" and "outer" horizons of temporal experience resulting in an impairment of cognitive processing based on temporal order. Recently, a link between the loss rate of duration representations and certain serotonin system genotypes was detected: A higher loss rate was found for the carriers of genotypes known to augment serotoninergic transmission (Sysoeva et al. 2010). In line with an earlier study conducted with LSD

(Aronson et al. 1959), Wittmann et al. (2007) reported that participants underestimated intervals longer than 2.5 s in an interval reproduction task and had an impaired ability to synchronize tapping when inter-beat intervals were longer than 2 s. No impairments were found for shorter time intervals, indicating a potential interaction with cognitive processes such as working memory. Furthermore, individual tapping tempos were slowed down.

8 Non-pharmacological Predictors

Individual reactions to serotonergic hallucinogens vary widely, even when the experimental conditions are maintained constant (Dittrich 1994; Nichols and Chemel 2006). Out of 24 variables analyzed by Studerus et al. (2012), dose was found to be the most important factor predicting the acute response to psilocybin in a large sample of participants (n = 409 trials); however, a large proportion of inter- and intraindividual differences in reactions were also influenced by non-pharmacological variables (Studerus et al. 2012; Grof 1975). Within these factors, an important distinction has been made between "set" and "setting" (Grof 1975; Leary et al. 1963). Set refers to expectations of the subject, their personality structure, and current mood state, whereas setting comprises the specific circumstances in which the drug is administered, e.g., the physical, social, and cultural environment (Leary et al. 1963; Grof 1975). While the earliest academic research on serotonergic hallucinogens was conducted without considering the influence of set and setting, later research included more preparation and interpersonal support and consequently reported fewer adverse psychological reactions (Johnson et al. 2008). The specific effects of set and setting variables are discussed below.

8.1 Demographic Variables

In line with early studies investigating the effects of LSD (Hyde 1960), Studerus et al. (2012) showed that the effects of psilocybin were not moderated by gender. Additionally, neither body mass index nor years of education predicted the psilocybin reaction in the study (Studerus et al. 2011). However, this and previous studies found that older subjects reported less impairment of control and cognition and tended to experience a more blissful state compared with younger subjects, and there was an inverse relationship between age and the likelihood of having an unpleasant and/or anxious reaction to psilocybin or LSD (Hyde 1960; Studerus et al. 2012).

8.2 Personality

Many studies suggest that responses to serotonergic hallucinogens are dependent—at least to some degree—on personality structure (e.g., Fischer et al. 1968; Hemsley and Ward 1985; Dittrich 1994; Lienert and Netter 1996; Thatcher et al. 1971; Silverman 1971). Fischer et al. (1965a, b, 1968; Fischer 1971) reported that taste-sensitive subjects who also tended to be intuitive and introversive were more sensitive to the effects of psilocybin, whereas taste-insensitive, taste-extroversive subjects require larger doses to elicit comparable effects. The authors described the psilocybin-sensitive subjects as being "variable subjects" or "minimizers" with high creativity scores, large handwriting area, large standard deviation on the handwriting area, large retest variance on perceptual tasks, and fast reaction time. In contrast, insensitive subjects were reported to be "stable subjects" or "maximizers" with a small standard deviation and handwriting area, small retest variability, and low creativity scores (Thatcher et al. 1971; Gwynne et al. 1969; Fischer et al. 1969b). Such findings are in line with observations by Leuner (1962) who reported that rational, down to earth, intellectual predispositioned individuals tended to be resistant to hallucinogen-induced intoxication. Leuner concluded that scholarly individuals in particular required higher doses, whereas individuals who were more emotionally guided (e.g., artists or musicians) required lower doses of LSD to produce an experience of the same intensity (Leuner 1962). Additionally, studies conducted using LSD also found that more intense reactions occurred in individuals with introversive or histrionically structured personalities compared with those having extroversive or compulsively structured personalities (Hyde 1960; Buckman 1967; Grof 1975). However, Dittrich (1994) reported that an optimistic, extroversive attitude toward life predicted scores in the oceanic boundlessness (OBN) and visual alteration (VIR) dimensions and the overall intensity of the altered state of consciousness (ASC). This finding is in line with a pilot study conducted by Thatcher et al. (1971), who found that certain extroversive subjects were more strongly affected by psilocybin. A recent study added to these data by showing that subjects who were more sociable (i.e., outgoing and extroverted) reported lower scores on items assessing spiritual experience and higher scores on items assessing audio-visual Synesthesia (Studerus et al. 2012). Furthermore, in the study by Dittrich (1994), a high esthetic sensibility and a non-dogmatic religiosity also predicted the experience of oceanic boundlessness (OBN) and the overall ASC intensity. A high esthetic sensibility was also associated with the occurrence of visual alterations. Furthermore, rigid and emotional unstable subjects were reported to be more at risk for anxious reactions to serotonergic hallucinogens (Lamparter and Dittrich 1994; Dittrich 1994). Several studies found moderate-to-strong correlations between measures of neuroticism and the occurrence of anxious reactions to serotonergic hallucinogens (Dittrich 1994; Hemsley and Ward 1985; Lienert and Netter 1996). Nonetheless, a recent study conducted with a large sample did not detect any statistically significant relationship between Neuroticism–Anxiety scores and negative reactions to psilocybin (Studerus et al. 2012). However, subjects with very high Neuroticism scores (i.e., scores that were more than two standard deviations above the mean) were excluded from the latter study during the initial screening. Furthermore, the personality trait Absorption (i.e., an individual's openness to a variety of cognitive, perceptual, and imaginary experiences, as well as vivid imagery, synesthesia, and intense involvement in aesthetics and nature) was found to be the second most important predictor after dosing of psilocybin response. The Absorption trait was highly positively correlated with the overall level of consciousness alteration and strongly predicted the occurrence of mystical-type experiences and the intensity of visual effects induced by psilocybin (Studerus et al. 2012). In contrast to previous studies (e.g., Buckman 1967; Dittrich 1994; Fischer et al. 1968), other than absorption, the only personality trait that accurately predicted psilocybin responses was the factor Sociability (i.e., being outgoing and extroverted). Other personality traits made only marginal contributions to the prediction of psilocybin responses (Studerus et al. 2012).

8.3 Mood

Early studies identified mood as an important predictor of the experience induced by hallucinogens (Johnson et al. 2008). For example, Metzner et al. (1965) reported that the best predictor for mood during a psilocybin session was mood prior to the session, e.g., a presession negative mood often resulted in anxious or negative experiences during the session. Dittrich (1994) found that emotional lability, measured as a state variable, predicted higher values on the APZ anxious ego-dissolution (AED) dimension after administration of DMT. Furthermore, Dittrich found that the presence of feelings of calmness in subjects prior to administration of hallucinogens correlated positively with their experience of oceanic boundlessness and visual alterations, as well as with the overall intensity of the ASC (Dittrich 1994). These findings are in line with those of a study recently conducted by Studerus et al. (2012), which found that current mood state and the presence of psychological distress during the past four weeks preceding drug intake were more important for predicting psilocybin responses compared with personality factors. In particular, this study found that emotional excitability shortly before drug intake predicted the occurrence of anxious reactions to psilocybin better than the trait anxiety–depressiveness. Furthermore, measures of performance-related activity (i.e., go-getting, avid, active, and energetic items on the EWL affective mood rating scale) were among the most important predictors of the overall level of consciousness alteration, as well as experiences of oceanic boundlessness (OBN) and visual reconstructualization (VIR) (Studerus et al. 2012). The authors suggested that the items assessing performance-related activity were strongly associated with positive mood and general optimism. Therefore, the authors concluded that being in an emotionally excitable and active state prior to drug intake and having experienced few psychological problems for several preceding weeks increased the intensity of pleasurable and visual experiences (Studerus et al. 2012). Other factors closely related to mood are peer support and the attitude toward the experiment and the investigator (Johnson et al. 2008). Leuner (1962) reported that the intensity of the response to LSD was influenced by how participants felt regarding the study and the study personnel. Furthermore, peer support and concomitant emotional support provide a benign interpretive context for the hallucinogen experience and therefore can reduce adverse psychological effects (Dunsmore and Kaplan 1997; Leary et al. 1963; Johnson et al. 2008).

8.4 Expectations

The expectations of subjects regarding the effects of hallucinogens can reportedly influence the nature of experience (Dittrich 1994; Leary et al. 1963; Metzner et al. 1965; ten Berge 2002; Unger 1963; Savage et al. 1966) and are probably closely linked to mood preceding the session. For example, although most studies investigating the potential for psychedelic drugs to enhance creativity showed mixed results (see above), Harman et al. (1966) showed that LSD and mescaline enhanced performance in creativity tests that were made in a setting where the expectation that the drug would enhance creativity was reinforced. Additionally, the expectation and attitude of the experimenter or therapist can reportedly influence the hallucinogen-induced experience, probably by inducing similar expectations in the subjects (Savage et al. 1966; Unger 1963). In particular, anxiety in the experimental staff may lead to anxiety-ridden experiences for subjects, whereas calm experimenters may reduce anxiety in participants and patients (Unger 1963). Johnson et al. (2008) recommend careful preparation of participants, which also involves the discussion of expectations—including those that are potentially unrealistic—as well as the study procedures and timing in order to avoid surprises during the session.

8.5 Genetic Factors

Although to date no investigations have directly assessed the influence of genetics on the reaction to psilocybin, several studies suggest that the hallucinogen-induced experience might, at least to some extent, be influenced by genetic factors. It is well established that many effects of serotonergic hallucinogens are attributable to excessive 5-HT2A receptor activation (Quednow et al. 2012; Kometer et al. 2012; 2013; Passie et al. 2002; Vollenweider et al. 1998). The 5-HT2A receptor gene resides on chromosome 13q14–q21; several different gene variants that are associated with functional and clinically relevant effects have been identified (Benedetti et al. 2008; Serretti et al. 2007). These 5-HT2A polymorphisms may contribute to inter-individual differences in response to serotonergic hallucinogens. In particular, Ott et al. (2005) demonstrated that a significant association exists between a T102C polymorphism of the 5-HT2A receptor and the personality trait of Absorption.

Participants with the T/T genotype of the T102C 5-HT2A polymorphism, implying a stronger binding potential of the 5-HT2A receptor, had higher absorption scores. As discussed above, the personality trait of Absorption was found to be an important predictor of psilocybin response (Studerus et al. 2012). Therefore, polymorphisms related to differences in the binding potential of the 5-HT2A receptor, such as the T102C polymorphism, may also contribute to variations in the responsiveness of subjects to hallucinogenic drugs. Furthermore, Sysoeva et al. (2010) reported that the loss rate of duration representation, which has been shown to be altered after administration of psilocybin, is higher for carriers of genotypes that are characterized by enhanced serotonin transmission (e.g., the 5HTLPR SS polymorphism compared to the LL genotype, the "low expression" variant of the MAOA VNTR gene compared to the "high-expression" variant, and the T/T genotype of the T102C polymorphism compared to the C/C genotype). Additionally, further investigations showed an association between T102C polymorphisms and sensorimotor gating (Quednow et al. 2008, 2009), thereby providing a possible link between genetic differences and hallucinogen-induced deficits in prepulse inhibition and cognition. In sum, there is evidence that a common genetic basis may exist for inter-individual differences in serotonergic neurotransmission and for differences in response to hallucinogens. However, the association between these polymorphisms and the phenomenology of hallucinogen-induced experiences has not yet been tested directly and will likely be the subject of further studies.

8.6 Physical Environment

The physical environment in which hallucinogens are administered has been reported to greatly influence the experience and probably has effects on expectations and mood before drug administration (Johnson et al. 2008; Strassman 2001; ten Berge 2002; Grof 1975). Johnson et al. (2008) recommend an aesthetically pleasing and comfortable environment and suggested that "cold" and overly clinical surroundings should be avoided in hallucinogen research in order to reduce the risk of anxious reactions. The environment can directly influence perceptual changes and the possibility of disorientation occurring under the influence of hallucinogens (Johnson et al. 2008; Strassman 2001). Relevant features of the environment include, for example, the furniture of the experimental room and its temperature, odors and fragrances, as well as acoustic stimuli such as music or the ringing of a telephone, which can trigger markedly different experiences (Grof 1975). Therefore, using a safe, familiar environment that minimizes the risk of surprises (e.g., meeting new people, receiving unexpected phone calls) is encouraged (Johnson et al. 2008). These reports are corroborated by Studerus et al. (2011) who investigated the effects of psilocybin administration in a rather technical PET environment versus laboratory rooms furnished in a comfortable, aesthetically pleasing manner. The latter study reported that the PET environment was associated with more anxious reactions; the absence of distractions in the scanner environment—subjects did not have to interact with peers or complete tasks, for example—may have allowed participants to concentrate on the experience, potentially causing increased confrontation with their inner fears (Studerus et al. 2011). However, even in the PET environment, the percentage of subjects experiencing a strong anxious reaction was still very low. Another factor that helped to create a positive environment was the presence and support of other people (Dunsmore and Kaplan 1997; Johnson et al. 2008; Grof 1975). The inclusion of people from both genders in the monitoring team is recommended to foster feelings of security in the research environment (Johnson et al. 2008; Grof 1975). As mentioned above, the attitude of the experimenter can strongly influence the participant's experience (Unger 1963; Savage et al. 1966). Hyde reported that impersonal, investigative, or hostile attitudes can arouse hostile paranoid responses (Hyde 1960). Alternatively, when the experimenter has the specific intent of making the experience positive, the majority of participants report a positive experience (Metzner et al. 1965; Unger 1963). Therefore, Johnson et al. (2008) emphasized the importance of having a friendly, welcoming, compassionate, and respectful atmosphere supported by all research staff members, as well as a trusting relationship between participant and monitors. Personal support has furthermore been reported to be very effective in reducing anxiety or fear during the course of action of hallucinogens (Johnson et al. 2008; Studerus et al. 2011).

8.7 Previous Experience

It has been reported that prior experience with substances inducing ASCs influences subsequent experiences and probably affects the mood and expectations preceding drug administration (Leuner 1962; Metzner et al. 1965). Various studies report that previous experience with classical hallucinogens is negatively associated with the occurrence of somatic symptoms (Leuner 1962; Studerus et al. 2012). While these symptoms may initially be perceived as disturbing, they recede into the background with more experience (Leuner 1962). Furthermore, Leuner (1962) noted that after several hallucinogen-induced experiences, participants would sometimes require only low doses to elicit experiences of the same intensity. Leuner (1962) named this phenomenon "paradoxical habituation." In line with previous studies (Leuner 1962; Metzner et al. 1965), Studerus et al. (2012) reported that hallucinogen-naïve subjects experienced slightly more visual alterations, disembodiment, and changed meaning of percepts. Furthermore, the intensity of concentration deficits was reported to diminish when subjects had previous experience with hallucinogens (Leuner 1962). However, complex, scenic hallucinations were found to be more likely to occur after repeated hallucinogen administration (Leuner 1962). Furthermore, Dittrich (1994) reported that familiarity with drug-induced ASCs correlated negatively with DMT-induced anxiety and general familiarity with ASCs predicted a higher score on the 5D-ADC oceanic boundlessness scale.

9 Long-Term Effects

Studerus et al. (2011) analyzed long-term follow-up data from 90 subjects that were collected 8–16 months after experimental sessions with psilocybin. The authors reported that the majority of subjects rated the psilocybin experience as enriching or very enriching (92%). In another follow-up study conducted 14 months after administration of a high dose of psilocybin, more than half of the participants even rated the psilocybin-induced experience as being among the five most personally meaningful experiences of their lives (Griffiths et al. 2008).

Studerus et al. (2011) reported that the subjective changes in attitude and personality included positive changes in items measuring the attitude toward ASCs (56% of subjects), relationship to nature (38%), and aesthetic experiences (37%). Enhanced appreciation of art and music was also reported by McGlothlin et al. (1967) in a follow-up assessment conducted six months after an LSD experience. Furthermore, participants engaged in more creative activities such as attending musical events, visiting museums, and buying records. However, no actual increase in creative ability was measured. This finding is in line with results obtained by Dobkin De Rios and Janiger (2003) who found that there was no overall increase in creativity but that certain aspects of artists' work were enhanced after an LSD experience—for example, there may be a tendency toward more expressionistic work, a sharpening of color, or a greater freedom from prescribed mental sets. Additionally, MacLean et al. (2010) reported that the personality trait openness of the NEO Personality Inventory remained significantly elevated one year after the session in participants who had a mystical experience after a high dose of psilocybin. Studerus et al. (2011) reported that a single dose of psilocybin in a non-therapeutic research setting did not significantly affect subsequent drug use patterns. Furthermore, detailed questions about possible flashback phenomena and spontaneous alterations of consciousness revealed that none of the subjects fulfilled the diagnostic criteria for hallucinogen persisting perception disorder (HPPD) in DSM-IV or flashbacks in ICD-10, and the subjects did not report experiencing visual phenomena reminiscent of the typical symptoms of HPPD (Studerus et al. 2011). This is in line with the findings of Johnson et al. (2008) who reported that the incidence of lasting perceptual abnormalities is much lower in research contexts compared to illicit recreational use due to careful screening and preparation of subjects.

10 Biological Models of ASCs

Numerous authors have proposed models to explain the biological underpinnings of ASCs in order to account for the symptoms experienced after administration of hallucinogens as well as the nature of human consciousness (Fischer 1971; Geyer and Vollenweider 2008; Tononi 2004; Vollenweider and Geyer 2001).

Early studies conducted by Fischer and colleagues led them to propose a psychophysiological model of ecstatic and meditative states that also addressed psychedelic experiences (Fischer 1971). According to their model, ecstatic states can be mapped on a continuum of increasing arousal ranging from normal perception to hallucinations. Hallucinogen-induced states are characterized by an increase in the amount of sensory information that is not offset by a corresponding increase in the rate of information processing. Consequently, as the amount of externally and internally generated stimuli increases, exteroception gradually shifts to interoception. Fischer supported his model by demonstrating elevated ratios of sensory-to-motor activity using various psychophysiological measurements (e.g., Fischer et al. 1965a, 1969a, 1970; Fischer and Kaelbling 1966; Fischer and Mead 1966). Furthermore, Fischer (1971) described the experience of subjective reality as an interpretation of CNS activity within a structure of learned constancies, i.e., a learned model of a world ordered and stabilized by self-programmed rules. Moving along the continuum causes an "unlearning" of constancies due to an arousal-induced transformation, which gives rise to the effects seen in the hallucinogen intoxication. With increasing levels of arousal, the "I" of the physical world gives way to the "Self" of the mental dimension and the level of external sensory information reaching the brain is gradually reduced.

More recent evidence from animal and human studies suggests that serotonergic hallucinogens disrupt information processing in cortico–striato–thalamocortical (CSTC) feedback loops, which have been implicated in gating the flow of sensory and sensorimotor information to the cortex (Vollenweider and Geyer 2001; Geyer and Vollenweider 2008). The so-called CSTC information processing model proposes that deficits in early information processing may underlie the alterations of perception, cognition, and the sense of self that occur in psychedelic states (Vollenweider and Geyer 2001). According to this model, the inability to filter, inhibit, and gate exteroceptive and interoceptive stimuli may lead to cortical sensory overload and subsequently to a breakdown of cognitive integrity, resulting in hallucinations and alterations of ego-functioning. Deficits in preattentive sensorimotor gating have repeatedly been reported after psilocybin and LSD administration (Quednow et al. 2012; Schmid et al. 2015; Vollenweider et al. 2007) and have been related to alterations in cognitive functioning (Quednow et al. 2012; Vollenweider et al. 2007). The CSTC model proposes that the thalamus plays a key role within the CSTC feedback loops for gating external and internal information to the cortex and, thereby, is crucially involved in the regulation of the level of awareness and attention (Geyer and Vollenweider 2008; Vollenweider et al. 2007). This view is in line with the information integration theory of consciousness (Tononi 2004), which proposed that the thalamus and thalamocortical systems play a key role in integrating information processing and the ensuing consciousness. Serotonergic hallucinogens may alter thalamocortical transmission by stimulation of 5-HT2A receptors located at multiple sites within the CSTC loops (Geyer and Vollenweider 2008; Vollenweider et al. 2007). Indeed, PET and SPECT neuroimaging studies have shown that oral administration of psilocybin, mescaline, and DMT can alter neuronal activity in frontomedial and frontolateral cortices ("hyperfrontality"), basal ganglia, and thalamus (Vollenweider et al. 1997c; Gouzoulis-Mayfrank et al. 1999; Hermle et al. 1993; Riba et al. 2006), effects that are variously correlated with different dimensions of the psychedelic state (see above). Nevertheless, a recent fMRI study reported a decrease in brain activity in the medial frontal cortex after intravenous administration of psilocybin (Carhart-Harris et al. 2012a). However, numerous factors (differences in the route of administration (i.v. vs. p.o.), the indices used to monitor neuronal activity (metabolic rate of glucose, CBF, BOLD), or the intensity and content of psilocybin-induced symptoms) could account for these apparent discrepancies. Much additional research is needed to gain a better insight in the relationship between subjective experience in psychedelic states and neuronal activity in humans, especially given the complexity and variability of psychedelic experiences.

More recently, an "entropic brain hypothesis" was proposed based on the effects of intravenous administration of psilocybin measured using fMRI and MEG techniques (Tagliazucchi et al. 2014; Carhart-Harris et al. 2012a, 2013, 2014; Muthukumaraswamy et al. 2013). The model proposes that various conscious states are related to entropy measures of brain activity by an inverted-U-shaped function. Whereas sedation and anesthesia reflect rigid, low-entropy states, psychedelic states are characterized by increased flexibility and high entropy, which is associated with more random brain activity and flexible cognition (Carhart-Harris et al. 2014). Those authors corroborated their model with studies showing that psilocybin produces an increase in BOLD signal variance, a greater diversity of connectivity motifs in a hippocampal/ACC network, decreased connectivity within the default mode network (DMN), a reduction in DMN-TPN anticorrelation, and desynchronization of cortical oscillatory rhythms (Carhart-Harris et al. 2012a, 2013; Muthukumaraswamy et al. 2013; Tagliazucchi et al. 2014). The authors concluded that a disorganization of brain activity underlies hallucinogen-induced ASCs (Carhart-Harris et al. 2014).

In summary, numerous models have been proposed emphasizing different aspects and focusing on different neurobiological correlates of ASC. These models are not mutually exclusive, but are probably not able to capture the diverse and dynamic effects induced by hallucinogens when considered separately. More research on all levels of the hallucinogenic action (from receptors over brain networks to psychological effects) is needed to develop a unified and comprehensive biological model of ASC.

Acknowledgements This article is dedicated to my dear friends—Dr. Dan Stoicescu and Dr. Bill Linton without whom much of this work would not have been possible (FXV).

References

Amabile TM (1982) Social psychology of creativity: a consensual assessment technique. J Pers Soc Psychol 43:997–1013

Aronson H, Silverstein AB, Klee GD (1959) Influence of lysergic acid diethylamide (LSD-25) on subjective time. AMA Arch Gen Psychiatry 1:469–472

Austin JH (1998) Zen and the brain: toward an understanding of meditation and consciousness. The MIT Press, Cambridge

Austin JH (2006) Zen-brain reflections: reviewing recent developments in meditation and states of consciousness. MIT Press, Cambridge, Massachusetts

Benedetti F, Barbini B, Bernasconi A, Fulgosi MC, Colombo C, Dallaspezia S, Gavinelli C, Marino E, Pirovano A, Radaelli D, Smeraldi E (2008) Serotonin 5-HT2A receptor gene variants influence antidepressant response to repeated total sleep deprivation in bipolar depression. Prog Neuropsychopharmacol Biol Psychiatry 32:1863–1866

Berlin L, Guthrie T, Weider A, Goodell H, Wolff HG (1955) Studies in human cerebral function: the effects of mescaline and lysergic acid on cerebral processes pertinent to creative activity. J Nerv Ment Dis 122:487–491

Bogenschutz MP, Forcehimes AA, Pommy JA, Wilcox CE, Barbosa P, Strassman RJ (2015) Psilocybin-assisted treatment for alcohol dependence: a proof-of-concept study. J Psychopharmacol 29:289–299

Brogaard B (2013) Serotonergic hyperactivity as a potential factor in developmental, acquired and drug-induced synesthesia. Front Hum Neurosci 7:657

Buckman J (1967) Theoretical aspects of L.S.D. therapy. Int J Soc Psychiatry 13:126–138

Caldwell DF, Domino EF (1967) Effect of LSD-25 in the rat on operant approach to a visual or auditory conditioned stimulus. Psychol Rep 20:199–205

Carhart-Harris RL, Erritzoe D, Williams T, Stone JM, Reed LJ, Colasanti A, Tyacke RJ, Leech R, Malizia AL, Murphy K, Hobden P, Evans J, Feilding A, Wise RG, Nutt DJ (2012a) Neural correlates of the psychedelic state as determined by fMRI studies with psilocybin. Proc Natl Acad Sci U S A 109:2138–2143

Carhart-Harris RL, Leech R, Williams TM, Erritzoe D, Abbasi N, Bargiotas T, Hobden P, Sharp DJ, Evans J, Feilding A, Wise RG, Nutt DJ (2012b) Implications for psychedelic-assisted psychotherapy: functional magnetic resonance imaging study with psilocybin. Br J Psychiatry 200:238–244

Carhart-Harris RL, Leech R, Erritzoe D, Williams TM, Stone JM, Evans J, Sharp DJ, Feilding A, Wise RG, Nutt DJ (2013) Functional connectivity measures after psilocybin inform a novel hypothesis of early psychosis. Schizophr Bull 39:1343–1351

Carhart-Harris RL, Leech R, Hellyer PJ, Shanahan M, Feilding A, Tagliazucchi E, Chialvo DR, Nutt D (2014) The entropic brain: a theory of conscious states informed by neuroimaging research with psychedelic drugs. Front Hum Neurosci 8:20

Carhart-Harris RL, Kaelen M, Whalley MG, Bolstridge M, Feilding A, Nutt DJ (2015) LSD enhances suggestibility in healthy volunteers. Psychopharmacology 232:785–794

Carter OL, Pettigrew JD, Burr DC, Alais D, Hasler F, Vollenweider FX (2004) Psilocybin impairs high-level but not low-level motion perception. NeuroReport 15:1947–1951

Carter OL, Burr DC, Pettigrew JD, Wallis GM, Hasler F, Vollenweider FX (2005) Using psilocybin to investigate the relationship between attention, working memory, and the serotonin 1A and 2A receptors. J Cogn Neurosci 17:1497–1508

Carter OL, Hasler F, Pettigrew JD, Wallis GM, Liu GB, Vollenweider FX (2007) Psilocybin links binocular rivalry switch rate to attention and subjective arousal levels in humans. Psychopharmacology 195:415–424

Dahlberg CC, Jaffe J (1979) The effects of LSD-25 and dextroamphetamine on the use of defensive language. J Clin Psychol 35:250–254

Díaz JL (2010) Sacred plants and visionary consciousness. Phenom Cogn Sci 9:159–170

Dittrich A (1985) Ätiologie-unabhängige Strukturen veränderter Wachbewusstseinszustände. Ferdinand Enke Verlag, Stuttgart
Dittrich A (1994) Psychological aspects of altered states of consciousness of the LSD type: measurements of their basic dimensions and prediction of individual differences. In: Pletscher A, Ladewig D (eds) 50 years of LSD. Current status and perspective of hallucinogens. Parthenon, New York
Dittrich A (1998) The standardized psychometric assessment of altered states of consciousness (ASCs) in humans. Pharmacopsychiatry 31(Suppl 2):80–84
Dittrich A, von Arx S, Staub S (1981) International study on altered states of consciousness (ISASC). Schweiz Z Psychol und Anw 40:189–200
Dittrich A, von Arx S, Staub S (1985) International study on altered states of consciousness (ISASC). Summary of the results. Germ J Psych 9:319–339
Dobkin De Rios M, Janiger O (2003) LSD, spirituality and the creative process. Park Street Press, Rochester, Vermont
Dunsmore MW, Kaplan HB (1997) Peer support, perceptions of adverse effects, and the hallucinogenic drug experience. Appl Behav Sci Rev 5:219–230
Fischer R (1967) The biological fabric of time. Ann N Y Acad Sci 138:440–488
Fischer R (1971) A cartography of the ecstatic and meditative states. Science 174:897–904
Fischer R, Kaelbling R (1966) Increase in taste acuity with sympathetic stimulation: the relation of a just-noticeable taste difference to systemic psychotropic drug dose. Recent Adv Biol Psychiatry 9:183–195
Fischer R, Landon GM (1972) On the arousal state-dependent recall of 'subconscious' experience: stateboundness. Br J Psychiatry 120:159–172
Fischer R, Mead EL (1966) Time contraction and psychomotor performance produced by "Psilocybin". Nature 209:433–434
Fischer R, Rockey MA (1967) A heuristic model of creativity. Experientia 23:150–151
Fischer R, Scheib J (1971) Creative performance and the hallucinogenic drug-induced creative experience or one man's brain-damage is another's creativity. Confin Psychiatr 14:174–202
Fischer R, Georgi F, Weber R (1951) Psychophysical correlations. VIII. Experimental tests in schizophrenia: lysergic acid diethylamide and mescaline. Schweiz Med Wochenschr 81: 817–819; contd
Fischer R, Griffin F, Archer RC, Zinsmeister SC, Jastram PS (1965a) Weber ratio in gustatory chemoreception; an indicator of systemic (drug) reactivity. Nature 207:1049–1053
Fischer R, Knopp W, Griffin F (1965b) Taste sensitivity and the appearance of phenothiazine-tranquilizer induced extrapyramidal symptoms. Arzneimittelforschung 15:1379–1382
Fischer R, Marks PA, Hill RM, Rockey MA (1968) Personality structure as the main determinant of drug induced (model) psychoses. Nature 218:296–298
Fischer R, Hill RM, Warshay D (1969a) Effects of psychodysleptic drug psilocybin on visual perception. Changes in brightness preference. Experientia 25:166–169
Fischer R, Thatcher K, Kappeler T, Wisecup P (1969b) Unity and covariance of perception and behavior. Perceptual variability: a predictor of psychotomimetic drug-induced behavior. Arzneimittelforschung 19:1941–1945
Fischer R, Hill R, Thatcher K, Scheib J (1970) Psilocybin-induced contraction of nearby visual space. Agents Actions 1:190–197
Forman RKC (1990) The Problem of pure consciousness. Mysticism and philosophy. Oxford University Press, New York
Gasser P, Holstein D, Michel Y, Doblin R, Yazar-Klosinski B, Passie T, Brenneisen R (2014) Safety and efficacy of lysergic acid diethylamide-assisted psychotherapy for anxiety associated with life-threatening diseases. J Nerv Ment Dis 202:513–520
Gasser P, Kirchner K, Passie T (2015) LSD-assisted psychotherapy for anxiety associated with a life-threatening disease: a qualitative study of acute and sustained subjective effects. J Psychopharmacol 29:57–68
Geyer MA, Vollenweider FX (2008) Serotonin research: contributions to understanding psychoses. Trends Pharmacol Sci 29:445–453

Gouzoulis E, Thelen B, Schreckenberger M, Kovar KA, Spitzer M, Sass H (1997) Human experimental neurometabolic and neuropsychological studies with psilocybin, d-methamphetamine and 3,4-methylenedioxyethylamphetamine (MDE). 20th symposium of AGNP nuremberg, 8–11 Oct 97. Pharmacopsychiat 30:171

Gouzoulis-Mayfrank E, Habermeyer E, Hermle L, Steinmeyer AM, Kunert HJ, Sass H (1998a) Hallucinogenic drug induced states resemble acute endogenous psychoses: results of an empirical study. Eur Psychiatry 13:399–406

Gouzoulis-Mayfrank E, Thelen B, Maier S, Habermeyer E, Spitzer M, Sass H (1998b) Modulation of semantic priming effects by psilocybin, MDE (ecstasy), and d-methamphetamine. Neurol Psychiatry Brain Res 6:19–28

Gouzoulis-Mayfrank E, Schreckenberger M, Sabri O, Arning C, Thelen B, Spitzer M, Kovar KA, Hermle L, Bull U, Sass H (1999) Neurometabolic effects of psilocybin, 3,4-methylenedioxyethylamphetamine (MDE) and d-methamphetamine in healthy volunteers. A double-blind, placebo-controlled PET study with [18F]FDG. Neuropsychopharmacology 20:565–581

Gouzoulis-Mayfrank E, Thelen B, Maier S, Heekeren K, Kovar KA, Sass H, Spitzer M (2002) Effects of the hallucinogen psilocybin on covert orienting of visual attention in humans. Neuropsychobiology 45:205–212

Griffiths RR, Richards WA, McCann U, Jesse R (2006) Psilocybin can occasion mystical-type experiences having substantial and sustained personal meaning and spiritual significance. Psychopharmacology (Berl) 187:268–283; discussion 284-92

Griffiths R, Richards W, Johnson M, McCann U, Jesse R (2008) Mystical-type experiences occasioned by psilocybin mediate the attribution of personal meaning and spiritual significance 14 months later. J Psychopharmacol 22:621–632

Grob CS, Danforth AL, Chopra GS, Hagerty M, McKay CR, Halberstadt AL, Greer GR (2011) Pilot study of psilocybin treatment for anxiety in patients with advanced-stage cancer. Arch Gen Psychiatry 68:71–78

Grof S (1975) Realms of the human unconscious. Observations from LSD research. The Viking Press, New York

Gwynne PH, Fischer R, Hill R (1969) Hypnotic induction of the interference of psilocybin with optically induced spatial distortion. Pharmakopsychiatrie 2:223–234

Hanks JB, Gonzalez-Maeso J (2013) Animal models of serotonergic psychedelics. ACS Chem Neurosci 4:33–42

Harman WW, McKim RH, Mogar RE, Fadiman J, Stolaroff MJ (1966) Psychedelic agents in creative problem-solving: a pilot study. Psychol Rep 19:211–227

Hasler F, Grimberg U, Benz MA, Huber T, Vollenweider FX (2004) Acute psychological and physiological effects of psilocybin in healthy humans: a double-blind, placebo-controlled dose-effect study. Psychopharmacology 172:145–156

Heimann H (1963) Observations on disturbed time perception in model psychosis. Schweiz Med Wochenschr 93:1703–1705

Heimann H (1994) Experience of time and space in model psychoses. In: Pletscher A, Ladewig D (eds) 50 years of LSD. Current status and perspectives of hallucinogens. The Parthenon Publishing Group, New York

Hemsley DR, Ward ES (1985) Individual differences in reaction to the abuse of LSD. Person Individ Diff 6:515–517

Hermle L, Spitzer M, Borchardt D, Gouzoulis E (1992) Beziehungen der Modell- bzw. Drogenpsychosen zu schizophrenen Erkrankungen. Fortschr Neurol Psychiatr 60:383–392

Hill RM, Fischer R, Warshay D (1969) Effects of excitatory and tranquilizing drugs on visual perception. Spatial distortion thresholds. Experientia 25:171–172

Hofmann A (1968) Psychotomimetic agents. In: Burger A (ed) Chemical constitution and pharmacodynamic actions. M. Dekker, New York

Hyde RW (1960) Psychological and social determinants of drug action. In: Sarwer-Foner GJ (ed) The dynamics of psychiatric drug therapy. Thomas, Springfield, IL

Johnson M, Richards W, Griffiths R (2008) Human hallucinogen research: guidelines for safety. J Psychopharmacol 22:603–620

Kaelen M, Roseman L, Kahan J, Santos-Ribeiro A, Orban C, Lorenz R, Barett FS, Bolstridge M, Williams T, Williams L, Wall MB, Feidling A, Muthukumaraswamy S, Nutt DJ, Carhart-Harris R (2016) LSD modulates music-induced imagery via changes in parahippocampal connectivity. Eur Neuropsychopharmacol 26:1099–1109

Kast EC, Collins VJ (1964) Study of lysergic acid diethylamide as an analgesic agent. Anesth Analg 43:285–291

Kenna JC, Sedman G (1964) The subjective experience of time during lysergic acid diethylamide (Lsd-25) intoxication. Psychopharmacologia 5:280–288

Klüver H (1966) Mescal and mechanisms of hallucinations. University of Chicago Press, Chicago

Knoll M, Kugler J, Hofer O, Lawder SD (1963) Effects of chemical stimulation of electrically-induced phosphenes on their bandwidth, shape, number and intensity. Confin Neurol 23:201–226

Kometer M, Cahn BR, Andel D, Carter OL, Vollenweider FX (2011) The 5-HT2A/1A agonist psilocybin disrupts modal object completion associated with visual hallucinations. Biol Psychiatry 69:399–406

Kometer M, Schmidt A, Bachmann R, Studerus E, Seifritz E, Vollenweider FX (2012) Psilocybin biases facial recognition, goal-directed behavior, and mood state toward positive relative to negative emotions through different serotonergic subreceptors. Biol Psychiatry 72:898–906

Kometer M, Schmidt A, Jancke L, Vollenweider FX (2013) Activation of serotonin 2A receptors underlies the psilocybin-induced effects on alpha oscillations, N170 visual-evoked potentials, and visual hallucinations. J Neurosci 33:10544–10551

Kraehenmann R, Preller KH, Scheidegger M, Pokorny T, Bosch OG, Seifritz E, Vollenweider FX (2014) Psilocybin-induced decrease in amygdala reactivity correlates with enhanced positive mood in healthy volunteers. Biol Psychiatry

Krippner S (1977) Research in creativity and psychedelic drugs. Int J Clin Exp Hypn 25:274–290

Krus DM, Wapner S, Freeman H, Casey TM (1963) Differential behavioral responsivity to LSD-25, study in normal and schizophrenic adults. Arch Gen Psychiatry 8:557–563

Krus DM, Resnick O, Raskin M (1966) Apparent eye level test. Its background and use in psychopharmacology. Arch Gen Psychiatry 14:419–428

Lamparter D, Dittrich A (1994) Differentielle Psychologie aussergewöhnlicher Bewusstseinszustände - Literaturübersicht und methodische Probleme. In: Dittrich A, Hofmann A, Leuner H (eds) Welten des Bewusstseins. Band 3. Experimentelle Psychologie, Neurobiologie und Chemie. VWB, Berlin

Leary T, Litwin GH, Metzner R (1963) Reactions to psilocybin administered in a supportive environment. J Nerv Ment Dis 137:561–573

Leuner H (1962) Die experimentelle psychose. Springer, Berlin/Göttingen/Heidelberg

Leuner H (1971) Halluzinogene in der Psychotherapie [Hallucinogens in psychotherapy]. Pharmacopsychiatry 4:333–351

Leuner H (1981) Halluzinogene. Psychische Grenzzustände in Forschung und Therapie. Hans Huber, Bern

Leuner H, Pletscher A, Ladewig D (1994) Hallucinogens as an aid in psychotherapy: basic principles and results. In: 50 years of LSD: current status and perspectives of hallucinogen research. The Parthenon Publishing Group, New York

Lienert GA, Netter P (1996) LSD response in Eysenckian trait types identified by polypredictive CFA. Personality Individ Differ 21:845–850

Ludwig AM (1966) Altered states of consciousness. Arch Gen Psychiatry 15:225–234

Maclean KA, Johnson MW, Griffiths RR (2010) Mystical experiences occasioned by the hallucinogen psilocybin lead to increases in the personality domain of openness. J Psychopharmacol 25:1453–1461

MacLean KA, Leoutsakos JM, Johnson MW, Griffiths RR (2012) Factor analysis of the mystical experience questionnaire: a study of experiences occasioned by the hallucinogen psilocybin. J Sci Study Relig 51:721–737

Masters R, Houston J (2000) The varieties of psychedelic experience. Park Street Press, Rochester
McGlothlin W, Cohen S, McGlothlin MS (1967) Long lasting effects of LSD on normals. Arch Gen Psychiatry 17:521–532
Metzinger T (2009) The ego tunnel. The science of the mind and the myth of the self. Basic Books, New York
Metzner R, Litwin GH, Weil GM (1965) The relation of expectation and mood to psilocybin reactions: a questionnaire study. Psychedelic Rev 5:3–39
Moreau De Tours JJ (1973) Hashish and mental illness. Raven Press, New York
Moreno FA, Delgado PL (1997) Hallucinogen-induced relief of obsessions and compulsions. Am J Psychiatry 154:1037–1038
Muthukumaraswamy SD, Carhart-Harris RL, Moran RJ, Brookes MJ, Williams TM, Errtizoe D, Sessa B, Papadopoulos A, Bolstridge M, Singh KD, Feilding A, Friston KJ, Nutt DJ (2013) Broadband cortical desynchronization underlies the human psychedelic state. J Neurosci 33:15171–15183
Natale M, Dahlberg CC, Jaffe J (1978) Effect of psychotomimetics (LSD and dextroamphetamine) on the use of primary- and secondary-process language. J Consult Clin Psychol 46:352–353
Nichols DE (2004) Hallucinogens. Pharmacol Ther 101:131–181
Nichols DE, Chemel BR (2006) The neuropharmacology of religious experience: hallucinogens and the experience of the divine. In: McNamara P (ed) Where God and science meet. Praeger, Westport, CT
Northoff G (2011) Self and brain: what is self-related processing? Trends Cogn Sci 15:186–187
Osmond H (1957) A review of the clinical effects of psychotomimetic agents. Ann N Y Acad Sci 66:418–434
Ott U, Reuter M, Hennig J, Vaitl D (2005) Evidence for a common biological basis of the Absorption trait, hallucinogen effects, and positive symptoms: epistasis between 5-HT2a and COMT polymorphisms. Am J Med Genet B Neuropsychiatr Genet 137B:29–32
Pahnke WN (1963) Drugs and mysticism: an analysis of the relationship between psychedelic drugs and the mystical consciousness. Harvard University Press, Cambridge
Passie T, Seifert J, Schneider U, Emrich HM (2002) The pharmacology of psilocybin. Addict Biol 7:357–364
Pekala RJ, Steinberg J, Kumar VK (1986) Measurement of phenomenological experience: pheneomenology of consciousness inventory. Percept Mot Skills 62:983–989
Quednow BB, Kuhn KU, Mossner R, Schwab SG, Schuhmacher A, Maier W, Wagner M (2008) Sensorimotor gating of schizophrenia patients is influenced by 5-HT2A receptor polymorphisms. Biol Psychiatry 64:434–437
Quednow BB, Schmechtig A, Ettinger U, Petrovsky N, Collier DA, Vollenweider FX, Wagner M, Kumari V (2009) Sensorimotor gating depends on polymorphisms of the serotonin-2A receptor and catechol-O-methyltransferase, but not on neuregulin-1 Arg38Gln genotype: a replication study. Biol Psychiatry 66:614–620
Quednow BB, Kometer M, Geyer MA, Vollenweider FX (2012) Psilocybin-induced deficits in automatic and controlled inhibition are attenuated by ketanserin in healthy human volunteers. Neuropsychopharmacology 37:630–640
Riba J, Rodriguez-Fornells A, Strassman RJ, Barbanoj MJ (2001) Psychometric assessment of the hallucinogen rating scale. Drug Alcohol Depend 62:215–223
Riba J, Romero S, Grasa E, Mena E, Carrio I, Barbanoj MJ (2006) Increased frontal and paralimbic activation following ayahuasca, the pan-amazonian inebriant. Psychopharmacology
Sandison RA, Whitelaw JD (1957) Further studies in the therapeutic value of lysergic acid diethylamide in mental illness. J Ment Sci 103:332–343
Savage C, Fadiman J, Mogar R, Allen MH (1966) The effects of psychedelic (LSD) therapy on values, personality, and behavior. Int J Neuropsychiatry 2:241–254
Schmid Y, Enzler F, Gasser P, Grouzmann E, Preller KH, Vollenweider FX, Brenneisen R, Müller F, Borgwardt S, Liechti ME (2015) Acute effects of LSD in healthy subjects. Biol Psychiatry 78:544–553

Schmidt A, Bachmann R, Kometer M, Csomor PA, Stephan KE, Seifritz E, Vollenweider FX (2012) Mismatch negativity encoding of prediction errors predicts S-ketamine-induced cognitive impairments. Neuropsychopharmacology 37:865–875

Schmidt A, Kometer M, Bachmann R, Seifritz E, Vollenweider F (2013) The NMDA antagonist ketamine and the 5-HT agonist psilocybin produce dissociable effects on structural encoding of emotional face expressions. Psychopharmacology 225:227–239

Schultes RE, Hofmann A (1980) The botany and chemistry of hallucinogens. Charles C. Thomas, Springfield

Serretti A, Drago A, de Ronchi D (2007) HTR2A gene variants and psychiatric disorders: a review of current literature and selection of SNPs for future studies. Curr Med Chem 14:2053–2069

Sessa B (2008) Is it time to revisit the role of psychedelic drugs in enhancing human creativity? J Psychopharmacol 22:821–827

Shanon B (2002) Ayahuasca visualizations. a structural typology. J Conscious Stud 9:3–30

Siegel RK (1977) Hallucinations. Sci Am 237:132–140

Siegel RK, Jarvik ME (1975) Drug-induced hallucinations in animals and man. In: Siegel RK, West LJ (eds) Hallucinations: behavior, experience and theory. Wiley, New York

Silverman J (1971) Research with psychedelics. Some biopsychological concepts and possible clinical applications. Arch Gen Psychiatry 25:498–510

Sinke C, Halpern JH, Zedler M, Neufeld J, Emrich HM, Passie T (2012) Genuine and drug-induced synesthesia: a comparison. Conscious Cogn 21:1419–1434

Hermle L, Fnfgeld M, Oepen G, Botsch H, Borchard D, Gouzoulis E, Fehrenbach RA, Spitzer, M (1993) Mescaline-induced psychopathological, neuropsychological, and neurometabolic effects in normal subjects: experimental psychosis as a tool for psychiatric research. Biol Psychiatry 32:976–991

Spitzer M, Braun U, Maier S, Hermle L, Maher BA (1993) Indirect semantic priming in schizophrenic patients. Schizophr Res 11:71–80

Spitzer M, Thimm M, Hermle L, Holzmann P, Kovar KA, Heimann H, Gouzoulis-Mayfrank E, Kischka U, Schneider F (1996) Increased activation of indirect semantic associations under psilocybin. Biol Psychiatry 39:1055–1057

Stace WT (1961) Mysticism and philosphy. Macmillan, London

Strassman RJ (2001) DMT: the spirit molecule. Park Street Press, Rochester, VT

Strassman RJ, Qualls CR, Uhlenhuth EH, Kellner R (1994) Dose-response study of N, N-dimethyltryptamine in humans—II. Subjective effects and preliminary results of a new rating scale. Arch Gen Psychiatry 51:98–108

Studerus E, Gamma A, Vollenweider FX (2010) Psychometric evaluation of the altered states of consciousness rating scale (OAV). PLoS One 5:e12412

Studerus E, Kometer M, Hasler F, Vollenweider FX (2011) Acute, subacute and long-term subjective effects of psilocybin in healthy humans: a pooled analysis of experimental studies. J Psychopharmacol 25:1434–1452

Studerus E, Gamma A, Kometer M, Vollenweider FX (2012) Prediction of psilocybin response in healthy volunteers. PLoS One 7:e30800

Sysoeva OV, Tonevitsky AG, Wackermann J (2010) Genetic determinants of time perception mediated by the serotonergic system. PLoS One 5

Tagliazucchi E, Carhart-Harris R, Leech R, Nutt D, Chialvo DR (2014) Enhanced repertoire of brain dynamical states during the psychedelic experience. Hum Brain Mapp 35:5442–5456

ten Berge J (2002) Jekyll and Hyde revisited: paradoxes in the appreciation of drug experiences and their effects on creativity. J Psychoactive Drugs 34:249–262

Thatcher K, Wiederholt WC, Fischer R (1971) An electroencephalographic analysis of personality-dependent performance under psilocybin. Agents Actions 2:21–26

Tononi G (2004) An information integration theory of consciousness. BMC Neurosci 5:42

Umbricht D, Vollenweider FX, Schmid L, Grubel C, Skrabo A, Huber T, Koller R (2003) Effects of the 5-HT2A agonist psilocybin on mismatch negativity generation and AX-continuous performance task: implications for the neuropharmacology of cognitive deficits in schizophrenia. Neuropsychopharmacology 28:170–181

Unger SM (1963) Mescaline, LSD, psilocybin, and personality change. Psychiatry 26:111–125
Vollenweider FX (1998) Advances and pathophysiological models of hallucinogen drug actions in humans: a preamble to schizophrenia research. Pharmacopsychiat 31:92–103
Vollenweider FX (2008) Neurobiologie der Halluzinogenerfahrung. In: Jungaberle H, Gasser P, Weinhold J, Verres R (eds) Therapie mit psychoaktiven Substanzen. Verlag Hans Huber, Hogrefe AG, Göttingen
Vollenweider FX, Geyer MA (2001) A systems model of altered consciousness: integrating natural and drug-induced psychoses. Brain Res Bull 56:495–507
Vollenweider FX, Kometer M (2010) The neurobiology of psychedelic drugs: implications for the treatment of mood disorders. Nat Rev Neurosci 11:642–651
Vollenweider FX, Leenders KL, Oye I, Hell D, Angst J (1997a) Differential psychopathology and patterns of cerebral glucose utilisation produced by (S)- and (R)-ketamine in healthy volunteers using positron emission tomography (PET). Eur Neuropsychopharmacol 7:25–38
Vollenweider FX, Leenders KL, Scharfetter C, Antonini A, Maguire P, Missimer J, Angst J (1997b) Metabolic hyperfrontality and psychopathology in the ketamine model of psychosis using positron emission tomography (PET) and [F-18]-fluorodeoxyglocose (FDG). Eur Neuropsychopharmacol 7:9–24
Vollenweider FX, Leenders KL, Scharfetter C, Maguire P, Stadelmann O, Angst J (1997c) Positron emission tomography and fluorodeoxyglucose studies of metabolic hyperfrontality and psychopathology in the psilocybin model of psychosis. Neuropsychopharmacology 16:357–372
Vollenweider FX, Vollenweider-Scherpenhuyzen MFI, Babler A, Vogel H, Hell D (1998) Psilocybin induces schizophrenia-like psychosis in humans via a serotonin-2 agonist action. NeuroReport 9:3897–3902
Vollenweider FX, Csomor PA, Knappe B, Geyer MA, Quednow BB (2007) The effects of the preferential 5 HT2A agonist psilocybin on prepulse inhibition of startle in healthy human volunteers depend on interstimulus interval. Neuropsychopharmacology 32:1876–1887
Wackermann J, Wittmann M, Hasler F, Vollenweider FX (2008) Effects of varied doses of psilocybin on time interval reproduction in human subjects. Neurosci Lett 435:51–55
Wapner S, Krus DM (1960) Effects of lysergic acid diethylamide, and differences between normals and schizophrenics on the stroop color-word test. J Neuropsychiatr 2:76–81
Weber K (1967) Alteration of music perception in experimental psychosis (psilocybin). Confin Psychiatr 10:139–176
Wittmann M, Carter O, Hasler F, Cahn BR, Grimberg U, Spring P, Hell D, Flohr H, Vollenweider FX (2007) Effects of psilocybin on time perception and temporal control of behaviour in humans. J Psychopharmacol 21:50–64
Zegans LS, Pollard JC, Brown D (1967) The effects of LSD-25 on creativity and tolerance to regression. Arch Gen Psychiatry 16:740–749

Serotonergic Hallucinogen-Induced Visual Perceptual Alterations

Michael Kometer and Franz X. Vollenweider

Abstract Serotonergic hallucinogens, such as lysergic acid diethylamide (LSD), psilocybin, and *N*,*N*-dimethyltryptamine (DMT), are famous for their capacity to temporally and profoundly alter an individual's visual experiences. These visual alterations show consistent attributes despite large inter- and intra-individual variances. Many reports document a common perception of colors as more saturated, with increased brightness and contrast in the environment ("Visual Intensifications"). Environmental objects might be altered in size ("Visual illusions") or take on a modified and special meaning for the subject ("Altered self-reference"). Subjects may perceive light flashes or geometrical figures containing recurrent patterns ("Elementary imagery and hallucinations") influenced by auditory stimuli ("Audiovisual synesthesia"), or they may envision images of people, animals, or landscapes ("Complex imagery and hallucinations") without any physical stimuli supporting their percepts. This wide assortment of visual phenomena suggests that one single neuropsychopharmacological mechanism is unlikely to explain such vast phenomenological diversity. Starting with mechanisms that act at the cellular level, the key role of 5-HT2A receptor activation and the subsequent increased cortical excitation will be considered. Next, it will be shown that area specific anatomical and dynamical features link increased excitation to the specific visual contents of hallucinations. The decrease of alpha oscillations by hallucinogens will then be introduced as a systemic mechanism for amplifying internal-driven excitation that overwhelms stimulus-induced excitations. Finally, the hallucinogen-induced parallel decrease of the N170 visual evoked potential and increased medial P1 potential will be discussed as key mechanisms for inducing a dysbalance between global

M. Kometer (✉) · F.X. Vollenweider
Neuropsychopharmacology and Brain Imaging Research Unit, Heffter Research Center Zurich, Department of Psychiatry, Psychotherapy and Psychosomatics, Psychiatric University Hospital, University of Zurich, Lenggstrasse 31, 8032 Zurich, Switzerland
e-mail: mikometer@gmail.com

F.X. Vollenweider
e-mail: vollen@bli.uzh.ch

Curr Topics Behav Neurosci (2016) 36: 257–282
DOI 10.1007/7854_2016_461
Published Online: 30 November 2016

integration and early visual gain that may explain several hallucinogen-induced visual experiences, including visual hallucinations, illusions, and intensifications.

Keywords Hallucination · Imagery · Hallucinogen · Psilocybin · LSD · Ayahuasca

Contents

1 Introduction

The dramatic visionary impact of serotonergic hallucinogens has prompted their use across various cultures and throughout time, discerned in rock paintings and symbolic cultural materials dating back thousands of years. Notably, these visual expressions frequently resemble forms that are characteristics of the elementary visual hallucinations seen in drug-induced states (Kroeber 1925; Lewis-Williams et al. 1988; Winkelman 2002). In line with this idea, several visual artists have claimed that serotonergic hallucinogens greatly increased their visual creativity, allowing a connection between the artist and comparatively subconscious, internal aspects of vision (Berlin et al. 1955; Krippner 1985; Janiger and de Rios 1989; Grey and Wilber 2001). The strong link between hallucinogens-induced visual percepts, emotions, and autobiographic memory has further been exploited in psycholytic or psychedelic therapy in western society (Leuner 1967; Grof 1973; Grinspoon and Bakalar 1986; Vollenweider and Kometer 2010; Carhart-Harris et al. 2012). Similarly, shamans have been taken advantages of hallucinogen-induced imagery/hallucinations to influence illness processes since a long time (Achterberg 1987; Mercante 2006; Achterberg 2013). Finally, hallucinogen-induced visual perceptual alterations have been assessed using modern brain imaging techniques to elucidate the neuropsychopharmacological mechanisms of visual perceptual alteration and thereby gain insights into functional properties of the visual system and the pathophysiology of visual hallucinations (Kometer et al. 2011, 2013).

Given the cross-cultural influence of the visionary properties of serotonergic hallucinogens, Sect. 2 will review the phenomenology and psychology of drug-induced perceptual alterations, as well as their cultural determinants. Next, Sect. 3 will introduce several neural mechanisms that potentially underlie the rich visual phenomenology of serotonergic hallucinogens, ranging from mechanisms with actions at the cellular level to mechanisms with actions at the whole-brain level.

2 Phenomenology and Psychology

2.1 *Visual Illusions, Distortions, and Intensifications*

Serotonergic hallucinogens induce several visual percepts that are driven by the environment, but are characterized by an increased mismatch to the actual physical constitution of the subject's surroundings. This phenomenon can be appreciated by the altered perception of visual objects as decreased or increased in their perceived versus actual size (Dittrich 1998), or with modified angles (Díaz 2010). Furthermore, objects may rhythmically move or vibrate along their edges (Díaz 2010), which is apparently paralleled by a diminished performance in higher level motion recognition tasks for subjects under the influence of hallucinogens (Carter et al. 2004). Moreover, a subject's discernment of visual space is reportedly transformed after the administration of the serotonergic hallucinogen psilocybin, as evidenced by an augmented misjudgment regarding visual depth and the visual vertical or horizontal in tiled body positions (Hill et al. 1968; Fischer et al. 1970; Hill and Fischer 1973).

In addition to these alterations in complex, integrative visual processes (e.g., object recognition and formation of visual space), the perception of elementary visual features (brightness, color saturation, visual contrast) can be subjectively increased by serotonergic hallucinogens (Klüver 1942; Rümmele and Gnirss 1961; Klüver 1966; Dittrich 1998; Díaz 2010). Together with this subjective increase, a hallucinogen-intoxicated individual will show a preference for dimmer light in the room while experiencing the peak of the drug experience (Fischer et al. 1969). Following the administration of hallucinogens, subjects further reported that they could see more colors than usual due to flickering lights or visualization of after-images (Hartman and Hollister 1963).

On the other hand, the accuracy for detecting low-level features in behavioral tests remains mostly unaltered by serotonergic hallucinogens. For example, the threshold for discriminating near-threshold light stimuli was only slightly increased in early investigations of drug effects (Blough 1957; Carlson 1958), whereas color discrimination accuracy was either unaltered (Edwards and Cohen 1961; Hollister and Hartman 1962) or slightly decreased (Hartman and Hollister 1963). The contrast sensitivity for drifting gratings also stayed the same (Carter et al. 2004). The only clear exception to this detection pattern was an enhanced discrimination accuracy for flickering lights after administration of low-dose serotonergic hallucinogens (Becker et al. 1967). This exception implies that the formation of visual hallucinations and/or cognitive impairments at higher drug doses can prevent the detection of any increased accuracy in visual low-level behavioral tasks. Nonetheless, the overall picture suggests that increasing doses of serotonergic hallucinogens typically induce an augmented mismatch between subjective visual experiences and visual percept accuracy.

2.2 *Imagery and Hallucinations*

2.2.1 On the Distinctions Between Imagery and Hallucinations

Serotonergic hallucinogens induce several types of visual experiences ranging from imagery and pseudohallucinations to ideal hallucinations that are commonly defined by the absence of sensory input supporting these percepts. Distinctions between these categories were based on visual attributes such as vividness, intensity, appraisal, emotional reaction, volitional control, and sense of reality (Horowitz 1975), but no agreement emerged and instead a continuity between these categories has been proposed (Seitz and Molholm 1947; Horowitz 1975). Most typically, imagery lacks the vividness and intensity of hallucinations, but is more under the control of the subjects. The vividness of imagery is frequently reported to be increased by serotonergic hallucinogens (Dittrich 1998; Studerus et al. 2010), leading to a stage where the visual effects are better characterized as hallucinations. Pseudohallucinations are thought to be as vivid as ideal hallucinations, but are recognized to be self-produced. Although hallucinogen-induced percepts can usually be recognized as self-produced at moderate doses, this capacity seems to diminish with increasing dose (Shanon 2002; Rolland et al. 2014), leading first to a situation in which more time is required to differentiate between self- and external-induced percepts until finally this distinction is no longer possible (Cott and Rock 2008; Luke 2011). Thus, given the continuity between these categories, they will not be strictly differentiated throughout this Chapter. Instead, it seems more useful to note a progression of these visual experiences toward more vividness, intensity, and sensed reality with increasing doses (Shanon 2002). A clearer distinction can be made in terms of the content of imagery/hallucinations, which can either be composed of elementary or complex visual features.

2.2.2 Elementary Imagery and Hallucinations

Elementary imagery and hallucinations comprise visual experiences ranging from single-light flashes (known as phosphenes) to more repetitive visual elements with distinct boundaries, and then on to complete geometric images. Light flashes typically occur as single elements in the earliest stages of hallucinogen intoxication, while in later stages, the various elements multiply and may form more discrete boundaries (Shanon 2002). However, the most elaborated elementary hallucinations correspond to visions of geometrical figures, described by the intoxicated subject with words such as transparent oriental rug, wallpaper design, filigreed object of art, cobweb-like figure, spiral and prism. These elementary hallucinations are highly typical for serotonergic hallucinogens, and have been documented since the late eighteenth century (Lewin 1886; Mitchell 1896; Mooney 1896; Prentiss and Morgan 1896).

The first systematic analysis of hallucinogen-induced geometrical images was conducted by Heinrich Klüver by administrating mescaline-containing peyote cacti to several subjects, including himself (Klüver 1928, 1942). Despite large inter- and intra-individual differences in the descriptions of the ensuing geometric figures, recurring patterns with a remarkable uniformity were seen across subjects (Fig. 1). Klüver called these patterns "form constants" and categorized them into four classes, as follows: (1) lattices (including gratings, fretworks, honeycombs, filigrees, and chessboard designs), (2) cobwebs, (3) tunnels (including alleys, funnels, cones and vessels), and (4) spirals.

Almost five decades later, Siegel and coworkers assessed the consistency of this categorization scheme across different psychedelic substances in a study employing European subjects (Siegel and Jarvik 1975). To this end, four of the subjects included in the study were trained to categorize their visual experiences by repeated presentation of example visual stimuli for each of the four Klüver form constants (lattices, cobwebs, tunnels and spirals), as well as four additional investigator-defined categories. Furthermore, the four subjects were trained to categorize eight different groupings of color, movement and action patterns. The subjects then received either LSD (50 and 100 μg), 2-bromo-LSD (also known as BOL; 50 and 100 μg), psilocybin (10 and 20 mg), mescaline (200 and 300 mg), delta-9-tetrahydrocannabinol (THC) (10 and 20 mg), phenobarbital (30 and 60 mg), or D-amphetamine (5 and 15 mg) in a series of single-blind weekly test sessions. Consequently, they continually reported their eyes-open visual experiences in a completely dark chamber.

Interestingly, classical hallucinogens, including psilocybin, mescaline and LSD, frequently induced form constants of the lattice and tunnel types. Furthermore, hallucinogen-induced visual experiences were dominated by red, orange, or yellow colors, while blue colors were most commonly observed after the administration of delta-9-THC. Lastly, explosive and/or rotational motions were most frequently reported after the administration of classical hallucinogens, followed by pulsating motions. Thus, visual experiences showed marked consistency across various

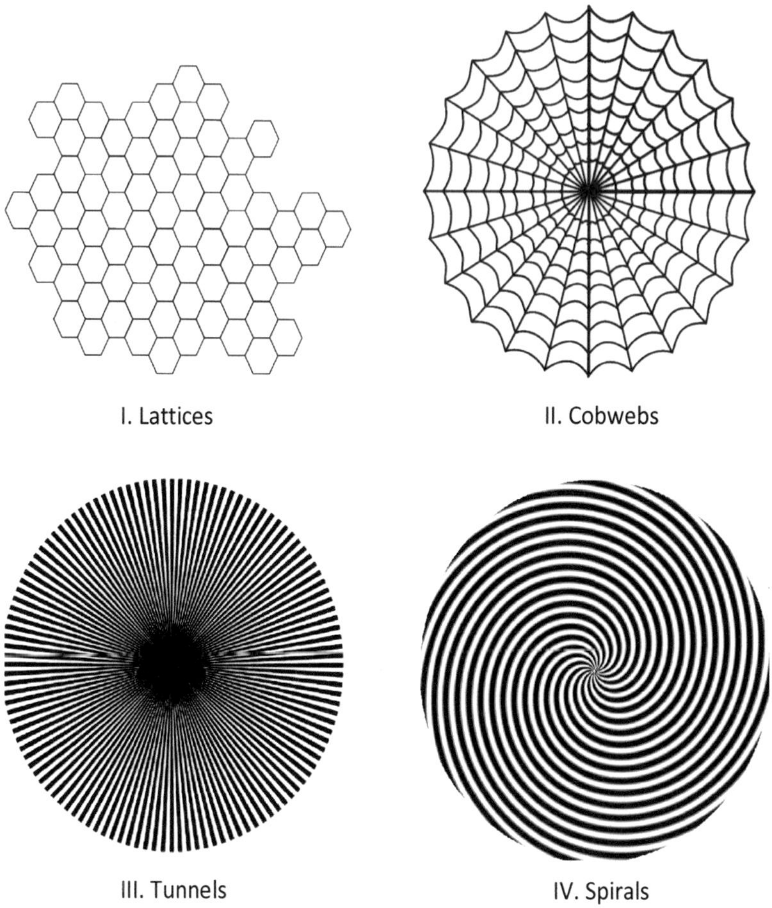

Fig. 1 Sample stimuli for the four hallucinatory form constants: (*I*) lattices (including gratings, fretworks, honeycombs, filigrees, and chessboard designs), (*II*) cobwebs, (*III*) tunnels (including alleys, funnels, cones, and vessels) and (*IV*) spirals.

classical hallucinogens, not only in terms of specific form constants, but also in terms of the color and movement categories.

Siegel went on to explore whether these consistencies in form, movement, and color are consistent across different cultures. Intriguingly, tunnels and funnels have been habitually reported by curanderos (folks healers or shamans) when using ayahuasca in medical and spiritual rites (Naranjo 1973). Furthermore, Tukano Indians in the Amazon region of Colombia often decorate their homes and pottery with geometrical paintings of images seen during ayahuasca rituals, including curves, spirals, lattices, and the sun (Reichel-Dolmatoff 1972). To further test the concept of cultural consistency, Siegel and colleagues conducted a small study in Mexico with four male Huichols, who ingested the equivalent of ~200 mg of mescaline in the form of a peyote button suspension during a traditional ceremony (Siegel and Jarvik 1975). Four hours after ingestion, the subjects made a total of 68 reports about simple forms, colors, and movement patterns, and a total of 27 reports referring to complex scenes. The predominant reported form was a lattice tunnel, and the predominant perceived movement was an explosive motion toward the subjects. These observations provide support for the hypothesis of a cross-cultural consistency of elementary hallucinations. However, blue was more often experienced by the Huichols than by the four European subjects who received mescaline in the tests described above.

2.2.3 Complex Imagery and Hallucinations

Complex imagery and hallucinations include on the one hand visual images of people, animals, or entities, and on the other hand, visions of whole scenes and landscapes. Hence, these hallucinations are usually composed of non-repetitive, figurative visual features, and they transport more semantic content than elementary hallucinations. Complex hallucinations are not reported as frequently as elementary hallucinations (Studerus et al. 2011; Kometer et al. 2012, 2013), and some people never seem to experience complex hallucinations following hallucinogenic drug administration (Shanon 2002).

Complex hallucinations usually appear after the first elementary hallucinations are observed (Butterworth 1967; Siegel and Jarvik 1975), and they are regarded in the shamanic tradition as higher stages of visioning (Reichel-Dolmatoff 1975; Lewis-Williams et al. 1988). At low drug doses, complex hallucinations only occur in the closed-eyes state or in complete darkness. However, with increasing drug doses, they are first seen with opened eyes in a dimmed environment or at the periphery of the visual field (Siegel and Jarvik 1975). At high drug doses and particularly with DMT, complex hallucinations are also observed with fully opened eyes in an undimmed environment (Shanon 2002; Cott and Rock 2008). Indeed, under these conditions, the subject fails to experience a strong distinction between the eyes-open and eyes-closed states, but most subjects nevertheless prefer to close their eyes to prevent external stimuli-mediated interruptions of internally driven percepts (Shanon 2002). Furthermore, at higher drug doses, complex hallucinations are increasingly experienced as having strong prevalence and independence. This phenomenon seems especially pronounced with DMT (Luke 2011).

The content of complex hallucinations is far-reaching and can largely differ between and within subjects. The scenes perceived under the influence of serotonergic hallucinogens encompass all-inclusive, progressively developing, visualized scenarios, varying from brief glimpses and snapshots to full-fledged panoramas viewed as in a film or theater (Shanon 2002). For instance, subjects have described incredible and beautiful landscapes, as well as futuristic cities (Shanon 2002). The visual images seen in the hallucinogen-induced states can include objects, people, human faces, and animals encountered in the visual environment, such as the anaconda or jaguar (Reichel-Dolmatoff 1972, 1975; Siegel and Jarvik 1975). Interestingly, these animals are a common aspect of ayahuasca-induced visions in shamanic rituals (Reichel-Dolmatoff 1972, 1975; Winkelman 2002), perhaps because certain animals have a special meaning for the affected individuals within daily life and during healing rituals (Harner et al. 1990; Saunders 1994; Winkelman 2002; Shepard 2004).

2.3 Audiovisual Synesthesia

Visual percepts observed in the hallucinogen-induced state can also be driven by stimulation of nonvisual sensory modalities a phenomenon termed synesthesia (Ellis 1898; Klüver 1966; Dittrich 1998; Shanon 2002; Studerus et al. 2011; Kometer et al. 2012; Brogaard 2013; Kometer et al. 2013; Luke and Terhune 2013). Most often, these visual percepts are modulated by auditory stimuli, such as the sound of music (Studerus et al. 2011; Luke and Terhune 2013). Only rarely are they reported to be induced by haptic, kinesthetic, or algesic stimuli (Klüver 1966; Luke and Terhune 2013). In agreement with these observations, intensified experiences of color and brightness were documented ın an early study during the presentation of auditory tones before versus after drug administration (Hartman and Hollister 1963). However, early behavioral studies often suffered from methodological problems, including lack of a placebo control, absence of a double-blind design, and a lack of randomized group assignments. Accordingly, further studies are required to assess the capacity of hallucinogens to induce synesthesia in more detail.

2.4 The Role of the Self in Visual Experiences

Hallucinogen-induced visual experiences are frequently described as having a deeply amended and personalized meaning for the subject, with profound individual significance (Dittrich 1998; Shanon 2002; Díaz 2010; Shanon 2010; Studerus et al. 2011; Froese et al. 2013). As described previously, the visual landscape may look new, and everything might seem as if it were viewed for the first time (Díaz 2010). Therefore, serotonergic hallucinogens change not only the visual percept per se, but also the unique relationship between the viewer and the visual percept. In line with this view, complex visual imagery and hallucinations can stem from autobiographic memory (Studerus et al. 2011) or can be characterized by a special psychological relationship to the current life situation of the subject (Shanon 2002). Hence, the appearance of visual hallucinations is strongly linked to the emotional state of the subject at the time of drug administration and thereby provide significant for the subject. Not rarely, visual hallucinations are described as exceedingly beautiful, surpassing anything ever seen, dreamt, or imagined (Shanon 2002). Such affirmative experiences are usually connected with intense positive emotions, while an anxious ego dissolution might be visualized by the subject in terms of terrifying images or scenarios. The strong association between autobiographic memory, emotions, and visual imagery/hallucinations has been exploited in psycholytic or psychedelic therapy, because this association provides a way to access and transform autobiographic memories and emotions (Leuner 1967; Grof 1973; Grinspoon and Bakalar 1986; Vollenweider and Kometer 2010; Carhart-Harris et al. 2012).

3 Neuropsychopharmacological Mechanisms

Serotonergic hallucinogens provoke a wide assortment of visual phenomena, suggesting that one single neuropsychopharmacological mechanism is unlikely to explain such vast phenomenological diversity. In this section, several partially overlapping, potential mechanisms of hallucinogen action will be discussed, ranging from mechanisms that act at the cellular level to mechanisms that act at the whole-brain level. Given that only a limited number of studies have specifically addressed the neural mechanisms underlying serotonergic drug-induced visual hallucinations, we will also consider the experimental evidence linking these mechanisms with the formation of visual perceptual alterations under various psychiatric conditions and during neurological disease states.

3.1 Primary Pharmacological Mechanism: 5-HT2A Receptor Activation

Serotonergic hallucinogens (e.g., psilocybin) display agonistic activity at several serotonergic receptors, including serotonin 5-HT2A, 5-HT2C, 5-HT1A, and 5-HT7 receptors (Nichols 2004; Sard et al. 2005; Ray 2010). However, the activation of 5-HT2A receptors seems to be primarily responsible for the psychedelic effects of these agents (Nichols 2004; Vollenweider and Kometer 2010). Early support for this view is provided by animal studies demonstrating that discriminative stimulatory hallucinogenic actions correlate with drug affinity at the 5-HT2A receptor (Glennon et al. 1983, 1984, Sanders-Bush et al. 1988). Furthermore, these actions can be blocked by the preferential 5-HT2A antagonists, ketanserin, and pirenperon (Colpaert et al. 1982; Leysen et al. 1982; Colpaert and Janssen 1983). More recently, hallucinogen-induced head shaking was used as an animal model of psychedelic drug action, and was found to be absent in transgenic mice lacking

5-HT2A receptors (González-Maeso et al. 2007) and in rats after administration of the 5-HT2A receptor antagonist M100907 (Schreiber et al. 1995).

In agreement with the crucial role of 5-HT2A receptors in serotonergic hallucinogen mechanisms of action, several investigations established that ketanserin can almost completely block the subjective psychedelic effects of psilocybin in humans (Vollenweider et al. 1998; Carter et al. 2005; Kometer et al. 2011; Quednow et al. 2012). These effects include elementary and complex hallucinations, audiovisual synesthesia and the altered significance of visual percepts (Kometer et al. 2012, 2013). Furthermore, the psilocybin-induced decrease in the visual evoked N170 potential, a marker of psilocybin-induced visual hallucinogenic activity in humans (discussed below) (Kometer et al. 2011, 2013), was also blocked by the administration of ketanserin (Kometer et al. 2013).

3.2 From 5-HT2A Receptor Activation to Increased Excitation

Activation of 5-HT2A receptors by serotonergic hallucinogens induces a robust increase in excitatory postsynaptic currents (EPSCs) of pyramidal neurons, predominantly within layer 5 of the frontal cortical area (Aghajanian and Marek 1997; Béïque et al. 2007, González-Maeso et al. 2007, 2008; Riga et al. 2014) and occipital cortex (Moreau et al. 2010). By contrast, inhibitory postsynaptic currents (IPSCs) are only weakly increased (Riga et al. 2014). The overall increase in excitation is abolished not only by administration of specific 5-HT2A receptor antagonists (Aghajanian and Marek 1997), but also by administration of α-amino-3-hydroxy-5-methyl-4-isoxazolepropionic acid (AMPA) receptor antagonists (Aghajanian and Marek 1997; Zhang and Marek 2008), metabotropic glutamate receptor (mGluR) agonists (Aghajanian and Marek 1997) and positive allosteric modulators of mGluR2 (Benneyworth et al. 2007). Taken together, these observations indicate that glutamatergic activity downstream of 5-HT2A receptor activation is strongly implicated in the mechanism of action of serotonergic hallucinogens. In keeping with this view, stimulation of postsynaptic 5-HT2A receptors on a subpopulation of glutamatergic cells in deep cortical layers increased glutamatergic recurrent network activity, resulting in an augmentation of EPSCs, mainly in layer V (Béïque et al. 2007; Aghajanian 2009; Moreau et al. 2010). Although the link between 5-HT2A receptor activation and increased excitation has been investigated in detail, the relationship between excitation and visual hallucinations has rarely been experimental investigated and will therefore be discussed by taking stimulation experiments and computational models into account.

3.3 From Increased Excitation to the Formation of Visual Hallucinations

3.3.1 Increased Excitation and Visual Hallucinations in Stimulation Experiments

The idea that increased excitation of cortical visual areas could lead to the formation of visual hallucinations derived in the nineteenth century from the observations that electrical stimulation of cortical areas in humans can induce several visual phenomena. Specifically, researchers observed that hallucinations of meaningful images were induced in some patients with epilepsies (Penfield and Rasmussen 1950; Penfield and Jasper 1954; Horowitz and Adams 1970) or schizophrenia (Ishibashi et al. 1964) by directly stimulating the temporal lobe. The content of these electrical-induced hallucinations seems to be related to what patients experienced immediate before surgery (Mahl et al. 1964). Furthermore, researchers found that visual phosphene, characterized by rather shapeless, nonspecific impressions of light, were induced by applying rectangular electric pulses to large electrodes placed on the cortical surface (Knoll 1958; Knoll et al. 1962a, b, 1963) or by applying magnetic fields (Thompson 1910; Dunlap 1911; Magnusson and Stevens 1911; Barlow et al. 1947; Seidel 1968). Since these early findings, it has been hypothesized than an increase in electrical or chemical processes underlies the formation of hallucinations and phosphenes.

More recent studies using electrical stimulation (Lee et al. 2000; Murphey et al. 2009; Jonas et al. 2013) or transcranial magnetic stimulation (TMS) (Hallett 2000; Kammer et al. 2001; Stewart et al. 2001) mostly confirm these early findings and extend them in several important ways. In TMS studies, it was found that phosphenes can be reliably induced by a certain stimulation protocol and head positioning (Hallett 2000; Kammer et al. 2001; Stewart et al. 2001). The threshold for experiencing these phosphenes differ between subjects and has even been taken as a marker for visual cortical excitability (Boroojerdi et al. 2000; Oliveri and Calvo 2003; Taylor et al. 2011). Interestingly, in subjects experiencing hallucinations this threshold was found to be decreased compared to non-hallucinators (Aurora et al. 1998, 2003; Taylor et al. 2011), providing further evidence that an increase in excitability underlies the formation of visual hallucination.

Electrical stimulation studies using modern tools to guide stimulation in various occipital, temporal, and parietal areas mostly found that the complexity of the visual phenomena tended to increase along the posterior–anterior axis (Lee et al. 2000; Jonas et al. 2013), however see also (Murphey et al. 2009). In addition, the probability of evoking visual phenomena was found to be generally higher in the right compared to the left hemisphere (Jonas et al. 2013). These anatomical observations are in line with the fMRI and leson studies suggesting that brain areas along the occipito-temporal cortex are hierarchically organized for processing increasingly complex visual features (Grill-Spector and Malach 2004). Thus, according to these findings, the content of the hallucinatory experience is to some extend driven by the cortical location of increased excitation. However, this association may not be strict because this association was not always found in stimulation studies (Murphey et al. 2009) and because each visual area is highly connected with areas providing top-down or bottom-up input (Grill-Spector and Malach 2004).

3.3.2 Increased Self-organized Excitation in Computational Models of Visual Hallucinations

Since the 1970s, a number of computational models have been proposed to explain how increased excitation leads to the formation of elementary, geometric hallucinations (Ermentrout and Cowan 1979; Bressloff et al. 2001, 2002; Gutkin et al. 2003; Rule et al. 2011; Billock and Tsou 2012; Butler et al. 2012). All these models are to some extent based on the idea of the Turing mechanism (Turing 1952), which explains morphogenesis, i.e., pattern formation during biological development, by reaction-diffusion systems. The idea of the turning mechanism has been transposed to the functioning of the brain in the Wilson–Cowan neural network equation (Wilson and Cowan 1973) and was fist applied to explain elementary geometrical hallucinations by Ermentrout and Cowan (Ermentrout and Cowan 1979). Specifically, they postulated a two-layer neural network model of excitatory and inhibitory neurons in primary visual cortex (V1). Similarly to the turing mechanism, this model contains two main elements that explain the formation of visual hallucinations: first, an asymmetry between two interacting mechanisms (excitation and inhibition) and second, a diffusion-like mechanism for spreading their influence. Within this model, a hallucinogen-induced increase in excitation destabilizes the resting state because of a distributional asymmetry between inhibition and excitation. As a consequence, spontaneous spatiotemporal patterns of activity emerge due to spreading of negative feedback by lateral interactions. These spatiotemporal patterns can be viewed as a self-organization process to reintroduce stability. In order to see what subjects would visually experience due to the emergent spatiotemporal patterns, a retinocortical mapping was applied, allowing neuronal activity in V1 to be transformed into retinal coordinates. Thus, the retinocortical mapping makes it possible to define the retinal input that would be required to induce these spatiotemporal patterns. Using this model Ermentrout and Cowan (1979) found that several patterns could be induced that resemble the form constant described by Klüver. This model did particularly well describing lattice patterns (Fig. 1), which are frequently induced by serotonergic hallucinogens.

Although this initial model was innovative, inspiring and influential, it was not able to produce all form constants and a large drawback was that it was not based on the actual neural architecture of the visual cortex. To overcome these problems, several authors (Bressloff et al. 2001, 2002; Butler et al. 2012) proposed models to explain geometrical elementary hallucinations based on the structure of V1. The model of Bressloff and colleagues incorporated the findings of anatomical and functional studies that the detection of local contours and oriented edges in visual input is mediated by structured connections between subgroups of V1 neurons that are organized in hypercolumns. This pattern of structural organization, combined with the neuronal Turing mechanism and increased cortical excitation, was able to generate form constants. Specifically, this model was able to describe some of the more complicated form constants, including cobwebs, honeycombs, and lattice (Bressloff et al. 2001, 2002).

Together, these mathematical models provide an appealing explanation for why increased excitation in V1 leads to the formation of specific elementary geometrical hallucinations. In line with these models, relatively simple visual features such as lines can be processed only within V1 (Grill-Spector and Malach 2004) and therefore V1 activity may be sufficient to explain form constants comprised of lines, such as lattices patterns. However, certain elementary form constants, as well as complex hallucinations, remain unexplained by these models and seem to require models incorporating higher level visual areas. Higher cortical areas may further be required to explain why these hallucinogen-induced experiences are often experienced as being meaningful (Froese et al. 2013), beautiful and detailed and why the interindividual predisposition for experiencing these visual hallucinations is associated with the personality trait absorption (Studerus et al. 2012). Thus, additional models incorporating higher level visual areas are required to explain, how increased self-organized excitation can explain a larger phenomenological range of serotonin hallucinogen-induced visual perceptual alterations. In the next section, alpha oscillations will be discussed as a systemic mechanism for regulating excitation across low- and high-level cortical visual areas, which could potentially be implicated in the formation of different type of visual perceptual alterations and hallucination.

3.3.3 Alpha Oscillations: Increased Spontaneous Excitation that Overwhelms Stimulus-Induced Excitability

Parieto-occipital alpha oscillations (8–12 Hz) regulate trough inhibition the excitability levels of neurons across the whole cortical visual system and thereby strongly influence visual perception (Foxe et al. 1998; Thut et al. 2006; Klimesch et al. 2007; Rihs et al. 2007; Romei et al. 2008a, b; Busch et al. 2009; Jensen and Mazaheri 2010; Romei et al. 2010; Klimesch 2011; Mathewson et al. 2011; Jensen et al. 2012). In line with this view, decreased alpha power levels were for instance found to be associated with increased neuronal firing rates (Haegens et al. 2011) and with a decreased threshold for perceiving visual stimuli (Ergenoglu et al. 2004; Thut et al. 2006; Hanslmayr et al. 2007; van Dijk et al. 2008) and TMS-induced phosphenes (Romei et al. 2008a, 2010). Furthermore, the likelihood of perceiving subliminal visual stimuli rhythmically varies with the phase of alpha oscillations (Busch et al. 2009; Spaak et al. 2014; VanRullen et al. 2014). Given this crucial role of alpha oscillations in modulating excitability through inhibition across cortical visual systems, a hallucinogen-induced decrease in alpha oscillations may not only be in line with the increased cortical excitation found in animals (Moreau et al.

2010), but may additionally explain a wide range of hallucinogen-induced phenomenology.

To address this idea, we recently assessed in healthy human subjects the effect of psilocybin (215 μg/kg vs. placebo) on posterior parieto-occipital alpha oscillations observed before and during presentation of simple visual stimuli (Kometer et al. 2013). A high level of parieto-occipital alpha power was seen in the placebo condition before the presentation of the stimuli, thus in the absence of any task-relevant visual input. This indicates a high level of inhibition reduces the excitability of the visual pathways in the absence of task-relevant visual input (Klimesch 2011; Palva and Palva 2011). Psilocybin strongly attenuated this high level of alpha power, suggesting that psilocybin increases the excitability of the visual pathway in the absence of externally presented stimuli. Thereby, spontaneous self-organized activity may gain perceptual quality, which could form the base for psilocybin-induced visual hallucinations. In line with this view, spontaneous self-organized background activity was found to resemble the neuronal activity seen by presenting simple visual geometric (Kenet et al. 2003). Therefore, it has been postulated that a inhibitory mechanism is necessary to prevent that the usually subliminal spontaneous neuronal activity leads to conscious percepts in the form of elementary visual hallucinations (Billock and Tsou 2007). Alpha oscillation may constitute this inhibitory mechanism, which was found to be attenuated by psilocybin, possible leading to a conscious perception of spontaneous neuronal activity in the form of hallucinations (Kometer et al. 2013).

Interestingly, alpha oscillations are not only implicated in regulating spontaneous internal-driven excitability, but also in controlling stimulus-induced excitation (Hanslmayr et al. 2009; Klimesch 2011). This leads to the question of whether the increased excitability seen in the absence of task-relevant input influences the excitation that is induced by the presentation of external visual stimuli? Such a stimulus-induced increase in excitation is seen by the strong decrease in alpha power around 200–400 ms after the presentation of the stimuli (Hanslmayr et al. 2009; Klimesch 2011). Interestingly, psilocybin was found to block this stimulus-induced reduction of alpha power (Kometer et al. 2013) and the lack of stimulus-induced alpha power reduction was further found to be due to the already attenuated prestimulus alpha power level (Kometer et al. 2013). Thus, by decreasing prestimulus alpha power, psilocybin seems to induce a dysbalance between the excitability that is seen in the absence of external visual input and the excitability that is induced by the presentation of the stimulus. Thus, psilocybin induces a processing mode, in which stimulus-driven cortical excitation is overwhelmed by spontaneous self-organizing neuronal excitation (Kometer et al. 2013). This psilocybin-induced shift away from stimulus-driven information processing toward internal-driven processing could well contribute to the formation of hallucinations, given the longstanding proposal that increased internal-driven information processing may lead to the formation of visual hallucinations (Horowitz 1975; Allen et al. 2008). Interestingly, a similar bias toward internal-driven information processing is seen at the single cell level. Specifically, the activation of 5-HT2A receptors was found to have an opposite effect on low and high neuronal firing rates

(Watakabe et al. 2009). That is, 5-HT2A receptor activation suppresses the activity of neurons with high firing rates (Watakabe et al. 2009), which are usually induced by external visual stimuli (Quiroga et al. 2005; Montemurro et al. 2008). By contrast, low firing rates, which may constitute stimulus-independent, internal-driven background activity, were found to be facilitated by 5-HT2A receptor activation (Watakabe et al. 2009). Thus, in line with our finding stimulus-independent background activity may overwhelm stimulus-induced processing. However, the strong effect of psilocybin on alpha oscillations may not only be implicated in the formation of visual hallucinations, but may further underlie a psilocybin-induced increase in distractibility (Carter et al. 2005) or decrease in working memory (Wittmann et al. 2007) due to the crucial role of alpha oscillations in dynamically adjusting spatial and temporal excitability parameters for optimizing processing for task demands (Capotosto et al. 2009; Zanto et al. 2011; Bonnefond and Jensen 2012; Hsu et al. 2014; Zumer et al. 2014). For instance, psilocybin may disrupt the possibility of increasing alpha oscillations in anticipation of distracting stimuli, which was found to be required to prevent interference of distracters with working memory maintenance (Bonnefond and Jensen 2012). Hence, taken together the decrease in alpha oscillations seems to amplify internal-driven excitation that overwhelms stimulus-induced excitations and in addition may induce cognitive impairments such as increased distractibility.

3.4 Dysbalance Between Early Low-Level and Late High-Level Visual Processing

3.4.1 Evidence from Phenomenological and Behavioral Studies

Converging lines of evidence form phenomenological and behavioral suggest that serotonergic hallucinogens differently modulate early low-level visual processing and late high-level visual processing. Specifically, phenomenological studies indicate that the perception of elementary visual features such as the brightness, the local contrast and the saturation of colors is subjectively increased by hallucinogens (Klüver 1942; Rümmele and Gnirss 1961; Klüver 1966; Dittrich 1998; Díaz 2010). These elementary visual features are typically processed fast (Proverbio and Zani 2002) and within low-level visual areas (Grill-Spector and Malach 2004). By contrast, the perception of whole objects, the construction of the visual space, and the detection of global motion patterns, which all require more time (Johnson and Olshausen 2003) and higher level visual areas (Grill-Spector and Malach 2004) to be processed, seems to be impaired by hallucinogens (Hill et al. 1968; Fischer et al. 1970; Hill and Fischer 1973; Dittrich 1998; Carter et al. 2004; Díaz 2010). Together, these findings suggest that hallucinogens impair late high-level processing, while increasing or having no effect on early low-level processing. In the following section, we will present studies that address this hypothesis in more detail by using high-density EEG recordings to assess the spatiotemporal dynamic of visual processing in psilocybin-induced states.

3.4.2 P1 Amplitude and V1 Activity: Increased Early Low-Level Processing

Using high-density EEG recordings psilocybin was found to dose-dependently increase the amplitude of the early visual evoked P1 potential selectively over the medial occipital electrode sites (Kometer et al. 2011, 2013). This psilocybin-induce increase, which was seen 100 ms after the presentation of simple visual stimuli was found by mathematical source reconstruction techniques to reflect increased activity in early visual area V1 (Kometer et al. 2011). Because processing of brightness has been associated with the medial P1 potential (Proverbio and Zani 2002) and with activity in V1 (Salminen-Vaparanta et al. 2013), this psilocybin-induced increase in the medial P1 potential may be the neuronal correlate of the often reported hallucinogen-induced increase in brightness perception (Kometer et al. 2011). Interestingly, this medial P1 increase induced by psilocybin was found to be driven by the hallucinogen-induced decrease in prestimulus alpha oscillations (Kometer et al. 2013). This suggests that the psilocybin-induced increase in visual cortical excitability before the presentation of the visual stimulus may have amplified the processing elementary visual features, such as the brightness, in early visual areas.

3.4.3 N170 Amplitude and Extrastriate Activity: Decreased Global Integration

In contrast to this initial increase in early visual cortex activity, during a later time frame ($\sim$150–190 ms after stimulus presentation) psilocybin dose-dependently decreased the visual N170 potential to the same simple visual stimuli (Kometer et al. 2011, 2013). This psilocybin-induced decrease of the N170 amplitude was localized in the lateral occipital complex (LOC) and the fusiform gyrus, which both belong to extrastriate, higher visual areas. Thus, this psilocybin-induced decrease of the N170 decrease is in line with the hypothesis that hallucinogens disrupt late higher level visual processing.

More specifically, the N170 potential has been implicated in global integrative processes, such as the structural encoding of emotional face expressions (Rossion et al. 2000; Bernasconi et al. 2013; Schmidt et al. 2013) or object recognition (Murray et al. 2002; Kometer et al. 2011, 2013; Knebel and Murray 2012). For instance, the N170 potential has been found to be crucial for object completion, which is the process of integrating local information into complex object representation and of interpolating missing parts of objects. This process is required due to the ambiguous and incomplete retinal information under partial occlusion or poor illumination conditions. In support of the proposed role of the N170 potential in object completion, the N170 potential was found to be higher for incomplete,

Kanizsa figures, compared to control figures (Murray et al. 2002, 2006; Kometer et al. 2011, 2013; Knebel and Murray 2012).

Using these Kanizsa figures, we found that psilocybin induced a more pronounced reduction of the N170 amplitude and activation of the LOC in response to Kanizsa figures compared to control figures (Kometer et al. 2011, 2013) (Fig. 2a). This indicates that psilocybin disrupts the neuronal processes of object completion. Given that object completion is crucial for perceiving coherent and meaningful structures in natural images (Lesher 1995), this disruption in object completion is likely to contribute to psilocybin-induced alterations in visual perceptual experiences. Interestingly, this contribution may be seen in the observation that subjective visual perceptual alterations first appear in dimmed environment (Siegel and Jarvik 1975); thus in lighting situations that require extensive object completion.

Most direct support for the view that the N170 potential decrease is associated with visual alterations derives from the finding that the reported intensity of subjective visual hallucinations correlated with decreases in the N170 amplitude in both the Kanizsa and the control conditions (Kometer et al. 2011, 2013). This association was equally seen for elementary and complex hallucinations, as well as audiovisual synethesia (Kometer et al. 2013) (Fig. 2c). Exploring this relationship in more detailed using mathematical source reconstruction techniques indicated that the psilocybin-induced decrease in the right-lateralized LOC and posterior parietal areas during the time frame of the N170 potential most strongly correlated with the intensity of visual perceptual alteration (Kometer et al. 2011) (Fig. 2b). This localization is in accord with the results of previous brain imaging studies reporting decrease extrastriate activation in response to external visual stimuli in hallucinating patients compared to patients without hallucinations (Howard et al. 1995; Ffytche et al. 1998; Oertel et al. 2007). Furthermore, decreased activation during the time frame of the N1/N170 potential has been associated in patient studies with the formation of visual (Spencer et al. 2004) and acoustic hallucinations (Tiihonen et al. 1992; Hubl et al. 2007). This strong association between the N1/N170 potential and hallucinations seems to be driven by the crucial role of this potential in global integration processes that are required to perceive coherent and meaningful structures in sensory input. Global integration is further important to differentiate internal-driven and external-driven sensory percepts, a process that is associated with the N1 potential (Ford et al. 2007; Heinks-Maldonado et al. 2007; Gentsch and Schütz-Bosbach 2011; Ford et al. 2013; Hubl et al. 2014). Taken together, the decrease of N170 potential by hallucinogens is a key mechanism underlying the formation of visual hallucinations due to the role of the N170 potential in global integration required for recognizing the meaning and the self-reference of visual percepts.

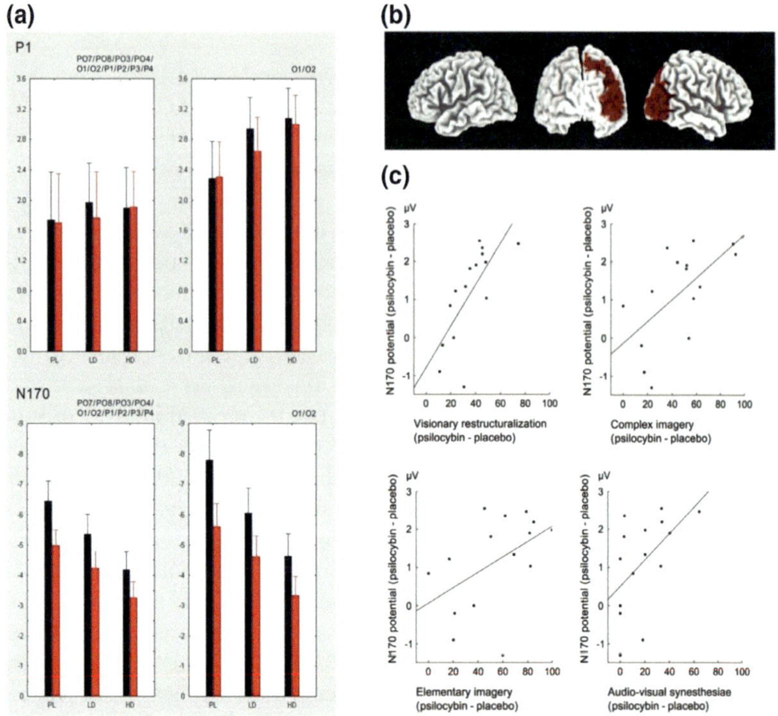

Fig. 2 **a** The bar graphs display the effect of placebo (PL), a low dose (LD) and a high dose (HD) of psilocybin on the amplitude of the P1 and N170 potential to kanizsa (*black*) and non-kanizsa figures (*red*) measured from 10 parieto-occipital electrodes sites (*left bar graphs*) and from the medial occipital electrode sides O1/O2 (*right bar graphs*) **b** *Red areas* display the psilocybin-induced decreases in current source density during the time period of the N170 potential that positively correlated with the intensity of psilocybin-induced visual hallucinations. [Figures 2a and 2b are reprinted from Kometer et al., The 5-HT2A/1A Agonist Psilocybin Disrupts Modal Object Completion Associated with Visual Hallucinations, page 399–406, Biological Psychiatry, Copyright (2011), with permission from Elsevier]. **c** The psilocybin-induced decrease of the N170 amplitude significantly correlates with the psilocybin-induced increase in visual restructuralization, complex imagery, elementary imagery and audiovisual synesthesiae as measured by the 5D-ASC questionnaire [Reprinted from Kometer et al., Activation of Serotonin 2A Receptors Underlies the Psilocybin-Induced Effects on α Oscillations, N170 Visual-Evoked Potentials, and Visual Hallucinations, page 10544–10551, The Journal of Neuroscience, Copyright (2013), with permission from Society for Neuroscience]

References

Achterberg J (1987) The shaman: master healer in the imaginary realm. In: Shamanism expanded view reality, pp 103–124

Achterberg J (2013) Imagery in healing: shamanism and modern medicine. Shambhala Publications, Boulder

Aghajanian GK (2009) Modeling "psychosis" in vitro by inducing disordered neuronal network activity in cortical brain slices. Psychopharmacology 206:575–585

Aghajanian GK, Marek GJ (1997) Serotonin induces excitatory postsynaptic potentials in apical dendrites of neocortical pyramidal cells. Neuropharmacology 36:589–599

Allen P, Larøi F, McGuire PK, Aleman A (2008) The hallucinating brain: a review of structural and functional neuroimaging studies of hallucinations. Neurosci Biobehav Rev 32:175–191

Aurora S, Ahmad B, Welch K, Bhardhwaj P, Ramadan N (1998) Transcranial magnetic stimulation confirms hyperexcitability of occipital cortex in migraine. Neurology 50:1111–1114

Aurora S, Welch K, Al-Sayed F (2003) The threshold for phosphenes is lower in migraine. Cephalalgia 23:258–263

Barlow HB, Kohn HI, Walsh EG (1947) Visual sensations aroused by magnetic fields. Am J Physiol 148:372–375

Becker DI, Appel J, Freedman D (1967) Some effects of lysergic acid diethylamide on visual discrimination in pigeons. Psychopharmacologia 11:354–364

Béïque JC, Imad M, Mladenovic L, Gingrich JA, Andrade R (2007) Mechanism of the 5-hydroxytryptamine 2A receptor-mediated facilitation of synaptic activity in prefrontal cortex. Proc Natl Acad Sci USA 104:9870–9875

Benneyworth MA, Xiang Z, Smith RL, Garcia EE, Conn PJ, Sanders-Bush E (2007) A selective positive allosteric modulator of metabotropic glutamate receptor subtype 2 blocks a hallucinogenic drug model of psychosis. Mol Pharmacol 72:477–484

Berlin L, Guthrie T, Weider A, Goodell H, Wolff HG (1955) Studies in human cerebral function: the effects of mescaline and lysergic acid on cerebral processes pertinent to creative activity. J Nerv Ment Dis 122:487–491

Bernasconi F, Schmidt A, Pokorny T, Kometer M, Seifritz E, Vollenweider FX (2013) Spatiotemporal brain dynamics of emotional face processing modulations induced by the serotonin 1A/2A receptor agonist psilocybin. Cereb Cortex 24:3221–3231

Billock VA, Tsou BH (2007) Neural interactions between flicker-induced self-organized visual hallucinations and physical stimuli. Proc Natl Acad Sci USA 104:8490–8495

Billock VA, Tsou BH (2012) Elementary visual hallucinations and their relationships to neural pattern-forming mechanisms. Psychol Bull 138:744

Blough DS (1957) Effect of lysergic acid diethylamide on absolute visual threshold of the pigeon. Science 126:304–305

Bonnefond M, Jensen O (2012) Alpha oscillations serve to protect working memory maintenance against anticipated distracters. Curr Biol 22:1969–1974

Boroojerdi B, Bushara KO, Corwell B, Immisch I, Battaglia F, Muellbacher W, Cohen LG (2000) Enhanced excitability of the human visual cortex induced by short-term light deprivation. Cereb Cortex 10:529–534

Bressloff PC, Cowan JD, Golubitsky M, Thomas PJ, Wiener MC (2001) Geometric visual hallucinations, Euclidean symmetry and the functional architecture of striate cortex. Philos Trans R Soc Lond B Biol Sci 356:299–330

Bressloff PC, Cowan JD, Golubitsky M, Thomas PJ, Wiener MC (2002) What geometric visual hallucinations tell us about the visual cortex. Neural Comput 14:473–491

Brogaard B (2013) Serotonergic hyperactivity as a potential factor in developmental, acquired and drug-induced synesthesia. Front Hum Neurosci 7:657

Busch NA, Dubois J, VanRullen R (2009) The phase of ongoing EEG oscillations predicts visual perception. J Neurosci 29:7869–7876

Butler TC, Benayoun M, Wallace E, van Drongelen W, Goldenfeld N, Cowan J (2012) Evolutionary constraints on visual cortex architecture from the dynamics of hallucinations. Proc Natl Acad Sci USA 109:606–609

Butterworth AT (1967) The psychotomimetic effect: a discussion of its unique nature and character. Existential Psychiatry Winter, p 9

Capotosto P, Babiloni C, Romani GL, Corbetta M (2009) Frontoparietal cortex controls spatial attention through modulation of anticipatory alpha rhythms. J Neurosci 29:5863–5872

Carhart-Harris RL, Leech R, Williams TM, Erritzoe D, Abbasi N, Bargiotas T, Hobden P, Sharp DJ, Evans J, Feilding A, Wise RG, Nutt DJ (2012) Implications for psychedelic-assisted psychotherapy: functional magnetic resonance imaging study with psilocybin. Br J Psychiatry 200:238–244

Carlson VR (1958) Effect of lysergic acid diethylamide (LSD-25) on the absolute visual threshold. J Comp Physiol Psychol 51:528

Carter OL, Pettigrew JD, Burr DC, Alais D, Hasler F, Vollenweider FX (2004) Psilocybin impairs high-level but not low-level motion perception. NeuroReport 15:1947–1951

Carter O, Burr D, Pettigrew J, Wallis G, Hasler F, Vollenweider F (2005) Using psilocybin to investigate the relationship between attention, working memory, and the serotonin 1A and 2A receptors. J Cogn Neurosci 17:1497–1508

Colpaert F, Janssen P (1983) A characterization of LSD-antagonist effects of pirenperone in the rat. Neuropharmacology 22:1001–1005

Colpaert FC, Niemegeers C, Janssen P (1982) A drug discrimination analysis of lysergic acid diethylamide (LSD): in vivo agonist and antagonist effects of purported 5-hydroxytryptamine antagonists and of pirenperone, a LSD-antagonist. J Pharmacol Exp Ther 221:206–214

Cott C, Rock A (2008) Phenomenology of N,N-Dimethyltryptamine use: a thematic analysis. J Sci Explor 22:359–370

Díaz JL (2010) Sacred plants and visionary consciousness. Phenomenol Cogn Sci 9:159–170

Dittrich A (1998) The standardized psychometric assessment of altered states of consciousness (ASCs) in humans. Pharmacopsychiatry 31:80–84

Dunlap K (1911) Visual sensations from the alternating magnetic field. Science 33:68–71

Edwards AE, Cohen S (1961) Visual illusion, tactile sensibility and reaction time under LSD-25. Psychopharmacologia 2:297–303

Ellis H (1898) Mescal: a new artificial paradise. US Government Printing Office

Ergenoglu T, Demiralp T, Bayraktaroglu Z, Ergen M, Beydagi H, Uresin Y (2004) Alpha rhythm of the EEG modulates visual detection performance in humans. Brain Res Cogn Brain Res 20:376–383

Ermentrout GB, Cowan JD (1979) A mathematical theory of visual hallucination patterns. Biol Cybern 34:137–150

Ffytche DH, Howard RJ, Brammer MJ, David A, Woodruff P, Williams S (1998) The anatomy of conscious vision: an fMRI study of visual hallucinations. Nat Neurosci 1:738–742

Fischer R, Hill R, Warshay D (1969) Effects of the psychodysleptic drug psilocybin on visual perception. Changes in brightness preference. Experientia 25:166–169

Fischer R, Hill R, Thatcher K, Scheib J (1970) Psilocybin-induced contraction of nearby visual space. Agents Actions 1:190–197

Ford JM, Gray M, Faustman WO, Roach BJ, Mathalon DH (2007) Dissecting corollary discharge dysfunction in schizophrenia. Psychophysiology 44:522–529

Ford JM, Mathalon DH, Roach BJ, Keedy SK, Reilly JL, Gershon ES, Sweeney JA (2013) Neurophysiological evidence of corollary discharge function during vocalization in psychotic patients and their nonpsychotic first-degree relatives. Schizophr Bull 39:1272–1280

Foxe JJ, Simpson GV, Ahlfors SP (1998) Parieto-occipital approximately 10 Hz activity reflects anticipatory state of visual attention mechanisms. NeuroReport 9:3929–3933

Froese T, Woodward A, Ikegami T (2013) Turing instabilities in biology, culture, and consciousness? On the enactive origins of symbolic material culture. Adapt Behav, 1059712313483145

Gentsch A, Schütz-Bosbach S (2011) I did it: unconscious expectation of sensory consequences modulates the experience of self-agency and its functional signature. J Cogn Neurosci 23:3817–3828
Glennon RA, Young R, Rosecrans JA (1983) Antagonism of the effects of the hallucinogen DOM and the purported 5-HT agonist quipazine by 5-HT2 antagonists. Eur J Pharmacol 91:189–196
Glennon RA, Young R, Hauck AE, McKenney J (1984) Structure-activity studies on amphetamine analogs using drug discrimination methodology. Pharmacol Biochem Behav 21:895–901
González-Maeso J, Weisstaub NV, Zhou M, Chan P, Ivic L, Ang R, Lira A, Bradley-Moore M, Ge Y, Zhou Q, Sealfon SC, Gingrich JA (2007) Hallucinogens recruit specific cortical 5-HT (2A) receptor-mediated signaling pathways to affect behavior. Neuron 53:439–452
González-Maeso J, Ang RL, Yuen T, Chan P, Weisstaub NV, López-Giménez JF, Zhou M, Okawa Y, Callado LF, Milligan G, Gingrich JA, Filizola M, Meana JJ, Sealfon SC (2008) Identification of a serotonin/glutamate receptor complex implicated in psychosis. Nature 452:93–97
Grey A, Wilber K (2001) The mission of art. Shambhala
Grill-Spector K, Malach R (2004) The human visual cortex. Annu Rev Neurosci 27:649–677
Grinspoon L, Bakalar JB (1986) Can drugs be used to enhance the psychotherapeutic process. Am J Psychother 40:393–404
Grof S (1973) Theoretical and empirical basis of transpersonal psychology and psychotherapy: Observations from LSD research. J Trans Pers Psychol 5:15
Gutkin B, Pinto D, Ermentrout B (2003) Mathematical neuroscience: from neurons to circuits to systems. J Physiol Paris 97:209–219
Haegens S, Nácher V, Luna R, Romo R, Jensen O (2011) α-Oscillations in the monkey sensorimotor network influence discrimination performance by rhythmical inhibition of neuronal spiking. Proc Natl Acad Sci USA 108:19377–19382
Hallett M (2000) Transcranial magnetic stimulation and the human brain. Nature 406:147–150
Hanslmayr S, Aslan A, Staudigl T, Klimesch W, Herrmann CS, Bäuml KH (2007) Prestimulus oscillations predict visual perception performance between and within subjects. Neuroimage 37:1465–1473
Hanslmayr S, Spitzer B, Bäuml KH (2009) Brain oscillations dissociate between semantic and nonsemantic encoding of episodic memories. Cereb Cortex 19:1631–1640
Harner MJ, Mishlove J, Bloch A (1990) The way of the shaman. Harper & Row San Francisco
Hartman AM, Hollister LE (1963) Effect of mescaline, lysergic acid diethylamide and psilocybin on color perception. Psychopharmacologia 4:441–451
Heinks-Maldonado TH, Mathalon DH, Houde JF, Gray M, Faustman WO, Ford JM (2007) Relationship of imprecise corollary discharge in schizophrenia to auditory hallucinations. Arch Gen Psychiatry 64:286–296
Hill RM, Fischer R (1973) Induction and extinction of psilocybin induced transformations of visual space. Pharmacopsychiatry 6:258–263
Hill R, Fischer R, Warshay D (1968) Effects of excitatory and tranquilizing drugs on visual perception. Am J Optom Archiv Am Acad Optom 45:454
Hollister LE, Hartman AM (1962) Mescaline, lysergic acid diethylamide and psilocybin: comparison of clinical syndromes, effects on color perception and biochemical measures. Compr Psychiatry 3:235–241
Horowitz MJ (1975) Hallucinations: an information-processing approach. Hallucinations: behavior, experience and theory, pp 163–196
Horowitz M, Adams J (1970) Hallucinations on brain stimulation: evidence for revision of the Penfield hypothesis. In: Origin and mechanisms of hallucinations. Springer, Berlin, pp 13–22
Howard R, Williams S, Bullmore E, Brammer M, Mellers J, Woodruff P, David A (1995) Cortical response to exogenous visual stimulation during visual hallucinations. Lancet 345:70
Hsu TY, Tseng P, Liang WK, Cheng SK, Juan CH (2014) Transcranial direct current stimulation over right posterior parietal cortex changes prestimulus alpha oscillation in visual short-term memory task. Neuroimage 98:306–313

Hubl D, Koenig T, Strik WK, Garcia LM, Dierks T (2007) Competition for neuronal resources: how hallucinations make themselves heard. Br J Psychiatry 190:57–62
Hubl D, Schneider RC, Kottlow M, Kindler J, Strik W, Dierks T, Koenig T (2014) Agency and ownership are independent components of 'sensing the self' in the auditory-verbal domain. Brain Topogr 27:1–11
Ishibashi T, Hori H, Endo K, Sato T (1964) Hallucinations produced by electrical stimulation of the temporal lobes in schizophrenic patients. Tohoku J Exp Med 82:124–139
Janiger O, de Rios MD (1989) LSD and creativity. J Psychoactive Drugs 21:129–134
Jensen O, Mazaheri A (2010) Shaping functional architecture by oscillatory alpha activity: gating by inhibition. Front Hum Neurosci 4:186
Jensen O, Bonnefond M, VanRullen R (2012) An oscillatory mechanism for prioritizing salient unattended stimuli. Trends Cogn Sci 16:200–206
Johnson JS, Olshausen BA (2003) Timecourse of neural signatures of object recognition. J Vis 3:4
Jonas J, Frismand S, Vignal JP, Colnat-Coulbois S, Koessler L, Vespignani H, Rossion B, Maillard L (2013) Right hemispheric dominance of visual phenomena evoked by intracerebral stimulation of the human visual cortex. Hum Brain Mapp 35:3360–3371
Kammer T, Beck S, Erb M, Grodd W (2001) The influence of current direction on phosphene thresholds evoked by transcranial magnetic stimulation. Clin Neurophysiol 112:2015–2021
Kenet T, Bibitchkov D, Tsodyks M, Grinvald A, Arieli A (2003) Spontaneously emerging cortical representations of visual attributes. Nature 425:954–956
Klimesch W (2011) Evoked alpha and early access to the knowledge system: the P1 inhibition timing hypothesis. Brain Res 1408:52–71
Klimesch W, Sauseng P, Hanslmayr S (2007) EEG alpha oscillations: the inhibition-timing hypothesis. Brain Res Rev 53:63–88
Klüver H (1928) Mescal: the divine plant and its psychological effects. Trubner & Company Limited
Klüver H (1942) Mechanisms of hallucinations. McGraw-Hill, New York
Klüver H (1966) Mescal and mechanisms of hallucinations. University of Chicago Press, Chicago
Knebel JF, Murray MM (2012) Towards a resolution of conflicting models of illusory contour processing in humans. Neuroimage 59:2808–2817
Knoll M (1958) Anregung geometrischer Figuren und anderer subjektiver Lichtmuster in elektrischen Feldern. Schweiz Z Psychol 17:110–126
Knoll M, Höfer O, Lawder S, Lawder U (1962a) Die Reproduzierbarkeit von elektrisch angeregten Lichterscheinungen (Phosphenen) bei zwei Versuchspersonen innerhalb von 6 Monaten. Biomedizinische Technik/Biomedical Engineering 7:235–242
Knoll M, Kugler J, Eichmeier J, Höfer O (1962b) Note on the spectroscopy of subjective light patterns. J Anal Psychol 7:55–70
Knoll M, Kugler J, Höfer O, Lawder S (1963) Effects of chemical stimulation of electrically-induced phosphenes on their bandwidth, shape, number and intensity. Stereotact Funct Neurosurg 23:201–226
Kometer M, Cahn BR, Andel D, Carter OL, Vollenweider FX (2011) The 5-HT2A/1A agonist psilocybin disrupts modal object completion associated with visual hallucinations. Biol Psychiatry 69:399–406
Kometer M, Schmidt A, Bachmann R, Studerus E, Seifritz E, Vollenweider FX (2012) Psilocybin biases facial recognition, goal-directed behavior, and mood state toward positive relative to negative emotions through different serotonergic subreceptors. Biol Psychiatry 72:898–906
Kometer M, Schmidt A, Jäncke L, Vollenweider FX (2013) Activation of serotonin 2A receptors underlies the psilocybin-induced effects on α oscillations, N170 visual-evoked potentials, and visual hallucinations. J Neurosci 33:10544–10551
Krippner S (1985) Psychedelic drugs and creativity. J Psychoactive Drugs 17:235–246
Kroeber AL (1925) Handbook of the Indians of California. Courier Dover Publications
Lee H, Hong S, Seo D, Tae W, Hong S (2000) Mapping of functional organization in human visual cortex Electrical cortical stimulation. Neurology 54:849–854

Lesher GW (1995) Illusory contours: toward a neurally based perceptual theory. Psychon Bull Rev 2:279–321

Leuner H (1967) Present state of psycholytic therapy and its possibilities. The use of LSD in psychotherapy and alcoholism, p 101

Lewin L (1886) Ueber Piper methysticum (kawa). Hirschwald

Lewis-Williams JD, Dowson TA, Bahn PG, Bandi H-G, Bednarik RG, Clegg J, Consens M, Davis W, Delluc B, Delluc G (1988) The signs of all times: entoptic phenomena in Upper Palaeolithic art [and comments and reply]. Curr Anthropol 29:201–245

Leysen J, Niemegeers C, Van Nueten J, Laduron P (1982) [3H] Ketanserin (R 41 468), a selective 3H-ligand for serotonin2 receptor binding sites. Binding properties, brain distribution, and functional role. Mol Pharmacol 21:301–314

Luke D (2011) Discarnate entities and dimethyltryptamine (DMT): psychopharmacology, phenomenology and ontology. J Soc Psychical Res 75:26

Luke DP, Terhune DB (2013) The induction of synaesthesia with chemical agents: a systematic review. Front Psychol 4:753

Magnusson C, Stevens H (1911) In our own experiments, we were concerned with the following points: I, To verify the results of Thompson and Dunlap. 2. To ascertain whether the magnetic field induced by the direct current gives a visual sensation. 3. To determine the threshold of the sensation in terms of ampere

Mahl GF, Rothenberg A, Delgado JM, Hamlin H (1964) Psychological responses in the human to intracerebral electrical stimulation. Psychosom Med 26:337–368

Mathewson KE, Lleras A, Beck DM, Fabiani M, Ro T, Gratton G (2011) Pulsed out of awareness: EEG alpha oscillations represent a pulsed-inhibition of ongoing cortical processing. Front Psychol 2:99

Mercante MS (2006) Images of healing: spontaneous mental imagery and healing process of the Barquinha, a Brazilian ayahuasca religious system. Saybrook Graduate School and Research Center, San Francisco

Mitchell SW (1896) Remarks on the effects of Anhelonium lewinii (the mescal button). Br Med J 2:1625

Montemurro MA, Rasch MJ, Murayama Y, Logothetis NK, Panzeri S (2008) Phase-of-firing coding of natural visual stimuli in primary visual cortex. Curr Biol 18:375–380

Mooney J (1896) The Mescal Plant and Ceremony. Ther Gazette 21:7–11

Moreau AW, Amar M, Le Roux N, Morel N, Fossier P (2010) Serotoninergic fine-tuning of the excitation-inhibition balance in rat visual cortical networks. Cereb Cortex 20:456–467

Murphey DK, Maunsell JH, Beauchamp MS, Yoshor D (2009) Perceiving electrical stimulation of identified human visual areas. Proc Natl Acad Sci 106:5389–5393

Murray MM, Wylie GR, Higgins BA, Javitt DC, Schroeder CE, Foxe JJ (2002) The spatiotemporal dynamics of illusory contour processing: combined high-density electrical mapping, source analysis, and functional magnetic resonance imaging. J Neurosci 22:5055–5073

Murray MM, Imber ML, Javitt DC, Foxe JJ (2006) Boundary completion is automatic and dissociable from shape discrimination. J Neurosci 26:12043–12054

Naranjo C (1973) Psychological aspects of the yage experience in an experimental setting. Hallucinogens Shamanism, p 190

Nichols DE (2004) Hallucinogens. Pharmacol Ther 101:131–181

Oertel V, Rotarska-Jagiela A, van de Ven VG, Haenschel C, Maurer K, Linden DE (2007) Visual hallucinations in schizophrenia investigated with functional magnetic resonance imaging. Psychiatry Res Neuroimaging 156:269–273

Oliveri M, Calvo G (2003) Increased visual cortical excitability in ecstasy users: a transcranial magnetic stimulation study. J Neurol Neurosurg Psychiatry 74:1136–1138

Palva S, Palva JM (2011) Functional roles of alpha-band phase synchronization in local and large-scale cortical networks. Front Psychol 2:204

Penfield W, Rasmussen T (1950) The cerebral cortex of man; a clinical study of localization of function. Macmillan, New York

Penfield W, Jasper H (1954) Epilepsy and the functional anatomy of the human brain. Little Brown & Co, Boston

Prentiss D, Morgan F (1896) Therapeutic uses of mescal buttons (Anhalonium Lewinii). Ther Gaz 20:4–7

Proverbio AM, Zani A (2002) Electrophysiological indexes of illusory contours perception in humans. Neuropsychologia 40:479–491

Quednow BB, Kometer M, Geyer MA, Vollenweider FX (2012) Psilocybin-induced deficits in automatic and controlled inhibition are attenuated by ketanserin in healthy human volunteers. Neuropsychopharmacology 37:630–640

Quiroga RQ, Reddy L, Kreiman G, Koch C, Fried I (2005) Invariant visual representation by single neurons in the human brain. Nature 435:1102–1107

Ray TS (2010) Psychedelics and the human receptorome. PLoS ONE 5:e9019

Reichel-Dolmatoff G (1972) The cultural context of an aboriginal hallucinogen: banisteriopsis caapi. Flesh of the gods: the ritual use of hallucinogens, pp 84–113

Reichel-Dolmatoff G (1975) The Shaman and the Jaguar. A study of narcotic drugs among the Indians of Colombia Med Hist 21:344–345

Riga MS, Soria G, Tudela R, Artigas F, Celada P (2014) The natural hallucinogen 5-MeO-DMT, component of *Ayahuasca*, disrupts cortical function in rats: reversal by antipsychotic drugs. Int J Neuropsychopharmacol 17:1–14

Rihs TA, Michel CM, Thut G (2007) Mechanisms of selective inhibition in visual spatial attention are indexed by alpha-band EEG synchronization. Eur J Neurosci 25:603–610

Rolland B, Jardri R, Amad A, Thomas P, Cottencin O, Bordet R (2014) Pharmacology of hallucinations: several mechanisms for one single symptom? BioMed Res Int 2014:9

Romei V, Rihs T, Brodbeck V, Thut G (2008a) Resting electroencephalogram alpha-power over posterior sites indexes baseline visual cortex excitability. NeuroReport 19:203–208

Romei V, Brodbeck V, Michel C, Amedi A, Pascual-Leone A, Thut G (2008b) Spontaneous fluctuations in posterior alpha-band EEG activity reflect variability in excitability of human visual areas. Cereb Cortex 18:2010–2018

Romei V, Gross J, Thut G (2010) On the role of prestimulus alpha rhythms over occipito-parietal areas in visual input regulation: correlation or causation? J Neurosci 30:8692–8697

Rossion B, Gauthier I, Tarr MJ, Despland P, Bruyer R, Linotte S, Crommelinck M (2000) The N170 occipito-temporal component is delayed and enhanced to inverted faces but not to inverted objects: an electrophysiological account of face-specific processes in the human brain. NeuroReport 11:69–72

Rule M, Stoffregen M, Ermentrout B (2011) A model for the origin and properties of flicker-induced geometric phosphenes. PLoS Comput Biol 7:e1002158

Rümmele W, Gnirss F (1961) Untersuchungen mit Psilocybin, einer psychotropen Substanz aus Psilocybe Mexicana. Schweiz Arch Neurol Neurochir Psychiatr 87:365–385

Salminen-Vaparanta N, Vanni S, Noreika V, Valiulis V, Móró L, Revonsuo A (2013) Subjective characteristics of TMS-induced phosphenes originating in human V1 and V2. Cereb Cortex 24:2751–2760

Sanders-Bush E, Burris KD, Knoth K (1988) Lysergic acid diethylamide and 2, 5-dimethoxy-4-methylamphetamine are partial agonists at serotonin receptors linked to phosphoinositide hydrolysis. J Pharmacol Exp Ther 246:924–928

Sard H, Kumaran G, Morency C, Roth BL, Toth BA, He P, Shuster L (2005) SAR of psilocybin analogs: discovery of a selective 5-HT 2C agonist. Bioorg Med Chem Lett 15:4555–4559

Saunders NJ (1994) Predators of culture: Jaguar symbolism and Mesoamerican elites. World Archaeol 26:104–117

Schmidt A, Kometer M, Bachmann R, Seifritz E, Vollenweider F (2013) The NMDA antagonist ketamine and the 5-HT agonist psilocybin produce dissociable effects on structural encoding of emotional face expressions. Psychopharmacology 225:227–239

Schreiber R, Brocco M, Audinot V, Gobert A, Veiga S, Millan MJ (1995) (1-(2, 5-dimethoxy-4 iodophenyl)-2-aminopropane)-induced head-twitches in the rat are mediated by

5-hydroxytryptamine (5-HT) 2A receptors: modulation by novel 5-HT2A/2C antagonists, D1 antagonists and 5-HT1A agonists. J Pharmacol Exp Ther 273:101–112

Seidel D (1968) Der Existenzbereich elektrisch und magnetischinduktiv angeregter subjektiver Lichterscheinungen (Phosphene) in Abhängigkeit von äußeren Reizparametern (Schluß). Biomedizinische Technik/Biomedical Engineering 13:208–211

Seitz PFD, Molholm HB (1947) Relation of mental imagery to hallucinations. Arch Neurol Psychiatry 57:469–480

Shanon B (2002) Ayahuasca visualizations a structural typology. J Conscious Stud 9:3–30

Shanon B (2010) The epistemics of ayahuasca visions. Phenomenol Cogn Sci 9:263–280

Shepard GH (2004) A sensory ecology of medicinal plant therapy in two Amazonian societies. Am Anthropol 106:252–266

Siegel RK, Jarvik ME (1975) Drug-induced hallucinations in animals and man. Hallucinations: Behavior, experience and theory, pp 163–195

Spaak E, de Lange FP, Jensen O (2014) Local entrainment of alpha oscillations by visual stimuli causes cyclic modulation of perception. J Neurosci 34:3536–3544

Spencer KM, Nestor PG, Perlmutter R, Niznikiewicz MA, Klump MC, Frumin M, Shenton ME, McCarley RW (2004) Neural synchrony indexes disordered perception and cognition in schizophrenia. Proc Natl Acad Sci USA 101:17288–17293

Stewart L, Walsh V, Rothwell J (2001) Motor and phosphene thresholds: a transcranial magnetic stimulation correlation study. Neuropsychologia 39:415–419

Studerus E, Gamma A, Vollenweider FX (2010) Psychometric evaluation of the altered states of consciousness rating scale (OAV). PLoS ONE 5:e12412

Studerus E, Kometer M, Hasler F, Vollenweider FX (2011) Acute, subacute and long-term subjective effects of psilocybin in healthy humans: a pooled analysis of experimental studies. J Psychopharmacol 25:1434–1452

Studerus E, Gamma A, Kometer M, Vollenweider FX (2012) Prediction of psilocybin response in healthy volunteers. PLoS ONE 7:e30800

Taylor JP, Firbank M, Barnett N, Pearce S, Livingstone A, Mosimann U, Eyre J, McKeith IG, O'Brien JT (2011) Visual hallucinations in dementia with Lewy bodies: transcranial magnetic stimulation study. Br J Psychiatry 199:492–500

Thompson SP (1910) A physiological effect of an alternating magnetic field. Proc Royal Soc Lon Ser B (Containing Papers of a Biological Character) 82:396–398

Thut G, Nietzel A, Brandt SA, Pascual-Leone A (2006) Alpha-band electroencephalographic activity over occipital cortex indexes visuospatial attention bias and predicts visual target detection. J Neurosci 26:9494–9502

Tiihonen J, Hari R, Naukkarinen H, Rimón R, Jousmäki V, Kajola M (1992) Modified activity of the human auditory cortex during auditory hallucinations. Am J Psychiatry 149:255–257

Turing AM (1952) The chemical basis of morphogenesis. Philos Trans R Soc Lond B Biol Sci 237:37–72

van Dijk H, Schoffelen JM, Oostenveld R, Jensen O (2008) Prestimulus oscillatory activity in the alpha band predicts visual discrimination ability. J Neurosci 28:1816–1823

Van Rullen R, Zoefel B, Ilhan B (2014) On the cyclic nature of perception in vision versus audition. Philos Trans Royal Soc B Biol Sci 369:20130214

Vollenweider FX, Kometer M (2010) The neurobiology of psychedelic drugs: implications for the treatment of mood disorders. Nat Rev Neurosci 11:642–651

Vollenweider FX, Vollenweider-Scherpenhuyzen MF, Bäbler A, Vogel H, Hell D (1998) Psilocybin induces schizophrenia-like psychosis in humans via a serotonin-2 agonist action. NeuroReport 9:3897–3902

Watakabe A, Komatsu Y, Sadakane O, Shimegi S, Takahata T, Higo N, Tochitani S, Hashikawa T, Naito T, Osaki H, Sakamoto H, Okamoto M, Ishikawa A, Hara S, Akasaki T, Sato H, Yamamori T (2009) Enriched expression of serotonin 1B and 2A receptor genes in macaque visual cortex and their bidirectional modulatory effects on neuronal responses. Cereb Cortex 19:1915–1928

Wilson HR, Cowan JD (1973) A mathematical theory of the functional dynamics of cortical and thalamic nervous tissue. Kybernetik 13:55–80

Winkelman M (2002) Shamanism as neurotheology and evolutionary psychology. Am Behav Sci 45:1875–1887

Wittmann M, Carter O, Hasler F, Cahn BR, Grimberg U, Spring P, Hell D, Flohr H, Vollenweider FX (2007) Effects of psilocybin on time perception and temporal control of behaviour in humans. J Psychopharmacol 21:50–64

Zanto TP, Rubens MT, Thangavel A, Gazzaley A (2011) Causal role of the prefrontal cortex in top-down modulation of visual processing and working memory. Nat Neurosci 14:656–661

Zhang C, Marek GJ (2008) AMPA receptor involvement in 5-hydroxytryptamine$_{2A}$ receptor-mediated pre-frontal cortical excitatory synaptic currents and DOI-induced head shakes. Prog Neuropsychopharmacol Biol Psychiatry 32:62–71

Zumer JM, Scheeringa R, Schoffelen J-M, Norris DG, Jensen O (2014) Occipital alpha activity during stimulus processing gates the information flow to object-selective cortex. PLoS Biol 12: e1001965

New World Tryptamine Hallucinogens and the Neuroscience of Ayahuasca

Dennis McKenna and Jordi Riba

Abstract New World indigenous peoples are noted for their sophisticated use of psychedelic plants in shamanic and ethnomedical practices. The use of psychedelic plant preparations among New World tribes is far more prevalent than in the Old World. Yet, although these preparations are botanically diverse, almost all are chemically similar in that their active principles are tryptamine derivatives, either DMT or related constituents. Part 1 of this paper provides an ethnopharmacological overview of the major tryptamine-containing New World hallucinogens. Part 2 focuses on ayahuasca and its effects on the human brain. Using complementary neurophysiological and neuroimaging techniques, we have identified brain areas involved in the cognitive effects induced by this complex botanical preparation. Initial SPECT data showed that ayahuasca modulated activity in higher order association areas of the brain. Increased blood perfusion was observed mainly in anterior brain regions encompassing the frontomedial and anterior cingulate cortices of the frontal lobes, and in the medial regions of the temporal lobes. On the other hand, applying spectral analysis and source location techniques to cortical electrical signals, we found changes in neuronal activity that predominated in more posterior sensory-selective areas of the brain. Now, using functional connectivity analysis of brain oscillations we have been able to reconcile these seemingly contradictory findings. By measuring transfer entropy, a metric based on information theory, we have shown that ayahuasca temporarily modifies the ordinary

D. McKenna
Director of Ethnopharmacology, Heffter Research Institute,
Santa Fe, NM, USA
e-mail: mcken031@umn.edu

J. Riba (✉)
Human Neuropsychopharmacology Research Group, Sant Pau Institute of Biomedical Research (IIB-Sant Pau), C/Sant Antoni María Claret, 167, 08025 Barcelona, Spain
e-mail: jriba@santpau.cat

J. Riba
Department of Pharmacology and Therapeutics, Universitat Autònoma de Barcelona (UAB), Barcelona, Spain

Curr Topics Behav Neurosci (2016) 36: 283–312
DOI 10.1007/7854_2016_472
Published Online: 02 April 2016

flow of information within the brain. We propose a model in which ayahuasca reduces top-down constraints and facilitates bottom-up information transfer. By simultaneously enhancing endogenous cortical excitability and reducing higher-order cognitive control, ayahuasca temporarily disrupts neural hierarchies allowing inner exploration and a new outlook on reality.

Keywords Tryptamine derivatives · Tryptamine hallucinogen · Hallucinogenic · New World · Shamanism · Botany · Chemistry · Ethnopharmacology

Contents

1 Botany, Chemistry, and Ethnopharmacology of New World Tryptamine Hallucinogens

1.1 Introduction

The indigenous cultures of the New World are infinitely more sophisticated than their Old World counterparts in their knowledge and utilization of vision-producing, or hallucinogenic, plants. Fewer than 100 genera have been identified as “major hallucinogens,” meaning that they form an important component of ethnomedical and ritual practices in one or more indigenous societies; fully sixty genera are used by New World aboriginal peoples, while only a dozen or so are utilized in Old World indigenous cultures (Schultes 1970a, b). This curious asymmetry in the ethnographic utilization of visionary plants has never been satisfactorily explained. One difficulty is that documentation of the antiquity of hallucinogenic plant use is drawn from sources that are much more recent than the actual practices, and archeological evidence, in the form of plant remains or artifacts used in drug rites, is rare to nonexistent. *Amanita muscaria*, the Fly Agaric mushroom, is at the root of shamanism, which has its origins among the Finno-Ugric peoples who are thought to have settled west of the Urals sometime after the end of the last Ice Age, between 8000 and 4200 BP. Linguistic analysis of Northern European languages indicates that the Uralic languages split into two branches about 4000 BP. Both of these branches contain similar words for "inebriation," and in some of these languages, the root "pang" signifies both intoxication and the *Amanita muscaria* mushroom. This evidence suggests, but does not prove, that *Amanita muscaria* was known to be an intoxicant prior to the linguistic split (Wasson 1968). There is evidence that earlier settlements existed in the region before the end of the last Ice Age and thus are more than 10,000 years old. There is no information on whether these cultures also used Amanita, and whether they had any cultural relation to the Finno-Ugric cultures. The dates for the ingression of Siberian or northern Asian people into the Americas are uncertain and cover a wide range. Discovery of human-worked mammoth bone artifacts in Bluefish Cave in the Canadian Yukon has been dated to 25,000–40,000 BP; this date currently stands as the best documented evidence of the earliest human habitation in the New World (Cinq-Mars and Morlan 1999).

It is likely, but impossible to prove, that the earliest immigrants into the North American continent may have brought with them both a tradition and a technology of psychoactive plant utilization; as these populations gradually expanded southward into the fecund rainforest ecosystems of Central and South America, they would have encountered an increasingly unfamiliar and biodiverse flora. Their penetration into these novel ecosystems, coupled with innate human curiosity, probably vastly expanded their access to plants utilized as food, as well as to floras rich in biodynamic plant species that included a greater variety of hallucinogens. Whether such migratory peoples already practiced shamanic traditions aided by divine intoxication with them, or discovered a rich new pallet of psychoactive species that fostered the development of purely New World practices, will probably never be known. What does seem certain, however, is that the innate human drive to access shamanic dimensions reached its fullest symbiotic expression in the tryptamine-saturated tropical rain forests of the New World. The legacy of that Paleolithic application of human ingenuity and curiosity to empirical psychopharmacological experimentation in an ecology of incredible chemical diversity persists to the present day in the many traditions involving the use of psychoactive plants that can still be found among New World aboriginal peoples.

While the ingenuity displayed by the New World Indians in discovering and utilizing psychoactive plants drawn from numerous families and genera is remarkable, perhaps equally remarkable is the fact that in most instances the active principles responsible for their psychoactive properties can be traced to chemical compounds known as tryptamines. With only three exceptions—the peyote cactus of North America and Mexico, the morning glories (*ololiuqui*) of Central America, and the columnar San Pedro cacti of the Andes—almost all of the "major" New World hallucinogens are derived from plants containing tryptamine derivatives. This section of this chapter presents an ethnopharmacologic overview of tryptamine-containing New World hallucinogens, including their botanical sources, chemistry, and pharmacology, and their geographic and ethnographic distribution.

1.2 Chemistry, Botany, and Pharmacology

Tryptamine derivatives are simple indole alkaloids, derived biosynthetically from tryptophan, an essential amino acid that is universally distributed in all plants and animals (although many animals, including man, cannot synthesize tryptophan de novo and must obtain it from dietary sources, which is why it is considered an "essential" amino acid). Decarboxylation of tryptophan by aromatic amino acid decarboxylase, an enzyme fundamental to basic metabolic processes in plants and animals, yields tryptamine, the structurally simplest of the tryptamine derivatives (Fig. 1).

Hydroxylation of tryptamine at position 5 on the indole ring yields 5-hydroxytryptamine (5-HT), also known as serotonin. Serotonin is widely distributed in plants where it functions as a defensive, irritant compound, e.g., in the leaves of nettles (*Urtica* spp.). Serotonin is also a major central nervous system neurotransmitter, and most hallucinogenic drugs are thought to be agonists at 5-HT_{2A} receptors, one of about fourteen subtypes of serotonin receptors. Tryptamine itself is not psychoactive, nor is serotonin, apart from a mild sedative effect; similar sedating and tranquilizing effects have been ascribed to tryptophan itself, undoubtedly because it is a precursor to serotonin in the central nervous system. A trivial chemical modification of tryptamine, viz. the addition of two

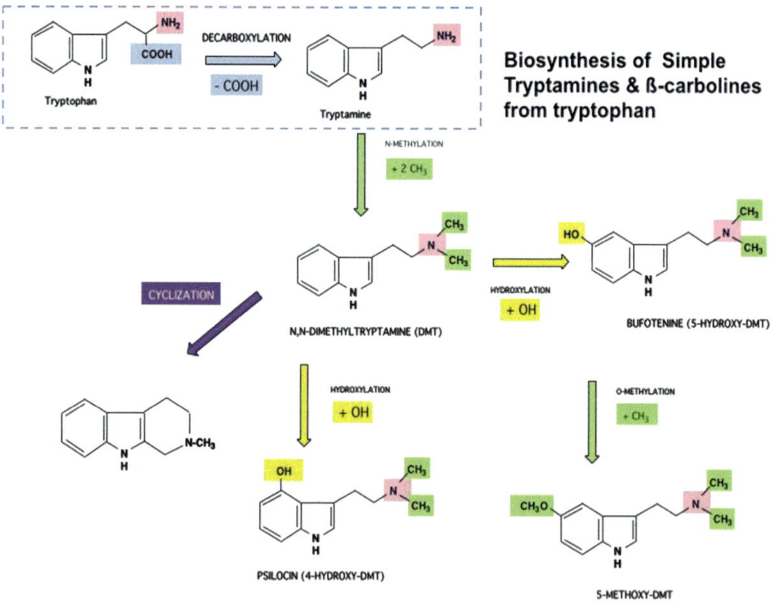

Fig. 1 Biosynthesis of simple tryptamines and ß-carbolines from tryptophan

methyl (CH_3) groups to the side chain nitrogen, yields the simplest of the hallucinogenic tryptamines, *N*,*N*-dimethyltryptamine (DMT). A closely related compound, bufotenine (5-hydroxy-*N*,*N*-dimethyltryptamine, 5-OH-DMT), can be derived from serotonin by a similar methylation reaction; bufotenine is so named because it is found in the skins of certain toads belonging to the genus *Bufo*. There is some controversy as to whether bufotenine is hallucinogenic or otherwise psychoactive, as it putatively does not cross the blood–brain barrier and purportedly produces only peripheral autonomic effects. However, careful self-bioassay experiments by Ott (2001) have established definitively that bufotenine is indeed psychoactive, although its activity is critically affected by dose and route of administration, as well as the chemical form of the alkaloid (salt or free base). Two further minor modifications of bufotenine can result in potent hallucinogens. *O*-Methylation of the hydroxy group of bufotenine yields 5-methoxy-*N*,*N*-dimethyltryptamine (5-MeO-DMT), a compound similar to DMT in that it is a short-acting hallucinogen, but that is approximately 10 times more potent, on a per weight basis. The 4-hydroxy regioisomer of bufotenine is psilocin (4-hydroxy-DMT), which is the active principle of hallucinogenic "magic" mushrooms, first described from Mexico. This compound is derived from psilocybin, which is simply the 4-phosphoryl ester of psilocin; psilocybin is converted to psilocin in the body and it is psilocin that is thought to be the active form of psilocybin. This completes the list of naturally occurring psychoactive tryptamine derivatives; two other classes of psychoactive indole compounds, however, are closely related to tryptamine derivatives and deserve to be mentioned here. One of these is the ß-carbolines, which, like tryptamines, are structurally simple and widely distributed in the plant kingdom. ß-carbolines have a tricyclic structure, with the tryptamine side chain incorporated into a third heterocyclic ring, while tryptamine derivatives have only two fused rings that comprise the indole nucleus. ß-carbolines are derived biosynthetically from tryptamine or other simple tryptamines and often are present in the same plants that contain hallucinogenic tryptamines. ß-carbolines can be important for the pharmacology of tryptamines, as will be discussed below. Lysergic acid derivatives are another class of naturally occurring psychoactive indoles that can be regarded, in some sense, as complex tryptamine derivatives. The extraordinarily potent hallucinogen LSD is a semisynthetic member of this class, but does not itself occur in nature. Other, less potent but definitely psychoactive lysergic acid derivatives are found in plants, notably the sacred morning glories of Mexico and Central America (members of the genera *Ipomoea*, *Rivea*, and *Turbina*) as well as in the well-known ergot fungus (*Claviceps purpurea* and other *Claviceps* spp.). Other than ergot fungi and the morning glory family (Convolvulaceae), however, lysergic acid derivatives are rare in nature and will not be discussed here further, since they do not strictly conform to the definition of simple tryptamine derivatives, that is, they are not indolealkylamines (Schultes and Hofmann 1981).

The simple hallucinogenic tryptamine derivatives, by which is meant here DMT and its derivatives (5-MeO-DMT, psilocin, psilocybin, and bufotenine), are widely distributed in nature, occurring in animals as well as plants and fungi (Smith 1977). Tryptamine derivatives have been reported from over 26 higher plant families; the families containing hallucinogenic tryptamine derivatives include the Aizoaceae, Apocynaceae, Poaceae, Fabaceae, Malpighiaceae, Myristicaceae, Pandanaceae, Rubiaceae, Rutaceae, and Urticaceae. Not all of the genera in which these compounds occur are used in shamanic traditions; in fact, it may be argued that the majority of them are not, since their hallucinogenic components were comparatively recent discoveries of modern science, and they were either overlooked or rejected by aboriginal psychopharmacologists. In fact, based on the known numbers of species in each genus reported to contain psychoactive tryptamines, these compounds are potentially present in over 4860 higher plant species! Since DMT is only two biosynthetic steps from tryptophan, these numbers are probably a serious underestimate. Psilocybin and psilocin are apparently restricted to higher fungi and are found primarily in mushrooms of the genera *Psilocybe*, *Stropharia*, and *Panaeolus*, although they have also been reported from other genera of basidiomycetes as well. These compounds have not been reported from any higher plant, although from a biosynthetic standpoint there is no a priori reason why they could not occur in a higher plant. Bufotenine and other tryptamine derivatives are principle constituents of the parotid gland secretions of New World toads belonging to the genus *Bufo*, for which the compound was named when first isolated. Curiously, neither DMT nor 5-MeO-DMT has been reported from any *Bufo* species, with a single exception: *Bufo alvarius*, which contains 5-MeO-DMT; rather staggering concentrations of up to 10% dry weight of the parotid gland have been reported by some investigators (Daly and Witkop 1971; Weil and Davis 1994).

The hallucinogenic tryptamines also display a unique pharmacology that bears importantly on their methods of utilization in the context of New World shamanic traditions. DMT and 5-MeO-DMT are both potent, extremely short-acting hallucinogens whose total duration of action is less than thirty minutes from "baseline" to "baseline." Neither of these derivatives is orally active, due to degradation by the enzyme monoamine oxidase (MAO), which is present in the liver and gut, as well as in the brain, of humans and other mammals. Thus, in order to experience the hallucinogenic effects of these compounds, they must be taken parenterally as a snuff or enema (although synthetic DMT can also be injected or smoked). If orally ingested, they must be protected from peripheral degradation by a monoamine oxidase inhibitor (MAOI). Traditional shamanic practitioners have ingeniously employed all three of these strategies. In contrast, psilocin, the active tryptamine in the "magic" mushrooms, is orally active and elicits an intense psychedelic experience lasting four to six hours. Psilocin is orally active by itself and does not require an MAOI for activation; as a result, no special preparation is required to experience the effects of hallucinogenic mushrooms. They are quite active when collected and eaten fresh, although they are often dried or stored in honey; these practices are a means of preservation, rather than a specific attempt to alter or activate the pharmacology. Psilocybin is readily converted to psilocin in the body and is considered to be a "pro-drug," i.e., the physiologically active form is psilocin.

1.3 Survey of New World Tryptamine Hallucinogens

1.3.1 Psilocybe Mushrooms

Just as DMT can be regarded, in some respects, as the prototype hallucinogen, the so-called magic mushrooms of Mexico qualify as the prototype of the New World hallucinogens. The Spanish conquistadores encountered a flourishing mycolatrous religion among the Aztecs at the time of the arrival of Cortés in the court of Moctezuma in 1519. Wasson (1980) has adduced evidence that the use of the magic mushrooms was spread throughout Mesoamerica at this time, and in some parts of the region, the practice may have dated well before the Christian era. Based on linguistic analyses of lexicons compiled by the Spanish missionaries, Wasson has presented convincing evidence that the inebriating mushrooms were known, not only to the Aztecs and Maya, but also to the Nahua, Otomi, Matlatzinca, Mazahua, Tarascan, Huastecan, Totomac, Mixe, Zoque, Mazatec, Zapotec, Chatino, Mixetec, and Chinantla linguistic groups. The discovery of "mushroom stones," carved effigies in the shape of mushrooms from the highlands of Guatemala, lends additional credence to his arguments. Some of these have been dated to the sixth century BC, indicating the considerable antiquity of the Mesoamerican mycolatry (Borhegyi 1961).

The Spanish missionaries, following close upon the coattails of Cortés, regarded the Aztec's ritual and religious use of *teonanácatl*, a name erroneously translated as "god's flesh" by one of their number, one Motolinía, as a particularly odious and blasphemous parody of the Christian Eucharist. They lost no time in pursuing the vigorous suppression of *teonanácatl* in all of its diabolical manifestations and, backed by the considerable intimidatory powers of the Spanish Inquisition, succeeded in driving the practice underground for the next 400 years (Wasson 1980). Despite the most energetic efforts to stamp out the use of the inebriating mushrooms in Mesoamerica, their employment in a religious and ritual context persisted into the twentieth century, when it was rediscovered and made known to the world by the famous team of R. Gordon and Valentina Wasson, in an article published in Life magazine in May, 1957 (Wasson and Wasson 1957). The repercussions of that event are still being felt, as this was really the first representation in popular media that such peculiar agents as psychedelic drugs even existed (LSD and peyote had been known for some decades but were familiar only to a few psychiatrists and literati). The Swiss chemist Albert Hofmann, the discoverer of LSD, shortly succeeded in isolating and characterizing the active principles of the magic mushrooms, psilocybin and psilocin, from material Wasson had brought to him from Mexico, thus marking yet another milestone in the history of psychedelic psychopharmacology (Heim and Hofmann 1958). In the early 1960s, the availability of these compounds from Sandoz for research purposes piqued the curiosity of one young Harvard psychologist, Dr. Timothy Leary, and the revelations he gained after a number of self-experiments with these novel substances led him to found the psychedelic movement that would shortly sweep North America and the world. Thus, it happened that a persecuted and reviled substance that had been at the heart of Mesoamerican religion since prehistoric times began a second career in the twentieth century. The possible clinical uses of psychedelics are only now being rediscovered after a hiatus of some forty years following their blanket prohibition in 1970. Of the many compounds that show promise for therapeutic use, psilocybin is one of the most promising; its lack of toxicity, short duration of action, and profound yet manageable psychedelic effects make it ideal for use in clinical settings.

The magic mushrooms encountered and collected by Wasson and his colleague, French mycologist Roger Heim, comprised about a dozen members of the genera *Psilocybe*, *Panaeolus*, and *Stropharia*, and were for the most part strict endemics, native only to the highlands of Central America. Subsequent work by ethnomycologists has shown that other species of psilocybin-containing mushrooms have a cosmopolitan and global distribution; in the tropics, this is exemplified by *Psilocybe cubensis* (formerly classified as *Stropharia cubensis*, one of the species collected by Wasson and Heim), while in temperate regions, the diminutive "Liberty Cap" mushroom, *Psilocybe semilanceata*, can be found in grassy meadows throughout North America and Europe. Guzmán et al. (2000) published a comprehensive list of the 186 known psilocybin-containing species in an Italian journal; the list can be found on the Erowid.org database (Erowid.org 2001a, b). The majority of the known species belong to the genus *Psilocybe*, but other genera reported to contain psilocybin include *Agrocybe*, *Conocybe*, *Copelandia*, *Panaeolina*, *Panaeolus*, *Galerina*, *Gymnopilus*, *Inocybe*, *Pluteus*, *Hypholoma*, *Gerronema*, and *Mycena*. Besides Mexico and Central America, another part of the New World that is particularly rich in psilocybin species is the Pacific Northwest of North America, where over 30 species are endemic. Curiously, although the recreational use of some of these species has become a popular pastime in recent years, there is no record that these mushrooms or their properties were known to any of the aboriginal groups who inhabited the Northwest Coast. These groups, including the Haida, Tlingit, Tsiamsham, and Salish, have never admitted knowledge of these species to ethnographers, although they possess a strong shamanic tradition and much of their traditional art could be characterized as "visionary" in nature. There is similarly no definitive evidence that psilocybin mushrooms had a place in shamanic practices in South America, although Schultes and Hofmann (2001) speculate that seventeenth-century reports by Jesuit missionaries referring to the use of a "tree fungus" for preparation of an intoxicating beverage by the Yurimaguas of the Peruvian Amazon may have referred to *Psilocybe yugensis*, a wood-growing species. In the same volume, they also mention the so-called telephone bell gods, anthropomorphic gold pectorals with dome-shaped ornaments on the head. These artifacts are reported from the Sinú region of northern Colombia and from the Calima region on the Pacific coast. The hemispherical ornaments, complete with a stem or stipe, are strongly suggestive of a mushroom effigy. Similar artifacts have been reported from Panama, Costa Rica, and Yucatan, suggesting that the prehistoric Mesoamerican mushroom cults may have extended as far south as modern-day Colombia. In any case, there is no evidence for contemporary use of psilocybin mushrooms in indigenous shamanic practices in South America.

1.3.2 *Anadenanthera* Snuffs

Now let us shift our focus southward, to the tryptamine-based hallucinogens that are endemic to the South American continent, for it is here that such plants and the sophisticated technologies used in employing them reached their fullest expression.

Among the many medicinal plants that Columbus encountered in his earliest visits to the New World, the intoxicating snuff prepared from the seeds of the Fabaceous tree *Anadenanthera peregrina* (formerly *Piptadenia peregrina*) may be considered the paradigm of hallucinogenic New World snuffs. While presumably Columbus observed the use of the snuff powder on his initial voyage, it was not until his second landing in the New World that he commissioned Friar Ramón Pané to undertake an ethnographic documentation of the use of *cohoba* (or *cogioba*), as it was known to the Taino people, the indigenous inhabitants of what is now Haiti and the Dominican Republic:

> The cogioba is a certain powder which they take sometimes to purge themselves, and for other effects which you will hear of later. They take it with a cane about a foot long and put one end in the nose and the other in the powder, and in this manner they draw it into themselves through the nose and this purges them thoroughly… [the *bohuti*, physician] takes a certain powder called *cohoba* snuffing it up his nose which intoxicates them so that they do not know what they do and in this condition they speak many things incoherently in which they say they are talking with the *cemis* and that by them they are informed how the sickness came upon them… (Wassén 1964).

Some scholars have asserted that the Taino word *cohoba* stood for tobacco, but Wassén cites evidence that the words *cohoba*, *cohobha*, *cahoba*, *cojoba-cogioba*, *cojioba*, *cohiba*, and *coiba* are equivalent and refer to a plant that was used by the medicine men to induce a state of trance, and that this was recognized as distinct from tobacco by the early chroniclers. To confuse matters further, the powdered *Anadenanthera* seeds may have been mixed with tobacco, at least on some occasions; the Jirara and Caquetio tribes of Venezuela, considered closely related to the West Indian Taino, commonly employ a mixture of tobacco and *Anadenanthera* snuff. The practice is also widespread among other snuff-using South American tribes (Wassén 1967).

Possibly as a result of this confusion of the two plants by the explorers in Columbus' party, as well as the rather more spectacular method of smoking tobacco that he observed among the peoples of the Antilles on his first voyage in 1492, Columbus took tobacco seeds back to Europe, but neglected to take seeds of *Anadenanthera*. Tobacco smoking quickly became a popular custom and diffused into Spanish society, and "within a few decades, there were more Spaniards converted to smoking than Indians converted to Christianity" (Emboden 1979). One is tempted to speculate how different our contemporary civilization might be had Columbus returned with the seeds of *cohoba*, rather than tobacco!

More modern ethnographic investigations have shown that the historical use of *cohoba* in the West Indies marks the easternmost boundary of the custom. The *Anadenanthera peregrina* used in that region were probably introduced cultivars from the South American mainland; the center of concentration of the species is the Orinoco valley of Colombia, Venezuela, and adjacent parts of Brazil, where it is known as *yopo*, or *niopo*. Archeological evidence in the form of carved snuff trays and snuffing tubes has placed the practice as far north as Costa Rica (Wassén 1967).

Further to the south, in the Atacama desert, another snuff, known as *vilca* or *huilca* in Peru and Bolivia and *cébil* in Northern Argentina, was similarly prepared from a different *Anadenanthera* species native to this region, *A. colubrina*. Many well-preserved grave sites containing snuffing implements, including carved snuff trays, snuffing tubes, and woven bags containing snuff powders, have been excavated in this area and dated to as early as 570 AD. A chemical and contextual analysis of the powders and implements recovered from these sites established the presence of tryptamine alkaloids in the snuff powders, thus confirming their identity as derived from *Anadenanthera* species (Torres et al. 1992). This evidence, combined with the archeological documentation of the antiquity of this practice in this region, has raised questions as to the true geographic origins of the practice, since it predates by approximately 1000 years any similar documented use of *Anadenanthera* snuffs in the Orinoco valley or the West Indies. It has been generally assumed among archeologists and ethnographers that the practice originated in the Orinoco valley and from there diffused north to Central America, east to the Antilles, and south to coastal Peru and Chile. This more recent evidence, however, suggests the possibility that the practice may have originated in the Atacama region and diffused north and eastward into the Orinoco basin, where the closely related native species, *A. peregrina*, was substituted for the more endemic southerly species, *A. colubrina*.

The excavation of paraphernalia, specifically snuff trays and snuffing tubes, enables estimation of the antiquity of the use of *Anadenanthera* snuffs. The oldest known snuffing implements have been dated to 1200 BC from an excavation by Junius Bird from the site of Huaca Prieta in the Chicama Valley on the central Peruvian coast (Torres 1995).

The primary indole alkaloid constituents of *Anadenanthera* species have been exhaustively reviewed by Torres and Repke (2006). The most abundant alkaloids reported from *A. peregrina* and *A. colubrina* are bufotenine, DMT, and 5-MeO-DMT. However, trace concentrations of a number of structurally related alkaloids, including *N*-methyltryptamine (NMT), 5-methoxy-*N*-methyltryptamine (5-MeO-NMT) DMT-*N*-oxide, serotonin, *N*-methylserotonin, bufotenine-*N*-oxide, and three ß-carbolines (2-methyl-tetrahydro-ß-carboline, 2-methyl-6-methoxy-tetrahydro-ß-carboline, and 1,2-dimethyl-6-methoxy-tetrahydro-ß-carboline) have been detected.

1.3.3 *Virola* Snuffs and Pastes

Ethnographers had long assumed that *Anadenanthera* species comprised the sole botanical source of psychotomimetic snuffs in use among indigenous peoples in the Amazon Basin, apart from tobacco snuff and the occasional use of coca as a snuff (Cooper 1949). However, subsequent investigations by ethnobotanist R.E. Schultes and toxicologist Bo Holmstedt in the 1950s and 1960s established that the use of *Anadenanthera* snuffs was less prevalent than formerly thought and that tribes belonging to the Waika groups in the upper Orinoco valley prepare an intoxicating snuff from the resin (sap) of several species of the genus *Virola*, in the Myristicaceae, or nutmeg family (Schultes and Holmstedt 1968). In addition to the Orinoco valley, the use of *Virola* snuffs is concentrated in the Colombian Vaupés and north of the Rio Negro in Brazil. Some overlap in the native nomenclature has undoubtedly contributed to the confusion regarding the distinction between *Virola* and *Anadenanthera* snuffs. Terms vary in different tribes, but *Virola* snuff is known as *yá-kee*, *yá-to*, and *paricá* in Colombia and Venezuela, and *epéna*, *ebene*, *paricá*, and *nyakwána* in Brasil. However, *paricá* may also refer to *Anadenanthera* snuff, and *epéna*, or *ebene* can be used as a general term for snuff (Schultes 1970a, b).

The principle species implicated in the preparation of hallucinogenic snuffs in the Colombian Amazon are *V. calophylla* and *V. calophylloidea*, while *V. theiodora* and *V. elongata* are the species utilized among the Waika, Paumari, and Taiwanos. There is considerable taxonomic confusion in the genus, however, and *V. theiodora* and *V. elongata* are regarded as equivalent by some taxonomists. *Virola cuspidata* and *V. rufula* have also been reported as snuffs. Occasionally, ashes or powdered leaves of other plants are used as admixtures to the snuffs. Among the Waika, one commonly employed admixture is the aromatic herb, *Justica pectoralis* var. *stenophylla*, which also is occasionally used as the sole ingredient of a snuff. Earlier reports of tryptamines in *Justicia* are apparently erroneous, although the plant does contain umbelliferone and other coumarins (Macrae and Towers 1984a). Interestingly, the resin of *Virola* species among the Waika is also occasionally used in the preparation of an arrow poison that is applied to darts used in hunting small animals. MacRae and Towers (1984b) investigated the possible mechanisms contributing to this activity in animal experiments. They found that, in their assays, extracts containing the tryptamine alkaloids were not highly toxic and did not interfere markedly with locomotion or motor activity. They isolated an alkaloid-free fraction containing lignans, however, and found that this fraction produced a marked inhibition of motor activity and apparent sedation of the test animals. They concluded that the lignans, rather than the tryptamines, were likely the agents responsible for the effectiveness of *Virola* arrow poison.

The reddish, resinous exudate of the inner cambial layer of the *Virola* species used as snuffs contains high concentrations of tryptamine alkaloids, including NMT, 5-MeO-NMT, DMT, and 5-MeO-DMT, of which the latter usually predominates, and often may be the sole constituent. Traces of ß-carbolines, including 2-methyl-1,2,3,4-tetrahydro-ß-carboline, 6-methoxy-2-methyl-1,2,3,4-tetrahydro-ß-carboline, and 6-methoxy-1,2-dimethyl-1,2,3,4-tetrahydro-ß-carboline have also been reported in some species (Holmstedt and Lindgren 1967). Both the source plants and the snuffs prepared from them exhibit considerable chemical variation.

In the Colombian Putumayo, marked on the north by the Rio Igaraparaná, and on the south by the Rio Yaguasyacu and Ampiyacu, the Bora, Witoto, and Muinane prepare an orally active hallucinogen from the resin of *Virola theiodora*, *V. elongata*, and *V. pavonis*(Schultes 1969). In this practice, the drug, which is known as *oo-koo-he* among the Witoto and *kú-ru-ku* among the Bora, is prepared by stripping the bark, collecting the resin, and concentrating it to a thick, syrupy consistency. This is then mixed with the ashes of other plants (usually *Gustavia peoppigiana* or *Theobroma* spp.) and rolled into pellets or boluses. Oral ingestion of two or three of these pellets is said to induce a rapid and violent intoxication, an effect that the author was able to confirm in self-experiments during fieldwork on the Rio Ampiyacu (McKenna et al. 1984a). Since the psychotomimetic tryptamines DMT and 5-MeO-DMT, the major alkaloids in the preparation, are not orally active unless activated by a monoamine oxidase inhibitor (MAOI), the documented oral activity of these *Virola* pellets raises some interesting pharmacological questions. Phytochemical analyses have shown that ß-carbolines, while acting as potent MAO inhibitors, are only present in trace concentrations in the *Virola* pellets and hence are unlikely to have any pharmacological significance. In vitro assay of extracts prepared from the pellets showed a somewhat weak MAOI activity, which moreover was shown to be due to the tryptamines alone; extracts from which the tryptamines were removed did not display any MAOI activity. At present, the pharmacological basis for the oral activity of the *Virola* pellets remains incompletely elucidated (McKenna et al. 1984a).

1.3.4 *Ayahuasca*

The orally active *Virola* pellets always had a restricted ethnographic distribution and the practice is now verging on extinction as a result of encroaching acculturation among the tribes that formerly used it. In contrast, the hallucinogenic beverage known variously as *ayahuasca*, *yagé*, *caapi*, *natema*, or *hoasca*, is the premier hallucinogen of the Amazon, and its use, far from dying out, is rapidly diffusing from aboriginal and mestizo society into mainstream South American (and global) culture. Like the *Virola* pellets, *ayahuasca* is also an orally active tryptamine-based hallucinogen, but its mechanism of action is relatively well understood.

Ayahuasca is prepared by boiling the bark or crushed stems of a Malpighiaceous jungle liana, *Banisteriopsis caapi*, together with various admixture plants, especially the leaves of *Psychotria viridis*, a member of the Rubiaceae. In the Colombian Putamayo and parts of Ecuador, the leaves of *Diplopterys cabrerana* (formerly classified as *Banisteriopsis rusbyana*) (Gates 1979), a liana in the same family as *Banisteriopsis*, are often substituted for those of *Psychotria viridis*. It is the admixture plants, *Psychotria* or *Diplopterys*, that contain the hallucinogenic alkaloid necessary for the activity; the leaves of both species contain substantial concentrations of DMT (Der Marderosian et al. 1968; Pinkley 1969). The *Banisteriopsis* liana, on the other hand, contains high concentrations of ß-carboline alkaloids, primarily harmine and tetrahydroharmine, with lesser amounts of harmaline (Rivier and Lindgren 1972; Schultes and Hofmann 1981). These compounds are potent peripheral MAO inhibitors, and it is the combination of DMT in the admixtures and the MAO-inhibiting ß-carbolines that provide the mechanism for the oral activity of this drink. The ß-carbolines are able to protect the DMT from degradation in the liver and gut, thus enabling it to cross the blood–brain barrier intact and exert its effect in the central nervous system (McKenna et al. 1984b).

Unlike the hallucinogenic snuffs or the *Virola* pellets, the custom of using *ayahuasca* has a widespread distribution among aboriginal groups in the Amazon, including the Guahibo, Jivaro, Colorado, Ingano, Siona, Kofan, Witoto, Tukano, Desana, Yakuna, and more than 20 others. In view of this widespread use, it is not surprising that the practice has diffused into mestizo society; in Peru and parts of Colombia and Ecuador, *ayahuasca* (or *yagé*, as it is known in the Colombian Putamayo) occupies a central position in the ethnomedical armamentarium of mestizo shamans. These practitioners consume the beverage themselves as a diagnostic and divinatory tool and also administer it to their patients as a panacea reliably able to cleanse both the body (it is often referred to as "la purga") and spirit. Regular consumption of *ayahuasca*, along with a special diet, sexual abstinence, and ingestion of other medicinal plants, also constitutes an essential part of shamanic training for a mestizo healer. Thus, *ayahuasca* is the primary "teacher" enabling the apprentice medicine man to learn about the curative properties of other plants (often by consuming them in the form of admixtures to *ayahuasca*), which are also conceived of as "plant teachers." (Luna 1984; McKenna et al. 1995). It is also through the medium of *ayahuasca* that the shaman acquires his "icaros," magical songs that are used in curing, and establishes alliances with his helping spirits, which may be conceived as animals, plants, or spirits (Luna 1984).

The origin of the use of *ayahuasca* by indigenous Amazonian peoples is lost in antiquity, and there is evidence that the practice was already centuries old by the time of the Columbian contact. Unlike snuffs, which leave unambiguous archeological evidence in the form of snuff trays and tubes, *ayahuasca* is consumed as a decoction, and there is no definitive link to ceramic or other vessels that may have been used to consume the beverage. Based on ambiguous evidence, Naranjo (1995) speculates that the earliest use of *ayahuasca* can be placed sometime between 500 BCE and 500 AD.

Whatever its historic context has been, in recent decades the ceremonial use of *ayahuasca* in a religious context has begun to diffuse from mestizo society into a wider cultural milieu. In Brazil, where it is known as *hoasca* or *Daime*, the beverage has become the central sacrament of several syncretic religious movements. The largest and most visible of these is the *Santo Daime* cult, which incorporates many elements of Christian liturgy in their practices and belief systems (Dale 1991) and the *União do Vegetal*, in which a collective spiritualism emphasizing ecology and harmony with nature plays a more prominent role. These cults have burgeoned from a few hundred members to thousands of members within the last two decades. The Brazilian government, recognizing that these are legitimate religious movements and perceiving little or no physical or moral detriment from their use of *ayahuasca*, has officially sanctioned *ayahuasca* by lifting legal restrictions against its sacramental use within a religious context (Erowid.org 2001a, b). In the USA, judicial rulings by the Supreme Court and the US District Court in Oregon have sanctioned the religious use of *ayahuasca* for practicing members of the UDV and Santo Daime churches (Erowid.org 2006; 2012).

1.3.5 *Bufo* Species and *Jurema* (*Mimosa* Species)

All of the New World tryptamine hallucinogens that we have discussed to this point —the psilocybin mushrooms, the *Anadenanthera* snuffs, the *Virola* preparations, and *ayahuasca*—have an extensive history and an association with New World shamanism that is Paleolithic in their origins. Their impact and influence on the cultures that utilized them is abundant and well documented, both in the ethnographic literature and in the art and iconography of the peoples who use the plants.

In the case of two other tryptamine hallucinogens, however, the information on their use in the New World is sparser, and as a result they are all the more fascinating; these are uncharted ethnopharmacological waters.

Bufo Species

It was mentioned above in the section on the distribution of the tryptamines in nature that bufotenine, as its name implies, was first isolated from the venom of toads of the genus *Bufo* and that in at least one instance (*Bufo alvarius*) the potent hallucinogen 5-MeO-DMT was a major ingredient of the venom. While it is true that the toad occupies a prominent position in Mayan, Aztec, and Olmec iconography (Kennedy 1982) and is often depicted together with mushrooms and stylizations of other "sacred" plants, there is no unambiguous proof that toad venom was used as a hallucinogen in Mesoamerica. A major source of controversy has been that the candidate species favored by most ethnographers has involved *Bufo marinus*, which is a highly toxic species that would require a rather sophisticated preparation if it were to be consumed safely. Davis and Weil have extensively reviewed the evidence for the hallucinogenic use of *B. marinus*, and they argue that *Bufo alvarius* is the more likely candidate to have been used, on both pharmacological and ethnographic evidence (Davis and Weil 1992, 1994).

Mimosa Species

Schultes reports that several tribes of eastern Brazil prepared the root of *Mimosa hostilis*, a scrubby, leguminous shrub native to the dry area, into a "miraculous drink" known as *ajuca* or *vinho de jurema*. Early reports of the *jurema* ceremony date back as far as 1788, and the practice is apparently ancient, having been practiced by a number of extinct tribes: Guegue, Acroa, Pimenteira, and Atanayé. A major application of the *jurema* ceremony at one time was apparently to prime the priests and warriors for going off to war (Schultes and Hofmann 1981).

In 1946, Brazilian chemist Oswaldo Gonçalves de Lima isolated a single alkaloid from the roots of a related species, *Mimosa tenuiflora* (*juremapreta*), which he named nigerine. Although this initial isolation was only partially pure, it was subsequently shown to be identical to DMT when Gonçalves de Lima provided a group of American researchers with a sample of *M. tenuiflora* root bark. These workers were able to unequivocally isolate DMT from the sample in 0.57% yield (Pachter et al. 1959). The history of this significant discovery—the first identification of DMT as a naturally occurring alkaloid—has been exhaustively reviewed by Ott (1998). Although as Ott points out, priority for the first **unequivocal** identification of DMT as a natural compound must be given to Fish et al. (1955) who reported DMT in the seeds of *Anadenanthera peregrina* (under its former name, *Piptadenia peregrina*). DMT is known to be orally inactive unless ingested with an MAOI, and in the traditional psychedelic brew *ayahuasca*, the DMT is protected from peripheral degradation by the ß-carboline alkaloids present in the bark of one of the plant components (the liana *Banisteriopsis caapi*). *Vinho de Jurema*, however, is prepared as a beverage without the use of ß-carboline containing admixture plants, and it has long been speculated that there must have been some long-forgotten admixture added to enable its oral activity. Ott (1998), however, conducted careful self-experiments using oral decoctions of *M. tenuiflora* (*Juremapreta*) and reported that it was potently psychoactive without the inclusion of any other admixture plants. Although the mechanism of its oral activity remains a mystery, the mystery may have been partially solved by the recent identification of another alkaloid, yuremamine, by J.C. Callaway and co-workers (Vepsäläinen et al. 2005).

Yuremamine, present in the stem bark of *Mimosa tenuiflora* at concentrations comparable to DMT, contains an unusual structure (Fig. 2) that incorporates the

Fig. 2 Structure of Yuremamine

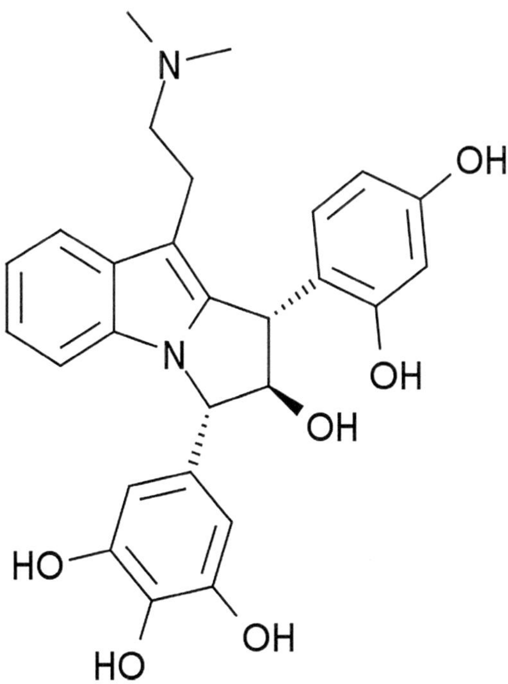

structure of DMT fused with phenolic moieties. The authors suggest that this novel compound may be active as an MAOI, and if confirmed this would account for the oral activity of this unusual preparation. Alternatively, cleavage of the D ring of the yuremamine molecule could free the DMT "caged" in the yuremamine structure. It is possible that the yuremamine is absorbed intact through the gut and the DMT subsequently becomes bioavailable through this mechanism. A third possibility is that yuremamine itself is hallucinogenic and thus accounts for the oral activity of traditional single-plant *jurema* preparations. Resolution of this question must await the isolation of sufficient quantities of pure yuremamine to permit a human bioassay. This interesting and little-known New World hallucinogen is yet another incompletely explored niche of ethnopsychopharmacology, awaiting the time and interest of some devoted investigator.

2 The Neuroscience of Ayahuasca

2.1 The Nuclear Medicine Approach

Using single photon emission tomography (SPECT), we conducted a neuroimaging study to assess the acute effects of a high ayahuasca dose in 15 healthy volunteers. We administered a dose of freeze dried ayahuasca equivalent to 1.0 mg DMT per kg body weight in one experimental session and a placebo in another session. A radiotracer was injected at the peak of the experience, one hour and forty minutes after ayahuasca intake. Subsequently, we obtained brain images showing regional cerebral blood flow at the time of injection (Riba et al. 2006).

As shown in Fig. 3, the statistical comparison between the images obtained after ayahuasca and the images obtained after a placebo revealed changes in a number of brain regions. The changes after ayahuasca were always increases in blood flow, and to our surprise, they were not found in low-level primary visual or auditory areas where we had expected changes based on the well-known effects of the ayahuasca on perception. Instead of effects on these hierarchically low sensory-selective regions, the increases occurred in regions placed higher in the information processing hierarchy, predominantly in anterior brain regions. We located significant clusters of activation in the medial aspects of the frontal lobe in an area encompassing parts of the anterior cingulate and medial frontal gyri. Increases were also observed in the medial temporal lobe (MTL) around the amygdala, hippocampus, and parahippocampal gyrus.

The medial frontal lobe plays a prominent role in cognitive control and in the binding of affective and cognitive processes, while the medial temporal lobe plays a role in emotional arousal and episodic memory. The pattern of activation we observed was in line with the findings by other researchers who had also administered serotonergic psychedelics and used the same nuclear medicine technique or the more advanced positron emission tomography (PET) technique. In 1992,

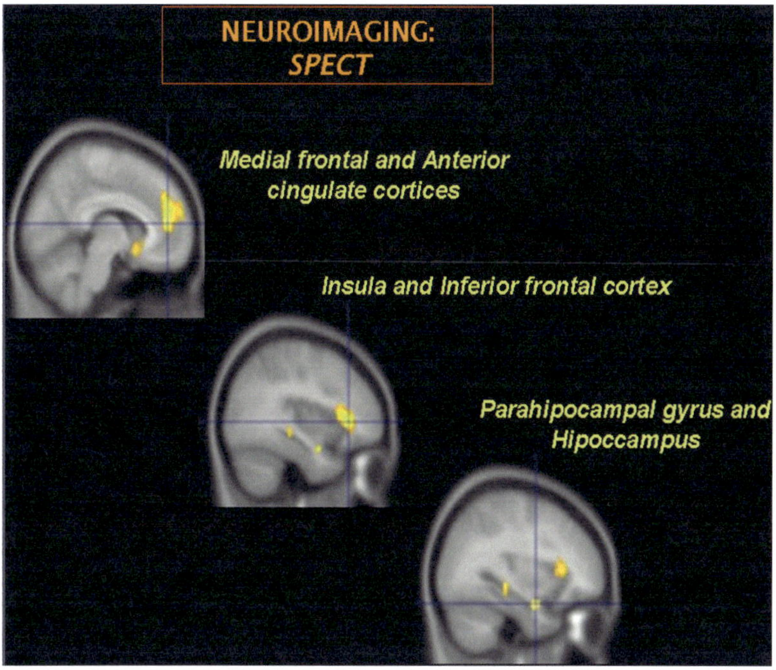

Fig. 3 Results of the neuroimaging SPECT study. The analysis showed significant areas of increased regional cerebral blood flow during the peak effects of ayahuasca. Significant clusters were located in: **a** the right anterior cingulate/right medial frontal gyrus; **b** the right insula/right inferior frontal gyrus; **c** the left insula/left inferior frontal gyrus; **d** the ventral anterior cingulate/subcallosal gyrus; and **e** the amygdala/parahippocampal gyrus. Results are shown at a p value of $p < 0.002$ uncorrected for an $n = 15$ subjects

Hermle and his team described a hyperfrontality pattern following the administration of mescaline (Hermle et al. 1992). The groups led by Vollenweider and by Gouzoulis-Mayfrank both observed increased fluorodeoxyglucose uptake in the medial prefrontal cortex after the administration of psilocybin (Vollenweider et al. 1997; Gouzoulis-Mayfrank et al. 1999). This converging evidence highlights the frontal cortex with its prominent role in executive function as a key target of psychedelic drugs.

2.2 *Spectral Analysis of Brain Electrical Activity*

In a previous study involving 18 participants and an ayahuasca dose equivalent to 0.85 mg DMT/kg body weight, we recorded the brain's spontaneous electrical activity (EEG) with sensors placed on the scalp (Riba et al. 2004). The spectral analysis of ayahuasca-induced changes in the EEG showed reductions in absolute power in all the classical frequency bands of the EEG. We subjected these power changes to intracerebral current density analysis in order to find their brain sources. The results showed only a partial overlap with the findings from the SPECT study. While there were reductions in the MTL and in the medial frontal lobe that matched the SPECT findings, we also found current density reductions in an extensive area in the posterior part of the brain, in the temporo-parieto-occipital junction that includes areas of the parietal, temporal, and occipital lobes. We found the effects specifically over the angular gyrus, the superior parietal lobule, the supramarginal gyrus, the precuneus, and the posterior cingulate cortex. We have replicated these findings more recently in another study involving the administration of two consecutive doses of ayahuasca, as shown in Fig. 4, and they have been independently corroborated by other researchers using magnetoencephalography (Muthukumaraswamy et al. 2013).

It is noteworthy that the brain areas identified using current density analysis correspond predominantly to association areas rather than primary sensory cortex. The temporo-parieto-occipital junction is involved in the secondary processing of visual and auditory information and has been found to play a role in the voluntary generation of visual imagery (Roland and Gulyás 1994). Additional support for a

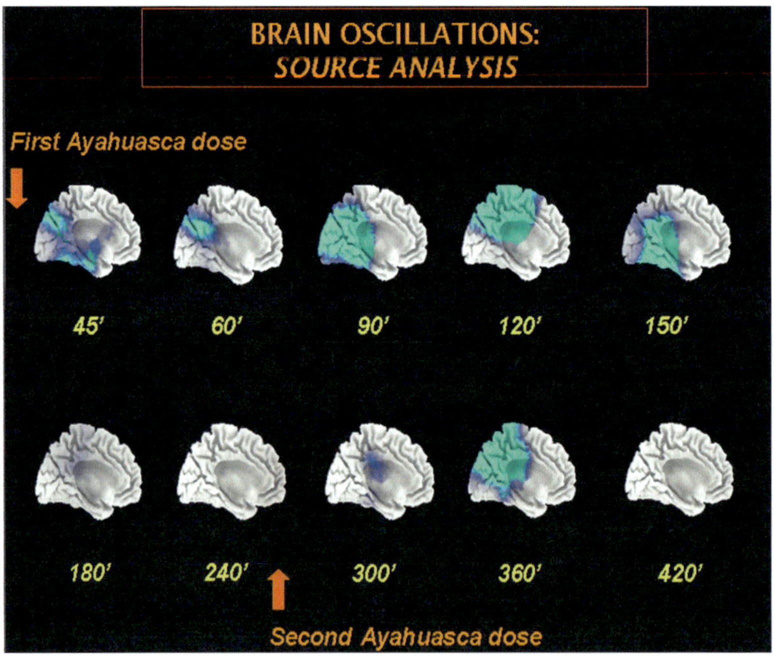

Fig. 4 Result of the current source density analysis. The figure shows areas of significant current density decreases in the alpha band of the electroencephalogram following two consecutive doses of 0.75 mg DMT/kg ayahuasca. *Blue* indicates significant decreases at $p < 0.05$ corrected as compared to placebo. *Note* The prominent decreases in posterior brain regions

role of these structures in ayahuasca-induced effects is provided by other authors who have postulated that, rather than requiring activation of the striate cortex, visual imagery is based on a more complex phenomenon involving the retrieval from memory of visual information stored in the temporal association cortex (Sakai and Miyashita 1994). An effect of ayahuasca at this level could explain phenomena such as synesthesia between the auditory and visual sensory modalities, given the lack of direct projections interconnecting primary sensory cortices or modality-specific areas (Mesulam 2000).

The targeted areas are characterized by their capacity to act as directories binding distributed components of sensory representations and associations (Mesulam 2000). They are believed to operate as gateways for integrating and accessing diffusely stored information. Increased excitability in these areas which intervene in higher-order processing and integration of information could underlie the complex cognitive modifications reported by users such as novel associations, insights, and revelations. It seems reasonable to assume that direct excitatory actions at these key structures can effectively modify the flow of information between the regions and consequently modify the ongoing mental activity.

2.3 *Structural Brain Modifications in Long-Term Users*

A recent study conducted in long-term ayahuasca users provides additional support for the involvement of higher-order association cortex in the effects of ayahuasca (Bouso et al., in press). We obtained high-definition structural images of the brains of 22 ayahuasca users and 22 controls matched for age, sex, years of education, and two intelligence measures (verbal and fluid IQ). We then compared cortical thickness in the two groups, and we also tested the participants for neuropsychological performance. The rationale behind this study was that pharmacological studies have shown that psychedelic 5-HT_{2A} agonists, such as DMT, stimulate neurotrophic factors (Gewirtz et al. 2002) and transcription factors (Frankel and Cunningham 2002; González-Maeso et al. 2007) associated with synaptic plasticity. Neurotrophic factors are small proteins that promote the survival and differentiation of neurons during development and influence neuronal plasticity in the adult brain. Transcription factors are proteins that bind to DNA and control gene transcription into RNA. Hence, transcription factors can stimulate or inhibit protein synthesis. By comparing the images of cortical structure, we expected to detect areas of the brain where structural changes such as increased dendritic arborization, enhanced vascularization, and glial cell proliferation might have occurred.

Interestingly, the analysis found differences in cortical thickness (CT) in anterior and posterior brain midline structures, specifically in the anterior and posterior cingulate cortices. Whereas CT had increased in the anterior cingulate cortex (ACC), thinning was observed in the posterior cingulate cortex (PCC) (see Fig. 5).

In the latter region, the degree of thinning was correlated with lifetime ayahuasca intake. Given that the study was cross-sectional rather than longitudinal, we cannot

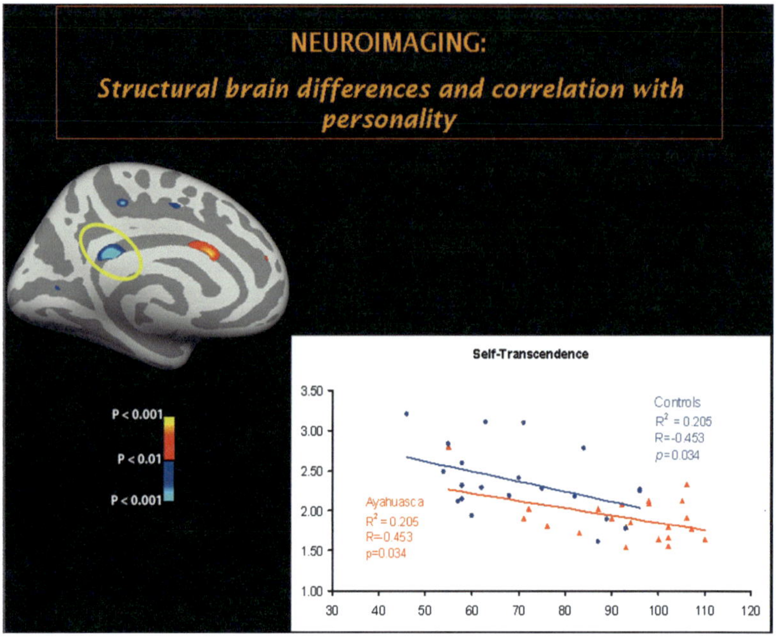

Fig. 5 Results of the structural analysis of T1 magnetic resonance images. The maps show areas of significant cortical thickness (CT) differences between ayahuasca users and controls displayed onto an inflated cortex. Regions with significantly lower CT in the ayahuasca group are shown in cool colors (*blue-cyan*), and regions with significantly higher CT appear as warm colors (*red-yellow*). *Note* CT in the posterior cingulate cortex shows an inverse correlation with the personality dimension of Self-transcendence. That is, participants with greater thinning scored higher on this personality trait

establish a causal relationship between the observed modifications and ayahuasca use, although the data indicate that they are closely related. It is of note that the structural changes were observed in the absence of any impairment of neuropsychological task performance. In fact, ayahuasca users performed better than controls in the two-back test (a measure of working memory), the Wisconsin Card Sorting Test (a measure of executive function) and in the task-switching task (a measure of set shifting) (Bouso et al., in press). Increased CT in the ACC, an area involved in attention and cognitive control, could explain the augmented performance in these tasks, a finding previously observed in experienced ayahuasca users (Bouso et al. 2012, 2013).

Another interesting aspect of the cortical thickness study was that changes in the PCC were associated with differences in personality between the two samples. The ayahuasca users scored higher in Self-Transcendence (ST) than controls, and scores showed a negative correlation with cortical thickness in the PCC. ST is a character dimension of the TCI personality questionnaire developed by Cloninger et al. (1993). It measures the individual's degree of religiousness and spirituality.

A possible role for the PCC mediating this facet of personality is of particular interest. The PCC is a key region within the default mode network (DMN), a series of functionally connected structures that have been associated with the intimate sense of self (Cavanna and Trimble 2006). Research in the 1960s showed that psychedelic experiences could be profound and lead to a more spiritual and less materialistic attitude. Our study suggested that these personality changes could have a neural basis and highlighted the involvement of the medial aspects of the frontal and parietal lobes in the (long-term) effects of ayahuasca and potentially in the effect of other psychedelics.

2.4 Functional Connectivity of Brain Oscillations

The structural study discussed above provided evidence of the involvement of both anterior and posterior brain regions in the effects of ayahuasca. In our most recent study, we corroborated this finding, reconciling the seemingly contradictory results of the initial SPECT and electrical source location studies. We assessed how ayahuasca modifies the normal flow of information within the brain during its acute effects. To do so, we studied the coupling of electrical signals using transfer entropy. Transfer entropy is a mathematical measure of functional connectivity based on the information theory that is model-free and takes into account both the linear and nonlinear components of signals. This measure can be applied to electrical brain oscillations; it identifies causal relationships and allows inferences regarding the directionality of information flow. Transfer entropy from y to x measures the amount of uncertainty reduced in the future values of x by taking into account the past values of y, as compared to when only the past values of x are used. Mathematically, this can be expressed as follows:

$$\mathrm{TE}_{y \to x} = \sum_{x_{n+1}} p(x_{n+1}, x_n, y_n) \log\left(\frac{p(x_{n+1}, x_n, y_n)p(x_n)}{p(x_n, y_n)p(x_{n+1}, x_n)}\right)$$

When the time series associated with spontaneous brain electrical activity was analyzed using TE, results showed significant ayahuasca-induced changes in the coupling of signals between anterior and posterior recording sites. Frontal sources decreased their influence over central, parietal, and occipital sites. At the same time, sources in posterior locations increased their influence over signals measured at anterior locations (see Fig. 6). These modifications were maximal at the time point when DMT plasma levels were highest and subjective effects most intense.

These findings indicate that ayahuasca modifies the functional coupling of oscillatory signals along the anterior-to-posterior axis. Given the asymmetric nature of transfer entropy, the results indicate a decrease in the predictability of activity in posterior areas based on the information available at anterior sites. They also indicate an increase in the predictability of activity in anterior areas when

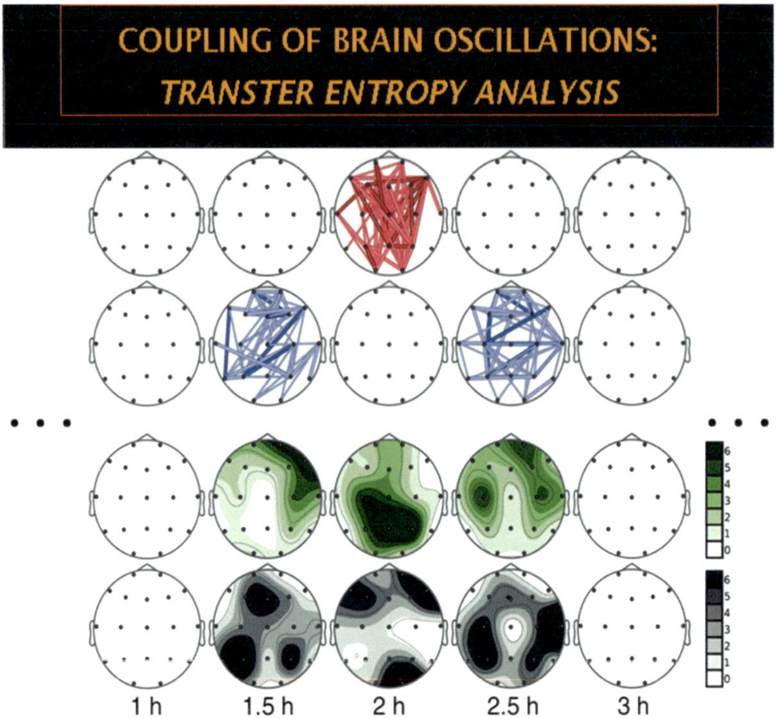

Fig. 6 Results of the transfer entropy (TE) analysis. The first two rows show maps of significant changes in connections between the corresponding electrodes. *Warm colors* indicate significant increases (first row) and *cold colors* (second row) indicate significant decreases. *Thick dark-colored* connections indicate statistically significant changes with p values lower than 0.01, whereas *thick* and *thin light-colored* connections indicate p values lower than 0.05 and 0.1, respectively. The third and fourth rows show the directionality maps depicting the number of outgoing (*in green*) or incoming (*in gray*) connections for each electrode. *Note* The decrease in anterior-to-posterior information transfers at 1.5 and 2.5 h, and the increase in posterior-to-anterior information transfers at 2 h

information at posterior sites is taken into account. Thus, the dynamics of the interaction between the higher-order frontal regions and the more sensory-selective posterior areas is modified. These results are in line with the findings in a functional MRI study in which functional connectivity between the frontal and parietal cortices was also found to be reversed by ayahuasca (de Araujo et al. 2012). Thus, under the influence of ayahuasca, the normal hierarchical structure regulating the flow of information is altered. Top-down or feedback control is reduced, and bottom-up or feedforward information is increased. This temporary disruption of normal information processing leads to a change in the "internal dialogue" and the experience of the world, as explained below.

2.5 A Model of Psychedelic Drug Effects on the Human Brain

Using the comprehensive data we have gathered using several techniques, we propose a model of how ayahuasca and, by extension, other serotonergic psychedelics work in the human brain. As shown above, these compounds target association cortex, which is responsible for the secondary processing of sensory information and, more importantly, higher-level areas binding information from different sensory modalities and data stored in memory. As we have seen from the TE analysis, ayahuasca transiently disrupts the hierarchies governing the flow of information during normal consciousness.

Classical models of brain function have focused mainly on the bottom-up or feedforward flow of information from primary sensory areas to the modality-specific association areas and from there to the multimodal association cortex that combines all incoming elements into a meaningful whole. However, more recent views also take into account top-down or feedback projections from hierarchically high to low nodes in the network (see Fig. 7a).

These models propose that top-down control also plays a significant role in the interpretation of sensory information. Thus, the experience of reality would involve feedforward and feedback loops, rendering the interpretation of incoming signals (both external and internal) dependent on previous knowledge and expectations (Friston 2005; Mesulam 2008). In this framework, each level in the hierarchy sends backward projections that modulate incoming information based on pre-established constraints. The whole network would be under the executive control of the frontal cortex.

We propose that the interaction of a psychedelic with this network will reduce top-down constraints and increase excitability in various levels of the hierarchy. In the modified state of awareness induced by ayahuasca, weak endogenous activity, be it sensory or amnestic, will be able to reach higher levels in the hierarchy and become consciously perceptible. This would explain the endogenous visual and auditory phenomena reported for psychedelics and the distortion of external stimuli. Even in the absence of strong external sensory input (eyes closed), visions will emerge due to increased activity in brain areas processing visual information. The higher excitability in multimodal brain areas such as the posterior association cortex, the cingulate, and the MTL (Riba et al. 2004, 2006) would explain the rapidly evolving modifications of thought content and the novel associations that stand as characteristic features of they psychedelic experience. Mismatch signals or "discrepancies" with predictions will be sent upstream and a constant updating of these predictions will be necessary in the brain's attempt to "make sense" of the experience. The novelty and spontaneity of the thought associations occurring, the facilitation of insight, and the new perspective gained into a given matter are dramatic effects of psychedelics. These sensations of novelty and deep meaning are sometimes so compelling that they are experienced as revelations. See, for example, the compilations by Metzner (1999), McKenna (2000).

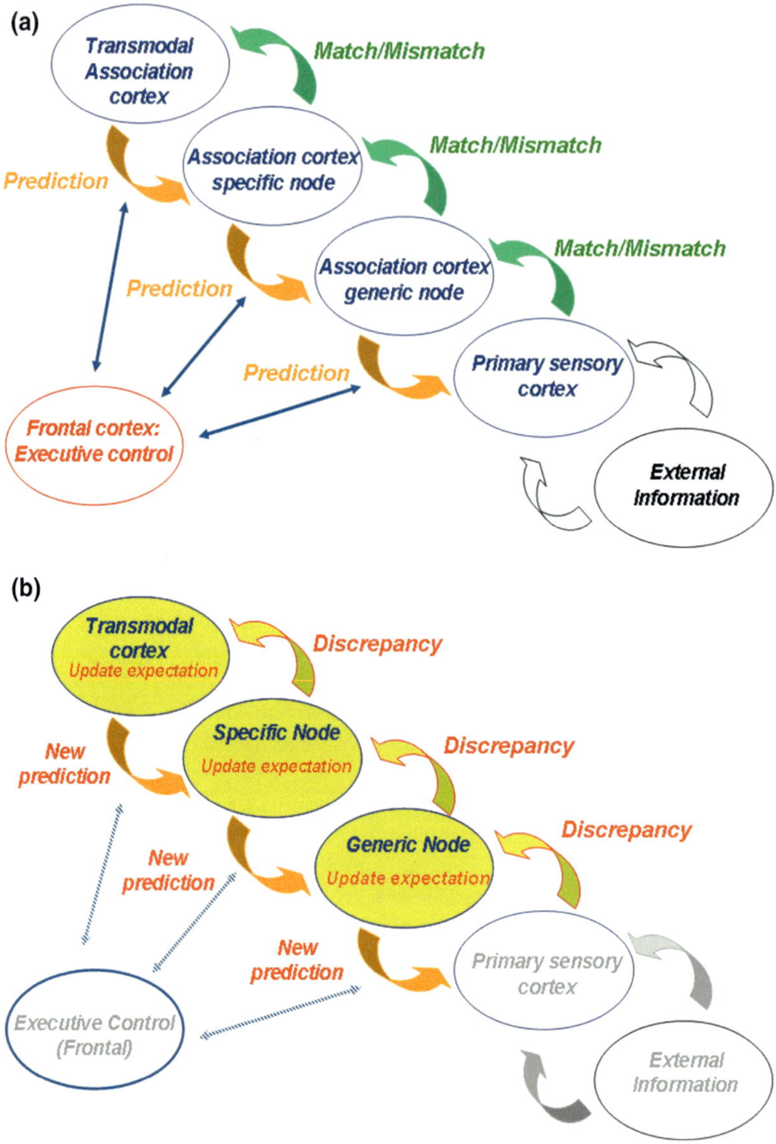
(a)
Transmodal Association cortex
Match/Mismatch
Association cortex specific node
Match/Mismatch
Prediction
Association cortex generic node
Match/Mismatch
Prediction
Prediction
Primary sensory cortex
Frontal cortex: Executive control
External Information
(b)
Transmodal cortex
Update expectation
Discrepancy
Specific Node
Update expectation
Discrepancy
New prediction
Generic Node
Update expectation
Discrepancy
New prediction
New prediction
Primary sensory cortex
Executive Control (Frontal)
External Information

◂**Fig. 7** Model of ayahuasca and other psychedelics on the human brain. **a** Bottom-up progress of external sensory information from sensory-specific primary cortex to secondary association and multimodal cortex under physiological conditions. Top-down constraints send inferences to lower-order nodes regarding the nature of the incoming information. Depending on the match or mismatch with the inference, an upward error message is sent and an updated prediction generated. The whole inference-prediction error is governed by executive control mechanisms at the frontal cortex. Adapted from Mesulam (2008). **b** Under ayahuasca and other psychedelics, secondary association and multimodal cortex shows enhanced excitability. The nature of information traveling up the hierarchy does not fit top-down predictions and an error message (discrepancy) is sent upwards. New updated predictions need to be generated. Iterations can be maintained even in the absence of incoming external information. A decrease in overall executive control exerted by the frontal cortex leads to a breakdown of "constancies" and a modification of internal dialogue. Elements of information can be associated in unusual ways and new "meaning" ascribed to sensory inputs and internal thoughts. The subjective state generated by the psychedelic is characterized by its novelty and sometimes by its overwhelming nature

Individual differences such as personality, mood, and prior experience with psychedelics will be part of each person's pre-established constraints and will consequently modulate the experience. The degree to which each person lets go of the cognitive grip exerted by frontal executive control will also influence the experience and could explain the common lack of effects reported by users when ayahuasca is taken for the first time. Directing attention to external cues such as the ritual and other participants or the desire to remain "in control" frequently leads to experiencing very weak effects or none at all. Typically, in subsequent sessions, the participant lets go and strong effects are finally experienced.

2.6 *Concluding Remarks*

The use of nuclear medicine and neurophysiological techniques has allowed us to identify brain structures targeted by ayahuasca and the changes in neural dynamics underlying the cognitive effects induced by the drug. By acting on key nodes of the association cortex, ayahuasca modifies the flow of information through the brain. The temporary modification of neural hierarchies induces dramatic changes in cognition. The capacity to provide a new outlook on internal and external reality constitutes the uniqueness of ayahuasca and other psychedelics and distinguishes them from all other psychotropic drugs.

References

Borhegyi SF (1961) Miniature mushroom stones from Guatemala. Am Antiq 26:498–504

Bouso JC, Palhano-Fontes F, Rodríguez-Fornells A, Ribeiro S, Sanches R, Crippa JA, Hallak JEC, de Araujo DB, Riba J (in press) Long-term use of psychedelic drugs is associated with differences in brain structure and personality in humans. Eur Neuropsychopharmacol

Bouso JC, González D, Fondevila S, Cutchet M, Fernández X, Ribeiro Barbosa PC, Alcázar-Córcoles MÁ, Araújo WS, Barbanoj MJ, Fábregas JM, Riba J (2012) Personality, psychopathology, life attitudes and neuropsychological performance among ritual users of ayahuasca: a longitudinal study. PLoS ONE 7:e42421

Bouso JC, Fábregas JM, Antonijoan RM, Rodríguez-Fornells A, Riba J (2013) Acute effects of ayahuasca on neuropsychological performance: differences in executive function between experienced and occasional users. Psychopharmacology 230:415–424

Cavanna AE, Trimble MR (2006) The precuneus: a review of its functional anatomy and behavioural correlates. Brain 129:564–583

Cinq-Mars J, Morlan RE (1999) Bluefish caves and old crow basin: a new rapport. In: Bonnichsen R, and Turnmire KL (eds) Ice age peoples of North America, Corvallis: Oregon State University Press for the Center for the Study of the First Americans, pp. 200–212

Cloninger CR, Svrakic DM, Przybeck TR (1993) A psychobiological model of temperament and character. Arch Gen Psychiatry 50:975–990

Cooper JM (1949) Stimulants and narcotics. In: Handbook of South American Indians, vol 5. Bureau of American Ethnology Bulletin #143. Smithsonian Institution, Washington

Dale GD (1991) The Santo Daime Doctrine: an interview with Alex Polari de Alverga. Shaman's Drum 22:30–41

Daly JW, Witkop B (1971) Chemistry and pharmacology of frog venoms. In: Bücherl W, Buckley EE (eds) Venomous animals and their venoms, vol 2. Academic Press

Davis W, Weil AT (1992) Identity of a new world psychoactive toad. Ancient Mesoamerica 3:51–59

Davis W, Weil AT (1994) *Bufo alvarius*: a potent hallucinogen of animal origin. J Ethnopharmacol 41:1–8

De Araujo DB, Ribeiro S, Cecchi GA, Carvalho FM, Sanchez TA, Pinto JP, de Martinis BS, Crippa JA, Hallak JEC, Santos AC (2012) Seeing with the eyes shut: neural basis of enhanced imagery following ayahuasca ingestion. Hum Brain Mapp 33:2550–2560

Der Marderosian AH, Pinkley HV, Dobbins MF (1968) Native use and occurrence of N, N-dimethyltryptamine in the leaves of *Banisteriopsis rusbyana*. Am J Pharm 140:137–147

Emboden W (1979) Narcotic plants. Collier Books, New York

Erowid.org (2001a) A list of the (186) known psilocybin mushrooms. https://www.erowid.org/plants/mushrooms/mushrooms_info12.shtml. Accessed 27 Dec 2014

Erowid.org (2001b) Legal status of ayahuasca in Brazil. http://www.erowid.org/chemicals/ayahuasca/ayahuasca_law6.shtml. Accessed 29 Dec 2014

Erowid.org (2006) UDV wins Supreme Court decision on preliminary injunction. http://www.erowid.org/chemicals/ayahuasca/ayahuasca_law22.shtml. Accessed 29 Dec 2014

Erowid.org (2012) Santo Daime wins court decision. http://www.erowid.org/chemicals/ayahuasca/ayahuasca_law24.shtml. Accessed 29 Dec 2014

Fish MS, Johnson NM, Horning EC (1955) *Piptadenia* alkaloids—indole bases of *Piptadenia peregrina* (L.) Benth and related species. J Am Chem Soc 77:5892–5895

Frankel PS, Cunningham KA (2002) The hallucinogen d-lysergic acid diethylamide (d-LSD) induces the immediate-early gene c-Fos in rat forebrain. Brain Res 958:251–260

Friston K (2005) A theory of cortical responses. Philos Trans R Soc Lond B Biol Sci 360:815–836

Gates B (1979) New names in *Banisteriopsis* and *Diplopterys* (Malpighiaceae) of the Guayana highland. Brittonia 31:108–109

Gewirtz JC, Chen AC, Terwilliger R, Duman RC, Marek GJ (2002) Modulation of DOI-induced increases in cortical BDNF expression by group II mGlu receptors. Pharmacol Biochem Behav 73:317–326

González-Maeso J, Weisstaub NV, Zhou M, Chan P, Ivic L, Ang R, Lira A, Bradley-Moore M, Ge Y, Zhou Q, Sealfon SC, Gingrich JA (2007) Hallucinogens recruit specific cortical 5-HT (2A) receptor-mediated signaling pathways to affect behavior. Neuron 53:439–452

Gouzoulis-Mayfrank E, Schreckenberger M, Sabri O, Arning C, Thelen B, Spitzer M, Kovar KA, Hermle L, Büll U, Sass H (1999) Neurometabolic effects of psilocybin, 3,4-methylenedioxyethylamphetamine (MDE) and d-methamphetamine in healthy volunteers. A double-blind, placebo-controlled PET study with [18F]FDG. Neuropsychopharmacol Off Publ Am Coll Neuropsychopharmacol 20:565–581

Guzmán G, Allen JW, Garrtz J (2000) A worldwide geographical distribution of the neurotropic fungi, an analysis and discussion. Annalidei Museocivico - Rovereto, Italia 14:189–280 (in English)

Heim R, Hofmann A (1958) Isolement de la Psilocybine à partir du Stropharia cubensis Earle et d'autrese spèces de champignons hallucinogens mexicains appartenant au genre Psilocybe. Extrait des Comptesrendus des séances de l'Académie des Sciences, t 247:557–561

Hermle L, Fünfgeld M, Oepen G, Botsch H, Borchardt D, Gouzoulis E, Fehrenbach RA, Spitzer M (1992) Mescaline-induced psychopathological, neuropsychological, and neurometabolic effects in normal subjects: experimental psychosis as a tool for psychiatric research. Biol Psychiatry 32:976–991

Holmstedt B, Lindgren JE (1967) Chemical constituents and pharmacology of South American snuffs. In: Efron DH, Holmstedt B, Kline NS (eds) Ethnopharmacologic search for psychoactive drugs, US Public Health Service Publication #1645. US Government Printing Office

Kennedy AB (1982) Ecce Bufo: the toad in nature and in Olmec iconography. Curr Anthropol 23:273–290

Luna LE (1984) The concept of plants as teachers among four mestizo shamans of Iquitos, northeastern Peru. J Ethnopharmacol 11:135–156

Macrae WD, Towers GH (1984a) *Justicia pectoralis*: a study of the basis for its use as a hallucinogenic snuff ingredient. J Ethnopharmacol 12:93–111

Macrae WD, Towers GH (1984b) An ethnopharmacological examination of *Virola elongata* bark: a South American arrow poison. J Ethnopharmacol 12:75–92

McKenna DJ (2000) An unusual experience with "hoasca": a lesson from the teacher. In: Ayahuasca Reader: encounters with the Amazon's sacred vine. Synergetic Press, Santa Fe, pp 154–157

McKenna DJ, Towers GH, Abbott FS (1984a) Monoamine oxidase inhibitors in South American hallucinogenic plants part II: constituents of orally active Myristicaceous hallucinogens. J Ethnopharmacol 12:179–211

McKenna DJ, Towers GH, Abbott FS (1984b) Monoamine oxidase inhibitors in South American hallucinogenic plants: tryptamine and ß-carboline constituents of ayahuasca. J Ethnopharmacol 10:195–223

McKenna DJ, Luna LE, Towers GH (1995) Biodynamic constituents in *ayahuasca* admixture plants: an uninvestigated folk pharmacopoeia. In: von Reis S, Schultes RE (eds) Ethnobotany: evolution of a discipline. Dioscorides Press, Portland

Mesulam MM (2000) Behavioral neuroanatomy: large-scale networks, association cortex, frontal syndromes, the limbic system, and hemispheric specializations. In: Mesulam MM (ed) Principles of behavioral and cognitive neurology. Oxford University Press, New York, pp 1–120

Mesulam M (2008) Representation, inference, and transcendent encoding in neurocognitive networks of the human brain. Ann Neurol 64:367–378

Metzner R (1999) Ayahuasca: hallucinogens, consciousness and the spirit of nature. Thunder's Mouth Press, New York

Muthukumaraswamy SD, Carhart-Harris RL, Moran RJ, Brookes MJ, Williams TM, Errtizoe D, Sessa B, Papadopoulos A, Bolstridge M, Singh KD, Feilding A, Friston KJ, Nutt DJ (2013) Broadband cortical desynchronization underlies the human psychedelic state. J Neurosci 33:15171–15183

Naranjo P (1995) Archeology and psychoactive plants. In: von Reis S, Schultes RE (eds) Ethnobotany: evolution of a discipline. Dioscorides Press, Portland

Ott J (1998) Pharmahuasca, anahuasca, *vinho de Jurema*: human pharmacology of oral DMT + harmine. In: Müller-Ebeling C (ed) Special: psychoactivity. Yearbook for ethnomedicine and the study of consciousness, 6/7 (1997–1998). Berlin, VWB

Ott J (2001) Pharmañopo-psychonautics: human intranasal, sublingual, intrarectal, pulmonary and oral pharmacology of bufotenine. J Psychoactive Drugs 33:273–281

Pachter IJ, Zacharius DE, Ribeiro O (1959) Indole alkaloids of *Acer saccharinum* (the silver maple), *Dictyloma incanescens*, *Piptadenia colubrina*, and *Mimosa hostilis*. J Org Chem 24:1285–1287

Pinkley HV (1969) Plant admixtures to *ayahuasca*, the South American hallucinogenic drink. Lloydia 32:305–314

Riba J, Anderer P, Jané F, Saletu B, Barbanoj MJ (2004) Effects of the South American psychoactive beverage ayahuasca on regional brain electrical activity in humans: a functional neuroimaging study using low-resolution electromagnetic tomography. Neuropsychobiology 50:89–101

Riba J, Romero S, Grasa E, Mena E, Carrió I, Barbanoj MJ (2006) Increased frontal and paralimbic activation following ayahuasca, the pan-Amazonian inebriant. Psychopharmacology 186:93–98

Rivier L, Lindgren J (1972) *Ayahuasca*, the South American hallucinogenic drink: ethnobotanical and chemical investigations. Econ Bot 29:101–129

Roland PE, Gulyás B (1994) Visual imagery and visual representation. Trends Neurosci 17:281–287; discussion 294–297

Sakai K, Miyashita Y (1994) Visual imagery: an interaction between memory retrieval and focal attention. Trends Neurosci 17:287–289

Schultes RE (1969) *Virola* as an orally administered hallucinogen. Harvard Bot Mus Leaflets 22:229–240

Schultes RE (1970a) The New World Indians and their hallucinogenic plants. Morris Arboretum Bull 21:3–14

Schultes RE (1970b) The plant kingdom and hallucinogens (part III). Bull Narc 22:25–53

Schultes RE, Hofmann A (1981) The botany and chemistry of hallucinogens, 2nd edn. Charles C. Thomas Publishers, Springfield, Illinois

Schultes RE, Hofmann A (2001) Plants of the gods: their sacred, healing, and hallucinogenic powers. Healing Arts Press

Schultes RE, Holmstedt B (1968) The vegetal ingredients of the Myristicaceous snuffs of the Northwest Amazon. Rhodora 70:113–160

Smith TA (1977) Tryptamine and related compounds in plants. Phytochemistry 16:171–175

Torres CM (1995) Archeological evidence for the antiquity of psychoactive plant use in the central Andes. Ann Mus Civ Reverero 11:291–326

Torres CM, Repke DB (2006) *Anadenanthera*: visionary plant of ancient South America. Routledge, London

Torres CM, Repke DB, Chan K, McKenna D, Llagostera A, Schultes RE (1992) Botanical, chemical, and contextual analysis of archaeological snuff powders from San Pedro de Atacama, Northern Chile. Curr Anthropol 32:640–649

Vepsäläinen JJ, Auriola S, Tukiainen M, Ropponen N, Callaway JC (2005) Isolation and characterization of yuremamine, a new phytoindole. Planta Med 71:1053–1057

Vollenweider FX, Leenders KL, Scharfetter C, Maguire P, Stadelmann O, Angst J (1997) Positron emission tomography and fluorodeoxyglucose studies of metabolic hyperfrontality and psychopathology in the psilocybin model of psychosis. Neuropsychopharmacology 16:357–372

Wassén SH (1964) Some general viewpoints in the study of native drugs especially from the West indies and South America. Ethnos 1–2:97–120

Wassén SH (1967) Anthropological survey of the use of South American snuffs. In: Efron DH, Holmstedt B, Kline NS (eds) Ethnopharmacologic search for psychoactive drugs, US Public Health Service Publication #1645. US Government Printing Office

Wasson RG (1968) Soma: divine mushroom of immortality. Harcourt, Brace, Jovanovich

Wasson RG (1980) The wondrous mushroom: mycolatry in Mesoamerica. McGraw-Hill Book Company, New York
Wasson RG, Wasson V (1957) Seeking the magic mushroom. Life Magazine, 13 May 1957
Weil AT, Davis W (1994) Bufo alvarius: a potent hallucinogen of animal origin. J Ethnopharmacol 41(1–2):1–8

Experimental Psychosis Research and Schizophrenia—Similarities and Dissimilarities in Psychopathology

Leo Hermle and Rainer Kraehenmann

Abstract The aim of experimental psychopathology is to delineate overlapping functional disorders of psychoneurobiologically-defined systems where a set of common symptoms may correspond to a variety of nosological entities. According to the vulnerability model of psychosis, experimental research needs to go beyond categories such as "schizophrenia". Prospective studies of the effects of psychoactive substances in normal control subjects offer several methodological advantages over routine clinical reviews of schizophrenic patients, especially in terms of standardization. Carefully designed studies utilizing a model psychosis paradigm are a step toward symptom-oriented research. Combining psychological and neurobiological techniques, the experimental psychopathological approach can provide us with a valuable tool for psychiatric research.

Keywords Experimental psychopathology · Hallucinogen induced model psychosis · Altered state of consciousness · Ego-/self disturbance · Key functions of model psychosis · Psychotoxic basic syndrome

---

L. Hermle (✉)
Christophsbad, Klinik für Psychiatrie und Psychotherapie, Faurndauer Straße 6-28, D-73035 Göppingen, Germany
e-mail: leo.hermle@christophsbad.de

R. Kraehenmann
Neuropsychopharmacology and Brain Imaging, Department of Psychiatry, Psychotherapy and Psychosomatics, University Hospital for Psychiatry Zurich, Lenggstrasse 31, P.O Box 1931, 8032 Zurich, Switzerland
e-mail: r.kraehenmann@bli.uzh.ch

Curr Topics Behav Neurosci (2017) 36: 313–332
DOI 10.1007/7854_2016_460
Published Online: 26 April 2017

Contents

1 Introduction

Most of the early research into experimental psychosis was performed based on the assumption that hallucinogen-induced altered states of consciousness (ASC) are similar to schizophrenia spectrum disorders and therefore could be used as a "model" for investigating psychosis-related phenomena (Geyer and Vollenweider 2008; Hermle et al. 1992b; Vollenweider 2008; Vollenweider and Geyer 2001). Clinical case studies suggest that hallucinogen-induced altered states of consciousness (ASC) and schizophrenia spectrum disorders share a set of common symptoms (Chapman 1966). However, there is only weak evidence to support the hypothesis that the mental phenomena characteristic of the hallucinogenic intoxication can also occur during the incipient and acute stages of schizophrenia (Gouzoulis-Mayfrank et al. 1998). The aim of experimental psychopathology is to delineate overlapping functional disorders of psychoneurobiologically defined systems where a set of common symptoms may correspond to a variety of nosological entities. Therefore, in this chapter, the authors will present evidence regarding the similarities and differences between the psychopathology of schizophrenia and of hallucinogen intoxication.

2 The Beginning of the Experimental Method in Psychiatry

The first systematic clinical investigations of the intoxication produced by mescaline were conducted at the Munich Psychiatric University Hospital. In 1913, Knauer administered 0.15–2 g mescaline subcutaneously to nine physicians (Knauer and Maloney 1913). Knauer was the first to notice that psychological factors can influence the hallucinogen intoxication, and he was able to observe that repeated administration of mescaline to the same test subject would sometimes result in very variable experiences. Although studies from that time period had a great heuristic value, they also suffered from several methodological weaknesses. For example, the experimental conditions were not standardized and scientists would use plant extracts of variable composition. In the 1920s, Kurt Beringer studied the effects of mescaline using a more systematic approach that included repeated tests in his subjects. Beringer was inspired by the German psychiatrist Emil Kraepelin, a pioneer in psychopharmacological research. For the first time, investigations of mescaline were conducted using a standardized design, subjects were given a fixed dose (400–600 mg) using a single route of administration (subcutaneous), and subjective experiences were assessed in great detail. The goal of these experiments was to characterize the range of psychopathological effects produced by mescaline in healthy subjects and to compare mescaline-induced symptoms with those known to occur in endogenous psychoses. Publication of Beringer's monograph *Der Meskalinrausch. Seine Geschichte und Erscheinungsweise* ["The mescaline inebriation. Its history and phenomenology"] in 1927 marked the culmination of mescaline research in Heidelberg. With his mescaline trials, Beringer believed he had found an artificial model of psychosis because the effects of mescaline were very similar to the symptoms of acute schizophreniform diseases.

What impressed Beringer the most was the close similarity between the effects of mescaline and the symptoms present during the incipient stage of schizophrenia. In particular, Beringer found that both conditions lead to a state characterized by extremely rich and intense subjective experiences, which were described as being highly significant by both the mescaline subjects and the schizophrenia patients. Beringer called these states *Primärerlebnis* ["primary experience"] and noted that they often have a brief duration and are distinct from other phases of the illness, which are typically characterized by the absence of positive symptoms and development of progressive mental deficits. According to Beringer, the most important symptoms of the primary experiences in the model psychoses and schizophrenia are disturbances of self-awareness, experiences of ego-fragmentation, abnormal affect, mystical states, and feelings of enlightenment and revelation (Beringer 1927).

In addition to focusing on the development of a comprehensive phenomenological description of the course of the mescaline intoxication, Beringer's investigations also attempted to define how personality influences the mescaline intoxication. According to his own statements, Beringer was not able to identify a clear relationship between personality factors and the response to mescaline. Beringer's principle aim was the search for an underlying disturbance that was common to both endogenous and drug-induced psychoses. By contrasting the symptoms occurring during primary experiences with those seen during the chronic phase of schizophrenia, Beringer stressed that there were both qualitative and quantitative differences. Furthermore, Beringer found that these two conditions may occur independently of each other and without one influencing the course of the other. Therefore, he concluded that the primary experiences have a biological basis and were unique entity within incipient schizophrenic illness and during mescaline intoxication.

3 Classification of the Subjective Effects of Hallucinogens

The complex phenomenology of the intoxication induced by mescaline and other hallucinogenic drugs has hampered development of a standardized nomenclature to describe and classify their effects. Depending on the perspective of the investigator, hallucinogenic drugs have been called *phantastics* (Lewin 1927) or *eidetics* (Hellpach 1941) to emphasize the visual hallucinations produced by these drugs; *hallucinogens* to describe pseudohallucinatory perceptual effects (Hoffer et al. 1967); *psychotics* (Becker 1949) and *psychotomimetics* (Leuner 1962) to describe the similarities between the hallucinogen intoxication and the symptoms of incipient or acute schizophrenia; *psychedelics* ("mind-manifesting") to describe the positive, beneficial effects produced by these drugs (Osmond 1957); and *psycholytics* to describe their therapeutically useful effects, such as their ability to facilitate access to repressed and unconscious spheres of personality (Sandison 1954).

4 The Common Nucleus of Hallucinogen-Induced Altered States of Consciousness

Despite their pharmacological heterogeneity, Ludwig found that all hallucinogens produce similar psychopathological syndromes that are characterized by a primary alteration of consciousness and by secondary changes in cognition, perception, and emotion (Ludwig 1966). Dittrich further elaborated on the common core dimensions of hallucinogen-induced states of consciousness (Dittrich 1996). Extending Ludwig's work, Dittrich (1998) identified three core dimensions that are consistently affected by hallucinogens: (1) *oceanic boundlessness*, (2) *dread of ego-dissolution*, and (3) *visionary restructuralization* (Fig. 1). These three core dimensions can be reliably measured by the APZ (*außergewöhnliche psychische Zustände*) questionnaire, which assesses altered states of consciousness (ASC).

Oceanic boundlessness (OSE) refers to positively experienced depersonalization and derealization, positive emotions, feelings of unity, and mystical experiences. *Dread of ego-dissolution* (AIA) refers to negatively experienced derealization and depersonalization, cognitive disturbances, catatonia, paranoid ideation, and loss of

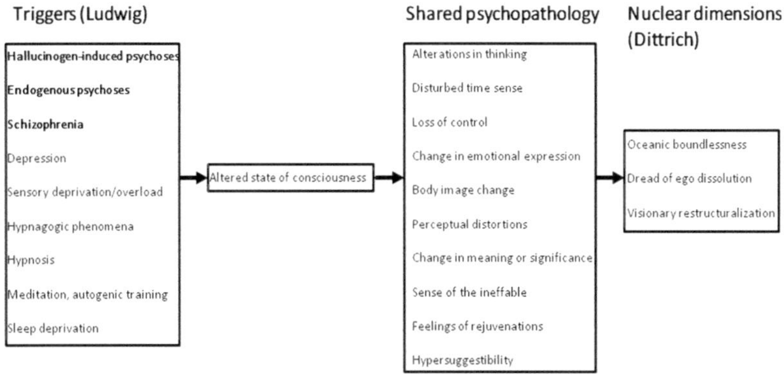

Fig. 1 The illustration shows that a variety of drug and non-drug conditions induce altered states of consciousness, leading to profound psychopathological changes. Altered states of consciousness consist of three core dimensions that have been found to be etiology-independent (Ludwig 1966; Dittrich 1996)

thought and body control; this indicates a very unpleasant state comparable to so-called bad trips. *Visionary restructuralization* (VR) refers to visual hallucinations, illusions, synesthesia, and changes in the meaning of perception. Vollenweider compared the effects of psilocybin (n = 99, 0.26 mg/kg p.o.), ketamine (n = 68, 0.012 mg/kg/min i.v.), and MDMA (n = 74, 1.5–1.7 mg/kg p.o.) on these three core dimensions in healthy volunteers (Vollenweider 2001). According to this comparison, both psilocybin and ketamine can produce ego-disintegration, an effect that is associated with either positive emotions or thought disorder and loss of self-control. Compared to psilocybin, however, *S*-(+)-ketamine and racemic ketamine produced higher levels of anxiety, thought disorder, and ego-disintegration. *S*-(+)-Ketamine and racemic ketamine also produced feelings of apathy, emotional withdrawal, and indifference—effects that are similar to the negative symptoms of schizophrenia. These findings support the contention that the effects of the dissociative anesthetics ketamine and PCP closely mimic the thought disturbances and cognitive deficits in schizophrenia (Table 1) (Krystal et al. 2000; Lahti et al. 2001; Vollenweider et al. 1997a, b, 1998b).

To determine whether the effects of hallucinogens are similar to the symptoms of endogenous psychosis, Gouzoulis-Mayfrank et al. (1998) compared 50 healthy controls, 93 patients with endogenous psychoses, and 7 hallucinogen-treated subjects using the APZ and a psychometrically improved version known as the OAV. The patients were examined after remission of their last psychotic episode and were directed to answer the two questionnaires with reference to the incipient phase of their psychotic disorder. Although there were significant differences between the scores of the psychotic patients and the healthy controls, there was little difference between the scores of patients with endogenous psychoses versus drug-treated subjects (Fig. 2). The results of this study support the hypothesis that there are phenomenological similarities between the experiences induced by hallucinogens and the incipient stages of endogenous psychoses. The results also show that the OSE, AIA, and VUS dimensions of ASC are sensitive to the subjective phenomena that occur during acute episodes of schizophrenia. As noted by Gouzoulis-Mayfrank and co-workers, it is remarkable that schizophrenia episodes include experiences in the OSE dimension because in clinical practice it would be easier to overlook pleasant OSE experiences compared to negative experiences in the AIA dimension.

Following the tradition of Bleuler (Benedetti 1995), Döhmann asked 105 psychiatrists to assess whether any of the 158 items in the APZ reflect symptoms that are typically observed in patients with schizophrenia (Döhmann 1978). According to the psychiatrists, items from the AIA dimension were most consistent with the symptomatology of schizophrenia, indicating that the illness is often be associated with anxiety, derealization, and depersonalization. However, Döhmann did not focus on first-episode schizophrenia, which could potentially explain the preponderance of items from the AIA dimension. Schröter-Rosendahl assessed self-ratings of psychedelic experiences in 40 first-episode schizophrenia patients (Schröter-Rosendahl 1980). About 50% of the self-reports contained items from the OSE dimension, indicating that schizophrenic patients experience pleasant states in addition to anxiety. Visual perceptual changes (items in the VUS dimension) were rarely experienced by the patients, demonstrating that a robust difference exists between the effects of hallucinogens and the symptoms of schizophrenia.

5 Administration of Hallucinogens to Schizophrenia Patients

In the early 1930s, psychiatrists began to experiment with mescaline in schizophrenia patients in order to investigate whether subjects could distinguish the effects of hallucinogens from the symptoms of endogenous schizophrenia. Zucker (1930) administered mescaline to nine schizophrenia patients at doses between 350 and 400 mg. These trials confirmed that the patients could clearly differentiate mescaline-induced effects from illness-related hallucinations. Alternatively, various other clinicians (Kant 1930; Stockings 1940; Condrau 1949) believed the intoxications induced by mescaline, and LSD could serve as valid models of psychosis. For example, Condrau (1949) conducted a total of 197 experiments with LSD in 30 schizophrenic patients and found that the hallucinations induced by LSD were not discriminable from the patients' spontaneous hallucinatory symptom. Nevertheless, many clinicians had observed that schizophrenic patients were resistant to the effects of LSD and that individual reactions could vary substantially. Hoch (1957), for instance, found that patients suffering primarily from negative symptoms showed only minimal reactions compared to normal subjects, whereas schizophrenics with positive symptoms sometimes showed very intense reactions to mescaline and LSD.

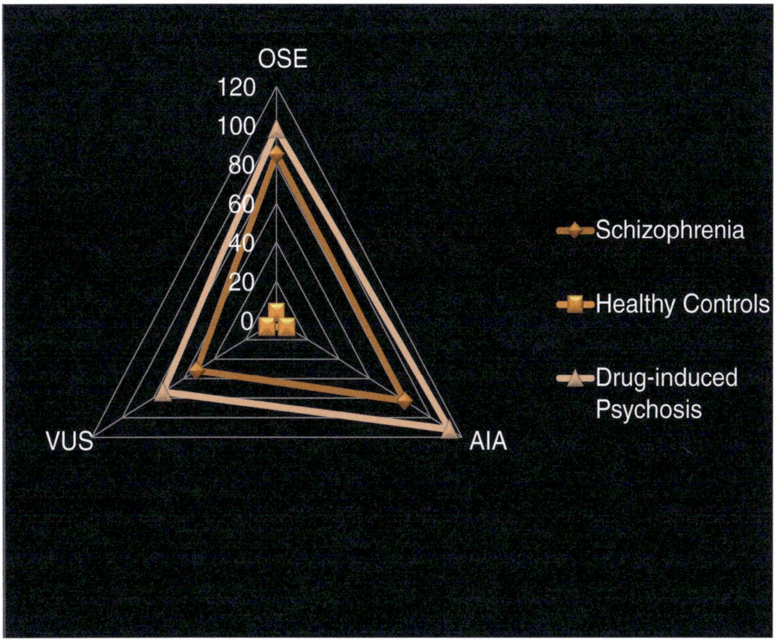

Fig. 2 Comparison of the scores on the three core dimensions of the altered states of consciousness scale (APZ) for schizophrenia patients (n = 93), healthy controls (n = 50), and hallucinogen intoxication (n = 7). Adapted from: Gouzoulis-Mayfrank et al. (1998)

Because there are now many schizophrenics who have used hallucinogens recreationally, comparing their experiences has provided important clues regarding the similarities and differences between the symptoms of schizophrenia and the effects of hallucinogen intoxication. Some patients report that their drug experience was completely different than their experience during the acute psychotic break. These results indicate that there are differences between the hallucinogen model psychosis and schizophrenia, but the difficulties associated with differential diagnosis and the high frequency of drug-induced schizophreniform psychoses argue for an overlap of pathogenesis. This is especially true if one assumes that a relatively non-specific mechanism is responsible for the development of schizophrenia. For example, according to the vulnerability stress model (Zubin and Spring 1977), only low doses of a hallucinogen would be required to induce psychotic reactions in susceptible individuals, making the old clinical dichotomy of provoked versus caused more of a continuum.

6 Similarities and Differences Between the Psychopathology of Psychosis and the Effects of Hallucinogens

6.1 Changes in Self-experience

The question of whether there is a close relationship between hallucinogen intoxication and endogenous psychosis—above and beyond their phenomenological similarities—is controversial and remains unresolved. The arguments for and against a common psychopathology will be summarized below for one specific disturbance: alterations of self-experience.

Self-awareness, also known as ego-consciousness, relates to the ability to conceive oneself as an individual, separate from others and from the environment, who can integrate and control their own thoughts, emotions, perceptions, and actions (Scharfetter 1995). Disorders of self-awareness, also known as self-disorders, comprise a variable set of symptoms and are nosologically unspecific, occurring in normal subjects after administration of hallucinogens as well as in patients suffering from schizophrenia, affective disorders, neurosis, or personality disorders. Self-disorders can also occur in healthy individuals, for example, during states of strong fatigue or of affective tension. There is substantial evidence that disrupted self-experience is a core disturbance of schizophrenia, which has been termed a self-disorder or an ipseity disturbance (*ipse* is Latin for "self") (Sass and Parnas 2003). Bleuler, for example, noted that schizophrenia patients experience disruptions of the minimal sense of self, including splitting of their self-image, alterations of bodily experience, and loss of thought control (Bleuler 1911). Schneider (1955) was convinced that self-disorders due to schizophrenia and other causes should be discriminable; hence, he argued that the self-disorders in schizophrenia may have an important differential diagnostic value. By systematically studying self-experience in schizophrenic patients, Scharfetter (1995) identified five fundamental dimensions of self-consciousness that can reliably distinguish patients with schizophrenia from non-schizophrenic patients (e.g., patients with borderline personality disorder or depression):

1. *Ego-identity*—changes of one's identity in respect to gestalt, physiognomy, gender, genealogy, and biography.
2. *Ego-demarcation*—loss of ego-boundaries and the ability to differentiate between self and non-self.
3. *Ego-consistency*—dissolution or destruction of a coherent, unified experience of the self, one's thoughts, and the external world.
4. *Ego-activity*—altered ability to perform self-determined actions, thoughts, feelings, and perception.
5. *Ego-vitality*—fear of one's death.

Table 1 Summary of the comparison of the dimensions of psychopathology during drug-induced hallucinations and schizophrenia (Vollenweider and Geyer 2001; Leuner 1962; Hollister 1961; Vollenweider 2001; Hermle et al. 1992a, b; Vollenweider et al. 1998a; Dumont et al. 2008)

	LSD, psilocybin, mescaline	Ketamine, PCP	MDMA	Amphetamine	Schizophrenia
Ego-/self-dimensions					
Ego-identity disturbance	+	+	?	(+)	+
Ego-demarcation disturbance	(+)	(+)	?	–	++
Ego-consistency disturbance	+	++	?	–	++
Ego-activity disturbance	+	++	?	(+)	+++
Ego-vitality disturbance	(+)	+	?	–	++
Reduction of reflecting self	+	++	?	–	+++
Psychedelic core dimensions					
Oceanic boundlessness	+++	+++	++	+	+++
Dread of ego-dissolution	+	++	(+)	+	+++
Visionary restructuralization	+++	++	+	(+)	(+)
Symptoms					
Visual hallucinations	+++	+(+)	–	–	+++ (acute) + (chronic)
Auditory hallucinations	(+)	(+)	–	–	+++
Delusions	(+)	(+)	–	–	+++
Negative symptoms	(+)	++	(+)	–	++

Mullen (2008) described experiences of passivity in schizophrenic patients, meaning they believed their internal mental processes were directed or influenced by outside forces or by other individuals. Beringer (1927) also wrote of a passivity syndrome in relation to the intoxication produced by mescaline. However, according to reports published by Arnold and Hoff (1953) and Weyl (1951), subjects using hallucinogens do not normally have the impression that their experiences are induced or controlled by outside forces. By contrast, Leuner (1962, p. 29), Savage (1955), and Grof (1967) noted that subjects treated with LSD and other hallucinogens sometimes believed their experiences were influenced or controlled by outside forces—in most cases, the investigators themselves or the investigational procedures.

To compare the changes of ego-functioning induced by drug intoxication and schizophrenia, Vollenweider and co-workers assessed first-episode schizophrenia patients as well as healthy volunteers administered psilocybin (15–20 mg p.o.), ketamine (0.02–0.03 mg/kg/min), or amphetamine (1.0 mg/kg p.o.) using Scharfetter's Ego Pathology Inventory (EPI) scale (Vollenweider et al. 1997a, b, 1998b; Vollenweider and Geyer 2001). At the doses studied, the dimensions *ego-identity* and *ego-consistency* were impaired to a similar degree in hallucinogen intoxication and schizophrenia. The dimensions of *ego-demarcation* and *ego-vitality*, however, were only slightly disrupted by hallucinogens while being severely impaired in schizophrenia (Table 1). Those authors concluded that the hallucinogen-induced model psychosis may be comparable to the symptoms of psychosis in schizophrenia patients, but with several important differences in the quality and quantity of effects on the ego-dimensions (Vollenweider and Geyer 2001). In particular, hallucinogen-treated subjects—unlike schizophrenic patients—typically recognized that their altered ego-functions were abnormal and retained insight into the fact that the changes were caused by a drug. The preservation of a residual self that is capable of reflection is the characteristic of hallucinogen effects and is only lost after administration of very high doses (Leuner 1962).

6.2 Altered Body Experience and Depersonalization as Special Cases of Self-disturbance in the Model Psychosis

There are numerous reports in the literature of altered body experience during hallucinogen intoxication (Beringer 1927; Arnold and Hoff 1953; Masters and Houston 2000; Savage and Cholden 1956; Heimann 1961; Stoll 1947). At the beginning of the mescaline intoxication, paresthesia-like sensations may occur, including sensations of vibration or electric shock, tingling, formication, and cold shivers (Beringer 1927). Moreover, there are reports of unpleasant muscular tension, feelings of heaviness in the limbs, and generalized muscle flaccidity. These so-called vegetative sensations appear to depend on the prevailing mood and personality of the subject (Hermle et al. 1992b). Alterations of body image may lead to disturbed self-experience and ego-dissolution. The dissolution of the boundary between the body and the outside world may be experienced as depersonalization phenomena (Stoll 1947). Body parts and even the entire body may be perceived as being separate from the self or may feel alien. Some subjects had out-of-body experiences where they felt that their consciousness was located outside of their physical body (Klee 1963; Hoffer et al. 1967; Savage 1955). It appears that these altered body experiences are often closely related to changes in perception or hallucinations. The perception of hallucinatory objects may lead to intensely experienced feelings that the self has merged with the cosmos and that the body has undergone a metamorphosis (Hermle et al. 1992b). Savage reported that bodily sensations during the hallucinogenic state may be intensely pleasurable (Savage and

Cholden 1956). Hence, the altered bodily experiences induced by hallucinogens have been described as both an intensification of the bodily self and a dissolution of the boundary between the body and the external world. Alterations of bodily experiences typically occur at the beginning of the intoxication and are paralleled by a more general change of self-experience where the drug state causes the subject to focus their attention inward (Heimann 1961).

6.3 Alteration of the Experience of Time and Space

Changes of the perception of three-dimensional space during the hallucinogen intoxication can vary widely. Objects seen may be perceived especially vividly, with increased differentiation of colors and brightness, as well as enhancement of stereoscopic vision (Beringer 1927, p. 39). Illusory motion and distortions of perspective, including micropsia, macropsia, and dysmegalopsia, are frequently reported (Beringer 1927, p. 43). In the context of these distortions of visual perception, the environment may be experienced as being abnormally large or threateningly small, and like the depersonalization experiences, the boundary between the body and the external world may be distorted. Alterations of the perception of time are frequently reported during hallucinogen intoxication (Beringer 1927; Becker 1949). Time is either perceived to be contracted, dilated, or the sense of time may completely dissolved. If the disturbed temporal experiences coincide with hallucinogen-induced euphoria, then they may lead to feelings of ecstatic exhilaration; alternatively, if accompanied by anxiety, the altered temporal experiences may intensify the negative feelings due to the impression that time has stopped (Becker 1949; Cohen 1968).

Changes in time perception have also been reported in patients with schizophrenia. For example, it has been shown that schizophrenic patients tend to overestimate elapsed time (Bonnot et al. 2011) or have less ability to judge correctly the temporal order of acoustic stimuli (Braus 2002).

6.4 Passivity Syndrome

Even some of the earliest studies noted that the hallucinogen intoxication can include elements of withdrawal and passivity that resemble the negative symptoms of schizophrenia (Hermle et al. 1992b). Some hallucinogen test subjects found themselves to be withdrawn and inactive, perceiving environmental stimuli in a passive and unfiltered manner, and at least partly incapable of controlling their thoughts, experiences, and behavior. At the peak of the mescaline intoxication, subjects may be so passive and inwardly focused on the experience that self-reflection or engagement with the external world is greatly reduced or impossible (Beringer 1927).

6.5 Self-loss

Self-loss refers to a very severe manifestation of self-disturbance. There may be feelings of loss of ownership of thoughts, perceptions, experiences, or emotions, or those things may feel alien in nature; additionally, the environment and time may no longer feel independent from the self or may completely cease to exist. These states are frequently accompanied by severe agitation and anxiety. Someone who takes a drug usually retains insight into the cause of the eventual changes, but those suffering from endogenous psychosis are not aware of the cause of their symptoms (Hermle et al. 1988). These responses appear to be an understandable reaction to changes of experiences that make the subject feel helpless. With further dose increases, an apocalyptic phase sets in (Conrad 1958, p. 104), which is characterized by loss of self-control and decay of the coherent self. As a consequence of these highly severe psychotic states, catatonic behavior may occur (Kraehenmann et al. 2010).

7 The Six Key Functions of the Model Psychosis Induced by Hallucinogens

In order to compare the phenomenology of schizophrenia and hallucinogen-induced experiences, Leuner (1962, p. 219) attempted to work out the key psychopathological features of the hallucinogen intoxication. Leuner regarded the so-called *psychotoxic basic syndrome* as the key functional substrate of the experiences produced by hallucinogens (Fig. 3). Importantly, there seems to be a uniform and consistent pattern of changes of psychological functioning that is common to all hallucinogenic substances. The *psychotoxic basic syndrome* is characterized by a regression of psychological functioning to ontogenetically earlier stages, as well as a shift from normal, waking consciousness to a so-called state of protopathic consciousness (Greek, proto-, "first, primitive" + pathos, "suffering, feeling") (Lienert 1959; Conrad 1948; Lohmar and Brudzinska 2011). The *psychotoxic basic syndrome* is similar to the *hypnagogic basic syndrome*. It is characterized by spontaneous symbolic visual phenomena and intensified mood and affect that occurs during hypnosis, before falling asleep, and during sensory deprivation (Leuner 1962). In the protopathic state, abstract thinking is impaired and thoughts are easily transformed into imagery (Klee 1963).

According to Leuner (1962), hallucinogenic states progress over time according to two basic patterns: (1) a "fluctuating-scenic" or "quasi-normal" course, and (2) a "stagnating fragmentary" or "extreme psychotic" course. During the quasi-normal course, subjects experience a continuous flow of imagery, accompanied by a significant emotional response. This experience is very similar to dreaming. If the level of emotional arousal peaks, there may a dissociation between conscious experience and emotional response. In that situation, the scenic course of intoxication may be

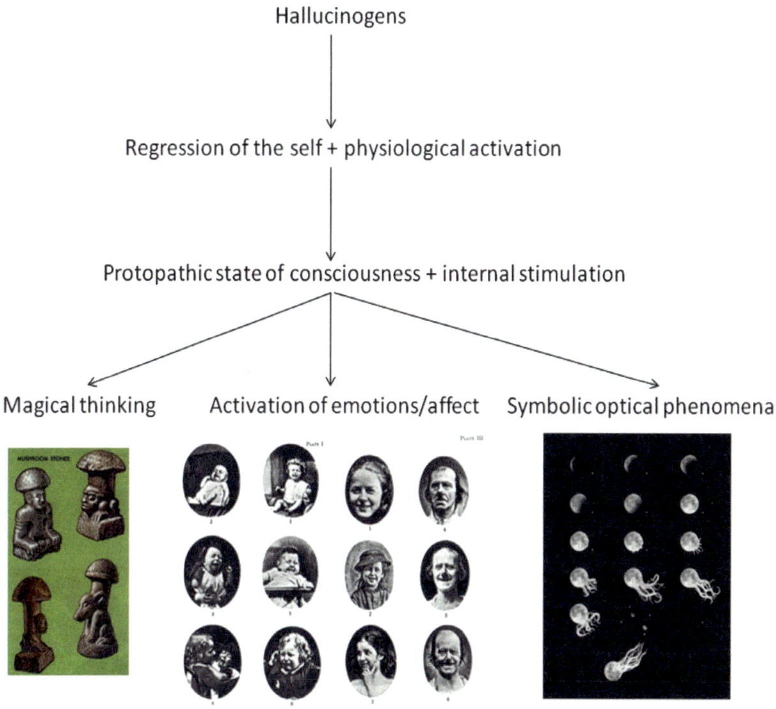

Fig. 3 Schematic illustration of the "psychotoxic basic syndrome" proposed by Leuner (1962) to explain the profound alterations of cognition, emotions, and perception produced by hallucinogens

interrupted and seemingly random fragmentary experiences may occur, often accompanied by inadequate affect and stereotypical motor behavior (a catatonic syndrome). Importantly, Leuner found that during an individual session, the hallucinogen intoxication may fluctuate between a quasi-normal course and an extreme psychotic course. Leuner also reported that even seemingly personality-neutral scenic visual hallucinations (e.g., images of animals, monsters, and mythical creatures) are often endowed with personal meaning. He categorized six key functions underlying the psychological effects of hallucinogens:

1. **Activation of intrapsychic processes**: On the one hand, hallucinogens activate physiological and psychological processes, including emotions and imagery; on the other hand, hallucinogens impair cognitive control; both effects lead to an enhancement of internally generated experiences;
2. **Dynamic overdrive of the psyche**: Mental hyperarousal and loss of control over thoughts and emotions further lead to a seemingly random succession of imagery, dissociation of affect, and motor reactions ranging from catatonia to hyperactivity;

3. **Transphenomenal dynamic governing system**: Individual memories are co-activated with associated thoughts and emotions, which can further drive and influence the hallucinogen experience and behavior;
4. **Dynamic reduction**: A gradual reduction of hyperarousal and dynamic overdrive in the course of time;
5. **Deep psychological affective dynamics**: expression of emotional themes via symbolic imagery, reliving of emotionally charged childhood memories, and cathartic abreaction; and
6. **Dynamic fixation**: The psychotic content of the experience is stagnant and fails to progress or change over time, potentially as a defense mechanism.

Therefore, Leuner revised Beringer's proposal that the experimental psychosis was not influenced by repressed material. Leuner found that it was even possible to identify symbolic content and repressed affective material during the course of extreme psychotic reactions (with catatonia and stupor), which is why it is possible to effectively perform psychotherapy in "psychotic" patients under the influence of hallucinogens.

8 Changes in Perception

Since the beginning of research into substances that mimic schizophrenia, it has been repeatedly discovered that hashish, mescaline, psilocybin, and LSD have a common effect: They intensify sensory perception and induce a wealth of subjective experiences (Beringer 1927; Leuner 1962; Stoll 1947; Joel and Fränkel 1926). The drug-induced alterations of perceptual processes can be categorized into three basic types (Boor 1956): (1) changes related to sensory input; (2) hallucinatory changes; and (3) alteration of the meaning of percepts. Changes related to sensory input (group 1) are characterized by only minimal alterations of their formal qualities (e.g., shape and color). Chromatopsia, micropsia, macropsia, and spatial distortions are examples of perceptual alterations belonging to this class. Hallucinations (group 2) are sensory perceptions that occur in the absence of external stimuli. Changes in the significance or meaning of sensory perceptions (group 3) are known to occur frequently in hallucinogen intoxication and incipient schizophrenia; these symptoms have been termed delusional perceptions. Delusional perception refers to the phenomenon where a patient believes that a normal percept has abnormal meaning, often with self-reference.

The degree to which the hallucinations and perceptual disturbances in schizophrenic patients overlap with the hallucinations induced by LSD and related substances has been addressed by several authors over the past few decades. It is believed that patients with schizophrenia primarily experience auditory hallucinations, whereas LSD-like drugs tend to cause visual hallucinations. These differences have frequently been cited as evidence that hallucinogen effects and schizophrenic symptomatology can be differentiated. However, others have noted that hallucinogens can sometimes produce acoustic hallucinations and that some patients with schizophrenia experience severe visual hallucinations (Chapman 1966; Hoffer and Osmond 1966; Small et al. 1966; Freedman and Chapman 1973; Young 1974; Winters 1975). These authors emphasized that the acute phase of schizophrenia is often associated with perceptual disturbances that are very similar to drug-induced hallucinations. According to Chapman (1966), delusions that develop in schizophrenia patients often occur in response to specific perceptual disturbances, making it difficult to differentiate between their hallucinations, illusory misperceptions, and delusional perceptions. Further, it has been argued that the difference between drug-induced hallucinations and the hallucinations in schizophrenia may to some degree be dependent on time rather than the underlying illness. For example, Winters (1975) postulated that the temporal stages of schizophrenia may be understood in analogy to a chronic state of hallucinogen-induced intoxication, where the typical schizophrenic symptoms such as multisensory perceptual disturbances and delusional ideation follow the initial visual hallucinations.

The extent to which the evaluation of perceptual disturbances relating to syndrome genesis may influence the classification of model psychoses was made clear by Hoffer and Osmond (1966). The authors noted that psychiatrists tend to ignore perceptual disorders in patients with schizophrenia, which they theorized occurs because Bleuler believed that perception is unaltered in schizophrenia (Bleuler 1911). By contrast, Hoffer and Osmond suggested that perceptual disturbances are a common occurrence in schizophrenia patients and should be viewed as a primary illness process and not as a secondary response (Hoffer and Osmond 1966; also see Kraehenmann et al. 2012). Previously, Beringer (1927), Stockings (1940), and Savage and Cholden (1956) noted that hallucinogen-like visual alterations may occur in the early stages of spontaneous psychotic states, although they believed that such phenomena would be easily overlooked. According to Süllwold and Huber (1986) and Klosterkötter et al. (1994), these subjective visual alterations occur during the prodromal phase of psychosis. Before the start of an acute psychotic episode, the prodromal symptoms become more intense and are often accompanied by fluctuating affect; eventually, the prodromal symptoms progress to classical psychotic symptomatology, such as delusional phenomena.

9 Inter- and Intra-individual Differences

The personality structure of test subjects, as well as differences in the integration of self-related functioning and affective responses, often influences the symptoms of the intoxication induced by hallucinogens (Leuner 1962, p. 43; Linton and Langs 1962; Ziolko 1966, p. 250; Grof 1967, p. 160). For example, subjects with compulsively structured personalities exhibit weak responses to moderate doses of hallucinogens (Langs 1967, p. 182; Leuner 1962, p. 44; Buckman 1967). Alternatively, subjects with histrionically structured personalities are highly sensitive to low doses of hallucinogens (Grof 1967, p. 161). Within individual subjects, there may also be substantial differences across test sessions (Sandison 1954; Leuner 1962, p. 446; Cohen 1968). Initially, test subjects are often disturbed by the vegetative symptoms produced by hallucinogens (e.g., hypertension, tachycardia, mydriasis, hyperthermia, tremor), but these symptoms often recede into the background. Likewise, the intensity of hallucinations, concentration deficits, and self-disturbances often gradually declines from session to session, giving way to well-arranged (so-called quasi-normal) experiences (Leuner 1962, p. 45). Leuner (1962, p. 45) described a "paradoxic habituation" where, after several hallucinogen experiments, test subjects would sometimes only need half the dose amount to be brought into an inebriation experience of similar intensity. The influence of *setting* appears to be of special importance here.

The effects of hallucinogens are extremely dose-dependent. Leuner (1962) divided hallucinogen effects into three dose ranges: At low doses (e.g., 25–80 μg LSD), hallucinogens tend to produce euphoria and a loosening of associations; medium doses (e.g., 80–200 μg LSD) produce a broad spectrum of psychopathological effects, including hallucinations and depersonalization; high doses (200–700 μg LSD) can provoke extreme disorientation, as well as loss of insight into the drug-induced nature of the experience, potentially resulting in extreme agitation or catatonia. The mystical experiences produced by hallucinogens are most likely to occur after administration of high doses.

10 Differential Typology of Intoxication and Psychosis

In the literature, the terms drug-induced intoxication and drug-induced psychosis are often used synonymously with regard to hallucinogen intake. Clarification of these terms, however, allows intoxications to be distinguished from psychoses. The terms *hallucinogen intoxication* and *hallucinogen-induced psychosis* are not uniformly defined and differentiated, which is a consequence of their very diverse psychopathological presentation. In order to define and evaluate these terms, it is necessary to understand the spectrum of effects produced by individual hallucinogens as well as the many factors that can influence the occurrence of intoxication and psychosis. Whereas intoxication may be directly traced back to the pharmacological effects of hallucinogens, a drug-induced psychosis can occur in the absence of recent hallucinogen use (ICD-10-CM: F16.9; F16.75).

11 Conclusions

Among psychiatrists, it is a widely held opinion that visual hallucinations predominate the hallucinogen intoxication, whereas auditory hallucinations predominate in schizophrenia. However, this view is not completely accurate, possibly due to unclear terminology; many of the so-called visual hallucinations occurring in hallucinogen intoxication correspond to complex perceptual disturbances that can only be described verbally with great difficulty (Spitzer 1988). Diverse changes in visual perception can also occur in the beginning phases of schizophrenia, although these symptoms often do undetected because most psychiatrists focus on the acoustic hallucinations that occur later in the developmental course of schizophrenia. Between the model psychosis produced by hallucinogens and the acute state of schizophrenia, there appears to be no basic difference in terms of the psychopathological phenomena that can occur. Therefore, we believe that the term model psychosis is well deserved and may be used as a tool to study "endogenous" psychoses (Bowers and Freedman 1966).

Experiments using such models in healthy subjects offer many advantages compared to the spontaneously developing psychoses. One major advantage is the fact that in normal subjects brain function is not impaired by a preexisting illness process. In addition, the findings before, during, and after remission of the hallucinogen intoxication may be compared using within-subject measures. Using this technique, it is possible to avoid the difficulties associated with the existence of large inter-individual differences in biological and psychological variables in schizophrenia patients. The aim of research is the identification of so-called linking variables that connect clinical psychopathological phenomena with the underlying biological factors (Callaway 1992).

The aforementioned considerations make it clear that states of hallucinogen-induced intoxication are not a specific model for functional endogenous psychosis; due to the relatively short duration of hallucinogen intoxication, the effects may at best resemble the symptoms of incipient psychosis. Depending on the dose, set (i.e., individual vulnerability to the effects of hallucinogens), and setting, the hallucinogen intoxication corresponds to the prodromal phase of schizophrenia in the sense of the "trema" according to Conrad (1958) and the type of dynamic instability according to Janzarik (1959). Sometimes, a fully developed psychosis (in the sense of dynamic overdrive according to Leuner (1962) and the "apophenia" or "apocalypse" according to Conrad (1958)) may develop, which may present as a catatonic state. The etiopathogenetic mechanisms of functional psychoses and of the hallucinogenic inebriation may be different, but in view of a putative common psychopathological final pathway, one may assume a similarity in their underlying neurobiological mechanisms. Therefore, analogical conclusions from the so-called experimental psychosis to naturally beginning psychotic illness processes may be justified. By combining psychological and neurobiological techniques, the experimental psychopathological approach can provide us with a valuable tool for psychiatric research (Sessa 2005).

References

Arnold OH, Hoff H (1953) Körperschemastörungen bei LSD 25. Wiener Zeitschrift für Nervenheilkunde und deren Grenzgebiete 6:259–274

Becker AM (1949) Zur Psychopathologie der Lysergsäurediäthylamidwirkung. Wiener Zeitschrift für Nervenheilkunde 2:402–440

Benedetti G (1995) Die Bleulersche Tradition der Schizophrenielehre und das Burghölzli als Stätte der Psychotherapie bei Schizophrenen. Schweizer Archiv für Neurologie und Psychiatrie 146:195–199

Beringer K (1927) Der Meskalinrausch: seine Geschichte und Erscheinungsweise. Springer, Berlin

Bleuler E (1911) Dementia praecox, oder, Gruppe der Schizophrenien. Deuticke, Leipzig

Bonnot O, Montalembert Md, Kermarrec S, Botbol M, Walter M, Coulon N (2011) Are impairments of time perception in schizophrenia a neglected phenomenon? J Physiol Paris 105:164–169

Boor Wd (1956) Pharmakopsychologie und Psychopathologie. Springer, Berlin

Bowers MB, Freedman DX (1966) Psychedelic experiences in acute psychoses. Arch Gen Psychiatry 15:240–248

Braus DF (2002) Wahrnehmen zeitlicher Relationen, neuronale Synchronisation und die Schizophrenien. Fortschritte Der Neurologie, Psychiatrie 70:591–600

Buckman J (1967) Theoretical aspects of LSD therapy. Int J Social Psychiatry 13(2):126–138.

Callaway E (1992) Psychopharmacology's need for linking variables. Biol Psychiatry 31:1–3

Chapman J (1966) The early symptoms of schizophrenia. Br J Psychiatry 112:225–251

Cohen S (1968) A quarter century of research with LSD. Charles C. Thomas, Illinois

Condrau G (1949) Klinische Erfahrungen an Geisteskranken mit Lysergsäure-Diäthylamid. Acta Psychiatr Scand 24:9–32

Conrad K (1948) Über differentiale und integrale Gestaltfunktion und den Begriff der Protopathie. Der Nervenarzt 19:315–323

Conrad K (1958) Die beginnende Schizophrenie: Versuch einer Gestaltanalyse des Wahns: Thieme Stuttgart

Dittrich A (1996) Ätiologie-unabhängige Strukturen veränderter Wachbewußtseinszustände: Ergebnisse empirischer Untersuchungen über Halluzinogene I. und II. Ordnung, sensorische Deprivation, hypnagoge Zustände, hypnotische Verfahren sowie Reizüberflutung. Stuttgart: VWB-Verlag für Wissen und Bildung

Dittrich A (1998) The standardized psychometric assessment of altered states of consciousness (ASCs) in humans. Pharmacopsychiatry 31:80–84

Döhmann M (1978) Eine empirische Untersuchung zum Schizophreniemodell der Bleuler-Schule. Dissertation, Universität Zürich

Dumont GJ, Wezenberg E, Valkenberg M, de Jong CA, Buitelaar JK, van Gerven JM, Verkes RJ (2008) Acute neuropsychological effects of MDMA and ethanol (co-) administration in healthy volunteers. Psychopharmacology 197:465–474

Freedman B, Chapman LJ (1973) Early subjective experiences in schizophrenic episodes. J Abnorm Psychol 82:46

Geyer MA, Vollenweider FX (2008) Serotonin research: contributions to understanding psychoses. Trends Pharmacol Sci 29:445–453

Gouzoulis-Mayfrank E, Habermeyer E, Hermle L, Steinmeyer Am, Kunert HJ, Sass H (1998) Hallucinogenic drug induced states resemble acute endogenous psychoses: results of an empirical study. Eur Psychiatry 13:399–406

Grof S (1967) Use of LSD in personality diagnostics and therapy of psychogenic disorders. Bobbs-Merrill, New York

Heimann H (1961) Ausdrucksphänomenologie der Modellpsychosen (Psilocybin). Eur Neurol 141:69–84.

Hellpach W (1941) Funktionelle Differenzierung der "psychischen Stimulantien". DMW-Deutsche Medizinische Wochenschrift 67:1358–1361

Hermle L, Fünfgeld M, Oepen G, Botsch H, Borchardt D, Gouzoulis E, Fehrenbach RA, Spitzer M (1992a) Mescaline-induced psychopathological, neuropsychological, and neurometabolic effects in normal subjects: experimental psychosis as a tool for psychiatric research. Biol Psychiatry 32:976–991

Hermle L, Oepen G, Spitzer M (1988) Zur Bedeutung der Modellpsychosen. Fortschritte der Neurologie Psychiatrie 56:48–58

Hermle L, Spitzer M, Borchardt D, Gouzoulis E (1992b) Beziehungen der Modell- bzw. Drogenpsychosen zu schizophrenen Erkrankungen. Fortschr Neurol Psychiatr 60:383–392

Hoch PH (1957) Remarks on LSD and Mescaline. J Nerv Ment Dis 125:442–443

Hoffer A, Osmond H (1966) Some psychological consequences of perceptual disorder and schizophrenia. Int J Neuropsychiatry 2:1–19

Hoffer A, Osmond H, Weckowicz T (1967) The hallucinogens. Academic Press, New York

Hollister LE (1961) Clinical, biochemical and psychologic effects of psilocybin. Arch Int Pharmacodyn Ther 130:42–52

Janzarik W (1959) Dynamische Grundkonstellationen in endogenen Psychosen. Springer, Berlin

Joel E, Fränkel F (1926) Der Haschisch-Rausch. J Mol Med 5:1707–1709

Kant F (1930) Über Reaktionsformen im Giftrausch. Eur Arch Psychiatry Clin Neurosci 91: 694–721

Klee GD (1963) Lysergic acid diethylamide (LSD-25) and ego functions. Arch Gen Psychiatry 8:461

Klosterkötter J, Albers E, Steinmeyer E, Hensen A (1994) Positive oder negative Symptome. Was ist brauchbarer für die Diagnose Schizophrenie? Der Nervenarzt, 444–453

Knauer A, Maloney WJ (1913) A preliminary note on the psychic action of mescalin, with special reference to the mechanism of visual hallucinations. J Nerv Ment Dis 40:425–436

Kraehenmann R, Küchenhoff B, Böker H, Schick M (2010) Katatones Dilemma unter Kombinationsbehandlung mit Lithium und Risperidon. Psychiatr Prax 37:306–309

Kraehenmann R, Vollenweider FX, Seifritz E, Kometer M (2012) Crowding deficits in the visual periphery of schizophrenia patients. PLoS ONE 7:e45884

Krystal JH, Bennett A, Abi-Saab D, Belger A, Karper LP, D'Souza DC, Lipschitz D, Abi-Dargham A, Charney DS (2000) Dissociation of ketamine effects on rule acquisition and rule implementation: possible relevance to NMDA receptor contributions to executive cognitive functions. Biol Psychiatry 47:137–143

Lahti AC, Weiler MA, Tamara Michaelidis BA, Parwani A, Tamminga CA (2001) Effects of ketamine in normal and schizophrenic volunteers. Neuropsychopharmacology 25:455–467

Langs RJ (1967) Stability of earliest memories under LSD-25 and placebo. J Nerv Ment Dis 144:171–184

Leuner H (1962) Die experimentelle Psychose. Ihre Psychopharmakologie, Phänomenologie und Dynamik in Beziehung zur Person: Versuch einer konditional-genetischen und funktionalen Psychopathologie der Psychose. Springer, Berlin.

Lewin L (1927) Phantastica. Stilke, Berlin

Lienert GA (1959) Changes in the factor structure of intelligence tests produced by D-lysergic acid diethylamide (LSD). Neuro-Psychopharmacol, 461–465 (Elsevier, Amsterdam)

Linton HB, Langs RJ (1962) Subjective reactions to lysergic acid diethylamide (LSD-25): measured by a questionnaire. Arch Gen Psychiatry 6:352–368

Lohmar D, Brudzinska J (2011) Founding psychoanalysis phenomenologically: phenomenological theory of subjectivity and the psychoanalytic experience. Springer, Berlin

Ludwig AM (1966) Altered states of consciousness. Arch Gen Psychiatry 15:225

Masters R, Houston J (2000) The varieties of psychedelic experience: the classic guide to the effects of LSD on the human psyche. Inner Traditions/Bear & Co, New York

Mullen P (2008) The mental state and states of mind. Cambridge University Press, Cambridge

Osmond H (1957) A review of the clinical effects of psychotomimetic agents. Ann N Y Acad Sci 66:418–434

Sandison RA (1954) Psychological aspects of the LSD treatment of the neuroses. Br J Psychiatry 100:508–515

Sass LA, Parnas J (2003) Schizophrenia, consciousness, and the self. Schizophr Bull 29:427–444

Savage C (1955) Variations in ego feeling induced by D-lysergic acid diethylamide (LSD-25). Psychoanal Rev, 1–16.

Savage C, Cholden L (1956) Schizophrenia and the model psychoses. J Clin Exp Psychopathol 17:405

Scharfetter C (1995) The self-experience of schizophrenics: empirical studies of the ego-self in schizophrenia, borderline disorders and depression. Private Publication, Zurich

Schneider K (1955) Klinische Psychopathologie. Thieme, Stuttgart

Schröter-Rosendahl C (1980) Beginnende Schizophrenie und verändertes Wachbewusstsein Gesunder: ein empirischer Vergleich mittels einer Selbstbeurteilungsskala. Dissertation, Universität Zürich

Sessa B (2005) Can psychedelics have a role in psychiatry once again? Br J Psychiatry 186: 457–458

Small IF, Small JG, Andersen JM (1966) Clinical characteristics of hallucinations of schizophrenia. Dis Nerv Syst 27:349

Spitzer M (1988) Halluzinationen: ein Beitrag zur allgemeinen und klinischen Psychopathologie. Springer, Berlin

Stockings GT (1940) A clinical study of the mescaline psychosis, with special reference to the mechanism of the genesis of schizophenic and other psychotic states. Br J Psychiatry 86:29–47

Stoll WA (1947) Lysergsäure-diäthylamid, ein Phantastikum aus der Mutterkorngruppe. Schweizer Archiv für Neurologie, Neurochirurgie und Psychiatrie 60:279

Süllwold L, Huber G (1986) Schizophrene Basisstörungen. Springer, Berlin

Vollenweider FX (2001) Brain mechanisms of hallucinogens and entactogens. Dialogues Clin Neurosci 3:265–279

Vollenweider FX (2008) Neurobiologie der Halluzinogen-Erfahrung. Hans Huber, Hogrefe AG, Bern

Vollenweider FX, Gamma A, Liechti M, Huber T (1998a) Psychological and cardiovascular effects and short-term sequelae of MDMA ("ecstasy") in MDMA-naive healthy volunteers. Neuropsychopharmacology 19:241–251

Vollenweider FX, Geyer MA (2001) A systems model of altered consciousness: integrating natural and drug-induced psychoses. Brain Res Bull 56:495–507

Vollenweider FX, Leenders KL, Scharfetter C, Antonini A, Maguire P, Missimer J, Angst J (1997a) Metabolic hyperfrontality and psychopathology in the ketamine model of psychosis using positron emission tomography (PET) and [< sup > 18 F] fluorodeoxyglucose (FDG). Eur Neuropsychopharmacol 7:9–24

Vollenweider FX, Leenders KL, Scharfetter C, Maguire P, Stadelmann O, Angst J (1997b) Positron emission tomography and fluorodeoxyglucose studies of metabolic hyperfrontality and psychopathology in the psilocybin model of psychosis. Neuropsychopharmacology 16:357–372

Vollenweider FX, Maguire RP, Leenders KL, Mathys K, Angst J (1998b) Effects of high amphetamine dose on mood and cerebral glucose metabolism in normal volunteers using positron emission tomography (PET). Psychiatry Research: Neuroimaging 83:149–162

Weyl B (1951) Versuch einer psychopathologischen analyse der LSD-Wirkung. Dissertation, Universität Freiburg

Winters WD (1975) The continuum of CNS excitatory states and hallucinosis. Hallucinations. Wiley, New York, pp 53–70.

Young BG (1974) A phenomenological comparison of LSD and schizophrenic states. Br J Psychiatry 124:64–74

Ziolko HU (1966) Persönlichkeitsabhängige Veränderungen in der experimentellen Psychose durch Lysergsäurediäthylamid. Arzneimittelforschung 16:249–251

Zubin J, Spring B (1977) Vulnerability: a new view of schizophrenia. J Abnorm Psychol 86:103

Zucker K (1930) Versuche mit Meskalin an Halluzinanten. Zeitschrift für die gesamte Neurologie und Psychiatrie 127:108–161

A Review of Hallucinogen Persisting Perception Disorder (HPPD) and an Exploratory Study of Subjects Claiming Symptoms of HPPD

John H. Halpern, Arturo G. Lerner and Torsten Passie

Abstract Hallucinogen persisting perception disorder (HPPD) is rarely encountered in clinical settings. It is described as a re-experiencing of some perceptual distortions induced while intoxicated and suggested to subsequently cause functional impairment or anxiety. Two forms exist: Type 1, which are brief "flashbacks," and Type 2 claimed to be chronic, waxing, and waning over months to years. A review of HPPD is presented. In addition, data from a comprehensive survey of 20 subjects reporting Type-2 HPPD-like symptoms are presented and evaluated. Dissociative Symptoms are consistently associated with HPPD. Results of the survey suggest that HPPD is in most cases due to a subtle over-activation of predominantly neural visual pathways that worsens anxiety after ingestion of arousal-altering drugs, including non-hallucinogenic substances. Individual or family histories of anxiety and pre-drug use complaints of tinnitus, eye floaters, and concentration problems may predict vulnerability for HPPD. Future research should take a broader outlook as many perceptual symptoms reported were not first experienced while intoxicated and are partially associated with pre-existing psychiatric comorbidity.

Keywords Hallucinogen Persisting Perceptual Disorder (HPPD) · Drug-induced flashback · Flashback · LSD · Hallucinogens · Posttraumatic Stress Disorder (PTSD) · Dissociation

---

J.H. Halpern (✉) · T. Passie
Laboratory for Integrative Psychiatry, Division of Alcohol and Drug Abuse, McLean Hospital, 115 Mill Street, Oaks Building, Belmont, MA 02478-1064, USA
e-mail: john_halpern@gmail.com

J.H. Halpern · T. Passie
Department of Psychiatry, Harvard Medical School, Boston, MA, USA

A.G. Lerner
Lev Hasharon Mental Health Center, Pardesya, Israel

A.G. Lerner
Sackler Medical School, Tel Aviv University, Tel Aviv, Israel

T. Passie
Senckenberg Institute for History and Ethics in Medicine, Goethe-University Frankfurt/Main, Frankfurt, Germany

Curr Topics Behav Neurosci (2016) 36: 333–360
DOI 10.1007/7854_2016_457
Published Online: 08 November 2016

Contents

1 Introduction

The use of hallucinogens in uncontrolled settings is widespread. In the USA, 19 % of adults by age 50 are estimated to have tried a hallucinogen (Johnston et al. 2013). Most such drug experiences occur without medically significant sequelae, but there have long been reports of subsequent perceptual effects in some users (Ellis 1898; Mayer-Gross 1931; Fischer 1971, 1976; Naditch 1974; Holsten 1976; Matefy et al. 1978).

1.1 Definitions of Hallucinogen Persisting Perceptual Disorder (HPPD)

The first formal description of "a repetition of the acute phase of the experience days or even weeks after the initial doses" emerged from a study of LSD-assisted psychotherapy (Sandison and Whitelaw 1957). Around 1970, the term "flashback" began to appear in the literature; as in Heaton and Victor (1976): "A flashback is the transient recurrence of psychedelic drug symptoms after the pharmacologic effects of such drugs have worn off and a period of relative normalcy has occurred." The ICD-10 (World Health Organization 1992) lists "F16.283 hallucinogen dependence with hallucinogen persisting perception disorder (HPPD) (flashbacks)" as temporary, short-lived re-experiences of aspects of the initial drug intoxication. Clinically meaningful impairment and/or suffering are required for its diagnosis. The DSM-V (American Psychiatric Association 2013) lists "HPPD (Flashbacks)" as a typically temporary re-experience of aspects of the drug intoxication. It also includes a subform involving long-term visual disturbances (Textbox 1). Identification of this form of HPPD is based on work by one psychiatrist researching this specific domain (Abraham 1982, 1983; Abraham and Aldridge 1993). Systematic and group studies exist for the brief temporary "flashback" type and some case studies for the chronic subtype; only a few studies have examined flashbacks in groups of users (Halpern and Pope 2003).

Textbox 1

DSM-V criteria for HPPD
(A) The reexperiencing, following cessation of use of a hallucinogen, of one or more of the perceptual symptoms that were experienced while intoxicated with the hallucinogen (e.g., geometric hallucinations, false perceptions of movement in the peripheral visual fields, flashes of color, intensified colors, trails of images of moving objects, positive afterimages, halos around objects, macropsia, and micropsia)
(B) The symptoms in Criterion A cause clinically significant distress or impairment in social, occupational, or other important areas of functioning
(C) The symptoms are not due to a general medical condition (e.g., anatomical lesions and infections of the brain, and visual epilepsies) and are not better accounted for by another mental disorder (e.g., delirium, dementia, and Schizophrenia) or hypnopompic hallucinations

To meet DSM-V criteria, hallucinogen use must precede the syndrome. The word "re-experiencing" in criterion A indicates that the symptom should resemble that of an actual hallucinogen-induced experience.

The requirement for "distress or impairment" in criterion B suggests that perceptual phenomena should be outside the range of normal experience. For example, seeing bright spots in front of one's eyes upon entering a dark room should not qualify.

Criterion C requires that alternative etiologies for perceptual changes be considered before diagnosing HPPD. DSM-V cites visual seizure, migraine, delirium, dementia, schizophrenia, hypnopompic hallucination, post-traumatic stress disorder (PTSD), and depersonalization and derealization associated with significant anxiety and depression as specific disorders to rule out. Finally, one must exclude other hallucinogen-induced disorders recognized by DSM-V, such as hallucinogen-induced psychotic, mood, or anxiety disorders.

1.2 Two Entities of HPPD

These definitions and other work (Hermle et al. 2008; Holland and Passie 2011) may warrant distinguishing two types of HPPD. We offer these tentatively to help define the findings of this article in terms of the literature. We note that Abraham's definition of HPPD with continuous visual perception disorders is still a subject of debate due to the absence of replication studies and methodological concerns (Halpern and Pope 2003; Studerus et al. 2011; Holland and Passie 2011).

1. *Type 1 HPPD* (consistent with the ICD-10 definition) consists of brief re-experiences of alterations in perception, mood, and/or consciousness, as previously experienced during a hallucinogenic intoxication. Brevity, infrequency, and intermittency of symptoms are signified in the concept of a "flashback" (ICD-10). "Flashbacks are sudden and unexpected re-experiences of aspects of a psychedelic drug trip that happened weeks, months, or even years before" (Matefy et al. 1979). Type 1 HPPD symptoms may be pleasurable and even controllable (Hasse and Waldmann 1971; Holsten 1976). They appear days to months after the hallucinogen-induced experience, sometimes without apparent cause. The subject is usually aware of the unreality of the experience. Often symptoms are visual increases in perceived color intensity, dimensionality, or vibrancy; illusory changes; and/or movement of a perceived object. The perception of time may be altered. Strong emotion felt during the drug experience may recur and in some cases ego boundaries can become diffuse. A significant element of the definition of HPPD in the ICD-10 is: "Flashbacks may be distinguished from psychotic disorders partly by their episodic nature, frequently of very short duration (seconds or minutes), and by their duplication (sometimes exact) of previous drug-related experiences." Relevant reviews concluded that "they are usually self-limited and diminish in duration, intensity and frequency with time…" (Strassman 1984; see also Horowitz 1969; Siegel and Jarvik 1975; Holland and Passie 2011).
2. *Type 2 HPPD* (consistent with Abraham (1983) and part of the definition in the DSM-V) entails constant or near-constant visual effects. These can include the following:
    1. Palinopsia: the persistent perception of an object removed from view;
    2. halos: a brightening glow or colored shining/shimmering surrounding objects;
    3. trails or akinetopsia: a series of discrete positive afterimages following in the wake of moving objects; and
    4. visual snow: a TV static-like graininess superimposed upon the visual field.

Symptoms may occur alone or in combination. Sound and other perceptions are unaffected. In most cases, visual phenomena are reported to be uncontrollable and disturbing, though some individuals regard them as enriching (Baggott et al. 2011). Claimed constant visual phenomena are often accompanied by mild-to-moderate depersonalization, derealization, anxiety, or depression (Holland and Passie 2011). These psychopathological states are claimed to trigger the occurrence and intensity of visual phenomena (Abraham 1982, 1983; Abraham and Duffy 1996, 2001) depending on the waxing and waning nature of current affect.

Interestingly, Type 2 HPPD was never clearly reported during the 1960s when millions of Americans took LSD on a regular basis with less knowledge about hallucinogens and more resultant complications. HPPD was not described in the comprehensive retrospective surveys of LSD use in psychotherapy in approximately 10,000 patients during the 1950–1960s (Cohen 1960; Malleson 1971; Passie 1997).

1.3 Prevalence

Data do not permit us to estimate, even crudely, HPPD's prevalence according to DSM-V or ICD-10 criteria. Although millions of doses of hallucinogens were consumed by millions of individuals since the 1960s (SAMHSA 2011), few large HPPD case series were reported. Horowitz (1969), Cohen (1960, 1977) estimate the incidence of Type 1 HPPD in a population of regular hallucinogen users in the 1960–1970s as 1:20. Type 2 HPPD, if it exists as a reliable and distinct entity, appears to be very rare (Hermle et al. 2008, 2015; Holland and Passie 2011). Grinspoon and Bakalar (1997) estimate that Type 2 HPPD occurs in 1 of 50,000 hallucinogen users. Baggott et al. (2011) collected data online in a Web-based questionnaire from 2455 individuals reporting visual experiences while drug-free that resembled a past hallucinogen intoxication. Most of these experiences were simple, non-disturbing "flashbacks," while 4.2 % found these visual phenomena significant enough to at least contemplate seeking treatment.

1.4 Reviews of Data and Theories on HPPD

Comprehensive reviews of the literature (Halpern and Pope 2003; Holland and Passie 2011) show that HPPD definitions vary broadly in the scientific literature. The disorder's clinical relevance and etiologies remain unclear. Causation may be linked to a complex set of triggers alone or in combination (see Fig. 1).

The known neurochemical activity of hallucinogens is poorly correlated with their physiological and cognitive effects (Brimblecombe and Pinder 1975; Nichols 2004; Passie and Halpern 2014). We have virtually no data on the processes occurring during the latency between drug effect and flashback or on what predisposing vulnerabilities may result in the two types of HPPD. HPPD may also easily be confused or misdiagnosed for some other ophthalmological, neurological, or psychopathological phenomena (see Materials and Methods below for a list). Several studies show HPPD-like experiences (intense memories, depersonalization, derealization, and over-intensification of perceptional phenomena) occur quite often in normal, healthy populations (Parish 1894; Shor 1960; Dixon 1963; Kokoszka 1992–1993). Even Abraham (1984) acknowledges that several non-LSD exposed individuals in his study on visual phenomenology of the LSD flashback (1983) described visual disturbances similar to those reporting LSD flashbacks (although with much less intensity and number of symptoms).

Holland and Passie's (2011) evaluation of proposed etiological models found that in every case, individuals reporting flashbacks experienced some elements relating to the original experience—level of arousal, music playing, environmental cues, ingesting the same kind of drug, time of day, and so forth. Therefore, different etiologies may apply to each specific case. But for every case, the formative causes of such associations may vary—sensitization effects, trauma and reaction patterns, state-dependent memory, psychophysical vulnerabilities, and more (Fig. 1).

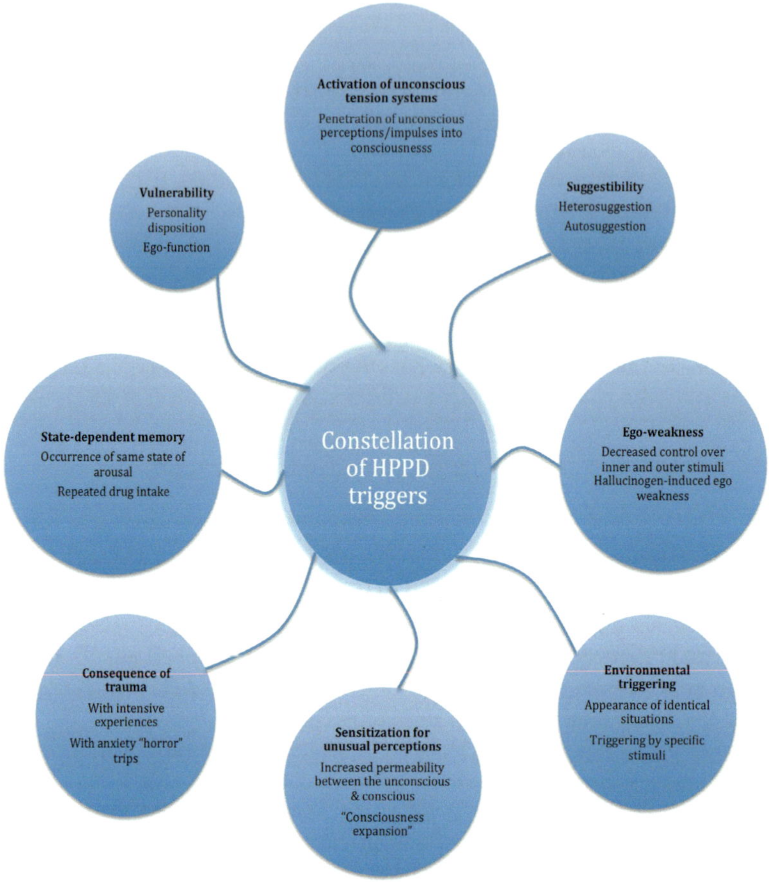

Fig. 1 Factors and triggers for the occurrence of HPPD phenomena (based on Holland and Passie 2011). According to the model of Holland and Passie (2011), a different pattern of factors (alone and/or in combination as well as their specific quantitative influences) contributes to the occurrence of HPPD in every instance

1.5 Risk Factors Associated with HPPD?

There are no recognized risk factors for HPPD (Halpern and Pope 2003). Some report a relationship between flashbacks and number of drug exposures (McGlothlin and Arnold 1971; Abraham 1983), but others have not (Horowitz 1969; Stanton and Bardoni 1972; Matefy et al. 1978). Abraham (1982, 1983) speculates there might be a very rare psychophysical vulnerability to a supposed toxic effect of LSD in Type 2 HPPD. In Type 1 HPPD, some older studies suggest pre-existing personality features (Naditch and Fenwick 1977), suggestibility

(Heaton and Victor 1976), or pre-existing psychopathology (Abraham and Duffy 1996) as possible contributing factors.

1.6 Treatment Options

No controlled treatment studies exist. Treatment of Type 1 HPPD is obviously brief; only very rarely will Type 1 HPPD lead to clinically relevant pathology. For those claiming Type 2 HPPD, improvements have been reported with sunglasses (Abraham 1983) and psychotherapy (Abraham et al. 1996). With Type 2 HPPD, antipsychotic drugs worsened some symptoms (Abraham and Mamen 1996; Morehead 1997, Lerner et al. 2002; Goldman et al. 2007). SSRIs worsened 4 cases documented by Markel et al. (1994), but other clinicians report improvements (Young 1997; Aldurra and Crayton 2001). Anti-seizure drugs and clonidine were also used with some success (Alarcon et al. 1982; Lerner et al. 2000). It is not easy to determine how best to treat HPPD given this literature. The widely variable, partially contradictory findings may require us to speculate on placebo effects, idiosyncratic neurochemistry, and spontaneous recovery, or perhaps more simply, an inadequately defined HPPD.

2 Materials and Methods

We conducted a study with a questionnaire specifically developed to identify prevalence and characteristics of self-reported altered perception experiences in hallucinogen users and to find relationships with drug use. Despite the obvious limitations of self-report questionnaires, a more carefully and thoroughly designed Web application seemed appropriate for delineating types of visual phenomena, triggering drug experiences, pre-existing medical conditions, general drug using habits, personality features, and more.

We sought individuals reporting persisting disorders of perception that started or worsened after a "triggering event," usually a drug intake, but other causes were explicitly not excluded. In this way, those who experienced problems prior to drug use, or even without a history of drug use, as well as those with a hallucinogen-related onset could complete the survey. Participants first gave informed consent as approved by the Institutional Review Board of McLean Hospital, assuring confidentiality. Survey software was LimeSurvey (v1.1; www.limesurvey.org) and was hosted on hospital servers for completion by adults aged 18 or older with access to the Internet. Recruitment occurred by word of mouth, postings about the survey at Web sites devoted either to the HPPD community or to those interested in the effects of hallucinogens, or from informing self-identified HPPD patients who had sought out the authors about the condition in response to the authors' prior publications on this condition. No compensation was provided for participation.

Subjects were asked about date of birth, marital status, level of education attained, employment, and family mental health and drug abuse. Then, a differential-diagnosis list was presented with yes/no buttons for disorders that might account for HPPD-type symptoms: HPPD, "visual snow," brain lesion, brain infection/ meningitis, seizure disorder in general, temporal lobe epilepsy, persistent migraine aura, schizophrenia or other psychotic disorder, bipolar disorder, PTSD, borderline personality disorder, conversion disorder, hypochondriasis, dissociative disorder (including Dissociative Identity Disorder), delirium, dementia, corneal or retinal disorder or damage to the eye in general, optic neuritis, multiple sclerosis, Charles Bonnet Syndrome, or Lyme disease.

The survey asked participants to rank the severity of disturbance across senses (vision, hearing, smelling, balance, touch, taste, and pain). Subjects were required to detail which drug, substances, or other triggering events they associate with their subsequent disorder. They were asked to quantify dose, frequency of use, drugs used in combination, prior drug use, drug use since HPPD-like symptoms commenced, and to list and rank which substances may worsen or improve the condition from an extensive list of drugs and drug categories. Subjects had to select a defined time interval from time of triggering event to time of development of HPPD-like symptoms. Subjects were asked about the presence of anxiety or panic before, during, and after the triggering event, including whether prior experiences with the same offending drug included anxiety or panic. Subjects listed the number of doctors, if any, they sought for treatment of their perception disorder. They were asked whether their condition made them contemplate or attempt suicide. They also had access to textboxes to write freely about their situations.

Based on clinical experience, subjects were asked about 21 forms of visual disturbance and 4 other symptoms (problematic concentration, communication, auditory hallucination, and tinnitus). For each statement, the participant had to declare whether or not the symptom presented occurred before the triggering drug event, during the worst episode of perception distortions, and in the last 30 days. For each of these three time points, subjects were asked to rank from 0 to 100 the severity of the symptom presented, the time duration and frequency the symptom would occur, and how much the particular symptom reminds them of what they experienced during the HPPD-triggering drug event.

Participants completed the 28-question Dissociative Experiences Scale (DES), a reliable, validated self-report measure quantifying the frequency of dissociative experiences (Bernstein-Carlson and Putnam 1986; Carlson and Putnam 1993). There are also three subscales evaluating forms of dissociation: amnestic (memory losses), absorption and imaginative involvement (preoccupations that distract from

Fig. 2 Photograph 1. Negative afterimage

present occurrences), and depersonalization and derealization (detachments from sense of self and/or mental function; sensations of unreality). Higher scores correlate with increased clinical severity of dissociation. Depending on population, normative scores range from 4 to just below 20. Scores greater than 20 are suggestive of PTSD or a dissociative disorder. Scores of 45 or more suggest severe conditions such as Dissociative Identity Disorder.

Finally, four photographs (see Figs. 2, 3, 4 and 5) and one short animation (http://www.youtube.com/watch?v=y63juPiMHu4) were presented, and subjects were asked to quantify how well they represented aspects of their symptoms.

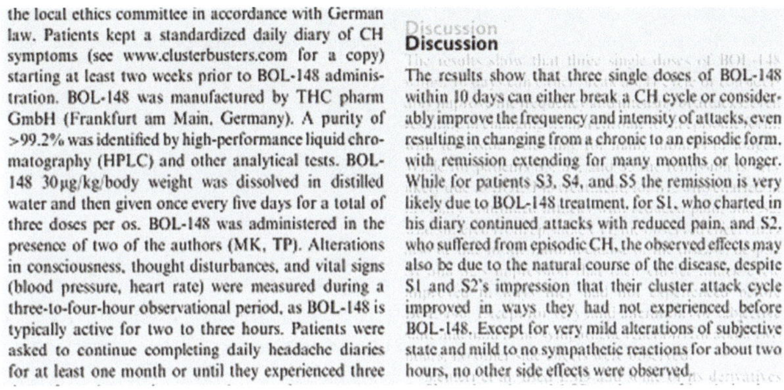

the local ethics committee in accordance with German law. Patients kept a standardized daily diary of CH symptoms (see www.clusterbusters.com for a copy) starting at least two weeks prior to BOL-148 administration. BOL-148 was manufactured by THC pharm GmbH (Frankfurt am Main, Germany). A purity of >99.2% was identified by high-performance liquid chromatography (HPLC) and other analytical tests. BOL-148 30 μg/kg/body weight was dissolved in distilled water and then given once every five days for a total of three doses per os. BOL-148 was administered in the presence of two of the authors (MK, TP). Alterations in consciousness, thought disturbances, and vital signs (blood pressure, heart rate) were measured during a three-to-four-hour observational period, as BOL-148 is typically active for two to three hours. Patients were asked to continue completing daily headache diaries for at least one month or until they experienced three

Discussion

The results show that three single doses of BOL-148 within 10 days can either break a CH cycle or considerably improve the frequency and intensity of attacks, even resulting in changing from a chronic to an episodic form, with remission extending for many months or longer. While for patients S3, S4, and S5 the remission is very likely due to BOL-148 treatment, for S1, who charted in his diary continued attacks with reduced pain, and S2, who suffered from episodic CH, the observed effects may also be due to the natural course of the disease, despite S1 and S2's impression that their cluster attack cycle improved in ways they had not experienced before BOL-148. Except for very mild alterations of subjective state and mild to no sympathetic reactions for about two hours, no other side effects were observed.

Fig. 3 Photograph 2. Ghosted text in right column (portion of photograph 1)

Fig. 4 Photograph 3. Positive afterimage

Fig. 5 Photograph 4. Repeating pattern only in snow (portion of photograph 3)

The photographs included a negative afterimage of a flower in a different color, a page of text with a ghosted second text over it, a positive afterimage, and a repeating geometric pattern superimposed upon a winter scene. The animation was a single photograph with a repeating loop of flickering grain to simulate "visual snow."

3 Results

Though subjects could save their answers and resume later (3 did), only 23 subjects completed the survey out of 67 who started it. The survey took 2 h to complete on average. Of the 23, two were healthy normals. One individual noted persisting perception problems after head trauma and temporal lobe epilepsy. These 3 individuals were excluded from further evaluation for not reporting Type 1 or Type 2 HPPD-like symptoms. One (Subject 5) reported "visual snow" since age 5 with no drug use but is included in the dataset because of HPPD-like symptoms.

Nineteen subjects (15M/4F) reported persisting perceptual disturbances triggered or worsened by past drug use. Sixteen were evaluated by physicians because of their disorder. Six were (co-)diagnosed with HPPD (all of the Type 2 variety), 3 persistent migraine aura, 2 psychotic disorders, 1 PTSD and 3 other anxiety disorders, 2 depression, 2 hypochondriasis (one of whom claims related eye injury and PTSD), and 3 dissociative disorders (one also having a history of psychosis). Family psychiatric histories included 6 reporting depression and/or anxiety in a primary relative.

In terms of demographics, mean age was 25.8 years (median 24.5, 18–40 range). Three were married and 17 single (2 cohabiting). Educational levels included 1 grade school graduate, 1 some high school, 3 high school graduates, 2 technical/vocational graduates and 1 who had started but not finished, 5 with some college, 6 college graduates, 1 with some graduate education, and 1 completed a master's degree. All participants self-identified as white. Prior-year income was 3 unemployed or on disability, 5 earning less than $10,000, 4 less than $25,000, 4 less than $50,000, 3 less than $75,000, and 1 more than $250,000.

In the survey, 17 complained of active symptoms consistent with Type 2 HPPD. When asked about perception disturbance symptoms (see Table 4), all 20 subjects noted they experienced some abnormality in the prior 30 days including all 20 reporting nighttime visual snow. All 20 report a chronic condition with 4 having symptoms of 1–6 months, 1 for less than 1 year, 4 for 1–2 years, and 11 for years longer. All 20 selected vision as their most significantly altered sense.

Other than Subject 5, all believe a drug triggered their perception disorder or markedly worsened visual symptoms associated with persisting migraine with aura (subjects 3 and 14). Seven subjects report symptoms starting after a single drug exposure. LSD was most commonly identified (12 of 19) and then psilocybin (4 of 19) (see Table 1). Four reported other substances: Subject 2, MDMA with alcohol; Subject 3, marijuana; Subject 13, 2C-I (2,5-dimethoxy-4-iodophenethylamine); and Subject 11, amphetamines, opiates, and an SSRI. Subjects 2, 3, and 13 had never tried LSD or psilocybin prior to their triggering experience. Subject 11 extensively used hallucinogens years before his disorder started (see Table 2), which occurred on his first day on the SSRI antidepressant citalopram, intensifying over the two weeks he continued to take it. Overall, subjects described an extensive range of drug and alcohol use histories. Twelve individuals' reports met criteria for a drug use disorder for one or more substances (see Table 2) prior to the start of persisting perception problems and 8 did after such problems started (Table 3). Subjects reported decreasing hallucinogen use after the start of their persisting perception disorder (with, e.g., 7 individuals admitting to use of LSD prior to the disorder and only one subject reporting subsequent LSD use).

Twelve claimed perception disturbances began during or within 24 h of the triggering drug experience (Table 1). Seven described disturbances starting one week to months after the experience (the rest of the individuals, as listed in Table 1, had symptoms start within 1 week or 1 month or longer). Five of the 7 felt they had no explanation for their condition other than this past drug exposure, yet only 4 of these 7 found perceptual disturbances reminiscent of the drug intoxication. In fact, only 8 of 19 subjects agreed to any degree that their symptoms are "exactly like" their triggering drug experience (see Table 4). However, all 19 recalled anxiety and/or panic reactions while on the drug. Of the 14 who described the intensity of their anxiety, 1 selected "mild," 1 "moderate," 5 "marked," and 7 "extreme." Three admit to psychiatric hospitalization because of their perception disorder.

Table 4 presents the 25 statements of symptoms (see also Textbox 2 for a typical HPPD history). Subject 5, with persisting migraine aura and complaints of visual snow, reported no history with hallucinogens or other drugs of abuse, yet she

Table 1 Drug use reported as contributing to triggering a persisting perception disorder

When did the altered perceptions start?	Subject #	Alcohol	Tobacco	MJ	Synthetic MJ	LSD	Psilocybin	2C	MDMA	Amphe-tamines	Opiates	Sedative-hypnotics	SSRI	Other
Within 3 months	1		Y				Y							
Within 24 h	2	Y							Y					
Almost immediately	3			Y										
Almost immediately[b]	4		Y	Y		Y	Y							
Started age 5. Never any drug use[a]	5													
Almost immediately	6					Y								
Almost immediately	7			Y		Y								
Almost immediately	8					Y								
Within 3 months[b]	9		Y	Y		Y			Y			Y	Y	
Almost immediately	10					Y								
Almost immediately	11									Y	Y		Y	
Within a month	12					Y[c]		Y[c]						

(continued)

Table 1 (continued)

When did the altered perceptions start?	Subject #	Alcohol	Tobacco	MJ	Synthetic MJ	LSD	Psilocybin	2C	MDMA	Amphe-tamines	Opiates	Sedative-hypnotics	SSRI	Other
Within 24 h	13							Y						
Within 24 h	14	Y	Y	Y		Y								
Within a month	15						Y		Y					
Almost immediately	16		Y	Y		Y								
Within a month	17	Y	Y	Y		Y		Y	Y					
A week after	18	Y	Y	Y	Y		Y		Y					LSA & Bromo-dragonfly
A week after	19	Y				Y								
Almost immediately	20	Y					Y							amoxicillin

Each subject who consumed one or more of the following drugs considered them to contribute or somehow be related to resultant disorders of perception. For some, there was a brief period of heavy polydrug use over a short time-span, for others they distinctly refer to a single drug event. Drugs that no one responded as having taken are not listed. We asked specifically about use of alcohol, tobacco, cannabis, synthetic cannabinoid analogs, 2-C series hallucinogens, DMT, mescaline, MDMA, 5-MeO-DIPT, alpha-methyltryptamine, Salvia divinorum, ketamine, PCP, dextromethorphan, cocaine, stimulant/amphetamines, inhalants, sedative-hypnotic/anxiolytics, traditional antipsychotics, atypical antipsychotics, SSRIs, TCAs, SNRIs, opiate antagonists, anticonvulsants, and Other. Subjects 3 and 14 have persisting migraine aura which pre-dated their drug use and found drug use worsened their visual symptoms. Shaded rows signify that the subject reported their perception disorder commenced after only a single drug exposure/event

Y = yes, drug consumed/contributed to disorder

[a]History of psychotic disorder

[b]Subject 5 with persisting migraine aura and no drug use associated with perception disturbances

[c]Subject is unsure if this was the drug consumed

Table 2 Drug use reported before persisting perception disorder started

Subject #	Drug																		
	Alcohol	Tobacco	MJ	Synthetic MJ	LSD	Psilocybin	2C	Mescaline	MDMA	S. divinorum	Ketamine	PCP	Cocaine	Amphetamines	Opiates	Inhalants	Sed/ Hypnotics	SSRI	SNRI
1	Y	Y	Y											Y					
2	Y		Y						Y				Y						
3	Y	Y	Y						Y				Y	Y	Y		Y		
4	Y	Y	Y																
5																			
6	Y	Y	Y																
7																			
8	Y	Y	Y														Y		
9		Y	Y			Y			Y				Y			Y			
10			Y																Y
11	Y	Y	Y		Y	Y		Y	Y		Y		Y	Y	Y				
12	Y				Y		Y[a]												
13							Y		Y					Y				Y	
14	Y	Y			Y				Y				Y						
15	Y		Y	Y	Y	Y	Y	Y	Y	Y				Y					
16	Y	Y	Y	Y	Y				Y	Y	Y		Y	Y		Y			
17	Y	Y	Y	Y	Y	Y			Y	Y									Y
18	Y	Y	Y																
19	Y		Y		Y							Y				Y			
20	Y																		

The following drugs were consumed BEFORE persisting perception problems started (or worsened for subjects 3 and 14) and such use is separate from association with the drug-triggering event ("Y" = yes, drug consumed; blank = drug not consumed). Any evidence of drug abuse or dependence results in a "**Y**" that is bolded. We asked specifically about use of alcohol, tobacco, cannabis ("MJ"), synthetic cannabinoid analogs ("synthetic MJ"), 2-C series hallucinogens, DMT, mescaline, MDMA, 5-MeO-DIPT, alpha-methyltryptamine, Salvia divinorum, ketamine, PCP, dextromethorphan, cocaine, stimulant/amphetamines, inhalants, sedative-hypnotic/anxiolytics, traditional antipsychotics, atypical antipsychotics, SSRIs, TCAs, SNRIs, opiate antagonists, anticonvulsants, and Other. Drugs that no one responded as having taken are not listed.

[a]Subject is unsure if this was the drug consumed.

Table 3 Drug use reported after persisting perception disorder started

Subject #	Drug																				
	Alcohol	Tobacco	MJ	Synthetic MJ	LSD	Psilocybin	2C	MDMA	S. divinorum	Ketamine	Cocaine	Amphetamines	Opiates	Inhalants	Sed/ Hypnotics	Atypical antipsychotics	SSRI	SNRI	Opioid antagonist	Anticonvulsant	Other
1	Y	Y	Y																		
2	Y							Y							Y						
3								Y					Y		Y						
4	Y	Y	Y																		
5	Y	Y													Y	Y	Y				
6	Y																				
7																					
8	Y	Y	Y																		
9	Y	Y	Y					Y						Y							
10															Y	Y	Y				
11	Y		Y										Y								
12																					
13		Y					Y	Y				Y									Mephedrone
14	Y	Y	Y									Y					Y				
15	Y		Y	Y					Y						Y	Y			Y		
16	Y	Y	Y	Y				Y		Y				Y	Y					Y	Mephedrone
17	Y	Y	Y	Y																	
18	Y	Y	Y			Y		Y				Y	Y		Y	Y	Y	Y			
19	Y		Y		Y						Y			Y	Y		Y				
20	Y														Y			Y			

The following drugs were consumed AFTER persisting perception problems started (or worsened for subjects 3 and 14) and such use is separate from association with the drug-triggering event ("Y" = yes, drug consumed; blank = drug not consumed). Any evidence of drug abuse or dependence results in a "**Y**" that is bolded. We asked specifically about use of alcohol, tobacco, cannabis ("MJ"), synthetic cannabinoid analogs ("synthetic MJ"), 2-C series hallucinogens, DMT, mescaline, MDMA, 5-MeO-DIPT, alpha-methyltryptamine, Salvia divinorum, ketamine, PCP, dextromethorphan, cocaine, stimulant/amphetamines, inhalants, sedative-hypnotic/anxiolytics, traditional antipsychotics, atypical antipsychotics, SSRIs, TCAs, SNRIs, opiate antagonists, anticonvulsants, and Other. Drugs that no one responded as having taken are not listed.

Table 4 Queried symptoms of persisting perception disorder

Symptom	# of subjects (max = 19) reporting symptom prior to drug use that triggered persisting perception disorder	Mean severity of symptom (0–100)	# of subjects (max = 20) reporting symptom during worst persisting perception episode	Mean severity of symptom (0–100)	# of subjects (max = 20) reporting symptom over the prior 30 days	Mean severity of symptom (0–100)	% of subjects who find symptom a little or more reminiscent of what was experienced during the drug intoxication(s) attributed to triggering persisting perception disorder (%)
Difficulty with color identification	1+	32	11++	43.5	9++	38.7	72.7
Seeing "halos" or "auras" around objects/people	2	14.5	18++*	69.8	18++*	58.7	88.2
Stationary objects appear to sway or move	1	27	15++*	68.5	15++*	48.6	100
See objects or faces when pressure placed to closed eyes	3+	7.3	11++*	56	10++*	39.6	80
Colored objects change in brightness	1	5	12++	58.3	12++	36.6	75
Macropsia	0	–	6	40.2	4	14.5	100
Micropsia	0	–	4*	35.3	2	17.5	100
Afterimage in other color	7++	4.7	18++*	67.2	18++*	52.1	88.2

(continued)

Table 4 (continued)

Symptom	# of subjects (max = 19) reporting symptom prior to drug use that triggered persisting perception disorder	Mean severity of symptom (0–100)	# of subjects (max = 20) reporting symptom during worst persisting perception episode	Mean severity of symptom (0–100)	# of subjects (max = 20) reporting symptom over the prior 30 days	Mean severity of symptom (0–100)	% of subjects who find symptom a little or more reminiscent of what was experienced during the drug intoxication(s) attributed to triggering persisting perception disorder (%)
Afterimage in same color	3++	4	15++*	71	15++*	59	85.7
Trailing image to moving object	3+	9.3	17++*	61.9	17++*	46.5	93.8
Seeing faces/objects in wood, clouds, trees	4++	11.3	7	54.4	6	32.6	71.4
“TV Static” (visual snow) projected over vision in DAYTIME	5+	16.8	19++*	69.4	19++*	62.4	66.7
“TV Static” (visual snow) projected over vision in NIGHTTIME	8	10.8	20++*	77.2	20++*	60.9	78.9
“Floaters” in field of vision	13++	23.5	18++*	69.6	18++*	59.3	70.6

(continued)

Table 4 (continued)

Symptom	# of subjects (max = 19) reporting symptom prior to drug use that triggered persisting perception disorder	Mean severity of symptom (0–100)	# of subjects (max = 20) reporting symptom during worst persisting perception episode	Mean severity of symptom (0–100)	# of subjects (max = 20) reporting symptom over the prior 30 days	Mean severity of symptom (0–100)	% of subjects who find symptom a little or more reminiscent of what was experienced during the drug intoxication(s) attributed to triggering persisting perception disorder (%)
Ghosted afterimage of text being read	1	20	16++*	76.4	15++	67.6	66.7
"Flash" of bright light appearing without explanation	2	2.5	11 + *	66 8	12 + *	50.7	72.7
Geometric patterns in field of vision whether eyes open or closed	2	12.5	13++*	56	12++*	40.8	100
Difficulty communicating thoughts	9+	8.8	18++*	63	18++*	43.2	88.2
Perceiving room as moving	1	1	16++*	49.4	13++*	38.8	93.3
Auditory hallucinations ("voices")	0	–	7++*	29.3	4++	16.8	66.7
Feel pressure in the head	4	6.5	16++*	58	15++*	34	13

(continued)

Table 4 (continued)

Symptom	# of subjects (max = 19) reporting symptom prior to drug use that triggered persisting perception disorder	Mean severity of symptom (0–100)	# of subjects (max = 20) reporting symptom during worst persisting perception episode	Mean severity of symptom (0–100)	# of subjects (max = 20) reporting symptom over the prior 30 days	Mean severity of symptom (0–100)	% of subjects who find symptom a little or more reminiscent of what was experienced during the drug intoxication(s) attributed to triggering persisting perception disorder (%)
Difficulty to light accommodation	6+	6.7	19++*	63	19++*	50.6	77.8
Trouble with concentration	14+	23	19++*	75.4	18++*	59.7	83.3
Faces appear distorted	0	–	11++*	41.2	7++	30.1	100
Tinnitus (ringing in ears)	12++	7.4	19++*	49.4	19++*	32	86.7

"+"/"++" = the # of subjects includes either one subject ("+") with a history of psychotic disorder or both subjects ("++") with that history, "*" = the # of subjects includes Subject 5 with persisting migraine aura and no drug use associated with perception disturbances; "+" and "*" is to show how symptoms broadly overlap between those with a history of psychosis, the individual with no drug use, and those reporting HPPD symptoms post-drug use reports symptoms similar to those who claim an association between their drug use and perceptual disorders.

Textbox 2

Typical history of type 2 HPPD (subject 1)
The long story short is I ate 2.5 g of strong shrooms. My family has a history of anxiety and depression, which I was not really aware of at the time. The trip itself started amazingly then took a turn for the worse when I got a stomachache. I bad tripped for a couple hours, but I told myself it would end at some point and mainly relegated myself to a chair and just chilled with my eyes closed. It was a bad setting with many people coming in and out of my residence, and this definitely made the vibe worse. Eventually, I felt better, and the trip was cool again…all ended well. Several months later, I started having trouble sleeping due to extremely bright closed-eye blotchy shapes that oozed around whenever I tried to sleep. Also at this time, I started having severe panic attacks that brought me to the hospital several times (only to be told that I was fine). The next 8 months was hell. I was convinced I was slowly going crazy, and that I would have to be committed. Then I found out about HPPD and have worked to beat it. At this point, I am happy. I have occasional anxiety, but my symptoms of depression and depersonalization have gone down drastically. In turn, my visual symptoms are less as well, although I still notice them every day (they don't bother me as much now). I have a ringing in my ears a lot of the time though, which is very bothersome. I have the sensation that I'm hearing sounds sometimes, but I usually chalk it up to being anxious and hypersensitive to my environment. Many times, I have found the source of the sound that I suspected to be not real (i.e. a beeping watch under some clothes, etc.) I find the fear of going crazy/hearing voices/seeing things that aren't there the permeating factor in many people with HPPD I've talked to, also including myself. Even though I've never actually heard or seen anything not real (besides HPPD visual phenomena which I don't consider "seeing" but rather "perceiving"), I have a worry that I soon will. I have largely gotten over this worry

To diagnose HPPD, symptoms must meet the DSM-V Criterion A: "The re-experiencing, following cessation of use of a hallucinogen, of one or more of the perceptual symptoms that were experienced while intoxicated with the hallucinogen." Of the 19 subjects associating their symptoms with drug use, only 5 symptoms were agreed by all who had that symptom that it felt somewhat or more like a re-experiencing of the triggering drug experience: "macropsia," "micropsia," "stationary objects appear to sway or move," "seeing geometric patterns in their field of vision whether eyes are open or closed," and "faces appear distorted." But all 20 subjects claim many of the 25 symptoms presented, including ones not part of their drug experience.

Some symptoms were experienced by most participants *prior* to the triggering event (of the 19 drug users): 12, mild tinnitus; 13, "floaters"; 14, trouble with concentration (Table 4). Symptoms experienced by at least 75 % of the 19 subjects *after* their disorder commenced were as follows: 15, head pressure; 15, ghosted afterimage of text being read; 15, stationary objects appear to sway or move; 15, afterimage seen in the same color, or, 18, in a different color; 17, trails; 18, halos or auras; 18, floaters; 18, difficulty communicating; 19, difficulty with light accommodation; 19, difficulty concentrating; 19, tinnitus; 19, daytime visual snow; 19, nighttime visual snow. Symptoms were always much more severe after the triggering drug event.

Of the 20 subjects, 6 had total Dissociative Experiences Scale scores over 20 with 2 greater than 45. Two scored above 20 (38.75 and 40) on the amnestic dissociation subscale, 9 scored at or greater than 20 (with 2 greater than 45) on the absorption and imaginative involvement subscale, and 8 scored greater than 20 (with 3 greater than 45) on the depersonalization and derealization subscale. Three subjects, as mentioned, were already aware of possible dissociative disorders and one additional subject self-identified as having problems with dissociation. Higher dissociative scores have been found in healthy individuals reporting perceptual anomalies (Wolfradt 1999).

Survey participants were asked to list substances that worsened (Table 5) or improved (Table 6) their symptoms. Cannabis was most commonly cited as worsening, whereas sedative-hypnotics ranked first for amelioration. Alcohol was second on both lists. SSRIs, atypical antipsychotics, and tobacco were also on both.

Responses to the 5 visual simulators are summarized in Table 7. Only the small photograph animation of daytime visual snow was chosen by all 20 participants as reflecting one element of their persisting perception disturbance and also had the highest score (0 to 10) on accuracy (7.9). Photographs 1 and 3 (Figs. 2 and 4) presented different afterimages with nearly as high a score on accuracy as the visual snow animation but 15–20 % of subjects stated they do not get those symptoms. Photograph 2 (Fig. 3) of ghosted text was identified as experienced by 95 % of subjects but had a much lower score on accuracy (5.6).

Table 5 Subjects that attribute worsening persisting perception problems to a specific drug or drug class

Drug	# of subjects (%)
Cannabis	12 (60 %)
Alcohol	6 (30 %)
MDMA	4 (20 %)
SSRIs	4 (20 %)
Stimulants	3 (15 %)
Atypical antipsychotics	3 (15 %)
Tobacco	2 (10 %)

Table 6 Subjects that attribute improved persisting perception problems to a specific drug or drug class

Drug	# of subjects (%)
Sedative-hypnotics	9 (45 %)
Alcohol	8 (40 %)
Opioids	2 (10 %)
Opioid antagonists	2 (10 %)
SSRIs	2 (10 %)
Atypical antipsychotics	2 (10 %)
Tobacco	2 (10 %)

Table 7 Photograph simulators. 0–10 score on how accurate it reflects an element of the perception disturbance

		# of Subjects ($N = 20$) who do not experience this as a symptom	Score 0–10 (avg.)
Photo 1	Negative afterimage	4	7.8
Photo 2	Ghosted text	1	5.6
Photo 3	Positive afterimage	3	7.3
Photo 4	Geometric pattern	7	3.9
Movie	Visual snow	0	7.9

10 = 100 % accurate and 0 = not accurate at all

4 Discussion

Only 23 of 67 subjects completed the survey between September 2010 and March 2012. Likely the time involved for survey completion kept a majority of survey initiators from finishing. It is impossible to know whether symptom severity interacted with completion rates. If there is a larger pool of individuals with HPPD-type symptoms, it is those who seek out medical and mental health attention whom we still need to be most aware of. The brief, time-limited effects described in Type 1 may be of such subclinical significance that none with this form elected to participate.

The DSM-V states that HPPD requires that the disturbances subsequent to hallucinogen use should be reminiscent of what was experienced during intoxication. Although all subjects reported primarily visual symptoms, by far not every disorder of vision detailed was also reminiscent of the triggering intoxication. One possibility is that hallucinogen use triggered subsequent disordered processing of vision beyond the alterations originally encountered. Another possibility is that those with Type 2 HPPD have a pre-existing set of subclinical symptoms that can be aggravated by various experiences, particularly by hallucinogens. Acute intoxication and later awareness of abnormal, "overactive" vision may alarm those with a pre-existing propensity for anxiety and may trigger states of more or less depersonalization in individuals with an appropriate predisposition.

If Type 2 "HPPD" symptoms are not only repetitions of a drug experience and/or existed prior to drug intoxication in milder intensity, this suggests that HPPD goes beyond hallucinogen use. The DSM-V criterion of re-experiencing focuses on drug exposure, but the constellation of symptoms is apparently more complex. One possibility is that hallucinogen use may generate symptoms not experienced during intoxication. The subjects might also be inaccurately recalling their histories, but it is also a possibility that some symptoms occurred well before drug use and that additional symptoms occur after, regardless of the drug experience. Our data suggest that there is a primary disorder of "overactive vision" prior to hallucinogen intake. Moreover, 7 of 19 subjects report their symptoms did not start "almost immediately" or "within 24 h" from exposure to the drug(s) they attribute to triggering their condition (Table 1): that a chronic disorder only in some starts weeks to months later, long after the drug has been excreted from the body, and that symptoms go beyond what the drug experience itself induced instead suggests a more subtle condition that remains poorly defined and understood, especially in respect to its causation.

Abraham (1982, 1983) postulated that a specific "LSD toxicity" that destroys some neurons of the visual system might be involved in HPPD. But many different substances, including non-hallucinogens, such as nefazodone (Kraus 1996; Horton and Trobe 1999), trazodone (Hughes and Lessell 1990), mirtazepine (Ihde-Scholl and Jefferson 2001), and others, can induce Type 2 HPPD-equivalent symptoms. One comprehensive review about HPPD concluded there is no consistent relationship between specific substances and the induction of HPPD: The range goes from alcohol and benzodiazepines to hallucinogens, cannabis, amphetamines, and inhalants (Holland and Passie 2011). Therefore, no single neuroreceptor system appears to be associated with the pathophysiology of Type 1 and Type 2 HPPD. It is evident from neuroimaging studies that the different drugs induce distinct alterations of brain activity, implying that different patterns of brain activity can lead to the same more or less lasting perceptual changes.

Even with a limited number of subjects, our data provide some tentative insight into who might be at risk for Type 2 HPPD. Those with individual or family histories of anxiety, who have pre-drug use complaints of tinnitus, visual floaters, and concentration problems, may be most susceptible for later development of persisting perception disorder (of Type 2 HPPD), particularly after LSD and or psilocybin. Our data also indicate that non-hallucinogenic substances can trigger HPPD symptoms.

Prominent anxiety during the drug intoxication, benzodiazepine anxiolysis appearing most helpful in reducing HPPD symptoms, and Dissociative Experiences Scale results (30 % reporting clinically significant pathology) together suggest that HPPD may be an anxiety disorder not unlike PTSD, where the triggering drug experience is the traumatic event. Indeed, in PTSD, DSM-V (Textbox 1) refers in Criterion B to an individual having intrusive recollections of the trauma, including the possibility of "Acting or feeling as if the traumatic event were recurring (includes a sense of reliving the experience, illusions, hallucinations, and dissociative flashback episodes, including those that occur upon awakening or when intoxicated)." With HPPD (Type 2 especially), the symptoms may wax and wane depending on environment and emotional state but symptoms may be described as durably present to some degree. As such, Type 2 HPPD might best be considered a disorder of "overactive vision," which, after worsening or the individual being more aware/concerned about the condition, renders clinically meaningful symptoms of anxiety.

In our subjects, claimed HPPD phenomena included symptoms that cannot be described as a "re-experiencing" of hallucinogen intoxication and Subject 5's HPPD-like symptoms, though indistinguishable from other subjects' descriptions, manifested despite declaring no history of any drug exposure whatsoever. As noted earlier, Type 1 and Type 2 HPPD have been reported after intoxication with alcohol, amphetamines, tobacco, and other substances. Subject 11's HPPD symptoms commenced years after last hallucinogen use but immediately after starting citalopram. Such delays in timing and the wide variety of triggering drugs do suggest a syndrome that is subclinical for most and which is aggravated perhaps most by hallucinogen exposure. As such, causation for HPPD has been linked to hallucinogens, but our findings combined with the reports of others, instead *suggest a kind of pre-existing neurological disorder or vulnerability of perception processing primarily in the visual domain that worsens with anxiety*. History of pre-morbid (to drug exposure) anxiety, family history of anxiety, and co-morbid dissociative phenomena (especially depersonalization and derealization) together suggest that the misprocessing of visual perception when drug-free takes on clinical significance for individuals who also meet criteria for an anxiety disorder (Passie et al. 2013). That so many substances are listed as worsening as well as ameliorating symptoms of HPPD (Tables 5 and 6) also suggests that this condition either may have more than one pathophysiological route to its expression and/or that indeed anxiety pushes individuals to latch on to whatever each discovers as useful for themselves.

4.1 Limitations of the Study

As with some other studies (Baggott et al. 2011), our findings are limited by a study design based on Internet survey of individuals not directly examined and without control and who may not be representative of the actual disorder. Yet, reports did show remarkable overlap across subjects, whether or not with an associated history of psychosis, dissociation, or even lifelong visual snow without any drug use. Moreover, the perceptual changes noted can briefly be experienced by most people without the disorder (simple examples of visual illusions we are all susceptible to can be explored at http://faculty.washington.edu/chudler/flash/nill.html). This may explain why, for example, 84.2 % of subjects experienced ghosted text as an afterimage in their worst HPPD-type event, but 95 % agreed that they experience the ghosted text displayed in our visual simulator (with a low score of 5.6 out of a possible 10 for how accurate it is for what they experience as HPPD-like). An additional limitation to the study of HPPD may indeed be that how we define the disorder is in need of revision.

4.2 Suggestions for Future Studies

In addition to careful screening for Type 1 versus Type 2 HPPD, future research would benefit from comparisons with healthy subjects or non-hallucinogen using patients with anxiety and depressive disorders. If possible, valid operationalized diagnostic procedures should be employed, including excluding other psychiatric, neurologic, ophthalmologic, and other medically relevant pre-existing conditions. It is especially important to exclude patients with a history of psychosis, which were sometimes consciously included (Abraham 1982), and dissociative disorders. Screening should include clinical evaluation for dissociative phenomena and disorders, pre-morbid visual disturbances, anxiety disorders, depression and dysthymia, psychotic disorders, and hypochondria.

Many endogenous and environmental etiologies have been created for HPPD, and they may account for the symptoms in a specific individual constellation in every single case, as proposed in the model of Holland and Passie (2011) (Fig. 1). A focus of future studies might be the validity of the Type 2 HPPD as initially proposed by Abraham (1982, 1983). More detailed examination of "HPPD" subjects is needed, especially of accompanying neurological and psychiatric disorders. One finding of our study is that anxiety and dissociation appear to be tightly connected to HPPD and may represent a significant vulnerability toward it, or even a partial model of its mechanisms.

Researchers as well as scientific journals have to be very careful about publishing case studies or case series because usually the descriptions in these are typically too crude for a scientific evaluation and may lead to inappropriate classification of psychological disturbances—usually without evaluating for further psychiatric/medical diagnoses. A scientific caveat is that such publications (together with a publication bias, preferring danger-related case stories about hallucinogens) may end up more as a science artifact than fact.

In conclusion, our results support the need for more rigorous research that goes beyond crude current definitions and case studies/series. It appears especially necessary to take into account the possibility that a subtle (pre-existing) over-activation of neural pathways for visual perception may be worsened and/or becomes a trigger for anxiety after the ingestion of an arousal-altering psychoactive drug. HPPD symptoms also appear to have a significant association with psychological trauma and dissociation. As revealed in our data, many perceptual symptoms are not consistent with the DSM-V Criterion A of HPPD that they are re-experiences of what transpired while hallucinogen intoxicated.

Acknowledgments The authors thank Vicka Corey, Ph.D. for numerous helpful comments and suggestions to an earlier draft of the manuscript.

Conflict of interest: The authors declare that there is no conflict of interest.

Funding: Supported in part by an unrestricted grant by the HPPD Foundation (www.hppdfoundation.org).

References

Abraham HD (1982) A chronic impairment of colour vision in users of LSD. Brit J Psychiatry 140:518–520

Abraham HD (1983) Visual phenomenology of the LSD flashback. Arch Gen Psychiatry 40:884–889

Abraham HD (1984) LSD flashbacks (In reply). Arch Gen Psychiatry 41:632–633

Abraham HD, Aldridge A (1993) Adverse consequences of LSD. Addiction 88:327–334

Abraham HD, Duffy FH (1996) Stable qEEG differences in post-LSD visual disorder by split half analysis: evidence for disinhibition. Psychiatry Res 67:173–187

Abraham HD, Duffy FH (2001) EEG coherence in post-LSD visual hallucinations. Psychiatry Res 107:151–163

Abraham HD, Mamen A (1996) LSD-like panic from risperidone in post-LSD visual disorder. J Clin Psychopharmacol 16:238–241

Abraham HD, Aldridge AM, Gogia P (1996) The psychopharmacology of hallucinogens. Neuropsychopharmacology 14:286–298

Alarcon RD, Dickinson WA, Dohn HH (1982) Flashback phenomena, clinical and diagnostic dilemmas. J Nerv Ment Dis 170:217–223

Aldurra G, Crayton JW (2001) Improvement of hallucinogen persisting perception disorder by treatment with a combination of fluoxetine and olanzapine: case report. J Clin Psychopharmacol 21:343–344

American Psychiatric Association (2013) Diagnostic and statistical manual of mental diseases (DSM-V), 5th edn. American Psychiatric Association Press, Washington, DC

Baggott MJ, Coyle JR, Erowid E, Erowid F, Robertson LC (2011) Abnormal visual experiences in individuals with histories of hallucinogen use, a web-based questionnaire. Drug Alcohol Depend 114:61–67

Bernstein-Carlson EB, Putnam FW (1986) Development, reliability, and validity of a dissociation scale. J Nerv Ment Dis 174:727–735

Brimblecombe RW, Pinder R (1975) Hallucinogenic agents. Wright-Scientechnica, Bristol

Carlson EB, Putnam FW (1993) An update on the dissociative experiences scale. Dissociation 6:16–27

Cohen S (1960) LSD, side effects and complications. J Nerv Ment Dis 130:30–40

Cohen S (1977) Flashbacks. Drug Abuse Alcohol Newslett 6:1–4

Dixon JC (1963) Depersonalization phenomena in a sample of college students. Br J Psychiatry 109:371–375

Ellis H (1898) Mescal: a new artificial paradise. Contemp Rev 73: 130–141 (reprinted in The Smithsonian Institution Annual Report 1898, 537–548)

Fischer R (1971) The "flashback": arousal-statebound recall of experience. J Psychedelic Drugs 3:31–39

Fischer R (1976) On the remembrance of things present: the state-bound and stage-bound nature of consciousness. Man-Environ Syst 6:131–136

Goldman S, Galarneau D, Friedman R (2007) New onset LSD flashback syndrome triggered by the initiation of SSRIs. Ochsner J 7:37–39

Grinspoon L, Bakalar J (1997) Psychedelics reconsidered, 3rd edn. The Lindesmith Center, New York

Halpern JH, Jr Pope HG (2003) Hallucinogen persisting perception disorder, what do we know after 50 years? Drug Alcohol Depen 69:109–119

Hasse HE, Waldmann H (1971) "Flashback", Spontane psychotische Episoden als Folgeerscheinung des Phantasticagebrauchs Jugendlicher. Arch Psychiat Nervenkr 214: 399–439

Heaton RK, Victor RG (1976) Personality characteristics associated with psychedelic flashbacks in natural and experimental settings. J Abnorm Psychol 85:83–90

Hermle L, Kovar KA, Hewer W, Ruchsow M (2008) Halluzinogen-induzierte psychische Störungen. Fortschr Neurol Psychiatr 76:334–342

Hermle L, Ruchsow M, Täschner KL (2015) Halluzinogen-induzierte Persistierende Wahrnehmungsstörung (HPPD) und Flashback-Phänomene-Differenzialdiagnose und Erklärungsmodelle. Fortschr Neurol Psychiatr 83:506–515

Holland D, Passie T (2011) Flashback-Phaenomene als Nachwirkung von Halluzinogeneinnahme VWB-Verlag, Berlin

Holsten F (1976) Flashbacks, clinical and social significance 1.5–4 years after the first admission. Tidsskr f d norske Lægefor 96:876–878

Horowitz MJ (1969) Flashbacks: recurrent intrusive images after the use of LSD. Am J Psychiatry 126:565–569

Horton JC, Trobe JD (1999) Akinetopsia from nefazodone toxicity. Am J Ophthalmol 128:530–531

Hughes MS, Lessell S (1990) Trazodone-induced palinopsia. Arch Ophthalmol 108:399–400

Ihde-Scholl T, Jefferson JW (2001) Mitrazapine-associated palinopsia. J Clin Psychiatry 62:373

Johnston LD, O'Malley PM, Bachman JG, Schulenberg JE (2013) Monitoring the future national survey results on drug use, 1975–2012. In: Volume II: college students and adults ages 19–50. Institute for Social Research, The University of Michigan, Ann Arbor

Kokoszka A (1992–1993) Occurrence of altered states of consciousness among students, profoundly and superficially altered states in wakefulness. Imagin Cogn Pers 12:231–247

Kokoszka H (1993) Flashbacks and schizophrenia. Irish J Psychol Med 10:37–39

Kraus RP (1996) Visual "trails" with nefazodone treatment. Am J Psychiatry 153:365–366

Lerner AG, Gelkopf M, Oyffe I, Finkel B, Katz S, Sigal M, Weizman A (2000) LSD-induced hallucinogen persisting perception disorder treatment with clonidine, an open pilot study. Int Clin Psychopharmacol 15:35–37

Lerner AG, Shufman E, Kodesh A, Kretzmer G, Sigal M (2002) LSD-induced Hallucinogen Persisting perception disorder with depressive features treated with reboxetine, case report. Is J Psychiatry Relat Sci 39:100–103

Malleson N (1971) Acute adverse reactions to LSD in clinical and experimental use in the United Kingdom. Brit J Psychiatry 118:229–230

Markel H, Lee A, Holmes RD, Domino EF (1994) LSD flashback syndrome exacerbated by selective serotonin reuptake inhibitor antidepressants in adolescents. L Ped 125:817–819

Matefy RE, Hayes C, Hirsch J (1978) Psychedelic drug flashbacks, subjective reports and biographical data. Addict Behav 3:165–178

Matefy RE, Hayes C, Hirsch J (1979) Psychedelic drug flashbacks, attentional deficits? J Abnorm Psychol 88:212–225

Mayer-Gross W (1931) Über Synästhesien im Meskalinrausch. In: Anschütz G (Ed) Farbe-Ton-Forschungen Bd. III, pp 266–277. Hamburg Psychologisch-ästhetische Forschungsgesellschaft, Hamburg

McGlothlin H, Arnold D (1971) LSD revisited: a ten-year follow-up of medical LSD use. Arch Gen Psychiatry 24:35–49

Morehead DB (1997) Exacerbation of hallucinogen-persisting perception disorder with risperidone. J Clin Psychopharmacol 17:327–328

Naditch MP (1974) Acute adverse reactions to psychoactive drugs, drug usage, and psychopathology. J Psychoactive Drugs 22:305–311

Naditch MP, Fenwick S (1977) LSD flashbacks and ego functioning. J Abnormal Psychol 86:352–359

Nichols DE (2004) Hallucinogens. Pharmacol Ther 101:131–181

Parish E (1894) Über die Trugwahrnehmungen mit besonderer Berücksichtigung der internationalen Enquete über Wachhalluzinationen bei Gesunden. Schr Ges psychol Forsch 7 (8):1–154

Passie T (1997) Psycholytic and psychedelic therapy Research: A complete international bibliography. Laurentius Publishers, Hannover

Passie T, Halpern JH (2014) The pharmacology of hallucinogens (Chapter 14). In: Ries RK, Fiellin DA (eds) Principles of addiction medicine, 5th edn. Lippincott Williams & Wilkins, NY

Passie T, Schneider U, Borsutzky M, Breyer R, Emrich HM, Bandelow B, Schmid-Ott G (2013) Impaired perceptual processing and conceptual cognition in patients with anxiety disorders: a pilot study with the binocular depth inversion paradigm. Psychol Health Med 18:363–374

Sandison RA, Whitelaw JD (1957) Further studies in the therapeutic value of lysergic acid diethylamide in mental illness. J Ment Sci 103:332–342

Shor RE (1960) The frequency of naturally occurring "hypnotic-like" experiences in the normal college population. Int J Clin Exp Hypn 8:151–156

Siegel RK, Jarvik ME (1975) Drug-induced hallucinations in animals and man. In: Siegel RK, West LJ (eds) Hallucinations: behavior, experience, and theory. Wiley, New York, pp 81–162
Stanton MD, Bardoni A (1972) Drug flashbacks, reported frequency in a military population. Am J Psychiatry 129:751–755
Strassman RJ (1984) Adverse reactions to psychedelic drugs, a review of the literature. J Nerv Ment Dis 172:577–595
Studerus E, Kometer M, Hasler F, Vollenweider FX (2011) Acute, subacute and long-term subjective effects of psilocybin in healthy humans: a pooled analysis of experimental studies. J Psychopharmacol 25:1434–1452
Substance Abuse and Mental Health Services Administration (SAMHSA) (2011) Results from the 2010 national survey on drug use and health: summary of national findings. Substance Abuse and Mental Health Services Administration, Rockville
Wolfradt U, Oubaid V, Straube ER, Bischoff N, Mischo J (1999) Thinking styles, schizotypal traits and anomalous experiences. Pers Indiv Differ 27:821–830
World Health Organization (1992) The ICD-10 classification of mental and behavioural disorders. World Health Organization, Geneva
Young CR (1997) Sertraline treatment of hallucinogen persisting perception disorder. J Clin Psychiatry 58:85

Therapeutic Applications of Classic Hallucinogens

Michael P. Bogenschutz and Stephen Ross

Abstract This chapter reviews what is known about the therapeutic uses of the serotonergic or classic hallucinogens, i.e., psychoactive drugs such as LSD and psilocybin that exert their effects primarily through agonist activity at serotonin 2A (5HT2A) receptors. Following a review of the history of human use and scientific study of these drugs, the data from clinical research are summarized, including extensive work on the use of classic hallucinogens in the treatment of alcoholism and other addictions, studies of the use of LSD and psilocybin to relieve distress concerning death, particularly in patients with advanced or terminal cancer, and more limited data concerning the use of classic hallucinogens to treat mood and anxiety disorders. A survey of possible mechanisms of clinically relevant effects is provided. The well-established safety of classic hallucinogens is reviewed. To provide a clinical perspective, case summaries are provided of two individuals who received treatment in recent controlled trials of psilocybin: one being treated for alcoholism, the other suffering from anxiety and depression related to fear of death due to a cancer diagnosis. Although promising early phase research conducted from the 1950s through the early 1970s was discontinued before firm conclusions could be reached concerning the efficacy of any of the classic hallucinogens for any clinical condition, the research that was conducted in that era strongly suggests that classic hallucinogens have clinically relevant effects, particularly in the case of LSD treatment of alcoholism. In the past decade, clinical trials have resumed investigating the effects of classic hallucinogens in the treatment of existential distress in the face of cancer, and in the treatment of addictions including alcoholism and nicotine addiction. The studies that have been completed to date are not sufficient to establish efficacy, but the outcomes have been very encouraging, and larger trials,

---

M.P. Bogenschutz (✉) · S. Ross
Department of Psychiatry, New York University Langone Medical Center, New York City, USA
e-mail: michael.bogenschutz@nyumc.org

S. Ross
e-mail: Stephen.Ross@nyumc.org

Curr Topics Behav Neurosci (2017) 36: 361–392
DOI 10.1007/7854_2016_464
Published Online: 18 May 2017

up to and including phase 3, are now underway or being planned. Although research has elucidated many of the acute neurobiological and psychological effects of classic hallucinogens on humans, animals, and in vitro systems, the mechanisms of clinically relevant persisting effects remain poorly understood.

Keywords Hallucinogens · Psychedelics · Review · Psychopharmacology · Psilocybin · LSD · Cancer · Addiction

Contents

1 Introduction

In this chapter, we review historical and recent literature on the clinical use of the classic serotonergic hallucinogens, i.e., those hallucinogens that are thought to exert their effects primarily through agonist activity at serotonin 2A (5HT2A) receptors. In order to maintain this focus, we will not attempt to cover several related topics. Clinical research on non-serotonergic hallucinogens such as ketamine will not be discussed. We also will not include discussion of MDMA and related compounds or iboga alkaloids. Although these drugs have clinically relevant effects and are subject to current clinical research, their effects and mechanisms of action are sufficiently distinct from those of the classic hallucinogens that they are best considered separately. Finally, we focus primarily on those indications for which meaningful data are available.

Although classic hallucinogens have been used by humans for at least 5 millennia (El-Seedi et al. 2005), western scientific study of hallucinogens dates to the late 1800s, when mescaline was isolated and its effects described by Arthur Heffter (Heffter 1898, 1896). In 1943, Albert Hofmann discovered the psychoactive effects of lysergic acid diethylamide (LSD) when he accidentally ingested traces of an ergot derivative he had synthesized, and confirmed the effects through self-experimentation (Hofmann 1979). He quickly recognized the therapeutic potential of LSD, as did other investigators (Busch and Johnson 1950). Following Gordon Wasson's report of hallucinogenic mushroom use by the Mazatec tribe in Mexico (Wasson 1957), Hofmann isolated psilocybin from samples of the mushrooms and synthesized it in the laboratory (Hofmann et al. 1958a, b).

Following the discovery or LSD, research on LSD and other classic hallucinogens expanded rapidly. Initially, the main focus of research was on the use of LSD as a model of psychosis (Rinkel et al. 1952). However, clinicians and clinical scientists soon began to explore the use of these unregulated chemicals in the treatment of alcohol and drug addiction, existential crisis related to death in the terminally ill, and various neurotic conditions being addressed in the context of psychodynamic psychotherapy. From the early 1950s to the mid-1960s, over 1000 scientific articles were published reporting on the treatment of over 40,000 patients with classic hallucinogens (Grinspoon and Balakar 1997).

Although classic hallucinogens were sometimes used in a strictly biological model of treatment, most clinicians and researchers combined the administration of hallucinogens with psychotherapy before, during, and/or after the drug experience, believing that the subjective experience during the drug's acute effects, and the successful integration of these experiences, was crucial to the therapeutic benefit of the treatment. The two prominent therapeutic models used in the 1950s through early 1970s were psycholytic and psychedelic therapy. Although both combined administration of hallucinogens with psychotherapy to achieve therapeutic change, they emphasized different processes in bringing about these changes (Grinspoon and Balakar 1997; Grof 2008). In the psycholytic method, low to moderate doses of hallucinogens were administered on multiple occasions to facilitate therapy that was based on traditional psychoanalytic principles, i.e., helping the patient to become aware of unconscious desires, emotions, attachments, and self-representations, and resolving intrapsychic conflicts (Leuner 1967; Buckman 1967). Therapy was conducted while the patient was under the influence of the drug. The psychedelic method used higher doses of LSD administered on no more than a few occasions, with the goal of occasioning a "peak-psychedelic" or mystical experience. These experiences are characterized by the phenomenology of unity (sense of oneness), transcendence of the ordinary experience of sense and time, sense of sacredness, sense of deep truth or ultimate meaning (noetic quality), deeply felt positive mood, and ineffability (Pahnke 1969). It was held that such experiences often facilitated lasting change in habitual patterns of thought, behavior, experience of emotion, and even personality (Hoffer 1967; Sherwood et al. 1962). Although these two models are conceptually distinct, some clinicians and investigators used both, or created hybrid models (Grof 2008; Masters and Houston 2000).

In reaction to the cultural upheaval of the 1960s and concern about the dangers of widespread illicit use of psychedelics during that era, clinical research on hallucinogens came to halt in the early 1970s. The Controlled Substances Act placed LSD, psilocybin, and related compounds in Schedule I, and human research was not allowed to continue. Basic research continued, and many additional compounds were and continue to be discovered, most of which have never been subjected to the most basic pharmacologic study, let alone human laboratory studies or clinical trials. Illicit use for recreational, therapeutic, and/or spiritual purposes continued.

After a hiatus of about two decades, human research on classic hallucinogens resumed in the early 1990s with Rick Strassman's studies of the subjective and physiological effects of intravenous dimethyltryptamine (DMT) on normal volunteers (Strassman and Qualls 1994). Since the beginning of the twenty-first century, there has been a sharp increase in research in this area. The safety of classic hallucinogens (particularly psilocybin) in clinical research settings has been well documented (Johnson et al. 2008). The prosocial and possibly beneficial use of hallucinogens in the context of organized religious activity has also attracted attention in recent years. A considerable body of work has characterized the acute effects of classic hallucinogens on physiology, cognition, emotion, and brain function. All of these factors have motivated the resumption of research into the possible clinical applications of these drugs.

2 Use of Classic Hallucinogens in the Treatment of Addiction

2.1 LSD

The use of LSD in the treatment of alcoholism was studied extensively in the 1950s through early 1970s [For reviews see Abuzzahab and Anderson (1971); Halpern (1996); Mangini (1998); Dyck (2006); Grinspoon and Balakar (1997)]. The earliest investigations in Saskatchewan, led by the pioneering psychedelic researcher Humphrey Osmond, were based on a model which held that the LSD experience mimicked the experience of delirium tremens, and that this experience of "hitting bottom" could induce abstinence in some cases (Smith 1958). However, based on their experience and the recommendations of A.M. Hubbard, the treatment was modified to facilitate the positive experience of self-surrender, consistent with the psychedelic model (Chwelos et al. 1959). Treatment of alcoholism with LSD became an accepted clinical treatment in Saskatchewan, and several thousand patients were treated by Dr. Osmond and colleagues using this model. Uncontrolled trials with severe, chronic alcoholic patients using a single high-dose LSD session had variable but generally encouraging results (Abuzzahab and Anderson 1971).

Controlled trials of LSD for alcoholism were underpowered with mixed results, leaving little to definitely be concluded from these studies regarding the clinical efficacy of LSD-assisted treatment for alcohol addiction (Abuzzahab and Anderson 1971; Halpern 1996; Grinspoon and Balakar 1997). However, a recent meta-analysis (Krebs and Johansen 2012) revealed consistent and clinically meaningful effects of LSD over control treatment in the six randomized trials of LSD for alcohol addiction that reported drinking outcomes (Smart et al. 1966; Hollister et al. 1969; Ludwig et al. 1969; Bowen et al. 1970; Pahnke et al. 1970; Tomsovic and Edwards 1970). Across the six studies, 325 participants received active treatment with LSD, and 211 received a control treatment. In all of the studies, LSD was administered in a single high-dose session, in doses ranging from about 210–800 mcg. At the first post-treatment follow-up, LSD-treated patients were more likely to show significant improvements in drinking outcomes (Odds Ratio = 1.96, 95% CI 1.36–2.84, $p = 0.0003$). Treatment effects remained significant at 6 months. The effect of LSD treatment was homogeneous across the six studies, in spite of great variability in the psychotherapeutic components of treatment and the control treatments employed. These promising findings strongly support renewed clinical investigation of LSD for the treatment of alcoholism.

Studies were also conducted using LSD as a component of treatment for opioid addiction. Seventy "post-narcotic drug addicts" at the U.S Public Health Service Hospital in Lexington, Kentucky received a single 2–3 h therapeutic session in which they were randomly assigned to receive one of five treatments: (1) insight-oriented psychotherapy, (2) hypnotherapy (psychotherapy conducted under hypnosis), (3) LSD with no psychotherapeutic intervention, (4) LSD with psychotherapy, or (5) LSD with hypnotherapy (Ludwig and Levine 1965). At 2-month follow-up, all groups showed significant improvement on a questionnaire designed to measure various dimensions of psychopathology, with greater improvement in the group that received LSD with hypnotherapy. Drug use behavior after discharge from the hospital was not investigated. The dose of LSD used in this study (0.2 mcg/kg, or 140 mcg for a 70 kg person) was lower than the doses used in the controlled alcohol trials summarized above. In another trial, Savage and McCabe randomly assigned 78 incarcerated male heroin addicts to outpatient treatment as usual versus psychedelic therapy followed by treatment as usual (Savage and McCabe 1973). The psychedelic therapy included a single high-dose (300–500 mcg) LSD session in the context of 4–6 weeks of residential treatment that included extensive psychotherapy before and after the session. Participants who received psychedelic therapy had better outcomes during the 12-month follow-up period (25% vs. 5% continuous abstinence). The design of this study did not separate the effects of the LSD from other aspects of the residential treatment.

2.2 Dipropyltryptamine

Dipropyltryptamine (DPT) is a classic hallucinogen, structurally very similar to DMT, which has effects lasting 1–6 h (depending on the dose) when given by intramuscular injection (Grof et al. 1973b). Grof et al. reported highly significant improvements in drinking behavior and other outcomes at 6 months among 47 participants who received between 1 and 6 (mean 1.9) DPT sessions using a psychedelic treatment model (Grof et al. 1973b). A randomized trial conducted by the same group found no difference in outcome between DPT treatment, "conventional treatment" (psychotherapy similar to that received by the DPT group, but without the DPT sessions), and "routine hospital treatment" (Rhead et al. 1977). However, this study suffered from methodological limitations including high attrition both during treatment and in the follow-up period (only 40% of the randomized sample was assessed at follow-up), as well as differential dropout among the groups.

2.3 Psilocybin

The effects of psilocybin were described over 50 years ago (Isbell 1959; Leary et al. 1963), and at least 2000 individuals received psilocybin in clinical studies using the *psychedelic* and *psycholytic* models of treatment (Passie 2005). However, other than very brief reports of psilocybin used in combination with LSD (Rydzyński et al. 1968; Rydzyński and Gruszczyński 1978), the first trial of psilocybin used to treat alcoholism was completed very recently (Bogenschutz et al. 2015a). In this proof-of-concept study, ten volunteers with DSM-IV alcohol dependence received orally administered psilocybin in 1 or 2 supervised sessions scheduled 4 weeks apart. Psilocybin was administered in the context of a 12-week manualized therapy program, in doses of 0.3 mg/kg and 0.4 mg/kg. Drinking decreased significantly following psilocybin administration, and gains remained significant during 36 weeks of follow-up. The intensity of self-reported effects during the first psilocybin session at week 4 strongly predicted improvement in drinking during weeks 5–8 ($r = 0.76$ to $r = 0.89$). A double-blind efficacy trial using a similar model of treatment is currently under way, a multicenter trial conducted at the University of New Mexico School of Medicine (NCT02061293) and the New York University School of Medicine.

In the first study of classic hallucinogens administered in the treatment of nicotine addiction, a recent open-label pilot study, utilizing a model of high-dose psilocybin in combination with cognitive behavioral therapy, found remarkably positive outcomes using psilocybin in the treatment of this addiction (Johnson et al. 2014). Fifteen nicotine-dependent smokers received psilocybin (20–30 mg/70 kg in 2–3 sessions) during a 15-week course of manualized therapy. At 6-month follow-up, 12/15 (80%) were abstinent based on self-report and verified by biological measures (urinary cotinine and breath carbon monoxide). This success rate far exceeds those seen in clinical trials of any currently available pharmacotherapy for tobacco addiction. Measures of mystical experience during psilocybin sessions and of spiritual significance and personal meaning attributed to the sessions were significantly correlated with smoking outcomes (Garcia-Romeu et al. 2014). Based on these very promising results, a comparative efficacy trial (psilocybin vs. nicotine replacement) is now being implemented at the Johns Hopkins School of Medicine (NCT01943994).

In addition, a pilot study is now under way to begin investigation of the effects of psilocybin-assisted psychotherapy in the treatment of cocaine dependence at the University of Alabama School of Medicine (NCT02037126). This study will be the first to use a classic hallucinogen in the treatment of this addiction.

2.4 *Mescaline and Peyote*

Mescaline, a phenylethylamine classical hallucinogen, has subjective effects very similar to the tryptamine classic hallucinogens described in this chapter (Wolbach et al. 1962). Mescaline was used interchangeably or in combination with LSD in some of the early work with the treatment of alcohol dependence (Smith 1958, 1959; Sherwood et al. 1962). However, it was never subjected to controlled trials. The peyote cactus (Lophophora williamsii), the San Pedro cactus, (Trichocereus pachanoi) and a number of other cacti contain mescaline in psychoactive quantities (Ogunbodede et al. 2010; Gabermann 1978). Peyote is used sacramentally by members of the Native American Church (NAC) (Stewart 1987) and the Huichol of northern Mexico (Meyerhoff 1974). Many authors have suggested that taking peyote in the context of NAC ceremonies helps alcoholics achieve and maintain sobriety (Kunitz and Levy 1994; Lu et al. 2009; Albaugh and Anderson 1974; Garrity 2000; Roy 1973), and NAC involvement has been integrated with addiction treatment (Albaugh and Anderson 1974). However, we were unable to find any quantitative data concerning alcohol use among NAC members, or outcomes of treatments including NAC involvement.

2.5 *Ayahuasca and DMT*

Ayahuasca is a hallucinogenic tea typically made by boiling the leaves of the shrub Psychotria viridis, containing the classic hallucinogen DMT, in combination with the vine Banisteriopsis caapi, containing beta carboline alkaloids (principally harmine, harmaline, and tetrahydroharmine), reversible MAO-A inhibitors that render DMT orally active by preventing GI degradation and may have significant effects of their own (McKenna et al. 1984; Callaway et al. 1996). Ayahuasca has been used by indigenous peoples of the Amazon basin for centuries, and is central to the religious practice of organized religions including the União do Vegetal and Santo

Daime (McKenna 2007). Decreased rates of alcohol misuse have been documented among members of both Brazilian and US religious groups using ayahuasca (Halpern et al. 2008; Doering-Silveira et al. 2005; Fabregas et al. 2010). Disapproval of alcohol within these religions may contribute to these low rates of substance use. Ayahuasca is currently being used in treatment centers in Peru and elsewhere for addiction and other conditions (Liester and Prickett 2012; Labate and Cavnar 2011), and many individuals have reported that ayahuasca has facilitated their recovery from addiction. However, with the exception of one observational study (Thomas et al. 2013), to our knowledge neither systematic outcome studies nor clinical trials have been conducted.

3 Use of Classic Hallucinogens to Ameliorate Distress Concerning Death

3.1 Prevalence and Impact of Psychiatric Disorders (Depression, Anxiety, Adjustment Disorder) and Pain in Cancer

Psychiatric disorders and psychological distress in individuals with cancer is common, clinically significant, and undertreated. Per year in the US, there are an estimated 1.6 million new cancer diagnoses and approximately 600,000 cancer-related deaths (Siegel et al. 2012). Additionally, in the US, there are approximately 10.5 million Americans with a current or past diagnosis of cancer, and approximately 40% will develop cancer at some point in their lives (Ries et al. 2007). The most common psychiatric disorders in patients with cancer are depressive, adjustment, and anxiety spectrum disorders, with rates of any psychiatric disorder in cancer patients as high as 40% (Zabora et al. 2001; Mitchell et al. 2011). Clinically relevant psychological distress in cancer patients is associated with a variety of poor outcomes if untreated. These include: medication non-adherence, increased emergency room visits and hospital stays, adverse medical outcomes, lower quality of life, decreased social function, increased disability, hastened desire for death, increased rates of suicide, and even decreased survival rates from the cancer (Partridge et al. 2003; Katon 2003; Brown et al. 2003; Kissane 2009; Bultz and Holland 2006; Li et al. 2012). Although a minority of patients with advanced or terminal cancer experience clinically relevant existential/spiritual distress, when it occurs its effects are highly consequential and associated with the following negative outcomes: increased pain perception, decreased quality of life, increased depressive and anxiety symptoms, increased healthcare visits, increased desire for hastened death, and increased suicidal ideation and behaviors (Puchalski 2012; LeMay and Wilson 2008). Pain is a common symptom associated with cancer syndromes, occurs in an estimated one-third to one-half of patients, and is often undertreated and contributes to a variety of adverse outcomes including fatigue, impaired function, desire for hastened death, and suicide (Carr et al. 2002; Jaiswal et al. 2014). Completed suicide is the worst possible outcome in patients with cancer diagnoses and a diagnosis of cancer increases the risk of suicide and in some estimates increases the risk by a factor of 2 (Allebeck and Bolund 1991; Levi et al. 1991; Yousaf et al. 2005; Storm et al. 1992).

When queried, patients with advanced or terminal cancer very commonly (as high as 90% in some samples) cite spiritual/existential themes as having significant importance to them with the following types of common themes identified: meaning, purpose, hope, seeking forgiveness, increased importance of relationships including wanting a closer connection with God or one's faith, thoughts of death (El Nawawi et al. 2012; Winkelman et al. 2011; LeMay and Wilson 2008). Despite the centrality of spirituality to the needs of patients with advanced or terminal cancer who are facing death, treatment providers do a poor job of diagnosing, referring, or treating this type of distress in patients with cancer (Institute of Medicine 2008; Puchalski 2012). Conversely, increased spiritual well-being in cancer patients has been associated with: increased hope/gratitude/positive outlook relative to a cancer diagnosis, increased quality of life as death approaches, decreased depression, decreased hopelessness, and decreased desire for hastened death (Taylor 2003; Gall and Cornblat 2002; Ferrell et al. 1998; Brady et al. 1999; Nelson et al. 2002; Breitbart et al. 2000; McClain et al. 2003; Greenstein and Breitbart 2000).

3.2 Spiritually/Existentially-Based Interventions for Cancer-Related Psychological Distress

Victor Frankl (a psychiatrist and holocaust survivor) and Dame Cicely Saunders (the founder of the modern hospice movement in Great Britain) were pioneers in highlighting the importance of spiritual and existential dimensions, both as symptoms of distress and as treatment targets, in patients with terminal cancer (Frankl 1984; Saunders 1988; LeMay and Wilson 2008). Spiritual or existential distress/pain lacks a consistent nosologic framework and has been described in various ways by prominent psycho-oncologists such as: severe distress associated with events that threaten the "intactness" of a person (Cassel 1982), pain caused by the threat of extinction of an individual and meaning of the self (Murata 2003), or mental turmoil experienced by those facing death accompanied by lack of meaning or purpose, powerlessness, hopelessness, remorse, a sense of futility, grief, isolation, loss of dignity, and demoralization (Kissane 2000; Clarke and Kissane 2002).

With limited evidence from randomized controlled trials in cancer populations, currently there are no well-established evidence-based treatment algorithms to guide the optimal pharmacologic or psychosocial treatment of cancer patients with depressive and anxiety spectrum disorders; rather, treatment guidelines have had to be derived from the available research conducted in cancer populations as well as from the more extensive research in psychiatric and other medical illness populations (Traeger et al. 2012; Li et al. 2012). Also, given the relationship between spiritual well-being and improved psychiatric outcomes in patients with advanced cancer (most importantly decreased depression, decreased hopelessness, decreased DHD and possibly decreased suicide), it would make sense to develop treatment (psychosocial, pharmacologic or combined pharmacologic–psychosocial) paradigms to specifically target such spiritual/existential distress when it manifests. While a handful of manualized existentially oriented psychotherapies have been developed to address the existential/spiritual issues faced by patients with advanced/terminal cancer, with some empirical support from clinical trials (LeMay and Wilson 2008), there are no current FDA approved pharmacologic interventions to treat this type of distress in cancer patients. However, historically and in the last 2 decades in the United States, psychedelic treatment models for end-of-life cancer distress have been empirically studied.

3.3 History of Psychedelic Therapy in Death and Dying

The first suggestions that classic hallucinogens could be useful for humans in the dying process came from Valentina Wasson and Aldous Huxley (Grof and Halifax 1977). Along with her husband Gordon Wasson, Valentina Wasson, a pediatrician and mycologist, was responsible for introducing psilocybin to western culture and medicine. After being the first known westerners to participate in a psilocybin ritual in Mexico with the Mazatec curandera Maria Sabina, the Wassons wrote about their experience in 1957 in Life Magazine (Wasson 1957). Valentina Wasson predicted that psilocybin would one day have medical utility for a variety of illnesses including terminal diseases. The famous writer Aldous Huxley also believed in the power of altered states and hallucinogens to help the dying. While his first wife, Maria, lay dying in 1955, he employed a hypnosis technique to alter her consciousness in an attempt to help with the dying experience and then in 1963 while dying of cancer himself, Huxley had his second wife Laura administer 100 mcg of LSD several hours before his death to facilitate his own dying process (Grof and Halifax 1977).

The first scientific exploration of classic hallucinogens as stand-alone pharmacologic interventions to help the dying in a medical setting occurred in the early 1960s with Eric Kast MD, an internist and psychiatrist at the University of Chicago Medical School. Kast was a specialist in pain medicine and became interested in studying LSD as a novel analgesic agent in patients with terminal cancer and pain syndromes by exploring LSD's ability to alter pain attention and perception (Grof and Halifax 1977). In his first paper on the topic, published in 1964, Kast conducted a blinded comparative efficacy trial of LSD (100 mcg orally) compared to Demerol 100 mg and dilaudid 2 mg in a sample of 50 gravely ill patients, mostly with terminal cancer but also including those with severe burns and infectious illnesses (Kast and Collins 1964). Kast reported that the LSD group had statistically significant reductions in pain compared to the 2 opioid groups, from 3 h post-dosing to up at least 19 h post-dosing. In addition, he noted that the patients in the LSD group appeared to display a type of detachment from the fear of dying that he thought was useful given their grave medical conditions. Based on this observation, he extended his LSD research by administering oral LSD 100 mcg to over 200 patients with terminal cancer and pain syndromes in an open-label design and paid increasing attention to psychological phenomenon (in addition to pain perception) such as sleep, affective changes, and attitudes toward death and dying. In further published research, Kast confirmed his earlier findings by observing: decreased pain perception acutely and lasting up to 2 weeks post-dosing, improved sleep and mood, improved communication between treatment provider and patients, enhanced morale and outlook on life, reports of mystical-type experiences ('happy, oceanic feelings'), enhanced philosophical and spiritual states, and decreased fear of cancer diagnoses and death (Kast 1966). It is important to note that Dr. Kast viewed LSD treatment in his patients as a type of pure pharmacologic intervention and did not account for or introduce the well-known components of set, setting, dose titration, and psychotherapeutic preparation and integration relative to the dosing sessions (Grof and Halifax 1977).

The other significant historical research utilizing classic hallucinogens to treat psychological distress associated with advanced or terminal cancer occurred from the early 1960s to the mid-1970s at the Spring Grove State Hospital in Maryland, a research affiliate of the Johns Hopkins School of Medicine and the University of Maryland School of Medicine at the time. Starting in 1963, a group of psychiatrists, psychologists, nurses, and social workers began examining LSD-assisted psychotherapy with alcoholics. In 1965, when one of the nurses on the research team became ill with metastatic breast cancer, one of the research psychologists (Sidney Wolf) suggested that a course of psychedelic therapy might help alleviate her anxiety associated with cancer. After she underwent LSD-assisted treatment and reported relief, the researchers at Spring Grove decided to embark on research utilizing LSD-assisted psychotherapy (a pharmacologic–psychosocial paradigm) to help patients with terminal cancer and psychological distress. This research project began in 1967 and was headed by the psychiatrists Stanislav Grof and Walter Pahnke (Pahnke et al. 1969). The treatment model utilized was significantly different from the pharmacologic only model developed by Dr. Kast. It was based on Dr. Grof's LSD research in administering moderate to high doses of LSD to normal volunteers (developed while doing research at the Psychiatric Research Institute in Prague, Czechoslovakia), paying careful attention to set, setting, as well as session preparatory psychotherapy and post-session integrative psychotherapy. The treatment model drew from psychoanalytic theory as well as transpersonal psychology, and Grof commented that "Many individuals who had the experience of death and rebirth sometimes accompanied by feelings of cosmic unity independently reported that their attitudes towards dying and their concepts of death underwent dramatic changes. Fear of their own physiological demise diminished, they became open to the possibility of consciousness existing after clinical death, and tended to view the process of dying as an adventure in consciousness rather than 'the ultimate biological disaster'" (Grof and Halifax 1977). A total of 60 patients participated in the open-label experimental trial with 44 of the patients receiving LSD (200–500 mcg orally) and 19 patients receiving dipropyltryptamine (DPT) (60–105 mg IM).

Systematic, longitudinal analyses performed on a subsample of 31 participants found statistically significant pre–post reductions in the following domains: depression, anxiety, pain, isolation, and fear of death; in addition, a global index of improvement broke down as follows (29% dramatically improved, 42% moderately improved, 23% unimproved, 6% worse) (Grof et al. 1973a).

Of note, from 1965 to 1970, a limited amount of open-label research studying LSD-assisted psychotherapy took place at the UCLA School of Medicine by the psychiatrist Sidney Cohen and the psychologist Gary Fisher. Dr. Cohen, a pioneering psychedelic researcher who also understood and articulated the risks associated with classic hallucinogens, commented: "Death must become a more human experience. To preserve the dignity of death and prevent the living from abandoning or distancing themselves from the dying is one of the great dilemmas of modern medicine" (Grof and Halifax 1977).

3.4 Phase II Trial: LSD-Assisted Psychotherapy for Anxiety Associated with Life-Threatening Disease

The study of LSD-assisted psychotherapy to treat psychosocial distress associated with life-threatening illnesses has resumed with a trial recently completed in Switzerland. In this double-blind, randomized, crossover study, 12 participants received the experimental condition (LSD 200 mcg orally) and the active control (LSD 20 mcg orally) in random order, in two medication administration sessions separated by 2 months, in conjunction with psychotherapy. Safety of LSD was supported in this small cohort with no reported serious adverse events. Compared to the active control, the experimental group had significant short-term (2 month follow-up) reductions in anxiety as measured by the State-Trait Anxiety Inventory (STAI) (Gasser et al. 2014). A recent 12-month follow-up study of participants from the above trial reported sustained reductions in anxiety (as measured by the STAI), increases in quality of life, and no reports of any adverse psychological or medical sequellae (Gasser et al. 2015).

3.5 Phase II and Phase III Trials: Psilocybin-Assisted Psychotherapy for Cancer-Related Psychological Distress

Randomized controlled trials utilizing psilocybin-assisted psychotherapy to treat psychosocial-spiritual distress (depression, anxiety, existential suffering) associated with advanced or terminal cancer have resumed in the United States within the last 2 decades at UCLA, NYU, and Johns Hopkins University (JHU). At all 3 sites, the treatment model was similar and was based on the Spring Grove psychedelic psychotherapy model, developed by Dr. Grof and colleagues. Important similarities among the three sites were as follows: (1) double-blind design methodology; (2) randomization among groups; (3) use of validated outcome measures (i.e., Hospital Anxiety and Depression Scale [HADS], Beck Depression Inventory, STAI); (4) use of exclusion criteria such as major unstable medical illnesses (e.g., cardiac/renal/hepatic failure) and major mental illness in particular psychotic spectrum illnesses (e.g., schizophrenia, bipolar I with psychotic features) or a family history of these disorders; (5) careful preparation of the participants for the experimental sessions by trained psychotherapists (usually as part of a dyad team of treatment providers) after a thorough life review and review of their cancer diagnosis narrative with an emphasis on spiritual/existential themes of distress; (6) conduct of the dosing sessions in a comfortable living room-like setting designed for maximal comfort and safety; (7) instructions to participants intended to increase the likelihood of the induction of mystical states of consciousness (participants instructed to lie down on a couch with eyeshades to reduce external visual distractions, and to focus on their inner experiences while a preselected music program played during the session); and (8) integration of the experience during the dosing sessions as part of post-integrative psychotherapy. The three studies differed in the following characteristics: (1) dose of psilocybin- UCLA (0.2 mg/kg), NYU (0.3 mg/kg); JHU (0.43 mg/kg); (2) single (UCLA, NYU) versus multiple (JHU) dosing schedules; (3) the active control (niacin at UCLA, NYU, low dose psilocybin at JHU); and (4) inclusion of nonterminally ill cancer patients (at NYU and JHU but not at UCLA).

The UCLA study results suggested an acute and short-term sustained antidepressant and anxiolytic effect associated with psilocybin versus placebo (Grob et al. 2011). Two other randomized controlled trials of psilocybin-assisted psychotherapy for patients with cancer and psychosocial distress (NYU $N = 29$; Johns Hopkins $N = 44$) have both recently finished and are at this time in the process of formal data analyses. Preliminary findings from both sites suggest treatment effects of the experimental psilocybin group versus the active placebo condition in terms of acute and sustained reductions in anxiety and depression as well as reports of highly salient mystical-type experiences having sustained personal and spiritual significance to participants (Ross et al. 2016; Griffiths et al. 2016).

4 Use of Classic Hallucinogens to Treat Mood and Anxiety Disorders

Prior to prohibition of clinical research with classic hallucinogens in the 1970s, LSD and, to a lesser extent, psilocybin and mescaline were used in the treatment of mood and anxiety disorders (anxiety neuroses in the parlance of the day) (Grinspoon and Balakar 1997). However, we are unaware of any published outcome data from trials of any of these disorders during that era.

Francisco Moreno and colleagues conducted the first clinical trial of a classic hallucinogen in the treatment of OCD (Moreno et al. 2006). In this study, nine participants with OCD received up to 4 doses of psilocybin at least 1 week apart. The active doses were 0.1 mg/kg, 0.2 mg/kg, and 0.3 mg/kg in ascending order, with a control dose of 0.025 mg/kg inserted in double-blind fashion at a random point in the sequence. Participants had significant pre–post decreases in OCD symptoms, as assessed with the Y-BOCS, lasting for 24 h or more. However, there was no significant effect of dose, i.e., higher doses did not result in more improvement than the putative control dose of 0.025 mg/kg. Because of the design of the study it is not possible to conclude whether there was a true treatment effect or only a time effect related to expectancy other sources of bias. Further controlled trials of the use of psilocybin-assisted therapeutics for OCD are clearly warranted.

Several researchers on these drugs have proposed the testing of psilocybin for this indication and proposed mechanisms by which such treatment could be effective (Baumeister et al. 2014; Vollenweider and Kometer 2010; Carhart-Harris et al. 2012b; Kraehenmann et al. 2014). Carhart-Harris et al. recently reported results of a pilot study in which 12 patients with treatment-resistant depression received psilocybin-assisted treatment including a low dose (10 mg orally) followed one week later by a high dose (25 mg orally) (Carhart-Harris et al. 2016). Depressive symptoms were markedly reduced following the high-dose session, and remained so at the final follow-up, with an effect size of $g = 2$ at three months..

5 Mechanisms of Clinically Relevant Effects

Detailed reviews of hallucinogen effects at all mechanistic levels are provided elsewhere in this volume. Very little is known about how any of these persisting effects might mediate lasting therapeutic benefits. Possibly relevant mechanisms reviewed in other publications (Bogenschutz and Pommy 2012; Bogenschutz and Johnson 2016) include downregulation of 5HT2A receptors, increased expression of neurotrophic factors resulting in increased neuroplastic potential, and persisting psychological changes including changes in personality. There is some evidence that the subjective experience during drug administration may be an important determinant of therapeutic effects. Recent studies in normal volunteers have demonstrated that self-reported ratings of the "mystical" quality of the psilocybin experience significantly predicts the lasting personal significance of the experience (Griffiths et al. 2008) and personality change (Maclean et al. 2011). Recent pilot work with psilocybin for alcohol and nicotine addiction demonstrated that strong mystical-type experiences were associated with greater improvement in substance use (Bogenschutz et al. 2015b; Garcia-Romeu et al. 2014). In the alcohol study, more general measures of the intensity of subjective effects (not limited to mystical-like effects) were associated with improvement in drinking (Bogenschutz et al. 2015b). In both of the two recently published trials of psilocybin-assisted treatment of depression and anxiety associated with a life-threatening cancer diagnosis, the degree of mystical experience during psilocybin administration was significantly correlated with improvement in mood and anxiety symptoms (Ross et al. 2016; Griffiths et al. 2016).

It is clear that the brain can be persistently (and sometimes permanently) altered and damaged by an overwhelming psychological event, as in the case of post-traumatic stress disorder (Karl et al. 2006). It has been hypothesized that under the right conditions a very intense experience occasioned by a classic hallucinogen can do the opposite, leading to lasting (and sometimes permanent) positive changes in the brain and in behavior (Garcia-Romeu et al. 2014).

6 Safety

As with all medications, there are risks associated with administration of classic hallucinogens. Within the context of clinical research, these risks are modest. In the reemergence of psychedelic research since the early 1990s, close to 1000 doses of psilocybin (ranging from low to high doses) have been administered safely in Europe and the United States at major academic medical centers (University Hospital of Psychiatry Zurich, Johns Hopkins University, NYU, UCLA, University of New Mexico, University of Arizona) with no reports of any treatment-related serious adverse events (SAEs), including no reported cases of prolonged psychosis or Hallucinogen Persisting Perception Disorder (HPPD) (see below sections on psychosis and HPPD) (Studerus et al. 2011); personal communications Roland Griffiths, Stephen Ross, Charles Grob, Michael Bogenschutz, Francisco Moreno 2014). In addition, approximately 200 doses of intravenous DMT were administered at the U of New Mexico to normal participants in the early 1990s without any SAEs, and LSD has now been administered to approximately two dozen research participants in Europe in the last year (Switzerland, England) without any SAEs reported (Gasser et al. 2014, 2015; Carhart-Harris et al. 2014a). The key commonality among all these studies is the careful attention to screening, set, setting, dose, and preparation and integration before and after drug administration.

6.1 Medical Toxicity

Classic hallucinogens possess remarkably low physiological toxicity and are not associated with end organ damage, carcinogenicity, teratogenicity, lasting neuropsychological deficits, or overdose fatalities (Johnson et al. 2008; Hasler et al. 2004; Cohen 1960; Halpern et al. 2005, 2008; Strassman 1984; Gable 2007; Barbosa et al. 2012). The classic hallucinogens produce sympathomimetic effects and can moderately increase pulse as well as diastolic and systolic blood pressure, although this has not been associated with cardiac, neurologic or other organ damage. (Hasler et al. 2004; Passie et al. 2002; Griffiths et al. 2006; Grob et al. 2011). Common physiological side effects of the classic hallucinogens include: mydriasis, blurry vision, dizziness, tremors, weakness, paresthesias, and increased deep tendon reflexes (Johnson et al. 2008).

6.2 Psychiatric Toxicity

6.2.1 Acute Effects

The acute psychological and behavioral effects of the classic hallucinogens are greatly influenced by set (personality and expectations of the individual), setting (environmental conditions and context of use) and dose, with the factors combining to influence the valence (positive or negative) of the experience. Affective changes can range from euphoric or ecstatic spiritual states to anxiety, terror and panic. Perception is intensified and amplified with alterations in time, space, and boundaries between self and others. Synesthesia (the mixing of various sensory stimuli, e.g., hearing colors) is common. Sensory illusions (e.g., walls breathing) are common, and frank hallucinations can occur, though less frequently. Thought processes are loosened, with effects ranging from increased creativity to thought disorder. Cognition is altered and can range from sudden and deeply felt insight ("noetic" effect) to confusion and disorientation (Wilkins et al. 2014). The sum total of the experience can range from positive mystical-type experiences associated with enduring positive changes in affect/cognition/behavior (Griffiths et al. 2008) to very unpleasant experiences dominated by fear and dysphoria. Severe adverse psychological experiences ('bad trips') tend to occur in poorly prepared individuals who use the substance in an uncontrolled setting and who have psychological risk factors (e.g., severe mental illness, recent trauma). These experiences typically include anxiety, panic, dysphoria, depersonalization, paranoid ideation, fear that the experience will never end, and fear of losing one's mind. Despite such adverse reactions, users usually retain insight into the fact that their symptoms are related to drug ingestion, and they usually respond to verbal reassurance. Classic hallucinogens can acutely engender frank psychosis marked by hallucinations, thought disorder and delusions, although this is rare in individuals without underlying psychotic spectrum illness (Wilkins et al. 2014). Such adverse psychological experiences can potentially lead to dangerous behavior toward self or others especially when used by vulnerable individuals without proper preparation and supervision. (Strassman 1984). In a review of LSD research conducted during the 1950s including thousands of research participants, the reported suicide attempt and completed suicide rates were very low at 1.2 and 0.4, respectively, per 1000 psychiatric patients, and no completed suicides or attempts among 1200 non-patients (Cohen 1960).

6.2.2 Prolonged Effects

Psychosis

The serotonergic hallucinogens have long provided evidence implicating the serotonergic system, as one of several neurotransmitter systems (the other main ones being the dopaminergic, glutamatergic, and endocannabinoid systems), in the pathophysiology of schizophrenia and related psychotic disorders (Murray et al. 2013; Beringer 1923; Halberstadt and Geyer 2013; Szara 1956). After LSD became widely used as a recreational drug in the 1960s, it was increasingly recognized that its use could trigger psychosis, a common example being first used by a teenager or young adult (during the age of typical onset of schizophrenia) who developed the onset of schizophrenia and went on to develop a chronic course of the illness. However, although it is the case that classic hallucinogen use can cause transient psychotic-like positive symptoms in normal volunteers and can provoke sustained psychosis in vulnerable people with psychotic spectrum illnesses (i.e., schizophrenia, schizoaffective disorder, bipolar disorder with psychotic features), there is little to no evidence linking classic hallucinogen use to prolonged psychosis in individuals without a psychotic diathesis (Ross and Peselow 2012). Estimates of the prevalence of LSD-induced psychosis as assessed by early psychedelic researchers and clinicians (many working with and administering LSD to psychiatric *inpatients*) were as follows from 2 reports: 0.8/1000 research volunteers (a single case out of 1250, where the volunteer was the identical twin of an individual with schizophrenia) and 1.8/1000 psychiatric patients (seven cases out of approximately 3850 patients) (Cohen 1960); and 0/170 research volunteers and 9/1000 psychiatric patients (37 out of 4300 patients) (Malleson 1971). A recent cross-sectional study evaluating data taken from years 2001–2004 of the National Survey on Drug Use and Health with a sample of 130,152 (representing a random sample of the US population living in households) did not find any significant associations between lifetime use of any psychedelic or past year use of LSD and increased rates of any psychiatric symptoms (including psychosis) or mental health outcomes (Krebs and Johansen 2013). Given that the classic hallucinogens are known to exacerbate psychosis in individuals with psychotic spectrum illnesses, in the modern era of psychedelic research in the last 20 years, all human trials at academic medical centers in the US and Europe have excluded individuals with psychotic spectrum illness or those who are at risk of such illnesses because of a known positive family history of psychotic illnesses. As mentioned above, no cases of persisting psychosis have been reported across these studies.

Hallucinogen Persisting Perception Disorder (HPPD)

Descriptions of persisting perceptual abnormalities ("flashbacks") following the use of the SHs were first described over 100 years ago with mescaline (Ellis 1898). The next reports of this condition occurred in the 1950s and 1960s, mostly in relation to

LSD ingestion, and described prolonged changes in normal perception that were similar in nature to the perceptual effects experienced under the influence of the hallucinogen, persisting subacutely (weeks to months), and chronically after use of the hallucinogen (Cooper 1955; Smart and Bateman 1967). In addition to the classic hallucinogens, cases of persisting perceptual disturbances have been reported and attributed to the use of MDMA, ketamine, and cannabis (Litjens et al. 2014). The symptoms can occur spontaneously or be triggered by stress, anxiety, exercise, or use of another drug (e.g., cannabis) (Abraham 1983). The term "flashback" has sometimes been used synonymously with HPPD but is distinguished from HPPD in that flashbacks do not necessarily persist or cause clinical distress or impairment, and can even be experienced as pleasurable by the individual (Lerner et al. 2002; Frecska and Luna 2006). Unlike flashbacks, HPPD tends to be recurrent with a continuous or intermittent/paroxysmal pattern of symptoms that is either a slowly reversible or a chronic permanent condition experienced as highly distressing by patients (Lerner et al. 2002). It is important to note that HPPD is not a psychotic spectrum illness, and patients have intact reality testing and awareness that the illusions and hallucinations are not real.

Although the exact prevalence of HPPD is unknown due to a lack of rigorous epidemiologic studies, it is thought to be a rare (but highly distressing) condition given the relatively few cases reported out of the tens of millions of doses of hallucinogens used over the last 50 years, and it appears to be significantly less common in research settings with careful screening and preparation (Studerus et al. 2011; Johnson et al. 2008; Halpern and Pope 2003). In two recent web-based surveys of hallucinogen users, visual distortions and illusions were quite commonly reported, but only 4–11% were distressed by these experiences (Baggott et al. 2011; Carhart-Harris and Nutt 2010). Data from the above-mentioned population study derived from the 2001–2004 National Survey on Drug Use and Health found no association between lifetime use of psychedelics or past year use of LSD and past year visual phenomena consistent with HPPD (Krebs and Johansen 2013).

Addictive Liability of the Serotonergic Hallucinogens

A commonality among all drugs that are capable of producing addiction is their ability to substantially increase extracellular DA levels in the nucleus accumbens, either directly by enhancing DA transmission through reuptake inhibition or facilitating presynaptic DA release (e.g., cocaine, amphetamine, MDMA), or by indirect GABAergic, cholinergic or glutamatergic mechanisms that affect DA-cell firing (e.g., alcohol, sedatives, opioids, cannabis, nicotine, NMDA antagonists such as PCP) (Baler and Volkow 2006).

In contrast to all other drugs of abuse, classic hallucinogens are not considered to be capable of producing sufficient reinforcing effects to produce addiction (O'Brien 2001). Animal models (i.e., self-administration, conditioned place preference) have failed to reliably demonstrate addictive liability of the classic hallucinogens, suggesting that they do not possess sufficient pharmacologic properties to initiate or maintain dependence (Nichols 2004). Almost all of the classic hallucinogens (with the exception of LSD) (Watts et al. 1995; Giacomelli et al. 1998) lack affinity for DA receptors or the DAT and do not directly affect dopaminergic transmission. Interestingly, despite the evidence that classic hallucinogens have been shown to increase DA transmission in striatal areas in humans, they fail to significantly activate the nucleus accumbens in PET imaging studies consistent with the lack of evidence linking classical hallucinogens with dependence syndromes (Vollenweider et al. 1998, 1999; Geyer and Vollenweider 2008). Although problematic or disordered use of classic hallucinogens certainly occurs, it is uncommon, and very rare in people over 25 years of age (Substance Abuse and Mental Health Services Administration 2013). The National Institute on Drug Abuse does not consider the classic hallucinogens drugs of "addiction" as they do not produce compulsive drug-seeking behavior or chronic addiction, and most recreational users decrease or stop their use over time (National Institute on Drug Abuse 2014).

7 Case Studies

Because the process of treatment with classic hallucinogens is so different from other pharmacotherapies, it may be useful to provide descriptions of the process in specific cases. Below we provide two brief case reports from the authors' clinical trials: one of a patient treated for alcohol dependence, the other of a cancer patient struggling depression and anxiety in relation to a cancer diagnosis. Details of these cases have been altered to obscure the identity of these individuals.

7.1 Psilocybin-Assisted Treatment of Alcoholism

C is a 59 year-old divorced mother of 2 who had struggled with alcohol since age 15. Her drinking had led to problems including recurrent physical violence, multiple arrests, poor work history, and intermittent homelessness. She had suffered severe abuse in the context of relationships with partners who also drank, including being beaten unconscious and suffering from intracranial bleeding on at least one occasion. She had made several past attempts to stop drinking, with little success. When she volunteered for the psilocybin trial, she had been sober for 11 days, and had been drinking 7 out of the past 84 days, an average of 16.9 standard drinks per drinking day.

During the preparatory phase of treatment in the study, she stated a goal of total abstinence, and rated the importance of abstinence as high and her readiness for abstinence as high, but her confidence in achieving it was low. She said that she wanted to understand why she drank, and hoped that this would help her stay sober. She listed God's will, forgiveness, humility, (to be) loved, and self-control as her most important values, and saw clearly that her drinking was in conflict with these values.

During her first psilocybin session, she reported that she experienced powerful feelings of sorrow and remorse regarding the course of her life, and particularly concerning her perceived failures as a parent as a result of her drinking. This experience was quite painful, and she believed that she was sobbing uncontrollably during much of this time, although she was actually lying quietly on the couch at the time. After the session, she felt a sense of relief, and said that she had been able to let go of these feelings and experience a sense of forgiveness. She was hopeful that the experience would help her stay sober, and had no desire to drink after the session.

C remained sober between the first and second psilocybin session. During the second session, she reported that she experienced a visual image of a small child lying "broken" on the floor. She realized that this child was her, and experienced herself as a 3-year-old child, devastated by abandonment by her father, an issue that she had not discussed in the preparatory sessions. After this, she began to perceive a white light, which she called "God's healing light," and felt a profound sense of love. She felt that she had been healed by this experience, and that she now felt "whole" and worthy of love.

In discussing these experiences afterwards, C said that she thought her drinking had been an attempt to escape the painful feelings of being unworthy of love, as well as the painful feelings of shame and loss related to her life as an alcoholic. She had avoided these feelings, believing that she would "fall apart" if she faced them. Following the sessions, she now felt that she was strong enough to face these feelings, and that she was a whole person, worthy of love. At her most recent follow-up, 5 months after the first psilocybin session, she remained abstinent and continued to feel that her life had been transformed, in spite of the unexpected death of a close family member during the interim.

7.2 *Psilocybin-Assisted Treatment of Cancer-Related Psychological and Existential Distress*

E is a 22-year-old woman, a first year law student, with recent diagnosis of leukemia. She had no prior medical or psychiatry history. She was treated successfully for her leukemia with chemotherapy and was told by her oncologist that she had a very good prognosis and was likely cured of her cancer. For the first 6 months following diagnosis and treatment, she did not experience any adverse psychological effects of the cancer diagnosis. However, at the 6-month point of remission, E started to become frightened that the cancer could recur. Even though she understood the good prognosis, she experienced increasing anxiety about the cancer returning and what that would mean for her. These thoughts continued to get worse to the point where E started experiencing panic attack-like events related to an acute fear of death. She had always thought of death as a distant event, something that she would not have to contend with for decades. However, the cancer provoked a kind of crisis and she began to feel that there was now a short distance between herself and death. She began to experience a dysphoric and anxious mood accompanied by anhedonia, poor concentration, preoccupation with death, and a desire for a hastened death because she felt that the end was imminent. From a Catholic faith background, she became angry at God for allowing her to get cancer and started for the first time in her life to have doubts about the existence of God. She stopped attending church and felt she had lost access to the nurturing parts of her religion and spirituality. She told her oncologist about these symptoms, and her oncologist referred her to the NYU Psilocybin Cancer Project. E was a casual user of alcohol and had no prior use of other drugs including hallucinogens and marijuana. She was intrigued by the use of psilocybin to treat anxiety but admitted that she was skeptical that it would actually do anything to decrease her anxiety or fear of death related to her cancer diagnosis. She felt it was worth a try given her level of distress and because she trusted her oncologist's judgment.

E underwent her first dosing session and happened to be randomized to the group that received psilocybin (0.3 mg/kg) first. On the day of the session, she reported having some trouble sleeping the night before and being somewhat apprehensive about the dosing session. She wanted to make sure that she would return to normative reality when the session ended. Her treatment dyad team did some deep breathing with her before the session and reassured her of all the ways the team would address any difficult or anxious psychological experiences. E received her psilocybin capsule at 9 AM. The treatment team checked in every hour to assess E's mental state. For the first 3 h, E reported nothing suggestive of a psychoactive experience (unusual for those that receive psilocybin whose effects usually are experienced in the first hour post-dosing). She told the dyad team that she was sure she must have gotten the placebo. At approximately 12:30, E sat up and took off her eye shades. She reported that the music sounded "trippy." When the dyad team inquired more, she said that she felt anxious and wondered if she would feel normal again. After verbal reassurance, E lay back down and placed the eye shades on. At approximately 2 PM, she sat up and reported the following visions: she journeyed back to parts of her childhood and came upon multiple scenes of her and her family (2 parents, and younger brother) joined together in laughter and love that made her feel a sense of unity and support; she reported a scene where she saw herself standing in front of a hospital and saw a cauldron of black, hot smoke circling inside of her body (which she interpreted as her cancer), followed by her boyfriend and family forming a loving circle around her which caused the black smoke inside of her to disappear outside of her body and into the sky (she interpreted this as a cathartic experience of the fear of cancer leaving her); she reported seeing her cancer-riddled body die in front of her eyes while the treatment providers and a transcendental entity together solved a complicated mathematical equation (she interpreted this as God and the treatment team working together to figure out the biological nature of her cancer-related anxiety). At the end of the experience, E declared that she felt "re-born" and freed from the anxiety and fear of death caused by her cancer diagnosis. Her scores on the main outcome measures of anxiety and depression (i.e., HADS, STAI, Beck Depression) dropped dramatically from high clinically significant values at baseline to almost zero, 7-h post-dosing with psilocybin. Her follow-up distress scores on all of these scales remained close to zero for the next 6-months until the final follow-up point. It is now 1.5 years after her treatment, and she continues to report no anxiety or distress associated with her cancer diagnosis. She is flourishing as a law student and reports a sense of calmness and psychological well-being that she continues to attribute to the psilocybin dosing experience. She has taken up a meditation practice and still can remember salient aspects of the experience that have stuck in her memory.

8 Discussion

Although classic hallucinogens have been used by humans for millennia, known to science for over a century, and subjected to extensive basic research for decades, the rigorous study of their clinical use is in its infancy. Promising early phase research conducted from the 1950s through the early 1970s was discontinued before any conclusions could be reached concerning the efficacy of any of the classic hallucinogens for any clinical condition. However, the research that was conducted in that era strongly suggests that classic hallucinogens have clinically relevant effects, particularly in the case of LSD treatment of alcoholism. In the past decade, clinical trials have resumed investigating the effects of classic hallucinogens in the treatment of existential distress in the face of cancer, and in the treatment of addictions including alcoholism and nicotine addiction. The studies that have been completed to date are small, and not sufficient to establish efficacy. However, they strongly suggest that the administration of psilocybin in the therapeutic context produces therapeutic benefit in both patients with addictions and cancer patients with existential distress. Only time will tell if hallucinogen-assisted treatments for addiction and cancer-related psychological/existential distress will prove to be safe and effective in large scale double-blind, placebo-controlled clinical trials. If the data supports their efficacy, they would constitute a novel psychopharmacologic–psychosocial treatment paradigm to treat these disorders.

The clinical safety of classic hallucinogens, well documented by extensive research in the 1950s through early 1970s, has been confirmed in the rigorous clinical work which began again in the 1990s. Although physical and psychological risks exist, they can be minimized through the safety procedures followed in recent clinical studies, including careful screening of participants, thorough training of therapists, close attention to dose, set, and setting, preparation, support and monitoring, and follow-up of participants, and predetermined procedures for dealing with medical and psychiatric emergencies (Johnson et al. 2008). Since the resumption of clinical research with classic hallucinogens in the 1990s to date, there have been no reported treatment-related serious adverse events or cases of persistent harm to participants. Although no medication is harmless, this fact is encouraging.

Future research must continue to make every effort to maximize safety. Should a classic hallucinogen be approved for clinical use in the future, it will also be critical that rigorous safety procedures be built into the approved conditions of clinical use.

Although research has elucidated many of the acute neurobiological and psychological effects of classic hallucinogens on humans, animals, and in vitro systems, the mechanisms of clinically relevant persisting effects remain poorly understood. At this point, the causal mechanisms with the strongest support are intensely memorable and personally meaningful experiences, particularly those of a mystical quality. Much more work is necessary to better understand the mechanisms by which such experiences, and their neurobiological correlates, can lead to persisting psychological, behavioral change, and neurobiological change.

Given the early stage of clinical research on classic hallucinogens, relative to other classes of drugs, many avenues of research could be explored that would advance the field. Among them, several topics appear particularly promising. First, well-designed and adequately powered efficacy trials should be conducted testing the efficacy of classic hallucinogens (particularly psilocybin and LSD) for the indications where the existing evidence strongly suggests efficacy: addiction to drugs including alcohol and nicotine, and existential crisis in the face of death among cancer patients or in other patients facing life-threatening illness. Phase II trials for psilocybin-assisted psychotherapy for cancer-related psychological distress were recently completed and a phase III trial for this same indication is currently being planned which potentially could lead to the historic partial rescheduling of psilocybin to become a prescribable medication. Furthermore, controlled trials are already under way testing the efficacy of psilocybin in the treatment of alcohol use disorders, tobacco addiction, and cocaine addiction.

Second, a broader range of classic hallucinogens and possible indications should be studied. The encouraging results that have been seen with classic hallucinogen treatment across alcohol, nicotine, and opioid dependence suggest the possibility that other chemical and behavioral addictions may also respond to treatment with these agents. While it cannot be assumed that addictions are interchangeable or that classic hallucinogen all have comparable efficacy in the treatment of addiction, it seems reasonable to pursue additional indications. One such possibility will be explored in a pending trial that will be the first to evaluate the effects of psilocybin-assisted treatment of cocaine dependence. Given the remarkable short-term efficacy of ketamine (an NMDA antagonist hallucinogen with biological and psychological effects overlapping those of classic hallucinogens), in the treatment of depression (Serafini et al. 2014), the efficacy of classic hallucinogens in depression should be studied as well (Baumeister et al. 2014).

Third, it will be important to investigate the mechanisms of action of classic hallucinogens in the treatment of specific disorders. Although neuroimaging studies have begun to characterize the acute effects of psilocybin on brain function (Tagliazucchi et al. 2014; Carhart-Harris et al. 2012a, 2013, 2014b; Kraehenmann et al. 2014), the persisting effects of classic hallucinogens on brain function have not been characterized. Changes in 5HT2A receptors, serotonin transporters, neurotrophic factors, and other relevant brain chemicals have not been measured in the context of clinical studies of classic hallucinogen treatments. Although the small trials that have been conducted have provided some information concerning the psychological processes that may underlie behavior change, future trials should continue to characterize both the acute medication effects and the longer term psychological changes that are associated with desired clinical outcomes.

References

Abraham HD (1983) Visual phenomenology of the LSD flashback. Arch Gen Psychiatry 40:884–889

Abuzzahab FS, Anderson BJ (1971) A review of LSD treatment in alcoholism. Int Pharmacopsychiatry 6:223–235

Albaugh BJ, Anderson PO (1974) Peyote in the treatment of alcoholism among American Indians. Am J Psychiatry 131:1247–1250

Allebeck P, Bolund C (1991) Suicides and suicide attempts in cancer patients. Psychol Med 21:979–984

Baggott MJ, Coyle JR, Erowid E, Erowid F, Robertson LC (2011) Abnormal visual experiences in individuals with histories of hallucinogen use: a Web-based questionnaire. Drug Alcohol Depend 114:61–67

Baler RD, Volkow ND (2006) Drug addiction: the neurobiology of disrupted self-control. Trends Mol Med 12:559–566

Barbosa PC, Mizumoto S, Bogenschutz MP, Strassman RJ (2012) Health status of ayahuasca users. Drug Test Anal 4:601–609

Baumeister D, Barnes G, Giaroli G, Tracy D (2014) Classical hallucinogens as antidepressants? A review of pharmacodynamics and putative clinical roles. Ther Adv Psychopharmacol 4:156–169

Beringer K (1923) Experimentelle Psychosen durch Mescalin. Zeitschrift für die gesamte Neurologie und Psychiatrie 84:426–433

Bogenschutz MP, Pommy JM (2012) Therapeutic mechanisms of classic hallucinogens in the treatment of addictions: from indirect evidence to testable hypotheses. Drug Test Anal 4:543–555

Bogenschutz MP, Johnson MW (2016) Classic hallucinogens in the treatment of addictions. Prog Neuropsychopharmacol Biol Psychiatry 64:250–258

Bogenschutz MP, Forcehimes AA, Pommy JA, Wilcox CE, Barbosa P, Strassman RJ (2015a). Psilocybin-assisted treatment for alcohol dependence: a proof-of-concept study. J Psychopharmacol 29(3):289–299

Bogenschutz MP, Forcehimes AA, Pommy JA, Wilcox CE, Barbosa PC, Strassman RJ (2015b) Psilocybin-assisted treatment for alcohol dependence: a proof-of-concept study. J Psychopharmacol 29:289–299

Bowen WT, Soskin RA, Chotlos JW (1970) Lysergic acid diethylamide as a variable in the hospital treatment of alcoholism: a follow-up study. J Nerv Ment Dis 150:111–118

Brady MJ, Peterman AH, Fitchett G, Mo M, Cella D (1999) A case for including spirituality in quality of life measurement in oncology. Psychooncology 8:417–428

Breitbart W, Rosenfeld B, Pessin H, Kaim M, Funesti-Esch J, Galietta M, Nelson CJ, Brescia R (2000) Depression, hopelessness, and desire for hastened death in terminally ill patients with cancer. JAMA 284:2907–2911

Brown KW, Levy AR, Rosberger Z, Edgar L (2003) Psychological distress and cancer survival: a follow-up 10 years after diagnosis. Psychosom Med 65:636–643

Buckman J (1967) Theoretical aspects of LSD therapy. In: Abramson HA (ed) The use of LSD in psychotherapy and alcoholism. Bobbs-Merrill, Indianapolis

Bultz BD, Holland JC (2006) Emotional distress in patients with cancer: the sixth vital sign. Community Oncology 3:311–314

Busch AK, Johnson WC (1950) L.S.D. 25 as an aid in psychotherapy; preliminary report of a new drug. Dis Nerv Syst 11:241–243

Callaway JC, Raymon LP, Hearn WL, McKenna DJ, Grob CS, Brito GS, Mash DC (1996) Quantitation of *N,N*-dimethyltryptamine and harmala alkaloids in human plasma after oral dosing with ayahuasca. J Anal Toxicol 20:492–497

Carhart-Harris RL, Nutt DJ (2010) User perceptions of the benefits and harms of hallucinogenic drug use: a web-based questionnaire study. J Subst Use 15:283–300

Carhart-Harris RL, Erritzoe D, Williams T, Stone JM, Reed LJ, Colasanti A, Tyacke RJ, Leech R, Malizia AL, Murphy K, Hobden P, Evans J, Feilding A, Wise RG, Nutt DJ (2012a) Neural correlates of the psychedelic state as determined by fMRI studies with psilocybin. Proc Natl Acad Sci USA 109:2138–2143

Carhart-Harris RL, Leech R, Williams TM, Erritzoe D, Abbasi N, Bargiotas T, Hobden P, Sharp DJ, Evans J, Feilding A, Wise RG, Nutt DJ (2012b) Implications for psychedelic-assisted psychotherapy: functional magnetic resonance imaging study with psilocybin. Br J Psychiatry 200:238–244

Carhart-Harris RL, Leech R, Erritzoe D, Williams TM, Stone JM, Evans J, Sharp DJ, Feilding A, Wise RG, Nutt DJ (2013) Functional connectivity measures after psilocybin inform a novel hypothesis of early psychosis. Schizophr Bull 39:1343–1351

Carhart-Harris RL, Kaelen M, Whalley MG, Bolstridge M, Feilding A, Nutt DJ (2014a) LSD enhances suggestibility in healthy volunteers. Psychopharmacology 232(4):785–794 (Berl)

Carhart-Harris RL, Leech R, Hellyer PJ, Shanahan M, Feilding A, Tagliazucchi E, Chialvo DR, Nutt D (2014b) The entropic brain: a theory of conscious states informed by neuroimaging research with psychedelic drugs. Front Hum Neurosci 8:20

Carhart-Harris RL, Bolstridge M, Rucker J, Day CM, Erritzoe D, Kaelen M, Bloomfield M, Rickard JA, Forbes B, Feilding A, Taylor D, Pilling S, Curran VH, Nutt DJ (2016) Psilocybin with psychological support for treatment-resistant depression: an open-label feasibility study. Lancet Psychiatry

Carr D, Goudas L, Lawrence D et al (2002) Management of cancer symptoms: pain, depression, and fatigue. Agency for Healthcare Research and Quality, Rockville

Cassel EJ (1982) The nature of suffering and the goals of medicine. N Engl J Med 306:639–645

Chwelos N, Blewett DB, Smith CM, Hoffer A (1959) Use of d-lysergic acid diethylamide in the treatment of alcoholism. Q J Stud Alcohol 20:577–590

Clarke DM, Kissane DW (2002) Demoralization: its phenomenology and importance. Aust N Z J Psychiatry 36:733–742

Cohen S (1960) Lysergic acid diethylamide: side effects and complications. J Nerv Ment Dis 130:30–40

Cooper HA (1955) Hallucinogenic drugs. Lancet 268:1078–1079

Doering-Silveira E, Grob CS, de Rios MD, Lopez E, Alonso LK, Tacla C, da Silveira DX (2005) Report on psychoactive drug use among adolescents using ayahuasca within a religious context. J Psychoactive Drugs 37:141–144

Dyck E (2006) 'Hitting highs at rock bottom': LSD treatment for alcoholism, 1950–1970. Soc Hist Med 19:313–329

El Nawawi NM, Balboni MJ, Balboni TA (2012) Palliative care and spiritual care: the crucial role of spiritual care in the care of patients with advanced illness. Curr Opin Support Palliat Care 6:269–274

Ellis H (1898) Mescal: a new artificial paradise. Contemp Rev 73:130–141

El-Seedi HR, de Smet PA, Beck O, Possnert G, Bruhn JG (2005) Prehistoric peyote use: alkaloid analysis and radiocarbon dating of archaeological specimens of Lophophora from Texas. J Ethnopharmacol 101:238–242

Fabregas JM, Gonzalez D, Fondevila S, Cutchet M, Fernandez X, Barbosa PC, Alcazar-Corcoles MA, Barbanoj MJ, Riba J, Bouso JC (2010) Assessment of addiction severity among ritual users of ayahuasca. Drug Alcohol Depend 111:257–261

Ferrell BR, Grant MM, Funk BM, Otis-Green SA, Garcia NJ (1998) Quality of life in breast cancer survivors: implications for developing support services. Oncol Nurs Forum 25:887–895

Frankl V (1984) Man's search for meaning. Simon & Schuster, New York

Frecska E, Luna LE (2006) The adverse effects of hallucinogens from intramural perspective. Neuropsychopharmacol Hung 8:189–200

Gabermann V (1978) Estimation of mescaline and pellotine in Lophophora coulter plants (Cactaceae) by means of the oscillographic polarography. Biokhimiia 43:246–251

Gable RS (2007) Risk assessment of ritual use of oral dimethyltryptamine (DMT) and harmala alkaloids. Addiction 102:24–34

Gall TL, Cornblat MW (2002) Breast cancer survivors give voice: a qualitative analysis of spiritual factors in long-term adjustment. Psychooncology 11:524–535

Garcia-Romeu A, Griffiths RR, Johnson MW (2014) Psilocybin-occasioned mystical experiences in the treatment of tobacco addiction. Curr Drug Abuse Rev 7:157–164

Garrity JF (2000) Jesus, peyote, and the holy people: alcohol abuse and the ethos of power in Navajo healing. Med Anthropol Q 14:521–542

Gasser P, Holstein D, Michel Y, Doblin R, Yazar-Klosinski B, Passie T, Brenneisen R (2014) Safety and efficacy of lysergic acid diethylamide-assisted psychotherapy for anxiety associated with life-threatening diseases. J Nerv Ment Dis 202:513

Gasser P, Kirchner K, Passie T (2015) LSD-assisted psychotherapy for anxiety associated with a life-threatening disease: a qualitative study of acute and sustained subjective effects. J Psychopharmacol 29:57–68

Geyer MA, Vollenweider FX (2008) Serotonin research: contributions to understanding psychoses. Trends Pharmacol Sci 29:445–453

Giacomelli S, Palmery M, Romanelli L, Cheng CY, Silvestrini B (1998) Lysergic acid diethylamide (LSD) is a partial agonist of D2 dopaminergic receptors and it potentiates dopamine-mediated prolactin secretion in lactotrophs in vitro. Life Sci 63:215–222

Greenstein M, Breitbart W (2000) Cancer and the experience of meaning: a group psychotherapy program for people with cancer. Am J Psychother 54:486–500

Griffiths RR, Richards WA, Mccann U, Jesse R (2006) Psilocybin can occasion mystical-type experiences having substantial and sustained personal meaning and spiritual significance. Psychopharmacology (Berl) 187:268–283 (discussion 284–292)

Griffiths R, Richards W, Johnson M, McCann U, Jesse R (2008) Mystical-type experiences occasioned by psilocybin mediate the attribution of personal meaning and spiritual significance 14 months later. J Psychopharmacol 22:621–632

Griffiths RR, Johnson MW, Carducci MA, Umbricht A, Richards WA, Richards BD, Cosimano MP, Klinedinst MA (2016) Psilocybin produces substantial and sustained decreases in depression and anxiety in patients with life-threatening cancer: a randomized double-blind trial. J Psychopharmacol 30:1181–1197

Grinspoon L, Balakar JB (1997) Psychedelic drugs reconsidered. The Lindesmith Center, New York

Grob CS, Danforth AL, Chopra GS, Hagerty M, McKay CR, Halberstadt AL, Greer GR (2011) Pilot study of psilocybin treatment for anxiety in patients with advanced-stage cancer. Arch Gen Psychiatry 68:71–78

Grof S (2008) LSD psychotherapy. Multidisciplinary Association for Psychedlic Studies, Ben Lomond

Grof S, Halifax J (1977) The human encounter with death. EP Dutton, New York

Grof S, Goodman LE, Richards WA, Kurland AA (1973a) LSD-assisted psychotherapy in patients with terminal cancer. Int Pharmacopsychiatry 8:129–144

Grof S, Soskin RA, Richards WA, Kurland AA (1973b) DPT as an adjunct in psychotherapy of alcoholics. Int Pharmacopsychiatry 8:104–115

Halberstadt AL, Geyer MA (2013) Serotonergic hallucinogens as translational models relevant to schizophrenia. Int J Neuropsychopharmacol 16:2165–2180

Halpern JH (1996) The use of hallucinogens in the treatment of addiction. Addict Res 4:177–189

Halpern JH, Pope HG (2003) Hallucinogen persisting perception disorder: what do we know after 50 years? Drug Alcohol Depend 69:109–119
Halpern JH, Sherwood AR, Hudson JI, Yurgelun-Todd D, Pope HG (2005) Psychological and cognitive effects of long-term peyote use among Native Americans. Biol Psychiatry 58:624–631
Halpern JH, Sherwood AR, Passie T, Blackwell KC, Ruttenber AJ (2008) Evidence of health and safety in American members of a religion who use a hallucinogenic sacrament. Med Sci Monit 14:SR15–SR22
Hasler F, Grimberg U, Benz MA, Huber T, Vollenweider FX (2004) Acute psychological and physiological effects of psilocybin in healthy humans: a double-blind, placebo-controlled dose-effect study. Psychopharmacology 172:145–156
Heffter A (1896) Ueber Cacteenalkaloïde. Ber Dtsch Chem Ges 29:216–227
Heffter A (1898) Ueber Pellote - Beiträge zur chemischen und pharmakologischen Kenntniss der Cacteen Zweite Mittheilung. Naunyn-Schmiedeberg's Archives of Pharmacology 40:385–429
Hoffer A (1967) A program for treatment of alcoholism: LSD, malvaria, and nicotinic acid. In: Abramson HA (ed) The use of LSD in psychotherapy and alcoholism. Bobbs-Merrill, Indianapolis
Hofmann A (1979) How LSD originated. J Psychedelic Drugs 11:53–60
Hofmann A, Frey A, Ott H, Petrzilka T, Troxler F (1958a) Konstitutionsaufklärung und Synthese von Psilocybin. Experientia 14:397–401
Hofmann A, Heim R, Brack A, Kobel H (1958b) Psilocybin ein psychotroper Wirkstoff aus dem mexikanischen Rauschpilz. Rev Mycologie 22:17–21
Hollister LE, Shelton J, Krieger G (1969) A controlled comparison of lysergic acid diethylamide (LSD) and dextroamphetmine in alcoholics. Am J Psychiatry 125:1352–1357
Institute of Medicine (2008) Cancer care for the whole patient: meeting psychosocial health needs. The National Academies Press, Washington, DC
Isbell H (1959) Comparison of the reactions induced by psilocybin and LSD-25 in man. Psychopharmacologia 1:29–38
Jaiswal R, Alici Y, Breitbart W (2014) A comprehensive review of palliative care in patients with cancer. Int Rev Psychiatry 26:87–101
Johnson M, Richards W, Griffiths R (2008) Human hallucinogen research: guidelines for safety. J Psychopharmacol 22:603–620
Johnson MW, Garcia-Romeu A, Cosimano MP, Griffiths RR (2014) Pilot study of the 5-HT2AR agonist psilocybin in the treatment of tobacco addiction. J Psychopharmacol 0269881114548296
Karl A, Schaefer M, Malta LS, Dorfel D, Rohleder N, Werner A (2006) A meta-analysis of structural brain abnormalities in PTSD. Neurosci Biobehav Rev 30:1004–1031
Kast E (1966) LSD and the dying patient. Chic Med Sch Q 26:80–87
Kast EC, Collins VJ (1964) Study of lysergic acid diethylamide as an analgesic agent. Anesth Analg 43:285–291
Katon WJ (2003) Clinical and health services relationships between major depression, depressive symptoms, and general medical illness. Biol Psychiatry 54:216–226
Kissane DW (2000) Psychospiritual and existential distress. The challenge for palliative care. Aust Fam Physician 29:1022–1025
Kissane D (2009) Beyond the psychotherapy and survival debate: the challenge of social disparity, depression and treatment adherence in psychosocial cancer care. Psychooncology 18:1–5
Kraehenmann R, Preller KH, Scheidegger M, Pokorny T, Bosch OG, Seifritz E, Vollenweider FX (2014). Psilocybin-induced decrease in amygdala reactivity correlates with enhanced positive mood in healthy volunteers. Biol Psychiatry 78(8):572–581
Krebs TS, Johansen PO (2012) Lysergic acid diethylamide (LSD) for alcoholism: meta-analysis of randomized controlled trials. J Psychopharmacol 26(7):994–1002
Krebs TS, Johansen PO (2013) Psychedelics and mental health: a population study. PLoS ONE 8: e63972

Kunitz SJ, Levy JE (1994) Drinking careers: a twenty-five year study of three Navajo populations. Yale University Press, New Haven

Labate BC, Cavnar C (2011) The expansion of the field of research on ayahuasca: Some reflections about the ayahuasca track at the 2010 MAPS "Psychedelic Science in the 21st Century" conference. Int J Drug Policy 22:174–178

Leary T, Litwin GH, Metzner R (1963) Reactions to psilocybin administered in a supportive environment. J Nerv Ment Dis 137:561–573

Lemay K, Wilson KG (2008) Treatment of existential distress in life threatening illness: a review of manualized interventions. Clin Psychol Rev 28:472–493

Lerner AG, Gelkopf M, Skladman I, Oyffe I, Finkel B, Sigal M, Weizman A (2002) Flashback and hallucinogen persisting perception disorder: clinical aspects and pharmacological treatment approach. Isr J Psychiatry Relat Sci 39:92–99

Leuner H (1967) Present state of psycholytic therapy and its possibilities. In: Abramson HA (ed) The use of LSD in psychotherapy and alcoholism. Bobbs-Merrill, Indianapolis

Levi F, Bulliard JL, la Vecchia C (1991) Suicide risk among incident cases of cancer in the Swiss Canton of Vaud. Oncology 48:44–47

Li M, Fitzgerald P, Rodin G (2012) Evidence-based treatment of depression in patients with cancer. J Clin Oncol 30:1187–1196

Liester MB, Prickett JI (2012) Hypotheses regarding the mechanisms of ayahuasca in the treatment of addictions. J Psychoactive Drugs 44:200–208

Litjens RP, Brunt TM, Alderliefste GJ, Westerink RH (2014) Hallucinogen persisting perception disorder and the serotonergic system: a comprehensive review including new MDMA-related clinical cases. Eur Neuropsychopharmacol 24:1309–1323

Lu L, Liu Y, Zhu W, Shi J, Liu Y, Ling W, Kosten T (2009) Traditional medicine in the treatment of drug addiction. Am J Drug Alcohol Abuse 35:11

Ludwig AM, Levine J (1965) A controlled comparison of five brief treatment techniques employing lsd, hypnosis, and psychotherapy. Am J Psychother 19:417–435

Ludwig A, Levine J, Stark L, Lazar R (1969) A clinical study of LSD treatment in alcoholism. Am J Psychiatry 126:59–69

Maclean KA, Johnson MW, Griffiths RR (2011) Mystical experiences occasioned by the hallucinogen psilocybin lead to increases in the personality domain of openness. J Psychopharmacol 25:1453–1461

Malleson N (1971) Acute adverse reactions to LSD in clinical and experimental use in the United Kingdom. Br J Psychiatry 118:229–230

Mangini M (1998) Treatment of alcoholism using psychedelic drugs: a review of the program of research. J Psychoactive Drugs 30:381–418

Masters R, Houston J (2000) The varieties of psychedelic experience: the classic guide to the effects of LSD on the human psyche. Park Street Press, Rochester

McClain CS, Rosenfeld B, Breitbart W (2003) Effect of spiritual well-being on end-of-life despair in terminally-ill cancer patients. Lancet 361:1603–1607

Mckenna DJ (2007) The healing vine: Ayahuasca as medicine in the 21st century. In: Winkelman MJ, Roberts TB (eds) Psychedelic medicine: New evidence for hallucinogenic substances as treatments, vol 1. Praeger Publishers/Greenwood Publishing Group, Westport

McKenna DJ, Towers GH, Abbott F (1984) Monoamine oxidase inhibitors in South American hallucinogenic plants: tryptamine and beta-carboline constituents of ayahuasca. J Ethnopharmacol 10:195–223

Meyerhoff B (1974) Peyote Hunt. Cornell, Ithaca

Mitchell AJ, Chan M, Bhatti H, Halton M, Grassi L, Johansen C, Meader N (2011) Prevalence of depression, anxiety, and adjustment disorder in oncological, haematological, and palliative-care settings: a meta-analysis of 94 interview-based studies. Lancet Oncol 12:160–174

Moreno FA, Wiegand CB, Taitano EK, Delgado PL (2006) Safety, tolerability, and efficacy of psilocybin in 9 patients with obsessive-compulsive disorder. J Clin Psychiatry 67:1735–1740

Murata H (2003) Spiritual pain and its care in patients with terminal cancer: construction of a conceptual framework by philosophical approach. Palliat Support Care 1:15–21

Murray RM, Paparelli A, Morrison PD, Marconi A, di Forti M (2013) What can we learn about schizophrenia from studying the human model, drug-induced psychosis? Am J Med Genet B Neuropsychiatr Genet 162B:661–670
National Institute on Drug Abuse (2014) Hallucinogens—LSD, peyote, psilocybin, and PCP [Online]. Accessed 12 Jan 2015
Nelson CJ, Rosenfeld B, Breitbart W, Galietta M (2002) Spirituality, religion, and depression in the terminally ill. Psychosomatics 43:213–220
Nichols DE (2004) Hallucinogens. Pharmacol Ther 101:131–181
O'brien CP (2001) Drug addiction and drug abuse. In: Hardman JG, Limbird LE, Molinoff PB, Ruddon RW, Gilman AG (eds) Goodman and Gilman's the Pharmacological Basis of Therapeutics. McGraw-Hill, New York
Ogunbodede O, McCombs D, Trout K, Daley P, Terry M (2010) New mescaline concentrations from 14 taxa/cultivars of Echinopsis spp. (Cactaceae) ("San Pedro") and their relevance to shamanic practice. J Ethnopharmacol 131:356–362
Pahnke WN (1969) Psychedelic drugs and mystical experience. Int Psychiatry Clin 5:149–162
Pahnke WN, Kurland AA, Goodman LE, Richards WA (1969) LSD-assisted psychotherapy with terminal cancer patients. Curr Psychiatr Ther 9:144–152
Pahnke WN, Kurland AA, Unger S, Savage C, Grof S (1970) The experimental use of psychedelic (LSD) psychotherapy. JAMA 212:1856–1863
Partridge AH, Wang PS, Winer EP, Avorn J (2003) Nonadherence to adjuvant tamoxifen therapy in women with primary breast cancer. J Clin Oncol 21:602–606
Passie T (2005). A history of the use of psilocybin in psychotherapy. In: Metzner R (ed) Sacred Mushrooms of Visions: Teonanacatl. Park Street Press, Rochester
Passie T, Seifert J, Schneider U, Emrich HM (2002) The pharmacology of psilocybin. Addict Biol 7:357–364
Puchalski CM (2012) Spirituality in the cancer trajectory. Ann Oncol 23(Suppl 3):49–55
Rhead JC, Soskin RA, Turek I, Richards WA, Yensen R, Kurland AA, Ota YO (1977) Psychedelic drug (DPT)-assisted psychotherapy with alcoholics: a controlled study. J Psychedelic Drugs 9:287–300
Ries LAG, Melbert D, Krapcho M, Stinchcomb DG, Howlader N, Horner MJ, Mariotto A, Miller BA, Feuer EJ, Altekruse S, Lewis DR, Clegg L, Eisner MP, Reichman M, Edwards BK (eds) (2007) SEER cancer statistics review, 1975–2005. National Cancer Institute, Bethesda
Rinkel M, Deshon HJ, Hyde RW, Solomon HC (1952) Experimental schizophrenia-like symptoms. Am J Psychiatry 108:572–578
Ross S, Peselow E (2012) Co-occurring psychotic and addictive disorders: neurobiology and diagnosis. Clin Neuropharmacol 35:235–243
Ross S, Bossis A, Guss J, Agin-Liebes G, Malone T, Cohen B, Mennenga SE, Belser A, Kalliontzi K, Babb J, Su Z, Corby P, Schmidt BL (2016) Rapid and sustained symptom reduction following psilocybin treatment for anxiety and depression in patients with life-threatening cancer: a randomized controlled trial. J Psychopharmacol 30:1165–1180
Roy C (1973) Indian peyotists and alcohol. Am J Psychiatry 130:329–330
Rydzyński Z, Gruszczyński W (1978) Treatment of alcoholism with psychotomimetic drugs. A follow-up study. Act Nerv Super (Praha) 20:81–82
Rydzyński Z, Cwyna RS, Grzelak L, Jagiello W (1968) Prelminary report on the experience with psychosomimetic drugs in the treatment of alcobolism. Act Nerv Super (Praha) 10:273
Saunders C (1988) Spiritual pain. J Palliat Care 4:29–32
Savage C, McCabe OL (1973) Residential psychedelic (LSD) therapy for the narcotic addict. A controlled study. Arch Gen Psychiatry 28:808–814
Serafini G, Howland RH, Rovedi F, Girardi P, Amore M (2014) The role of ketamine in treatment-resistant depression: a systematic review. Curr Neuropharmacol 12:444–461
Sherwood JN, Stolaroff MJ, Harman WW (1962) The psychedelic experience—a new concept in psychotherapy. J Neuropsychiatr 4:69–80

Siegel R, Naishadham D, Jemal A (2012) Cancer statistics, 2012. CA Cancer J Clin 62:10–29
Smart RG, Bateman K (1967) Unfavourable reactions to LSD: a review and analysis of the available case reports. Can Med Assoc J 97:1214–1221
Smart RG, Storm T, Baker EF, Solursh L (1966) A controlled study of lysergide in the treatment of alcoholism. 1. The effects on drinking behavior. Q J Stud Alcohol 27:469–482
Smith CM (1958) A new adjunct to the treatment of alcoholism: the hallucinogenic drugs. Q J Stud Alcohol 19:406–417
Smith CM (1959) Some reflections on the possible therapeutic effects of the hallucinogens; with special reference to alcoholism. Q J Stud Alcohol 20:292–301
Stewart O (1987) The peyote religion: a history. University of Oklahoma Press, Norman
Storm HH, Christensen N, Jensen OM (1992) Suicides among Danish patients with cancer: 1971 to 1986. Cancer 69:1507–1512
Strassman RJ (1984) Adverse reactions to psychedelic drugs. A review of the literature. J Nerv Ment Dis 172:577–595
Strassman RJ, Qualls CR (1994) Dose-response study of *N,N*-dimethyltryptamine in humans. I. Neuroendocrine, autonomic, and cardiovascular effects. Arch Gen Psychiatry 51:85–97
Studerus E, Kometer M, Hasler F, Vollenweider FX (2011) Acute, subacute and long-term subjective effects of psilocybin in healthy humans: a pooled analysis of experimental studies. J Psychopharmacol 25:1434–1452
Substance Abuse and Mental Health Services Administration (2013) Results from the 2012 national survey on drug use and health: summary of national findings. Substance Abuse and Mental Health Services Administration, Rockville
Szara S (1956) Dimethyltryptamine: its metabolism in man: the relation of its psychotic effect to serotonin metabolism. Experientia 12:441–442
Tagliazucchi E, Carhart-Harris R, Leech R, Nutt D, Chialvo DR (2014) Enhanced repertoire of brain dynamical states during the psychedelic experience. Hum Brain Mapp 35:5442–5456
Taylor EJ (2003) Spiritual needs of patients with cancer and family caregivers. Cancer Nurs 26:260–266
Thomas G, Lucas P, Capler NR, Tupper KW, Martin G (2013) Ayahuasca-assisted therapy for addiction: results from a preliminary observational study in Canada. Curr Drug Abuse Rev 6:30–42
Tomsovic M, Edwards RV (1970) Lysergide treatment of schizophrenic and nonschizophrenic alcoholics: a controlled evaluation. Q J Stud Alcohol 31:932–949
Traeger L, Greer JA, Fernandez-Robles C, Temel JS, Pirl WF (2012) Evidence-based treatment of anxiety in patients with cancer. J Clin Oncol 30:1197–1205
Vollenweider FX, Kometer M (2010) The neurobiology of psychedelic drugs: implications for the treatment of mood disorders. Nat Rev Neurosci 11:642–651
Vollenweider FX, Vollenweider-Scherpenhuyzen MF, Babler A, Vogel H, Hell D (1998) Psilocybin induces schizophrenia-like psychosis in humans via a serotonin-2 agonist action. NeuroReport 9:3897–3902
Vollenweider FX, Vontobel P, Hell D, Leenders KL (1999) 5-HT modulation of dopamine release in basal ganglia in psilocybin-induced psychosis in man—a PET study with [11C] raclopride. Neuropsychopharmacology 20:424–433
Wasson RG (1957) Seeking the magic mushroom. Life Mag 42:100
Watts VJ, Lawler CP, Fox DR, Neve KA, Nichols DE, Mailman RB (1995) LSD and structural analogs: pharmacological evaluation at D1 dopamine receptors. Psychopharmacology 118:401–409
Wilkins JN, Danovitch I, Gorelick DA (2014) Management of stimulant, hallucinogen, marijuana, phencyclidine, and club drug intoxication and withdrawal. In: Ries RA, Fiellin DA, Miller SA, Saitz R (eds) ASAM principles of addiction medicine, 5th edn. Wolters Kluwer
Winkelman WD, Lauderdale K, Balboni MJ, Phelps AC, Peteet JR, Block SD, Kachnic LA, Vanderweele TJ, Balboni TA (2011) The relationship of spiritual concerns to the quality of life of advanced cancer patients: preliminary findings. J Palliat Med 14:1022–1028

Wolbach AB Jr, Miner EJ, Isbell H (1962) Comparison of psilocin with psilocybin, mescaline and LSD-25. Psychopharmacologia 3:219–223

Yousaf U, Christensen ML, Engholm G, Storm HH (2005) Suicides among Danish cancer patients 1971–1999. Br J Cancer 92:995–1000

Zabora J, Brintzenhofeszoc K, Curbow B, Hooker C, Piantadosi S (2001) The prevalence of psychological distress by cancer site. Psychooncology 10:19–28

Classic Hallucinogens and Mystical Experiences: Phenomenology and Neural Correlates

Frederick S. Barrett and Roland R. Griffiths

Abstract This chapter begins with a brief review of descriptions and definitions of mystical-type experiences and the historical connection between classic hallucinogens and mystical experiences. The chapter then explores the empirical literature on experiences with classic hallucinogens in which claims about mystical or religious experiences have been made. A psychometrically validated questionnaire is described for the reliable measurement of mystical-type experiences occasioned by classic hallucinogens. Controlled laboratory studies show that under double-blind conditions that provide significant controls for expectancy bias, psilocybin can occasion complete mystical experiences in the majority of people studied. These effects are dose-dependent, specific to psilocybin compared to placebo or a psychoactive control substance, and have enduring impact on the moods, attitudes, and behaviors of participants as assessed by self-report of participants and ratings by community observers. Other studies suggest that enduring personal meaning in healthy volunteers and therapeutic outcomes in patients, including reduction and cessation of substance abuse behaviors and reduction of anxiety and depression in patients with a life-threatening cancer diagnosis, are related to the occurrence of mystical experiences during drug sessions. The final sections of the chapter draw parallels in human neuroscience research between the neural bases of experiences with classic hallucinogens and the neural bases of meditative practices for which claims of mystical-type experience are sometimes made. From these parallels, a functional neural model of mystical experience is proposed, based on changes in the default mode network of the brain that have been observed after the administration of classic hallucinogens and during meditation practices for which mystical-type claims have been made.

F.S. Barrett (✉) · R.R. Griffiths
Department of Psychiatry and Behavioral Sciences, Johns Hopkins University School of Medicine, 5510 Nathan Shock Drive, Baltimore, MD 21224, USA
e-mail: fbarrett@jhu.edu

R.R. Griffiths
Department of Neuroscience, Johns Hopkins University School of Medicine, 5510 Nathan Shock Drive, Baltimore, MD 21224, USA

Curr Topics Behav Neurosci (2017) 36: 393–430
DOI 10.1007/7854_2017_474
Published Online: 26 March 2017

Keywords Psilocybin · Hallucinogens · Meditation · Mystical experiences · Neural model · Default mode network · Medial prefrontal cortex · Posterior cingulate · Angular gyrus · Inferior parietal lobule

Contents

1 Introduction

Reports of mystical-type experiences date back many centuries (e.g., in the case of Rumi or St. Teresa of Avila), if not millennia (i.e., in the case of Plotinus) (Stace 1960a). Mystical experiences have occurred in the course of structured spiritual or religious practices as well as in cases in which there was no direct intention to have such an experience. Mystical experiences are uniquely interesting and important to study because they are sometimes associated with abrupt, substantial, and sustained changes in behavior and perception (Miller and C'de Baca 2001). Furthermore, the authoritative sense of interconnectedness and sacredness that defines such experiences suggest that mystical experiences may be foundational to the world's ethical and moral systems (Huxley 1947). Despite their apparent importance, the unpredictability and low probability of "naturally occurring" mystical-type experiences, whether they occur in religious or nonreligious contexts, has made them inherently difficult to study in controlled empirical research.

While there are countless reports of profound spiritual and mystical experiences that have occurred in the absence of psychoactive substances, historical evidence abounds for the role of psychoactive substances in ceremonial approaches to facilitating such experiences (Schultes et al. 2001). Further, descriptions of naturally occurring mystical experiences (Stace 1960b) are strikingly similar to profound spiritual experiences occasioned by hallucinogenic substances (Roberts 2001).

Experimental investigations have begun to utilize classic hallucinogens to study the reliability, characteristics, subjective nature, and behavioral consequences of mystical-type experiences (Pahnke 1963; Griffiths et al. 2006, 2011; Garcia-Romeu et al. 2015). Classic hallucinogens are a structurally diverse group of compounds that bind at 5-HT_{2A} serotonin receptors and produce a unique profile of changes in thoughts, emotions, and perceptions, often including profound alterations in the perception of reality, that are rarely experienced except in dreams, naturally occurring mystical experiences, and acute psychosis. The use of classic hallucinogens makes the study of mystical experiences more tractable because classic hallucinogens can be administered under double-blind conditions and can occasion mystical experiences with high probability (Griffiths et al. 2006, 2011). Classic hallucinogens allow for prospective and controlled exploration of such experiences, and provide a degree of neurobiological specificity and mechanistic understanding that is not possible in correlational or descriptive studies, or in reviews of present-day or historical case reports.

The following section of this chapter reviews descriptions and definitions of mystical-type experiences. Evidence of the historical connection between classic hallucinogens and mystical experiences is then presented. The chapter then reviews empirical evidence of mystical experiences occasioned by classic hallucinogens and the potential therapeutic benefits of such experiences. The chapter ends with an exposition of a functional neural model of mystical experience that is based on changes in the default mode network of the brain that have been observed after the administration of classic hallucinogens and during meditation practices which are sometimes associated with mystical-type experiences.

2 What Are Mystical Experiences?

> [Mystical experiences are] those peculiar states of consciousness in which the individual discovers himself to be one continuous process with God, with the Universe, with the Ground of Being, or whatever name he may use by cultural conditioning or personal preference for the ultimate and eternal reality (Watts 1970).

The theological and philosophical literatures have primarily taken descriptions of experiences as their source data for the exploration of mystical experiences. The theological literature on mystical experiences is based on accounts of prophets, saints, and the practices of mystic sects within religious traditions. Mystic sects exist within all major world religions. These sects are lineages of individuals who have organized around a spiritual practice within their culture and religion, with characteristic lifestyles, teachings, and traditions that instruct the individual and provide a supportive setting for the individual to achieve a deeper connection with the God of their understanding. In this fashion, mystics may be considered "empirical" theologians (Huxley 1947). Sufism is a mystical sect within the religion of Islam, while Kabbalists explore mysticism in the context of Judaism. Individuals from various religious traditions, including Christian (e.g., Meister Eckhart from the thirteenth century, St. Teresa of Avila and St. John of the Cross from the sixteenth century), Islamic (e.g., Rumi and Saadi from the thirteenth century), and Jewish (e.g., Maimonides of the twelfth century) traditions, are considered to be mystics. Some Hindu and Buddhist meditation practices that focus on non-dual experiences likely increase the probability of mystical-type experiences (Stace 1960a).

The most definitive philosophical treatise on mystical experiences was compiled by Stace (1960a). Stace identified, collated, and distilled descriptions of mystical experiences from a variety of sources. From this literature, he argued that mystical experiences have a "common core" of phenomenological features that are independent from the interpretation of those experiences. His thesis in this regard was that differences in the descriptions of mystical experiences given by mystics of different faiths amount not to a difference in the core experience, but rather a difference in the interpretational frame through which those experiences are viewed. The idea of a "common core" of mystical experience is consistent with the notion of a "perennial philosophy" in which an "immemorial and universal" substrate underlies all religions and spiritual paths and is reflected in "every religious tradition and in all the principal languages of Asia and Europe" (Huxley 1947).

A critical definitional feature of the mystical experience is a sense of unity, or the experience of becoming one with all that exists. Stace (1960a) described this mystical unity as "the apprehension of an ultimate nonsensuous unity in all things". Stace made a distinction between the qualitative nature of unity in extrovertive mystical experiences and the qualitative nature of unity in introvertive mystical experiences. Unity in the extrovertive mystical experience involves recognition of the oneness of all, in which one finds unity at the core of the inner subjectivity or inner reality of all things despite the diversity or apparent individual identity and separation of all things. Unity in the introvertive mystical experience, on the other hand, involves an experience of the complete dissolution of the self, loss of the notion of "I" and loss of all boundaries, such that there is no separation or individual identity. Introvertive mystical experience involves an experience of unity that is otherwise devoid of content, sometimes referred to as "the void". While both extrovertive and introvertive types are considered experiences of mystical unity, Stace considered extrovertive unity as an "incomplete kind of experience which finds its completion and fulfillment in the introvertive kind of experience" (Stace 1960a).

In addition to the experience of introvertive or extrovertive types of unity, Stace described six other dimensions of mystical experience: (1) *sacredness*: a sense that what is encountered is holy or sacred; (2) *noetic quality*: the experience is imbued with an aspect of meaning and a sense of encountering ultimate reality that is more real than usual everyday reality; (3) *deeply felt positive mood*: joy, ecstasy, blessedness, peace, tenderness, gentleness, tranquility, awe; (4) *ineffability*: the experience is difficult to put into words; (5) *paradoxicality*: to explain the experience, one seems to have to describe the coexistence of mutually exclusive states or concepts; (6) *transcendence of time and space*: Introvertive mystical experiences may have a nonspatial and nontemporal aspect, such that the traditional notions of time and space have no meaning.

As described in more detail later in this chapter, mystical experiences may be encountered following the ingestion of classic hallucinogens. The following is a written description of an experience reported by a volunteer who received 20 mg/70 kg of psilocybin in a study conducted at Johns Hopkins. Bold typeface indicates portions of the description that directly align with dimensions of mystical experience that were identified by Stace:

> In my mind's eye, I felt myself instinctively taking on the posture of prayer in my head. I was on my knees, hands clasped in front of me and I bowed to this force. I wasn't scared or threatened in any way. It was more about **reverence**. I was showing my respect. I was humbled and honored to be in this presence. This presence was a feeling, not something I saw or heard. I only felt it, but it felt **more real than any reality I have experienced**. And it was a familiar place too. One I had felt before. It was when I surrendered to this, that I felt like I let go. I was gone… or I should say this earthly part of me was. It was still on the couch in some sort of suspended animation awaiting my return. I was **in the void**. This void had a **strange and indescribable quality to it** in that there was nothing to it but this feeling of **unconditional and undying Love**. It felt like my soul was basking in the feeling of this space. I have no idea how long this lasted. **Time and space did not exist there** … it was all different manifestations of this Love feeling I found myself wrapped in.

Mystical experiences have been an active area of investigation in the experimental psychology literature, particularly within the psychology of religion (Hood 2009). Research in this domain has frequently used the Mysticism Scale, a psychometric instrument that codifies the descriptive definition of mystical experience provided by Stace (Hood 1975; Hood et al. 2001). The Hood Mysticism Scale is scored into three factors: extrovertive mysticism, introvertive mysticism, and religious interpretation. While the precise factor structure of items of the Hood Mysticism Scale may nominally vary between different cultures and different religious backgrounds (Chen et al. 2011a, b, 2012), the basic concepts that underlie the factor structure of this instrument are reliable and align with Stace's model. The extrovertive mysticism factor typically includes items related to external unity (unity in diversity) and inner subjectivity. The introvertive mysticism factor typically includes items related to internal unity (contentless unity), ego loss, ineffability, timelessness and spacelessness. The religious interpretation factor typically includes items that measure reverence or sacredness, noetic quality, and positive mood. Paradoxicality, though identified by Stace as a characteristic of mystical experiences, was not included as a criterion dimension in the Hood Mysticism Scale (Hood 1975).

The Hood Mysticism Scale has been used in survey research and to assess retrospective accounts of mystical experiences in other psychological studies. With tools such as the Mysticism Scale, Ralph Hood and colleagues have vigorously defended the thesis of the "common core" (Hood and Williamson 2000; Hood 2006; Anthony et al. 2010) by demonstrating the overall reliability and reproducibility of the factor structure of mystical experiences across various participant samples, across cultures, and across a number of different religious traditions. Iranian Muslims (Hood et al. 2001), American Christians (Hood et al. 2001), Chinese Christians (Chen et al. 2012), Chinese Buddhists (Chen et al. 2011a), and

Tibetan Buddhists (Chen et al. 2011b) were assessed using the Hood Mysticism Scale and the instrument was shown to validly measure mystical experiences in each sample. The common features of mystical experiences shared across cultures and religious traditions are consistent with idea that there is a common core to such religious experiences (i.e., the perennial philosophy; Huxley 1947).

It is important to acknowledge that the most extreme interpretation of the common core hypothesis, which holds that mystical experience is a direct encounter with the divine, and from which claims of a perennial philosophy are made, has been criticized by some scholars of religion. These scholars argue that the cross-religion and cross-cultural generality of such experiences is impossible from a constructionist position that asserts that all such experiences are necessarily and significantly shaped by language and culture, the differences in which obviate the potential for a common core to these experiences (Katz 1978, 1983; Proudfoot 1985; Sharf 1988). Although debating the conceptual extreme interpretations of mystical experience has provided a platform for academic scholarship, it may not be a productive strategy for advancing a scientific basis for exploring the immediate causes and consequences of such experiences (Hood 2003). The research discussed throughout this chapter is focused on an empirical description and analysis of mystical-type experiences, which we believe has documented impressive generality and replicability of such experiences.

Mystical experiences have not been restricted to any particular path, canon, or dogmatic viewpoint, and experiences identified as mystical have been encountered in secular contexts outside of the framework of any traditional religious or spiritual practice or interpretation (Hood et al. 1990). Experiences that fit the mystical-type framework have a wide range of apparent etiologies, including meditation and prayer (d'Aquili and Newberg 2000; Newberg and Iversen 2003; Newberg et al. 2010; Josipovic 2014), sensory deprivation/isolation (Hood et al. 1990), music listening by "deep listeners" (Gabrielsson 2010; Penman and Becker 2009), breathwork (Grof and Grof 2010), and ingestion of classic hallucinogens such as psilocybin (Griffiths et al. 2006, 2008, 2011).

While certain examples of mystical experience may be accompanied by quite dramatic and seemingly paranormal events (e.g. some of the experiences described by prophets in the Old Testament), these are exceptions rather than exemplars of mystical-type experiences. Likewise, abrupt experiences of "rapture" as described by mystics such as St. Teresa of Avila may be atypical (Stace 1960a). Rapture in the case of St. Teresa involved not only feelings of positive mood (ecstasy or pleasure) but also feelings of fear, and abnormal body changes (Underhill 1911 [2009]). Seeing visions and hearing voices are likewise not core elements of mystical experiences (Stace 1960a). Prophetic experiences per se, although heavily associated with classic religious and scriptural encounters with God, may not contain all the key features of mystical experience (Hood 2009).

Mystical experiences are not similar to the altered states of consciousness associated with intoxication of many common psychoactive drugs such alcohol or opiates (Aaronson and Osmond 1970). Nor are mystical experiences necessarily associated with religious experiences such as glossolalia, or speaking in tongues

(Newberg et al. 2006). Mystical experiences may be accompanied by spiritual insights, but the mystical experience is not, in and of itself, simply the experience of religious or spiritual insight. Mystical experiences are also not simply an interesting aesthetic and/or euphoric experience, an experience of an archetypal construct, or a psychodynamic or intellectual insight (Richards 2014). Mystical experiences are defined as a self-reported experience of unity accompanied by the additional dimensions of experience as outlined by Stace.

3 Historical Use of Indigenous Hallucinogens

The ingestion of naturally occurring classic hallucinogens in ceremonial contexts has a long history. While there is clear anthropological evidence of the ceremonial consumption of hallucinogenic plant and fungal matter in the past few centuries (Dobkin de Rios 1984; Guzmán 2008), there is speculation that ceremonial use of similar psychoactive compounds dates back many thousands of years (Schultes 1969; Westermeyer 1988; Wasson et al. 1998; Schultes et al. 2001). The reasons for ceremonial use of these substances by indigenous people included use for medicinal and divination purposes, but a prominent goal of ceremonial consumption of classic hallucinogens has also likely been to occasion primary spiritual experiences that fit a mystical-type description (Roberts 2001). Psychoactive plants and fungi for which there is substantive knowledge of ceremonial use include peyote, *ayahuasca*, and psilocybin mushrooms. While double-blind studies have not rigorously compared the psychoactive effects of these and other classic hallucinogens (including synthetic hallucinogens such as LSD), the alterations in consciousness produced by these compounds are generally considered to be quite similar, putatively because they share a common mechanism of action initiated through activation of the 5-HT_{2A} receptor and likely involving downstream glutamate effects (Nichols 2004, 2016; Vollenweider and Kometer 2010). Under appropriate set and setting conditions, these substances may produce mystical-type experiences.

Peyote is a cactus that contains the hallucinogenic alkaloid mescaline (3,4,5-trimethoxy-phenethylamine), which was used historically by Mexican indigenous people, including the Chichimeca, Huichol, and Tarahumara tribes, for thousands of years (El-Seedi et al. 2005; Schultes et al. 2001). Historically, peyote has been used both for medicinal and ceremonial purposes (Schultes et al. 2001). Peyote is currently used for religious purposes by the Native American Church (Halpern et al. 2005; Osmond 1970). Humphrey Osmond described an experience that he had with peyote as a participant in a ceremony of the Native American Church, and concluded: "We had wrestled with the angel. We had grappled with the Heavenly Father" and "Peyote acts not by emphasizing one's own self but by expanding it into the selves of others, with a deepening empathy or in-feeling. The self is dissolved and, in being dissolved, enriched" (Osmond 1970)

Psilocybin (4-phosphoryloxy-*N*,*N*-dimethyltryptamine), another hallucinogenic alkaloid, occurs naturally in mushrooms of the *Psilocybe* genus that are found on all continents (Guzman 1983). The ceremonial use of psilocybin-containing mushrooms by indigenous tribes in Mexico can be traced back at least to the fifteenth century. At least nine indigenous groups in Mexico have been identified that still consume psilocybin mushrooms in a ceremonial context (Guzmán 2008). R. Gordon Wasson used language suggestive of mystical-type experiences when describing his first exposure to psilocybin-containing mushrooms during a ceremony led by the now famous mushroom shaman, Maria Sabina: "… your soul is free, loses all sense of time … you know what the ineffable is, and what ecstasy means … the flight of the soul from the body" (Wasson et al. 1998).

DMT (*N,N*-dimethyltryptamine) is another hallucinogenic tryptamine alkaloid that has been used in various forms for medicinal, prophetic, and divination purposes by groups in South America since pre-Columbian times (Dobkin de Rios 1984; Schultes et al. 2001). *Ayahuasca* is a plant admixture made from two or more plants, including a plant (e.g. *Psychotria viridis*) that contains DMT and other plants that contain monoamine oxidase inhibitors (e.g., *Banisteriopsis caapi*). *Ayahuasca* is currently used for religious purposes in South America, the United States, and Europe by Uniao do Vegetal and the Santo Daime Church (Dobkin de Rios 1996; Grob et al. 1996; Tupper 2008). While not as widely used as *ayahuasca*, snuffs made from DMT-containing plants also have a history of ceremonial use in South America, with more frequent use in pre-Columbian than in modern times (Dobkin de Rios 1984).

4 Experimental Evidence of Mystical Experiences with Psilocybin

The classic hallucinogens produce a unique profile of cognitive, perceptual, and emotional changes that have similarities with various states of consciousness, including mystical experiences, psychoses, and liminal sleep states. During the first phase of modern human scientific research with classic hallucinogens, from roughly the 1950s to the 1970s, classic hallucinogens were primarily studied as models of psychosis and as therapeutic agents in a variety of disorders. In 1962, Walter Pahnke conducted the remarkable Good Friday experiment, administering either 30 mg psilocybin (10 subjects) or 200 mg nicotinic acid (10 subjects) with the intent of studying the incidence and character of psilocybin-induced mystical experiences (Pahnke 1963).

In an effort to maximize the effects of set and setting, Pahnke conducted the study with seminary students in a private chapel on Good Friday during the broadcast of the traditional Good Friday religious service. After the experience, and at a 6-month follow-up, participants completed a questionnaire that assessed eight dimensions of mystical experience that were based on the model of mystical experience developed by Stace (1960a): introvertive unity, extrovertive unity, transcendence of time and space, deeply felt positive mood, sacredness, objectivity (noetic quality), ineffability, and paradoxicality. In addition, the transiency of the experience was assessed for a total of nine dimensions of mystical experience.

Pahnke considered a "complete" mystical experience as one in which ratings of at least 60% of the total possible score for each of the nine categories of mystical experience were provided (Pahnke 1969). By this criterion, 30–40% of participants in the psilocybin condition achieved a complete mystical experience during the study, whereas none of the control participants who received nicotinic acid achieved a complete mystical experience (Pahnke 1967). In a 25-year follow-up to the Good Friday experiment, Doblin (1991) was able to contact 16 of the 20 original participants and collect additional retrospective ratings on the nine dimensions of mystical experience measured by Pahnke. Doblin found little change between the 6-month retrospective ratings and the 25-year retrospective ratings of mystical experience. Moreover, both Pahnke (at the 6-month follow-up) and Doblin (at the 25 years) found that participants in the psilocybin condition rated persisting positive changes in attitudes and behavior that they attributed to their psilocybin experience, while participants in the control condition indicated no such changes.

While groundbreaking, the Good Friday experiment had significant limitations, including limited generality due to the highly selective demographics of the participants (seminary students), conduct of the study in a group setting that allowed interactions among participants (thus resulting in nonindependence of individual subject data), explicit instructions to participants that some would and some would not receive psilocybin (thus creating powerful expectancy effects), and the fact that half of the researchers present during the study also received psilocybin. Not surprisingly, under these conditions, the blind was broken shortly after drug administration, which likely contributed to the assessed differences between groups (Doblin 1991; Wulff 1991; Smith 2000).

In a replication and extension of the Good Friday experiment, Griffiths and colleagues conducted a double-blind crossover comparative pharmacology study of psilocybin (30 mg/70 kg) and methylphenidate (40 mg/70 kg), which were administered in separate sessions to each of 36 participants individually, with at least two months between sessions (Griffiths et al. 2006, 2008). Participants in this study were well educated, psychiatrically and medically healthy, had no prior hallucinogen use, and represented a more general sample of the population than those used in the Good Friday experiment. The study reduced expectancy and group confounding effects by studying participants without personal histories of hallucinogen use, by studying only a single participant at a time, and by using an experimental design and instructions that obscured the range of drug conditions that would be administered as well as the total possible number of sessions. The study also utilized a stronger control condition (methylphenidate) than the Good Friday experiment (nicotinic acid). Methylphenidate and psilocybin can both induce strong subjective effects with some similarities, and with a reasonably similar time course. Nicotinic acid, in contrast, has a relatively short time course and a profile of subjective effects that is very different from psilocybin. Finally, in addition to using a revised and updated version of the mystical experience questionnaire used in the Good Friday experiment, Griffiths and colleagues used two psychometrically validated questionnaires that assessed mystical and spiritual effects (the Hood Mysticism Scale and the Spiritual Transcendence Scale) as well as ratings of changes in participant's attitudes and behavior by community observers (family members and friends of participants).

In this study, Griffiths and colleagues demonstrated a fairly high frequency of "complete mystical experiences" during psilocybin sessions (61% of participants), but not during methylphenidate sessions (11% of participants). The criterion for a complete mystical experience was a score of 60% of the total possible score on each dimension of the Mystical Experience Questionnaire, including scores on either internal or external unity factors (whichever was greater) and sacredness, noetic quality, transcendence of time and space, ineffability, and positive mood factors (Griffiths et al. 2006). This is a scoring system that is similar to the one used by Pahnke in the Good Friday Experiment. Two months after the session, most participants rated their psilocybin session as among the top five (71%) or single most (33%) spiritually significant experience of their lives, compared to 8% of participants who rated the methylphenidate experience to be among the top five spiritually significant experiences of their lives, with no one rating it as the single most. Ratings of positive attitudes about life and self, positive mood, positive behaviors, and positive social effects 2 months after psilocybin sessions were significantly greater than those provided 2 months after methylphenidate sessions. Negative ratings of these same dimensions were low and did not differ between the psilocybin and methylphenidate conditions. Further, community observers rated small but significant changes in participants' positive attitudes and behaviors 2 months after the psilocybin sessions, but no changes were found 2 months after methylphenidate sessions. In a 14-month follow-up report, 67% of participants rated their psilocybin session as among the top five most spiritually significant experiences of their lives, and 58% of participants rated their psilocybin session as among the top five most personally meaningful experiences of their lives (Griffiths et al. 2008). Ratings of positive behavior, mood, attitude, and social changes associated with the psilocybin session at the 14-month follow-up were not significantly different from those provided 2 months post session. Correlation and regression analyses indicated a central role of mystical experience assessed on the session day (but not intensity of psilocybin experience) in predicting the high ratings of spiritual significance and personal meaning assessed at 14 months (Griffiths et al. 2008). Figure 1 illustrates the significant relationship between mystical experience score and follow-up ratings of spiritual significance.

A further extension of this line of research was published by Griffiths and colleagues in 2011. This study utilized a double-blind placebo-controlled design that assessed the effects of placebo and a range of psilocybin doses on ratings of spiritual significance and meaningfulness of psilocybin sessions, and on the likelihood of complete mystical experience. Participants in this study were well educated, medically and psychiatrically healthy, and most were hallucinogen naïve. Participants received 5, 10, 20, and 30 mg/70 kg of psilocybin in separate sessions in either ascending or descending order, with at least one month between each session and a placebo session randomly placed within the sequence. Participants

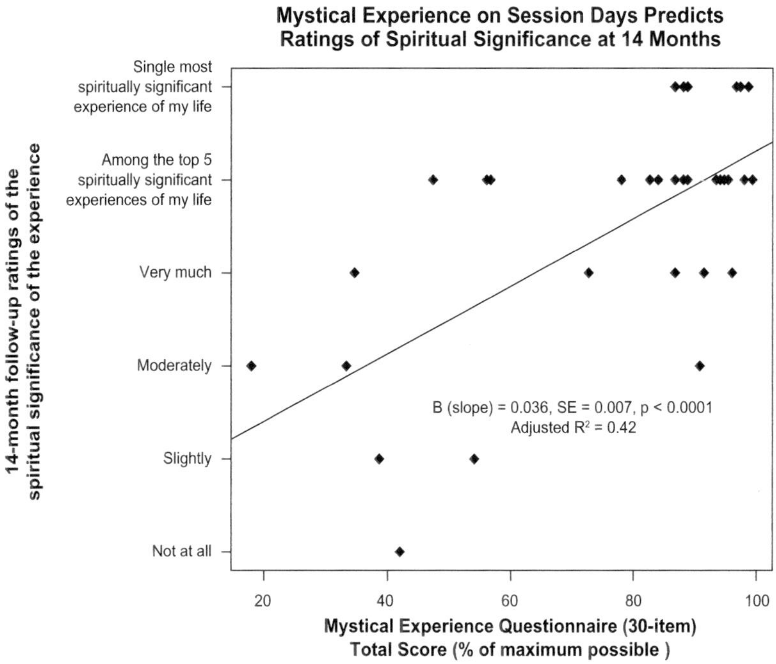

Fig. 1 Mystical experience on session days predicts ratings of spiritual significance at 14 months. Total score on the 30-item Mystical Experience Questionnaire (MEQ30) is expressed as the percentage of the maximum possible score. Data points represent individual participants (n = 36). B = slope of the regression of spiritual significance on mystical experience questionnaire total score. SE = standard error of the slope (B). Adjusted R^2 indicates the amount of variance in spiritual significance ratings that is explained by the mystical experience questionnaire total score. MEQ30 data have been rescored from MEQ43 responses reported in Griffiths et al. (2008)

were not informed of the ascending or descending nature of the dose sequence. 61% of participants met criteria for a "complete" mystical experience during the 20 mg/70 kg session, whereas 67% of participants met criteria for a "complete mystical experience during the 30 mg/70 kg session (Fig. 2, "Dose Effects Study", dark gray bars). Although "complete" mystical experience was coded in the original publication using items from the 7-factor, 43-item MEQ (MEQ43), the above percentages were derived using the psychometrically validated 4-factor, 30-item MEQ (MEQ30). The MEQ30 has been validated in both retrospective accounts of mystical experiences with psilocybin (MacLean et al. 2012) and in prospective, experimental laboratory studies with psilocybin (Barrett et al. 2015). The criteria for complete mystical experiences in the MEQ30 are a score of at least 60% of the total possible score on each of four factors of the MEQ30. The MEQ30 will be discussed in more detail later in this chapter.

Figure 2 ("Dose Effects Study") shows that the mean total score on the MEQ30 (light gray bars) as well as the percentage of volunteers who met criteria for a

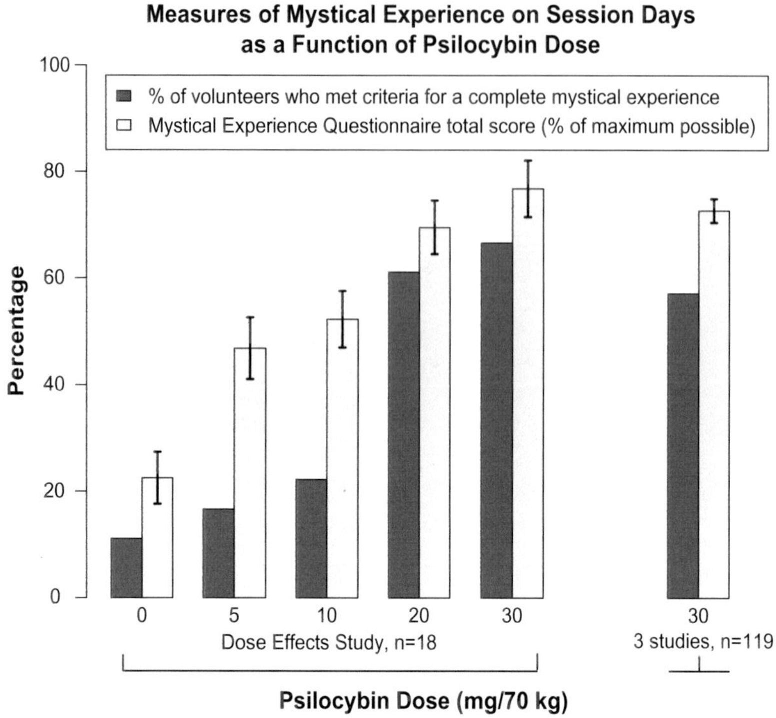

Fig. 2 Measures of mystical experience on session days as a function of psilocybin dose. "Dose Effects Study" data (5 pairs of *bars* at the *left*) show the dose-related effect of psilocybin on two measures of mystical experience (Griffiths et al. 2011). "3 studies" observations (*bars* at the *right*) show post-session ratings from the first high-dose (30 mg/70 kg) session administered in each of 3 studies (Griffiths et al. 2006, 2011, unpublished study in beginning meditators). *Dark bars* show the percentage of volunteers who met criteria for having a complete mystical experience. *Lighter bars* show the mean Mystical Experience Questionnaire total score expressed as a percentage of maximum possible score; brackets show ± 1 S.E.M. Categorizations of complete mystical experiences and calculations of total scores were based on the 30-item Mystical Experience Questionnaire (MEQ30)

complete mystical experience (dark gray bars) increased as a function of psilocybin dose. Overall, 72% of volunteers had a "complete" mystical experience during the 20 mg/70 kg session and/or the 30 mg/70 kg session (using MEQ30 scoring). Follow-up ratings 1 month after each session of the spiritual significance of the experience (Fig. 3, "Dose Effects Study", dark gray bars), as well as ratings of positive behavior change attributed to the experience (Fig. 3, "Dose Effects Study", light gray bars), increased in a dose-dependent fashion. Eighty-three percent of participants rated the session experiences after 20 and/or 30 mg/70 kg as among the five most spiritually significant experiences of their life; 61% also rated at least one of these as the single most spiritually significant experience of their life. Likewise,

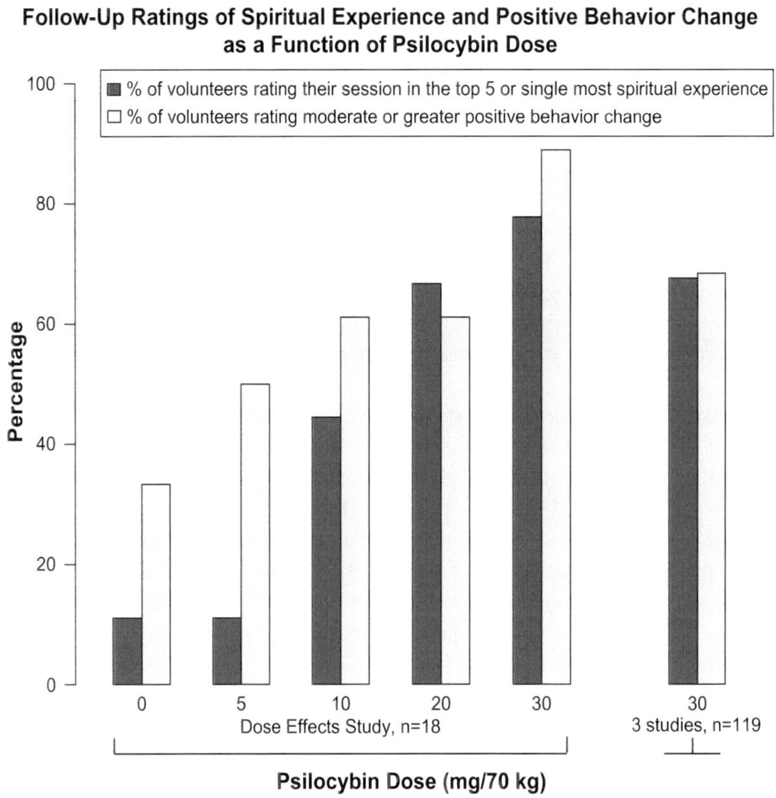

Fig. 3 Follow-up ratings of spiritual experience and positive behavior change as a function of psilocybin dose. "Dose Effects Study" data (5 pairs of *bars* at the *left*) show the dose-related effect of psilocybin on retrospective ratings assessed 1 month after sessions (Griffiths et al. 2011). "3 studies" observations (*bars* at the *right*) show retrospective ratings assessed 3–8 weeks after the first high-dose (30 mg/70 kg) session administered in each of three studies (Griffiths et al. 2006, 2011, unpublished study in beginning meditators). *Dark bars* show the percentage of volunteers who rated their session as within the top five or single most spiritual experience of their life. *Lighter bars* show the percentage of volunteers who rated a moderate or greater degree of positive behavior change at follow-up

1 month follow-up ratings of positive attitudes about life and self, positive behavior, positive social effects, and increased spirituality generally increased as a function of psilocybin dose. One month follow-up ratings after the 20 or 30 mg/70 kg sessions did not differ from follow-up ratings 14 months after study completion. Finally, compared to pre-study ratings, community observers rated significant positive change in the attitudes and behaviors of participants 3 to 4 weeks after the final session and 14 months after the final session.

In addition to showing psilocybin dose effects on mystical experience and follow-up ratings, the rightmost columns of Figs. 2 and 3 ("3 studies") present data from 119 volunteers after a high dose of psilocybin (30 mg/70 kg) (three studies: Griffiths et al. 2006, 2011 and an unpublished study in beginning meditators). Fifty-seven percent met criteria for a complete mystical experience (Fig. 2, "3 Studies"). More than 65% of high-dose sessions were retrospectively rated in the top five if not the single most personally meaningful (66%) or spiritually significant (68%) experience that volunteers have had in their lives, with most participants (70%) rating at least moderate positive behavior change that they attributed to their session experience (Fig. 3, "3 Studies"). The overall rate of complete mystical experiences achieved in these studies exceeds the rate achieved by Pahnke in the original Good Friday Experiment (30–40%; Pahnke 1967). Various factors likely contributed to this difference, with the most important being the greater volunteer preparation before and support during the sessions provided in the more recent studies.

Overall, the psilocybin studies reviewed in this section (Pahnke 1963; Griffiths et al. 2006, 2008, 2011, unpublished) make important contributions to the understanding of mystical experiences. These studies show that under double-blind conditions that provided significant controls for expectancy bias, psilocybin can occasion complete mystical experiences in the majority of people studied. These effects are dose-dependent, specific to psilocybin compared to a psychoactive control substance (methylphenidate), and have enduring impact on the moods, attitudes, and behaviors of participants as assessed by self-report of participants and ratings by community observers.

5 The Mystical Experience Questionnaire

A primary tool for the study of mystical experiences with classic hallucinogens is the Mystical Experience Questionnaire. The 43-item Mystical Experience Questionnaire (MEQ43), also known as the Pahnke–Richards Mystical Experience Questionnaire (Griffiths et al. 2006), was comprised of 43 items that were similar, but not identical, to the original item set used by Pahnke in the Good Friday experiment. The MEQ43 was scored into seven scales based on seven descriptive dimensions of mystical experience described by Stace (1960a): internal unity, external unity, noetic quality, sacredness, positive mood, transcendence of time and space, and a final factor that combined ineffability and paradoxicality scores. Items from a transiency scale used by Pahnke were not included. Thus, the MEQ43 dimensions are theoretically derived, not empirically derived. The MEQ43 was used in the psilocybin versus methylphenidate comparison study (Griffiths et al. 2006, 2008), the psilocybin dose effects study (Griffiths et al. 2011), and the psilocybin smoking cessation study (Johnson et al. 2014; Garcia-Romeu et al. 2015). The factor structure of this instrument was recently evaluated through analysis of retrospective accounts of mystical experiences with psilocybin-containing mushrooms, collected in two Internet surveys (MacLean et al. 2012) (n_1 = 1602, n_2 = 440). Psychometric analysis of the MEQ43 responses provided in these samples yielded a four-factor structure using a subset of 30 of the original 43 items. The four factors of the 30-item MEQ (MEQ30) are: (1) *mystical* (including items from the previously proposed internal unity, external unity, noetic quality, and sacredness factors); (2) *positive mood*; (3) *transcendence of time and space*; and (4) *ineffability*. The MEQ30 items and factor structure are presented in Fig. 4. This factor structure differs from that of the Hood Mysticism Scale, which has three factors: introvertive mysticism (typically including items measuring contentless unity, timelessness, and spacelessness), extrovertive mysticism (typically including items measuring unity in diversity and inner subjectivity), and religious interpretation (typically including items that measure sacredness, noetic quality, positive affect, paradoxicality, and ineffability). Although some items on the Hood Mysticism Scale were shown to load onto different factors in different subject populations (Chen et al. 2011a, b), the underlying conceptual framework of the three-factor structure of the items of the Hood Mysticism Scale has been repeatedly replicated. The four-factor structure for the MEQ30 was found to fit responses to the MEQ30 items better than two variations of a three-factor model that included the model proposed by Hood (MacLean et al. 2012).

The MEQ30 has more recently been validated using data collected after psilocybin administration in the context of prospective experimental laboratory psilocybin studies conducted at Johns Hopkins (Barrett et al. 2015). MEQ data from the first moderate to high-dose session (20 mg/70 kg or greater) for 184 volunteers from a total of 5 psilocybin studies (Griffiths et al. 2006, 2011; Johnson et al. 2014; Griffiths et al. 2016 and an unpublished study in beginning meditators) were pooled and submitted to confirmatory factor analysis. The analysis confirmed the internal validity of the MEQ30 by demonstrating acceptable fit of the four-factor structure of the MEQ30. This analysis also demonstrated the external validity of the instrument by showing that the four factors of the MEQ30 and a higher level MEQ total score significantly predicted subsequent ratings (from 3 to 8 weeks after psilocybin administration) of the meaningfulness, spiritual significance, impact on well-being, and positive change in behavior attributed to each psilocybin session, while also controlling for rated strength of drug effects (Barrett et al. 2015). Another recent study suggested a 2-factor solution to the MEQ30, based on ratings (N = 158) given by individuals who had just completed an ayahuasca ceremony, however this solution was not validated with either a separate or a large sample (Bouso et al. 2016). The limitations of this study include details of the set and setting as well as the actual dose of dimethyltryptamine consumed in any given instance. Without proper control of these elements of experience or experimental design, and without validation of the 2-factor structure with rigorous methods in separate samples, we must caution readers to treat the 2-factor structure tentatively (Barrett and Griffiths 2017).

The MEQ30 has also been utilized in studies of LSD and MDMA. A controlled laboratory experiment involving administration of psychedelics compared MEQ30 scores provided in reference to experiences with LSD (200 µg), MDMA (75 mg), methylphenidate (40 mg), and placebo (Liechti et al. 2016). While MEQ30 scores for those who received LSD were significantly higher than for those in any other

Revised 30-item Mystical Experience Questionnaire (MEQ30)

Factor 1: Mystical

Internal Unity

Feeling that you experienced eternity or infinity.

Freedom from the limitations of your personal self and feeling a unity or bond with what was felt to be greater than your personal self.

Experience of pure being and pure awareness (beyond the world of sense impressions).

Experience of oneness in relation to an "inner world" within.

Experience of the fusion of your personal self into a larger whole.

Experience of unity with ultimate reality.

External Unity

Experience of oneness or unity with objects and/or persons perceived in your surroundings.

Experience of the insight that "all is One."

Awareness of the life or living presence in all things.

Noetic Quality

Gain of insightful knowledge experienced at an intuitive level.

Certainty of encounter with ultimate reality (in the sense of being able to "know" and "see" what is really real at some point during your experience.

You are convinced now, as you look back on your experience, that in it you encountered ultimate reality (i.e., that you "knew" and "saw" what was really real).

Sacredness

Sense of being at a spiritual height.

Sense of reverence.

Feeling that you experienced something profoundly sacred and holy.

Factor 2: Positive Mood

Experience of amazement.

Feelings of tenderness and gentleness.

Feelings of peace and tranquility.

Experience of ecstasy.

Sense of awe or awesomeness.

Feelings of joy.

Factor 3: Transcendence of Time and Space

Loss of your usual sense of time.

Loss of your usual sense of space.

Loss of usual awareness of where you were.

Sense of being "outside of" time, beyond past and future.

Being in a realm with no space boundaries.

Experience of timelessness.

Factor 4: Ineffability

Sense that the experience cannot be described adequately in words.

Feeling that you could not do justice to your experience by describing it in words.

Feeling that it would be difficult to communicate your own experience to others who have not had similar experiences.

Fig. 4 Revised 30-item Mystical Experience Questionnaire (MEQ30). The MEQ30 is a psychometrically validated retrospective measure of mystical experience. The four factors of the psychometrically validated questionnaire are derived from a total of 30 items that probe seven dimensions (designated by *underlines*) of mystical experience that were identified by Stace (1960a). The Mystical factor is composed of 15 items probing four dimensions of the Stace model (internal unity, external unity, noetic quality, and sacredness). Positive Mood (6 items), Transcendence of Time and Space (6 items) and Ineffability (3 items) factors correspond to three separate dimensions of the Stace model

experimental condition, rates of complete mystical experience were lower for LSD (12.5%) than for previous reports of mystical experience after methylphenidate (23–33%) or psilocybin (up to 67% for a high dose of psilocybin; Barrett et al. 2015). It must be noted, however, that the set and setting involved in previous reports of mystical experience with methylphenidate and psilocybin (Barrett et al. 2015; Griffiths et al. 2006, 2011) were highly optimized to support the emergence of mystical experience, involving an experimental room modeled after a living room, a continuous musical accompaniment, art on the walls and soft lighting, and continual interpersonal preparation, support, and integration of the experience. The set and setting reported for LSD administration was significantly different, occurring in a standard hospital patient room with optional music listening (Liechti et al. 2016).

The MEQ30 and the Hood Mysticism Scale are psychometrically validated measures of mystical experience that have been derived from the same conceptual frame (Stace 1960a), but they differ in specific items and underlying factor structure. The MEQ30 and the Hood Mysticism Scale also differ in the timeframe over which the dimensions of mystical experiences are assessed. The MEQ30 assesses phenomena occurring during a single discrete experience, while the Hood Mysticism Scale typically assesses phenomena occurring over a lifetime. In addition, the MEQ30 has only been used in studies of classic hallucinogens (Griffiths et al. 2006, 2011; Johnson et al. 2014; Garcia-Romeu et al. 2015), whereas the Hood Mysticism Scale has been used most often in the context of survey research, in which the etiology of mystical experience is unknown, but generally not assumed to be due to classic hallucinogens. Future research is needed to validate the MEQ30 in assessing mystical experiences that occur in experimental and nonexperimental contexts in absence of drug administration. The psychometrically and experimentally validated MEQ30 may serve as a useful tool to facilitate investigation of both the determinants and consequences of mystical experience, including a wide variety of behavioral, pharmacologic, neurophysiological, genetic, personality, psychological, and therapeutic outcome measures.

While it has been suggested that the oceanic boundlessness (OAV) subscale of the Altered States of Consciousness (APZ; Dittrich 1998) questionnaire or the spiritual subscale of the derivative 5-Dimensional Altered States of Consciousness (5D-ASC; Dittrich et al. 2010; Studerus et al. 2010) questionnaire may provide a measure mystical experience (Majic et al. 2015), the items of these scales have only limited conceptual overlap with proposed (Stace 1960a) and empirically tested (Hood 1975; Griffiths et al. 2006; Hood 2009; Griffiths et al. 2011) models of mystical experience. Unity and positive mood (assessed in the “Blissful State” and “Experience of Unity” factors) are scored in the 11-dimension scoring of the 5D-ASC, and may be considered more suitable measures of some aspects of mystical experience. However, none of the proposed scales of 5D-ASC uniquely assess the constructs of ineffability or transcendence of time and space. While the experience oceanic boundlessness, unity, and positive mood may be necessary for a complete mystical experience, they may not be sufficient. One could expect to experience these effects after ingesting other drugs that would not be expected to reliably occasion a mystical experience. For instance, MDMA may be expected to increase feelings of unity and positive mood, but it may not be expected to occasion mystical experience (Lyvers and Meester 2012).

The newly developed Ego-Dissolution Inventory (EDI) measures a construct that is shown to correlate highly with a selected number of MEQ items that assess unity (Nour et al. 2016), however this scale does not measure positive affect, ineffability, or transcendence of time and space, nor does this scale assess other aspects of mystical experience that are assessed by items that load onto the Mystical factor of the MEQ30 (e.g., external unity, noetic quality, reverence, or sacredness). It is possible that one might report having experienced ego dissolution (e.g., might endorse items from the EDI including “All notion of self and identity dissolved away”, “I lost all sense of ego”, “I felt far less absorbed by my own issues and concerns”, and “I experienced a [dissolution or disintegration] of my ‘self’ or ego”) under the effects of anesthesia (e.g., propofol), but this experience would not include other dimensions of mystical experience described by Stace or included in either the MEQ30 or Hood Mysticism Scale. Further, some psychedelic experiences of ego dissolution are psychologically challenging, are devoid of positive affect, and may have enduring negative psychological effects (Carbonaro et al. 2016). As constructed, the EDI will not differentiate such experiences from mystical-type experiences.

A complete mystical experience, as described by Stace (1960a), which is cod ified in the MEQ30, includes not only just ego dissolution (as assessed by the EDI; Nour et al. 2016), unity (as assessed by the 5D-ASC), or positive mood (as assessed by the 5D-ASC), but also transcendence of space and time, ineffability, noetic quality, and reverence or sacredness. While other inventories have been used to measure and quantify important aspects of the subjective of psychedelics, such as experiences of unity and ego dissolution, none but the MEQ contains an array of items and subscales that adequately evaluates the construct of a complete mystical experience in relation to discrete psychedelic experiences.

6 Mystical Experiences, Classic Hallucinogens, and Therapeutic Interventions

Although mystical experiences are an intrinsically fascinating target of research in their own right in the study of consciousness, the correlates and consequences of mystical experiences may have relevance to therapeutic interventions. Research first conducted in the 1960s and continued in the present day suggests the possible efficacy of the classic hallucinogens in treating anxiety and depression (Grob et al. 2011; Carhart-Harris et al. 2016; Griffiths et al. 2016; Ross et al. 2016) and in treating addiction disorders such as alcohol dependence (Bogenschutz and Pommy 2012; Bogenschutz and Johnson 2016) and nicotine dependence (Johnson et al. 2014). Relevant to the present review, drug-occasioned mystical experiences have been suggested as a mediating mechanism underlying possible therapeutic effects (Richards et al. 1977; Garcia-Romeu et al. 2015; Griffiths et al. 2016; Ross et al. 2016).

A recent open-label, proof of concept study investigated the value of psilocybin as an adjunct to a smoking cessation intervention (Johnson et al. 2014). Up to three sessions with either a 20 mg/70 kg or 30 mg/70 kg dose of psilocybin were received by 15 participants. The authors reported a striking clinical outcome of successful smoking cessation in 80% of the sample (12/15 participants), with biologically verified abstinence 6 months after each participant's planned quit date. Seventy-three percent of participants rated at least one of their sessions as among the top five most spiritually significant experiences of their lives. Scores on a measure of individual mystical experience (the States of Consciousness Questionnaire, which contains the 43-item Mystical Experience Questionnaire, or MEQ43) correlated strongly, negatively, and significantly with a validated measure of cigarette craving (the Questionnaire on Smoking Urges). The MEQ43 data for this study have been rescored using the MEQ30 scoring, and are presented in Fig. 5. This suggests a link between strength of mystical experience during psilocybin sessions and clinical change in subjective effects that drive addictive behavior (Garcia-Romeu et al. 2015).

In a separate study, a similar finding was demonstrated in individuals with alcohol dependence (Bogenschutz et al. 2015). Ten volunteers were administered either a moderate (0.3 mg/kg) or high (0.4 mg/kg) dose of psilocybin in each of two experimental sessions during the course of treatment for alcohol dependence. While the exact number of participants who attained complete mystical experience was not reported, and while total scores on measures of mystical experience were lower than those attained in previous studies with psilocybin, patients in the study exhibited a significant improvement in drinking after their first psilocybin session and scores on ratings of psilocybin-occasioned mystical experience correlated strongly with change in drinking behavior.

Similar findings have been demonstrated in studies of the effect of psilocybin on anxiety and depression. A recent open-label pilot study consisting of an experimental session involving a 10 mg dose of psilocybin followed by a second experimental session involving a 25 mg dose of psilocybin reported sharp declines in depressive symptoms of patients with treatment-resistant depression at 1 week and 3 months after the second psilocybin session (Carhart-Harris et al. 2016). A separate series of investigations have demonstrated significant reduction in both anxiety and depression symptoms related to a life-threatening cancer diagnosis (Grob et al. 2011; Griffiths et al. 2016; Ross et al. 2016). Significant negative correlations were found between total mystical experience scores (scored using the MEQ30) and outcome measures assessed 5 weeks after a high-dose psilocybin session, including measures

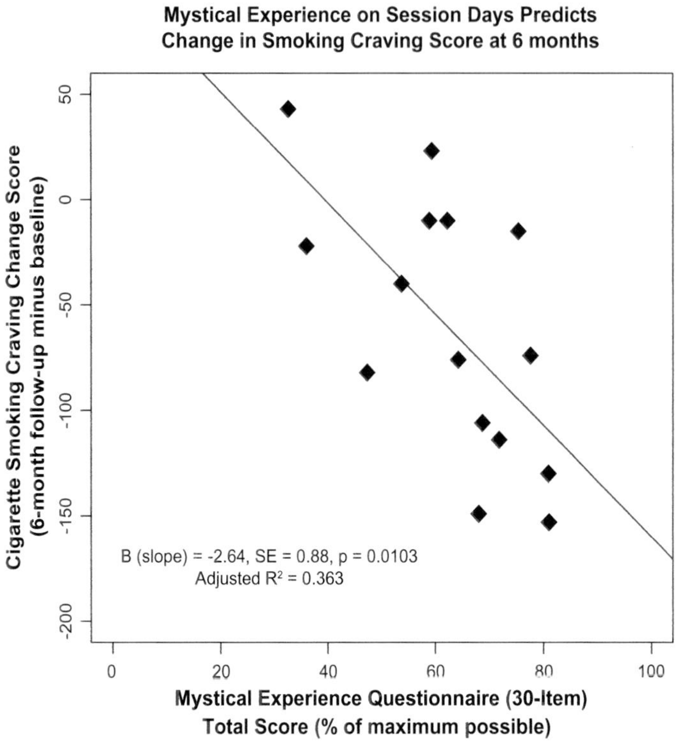

Fig. 5 Mystical experience scores on session days predicts change in smoking craving score at 6 months. *X*-axis: total score on the 30-item version of the Mystical Experience Questionnaire (MEQ30) expressed as the percentage of the maximum possible score. *Y*-axis: difference in score on the Questionnaire on Smoking Urges between study intake and 6-month follow-up. Data points represent individual participants (n = 15). B = slope of the regression of cigarette smoking craving change score on mystical experience questionnaire total score, SE = standard error of the slope (B). Adjusted R^2 indicates the amount of variance in spiritual significance ratings that is explained by mystical experience questionnaire total score. MEQ30 data have been rescored from MEQ43 responses reported in Garcia-Romeu et al. (2015)

of anxiety and depression symptoms. Figure 6 shows the relationship between total score on the MEQ30 and change from baseline to 5 weeks post-psilocybin on the Hamilton Anxiety Rating Scale (HAM-A). Importantly, total score on the MEQ30 was shown to mediate the effect of psilocybin on anxiety- and depression-related outcome measures (Griffiths et al. 2016; Ross et al. 2016). Initial findings from small-scale open-label studies (Carhart-Harris et al. 2016) along with findings from studies including more rigorous scientific controls (Grob et al. 2011; Griffiths et al. 2016; Ross et al. 2016) demonstrate the potential for therapeutic effects of psilocybin, and are suggestive of a link between mystical experience and therapeutic efficacy.

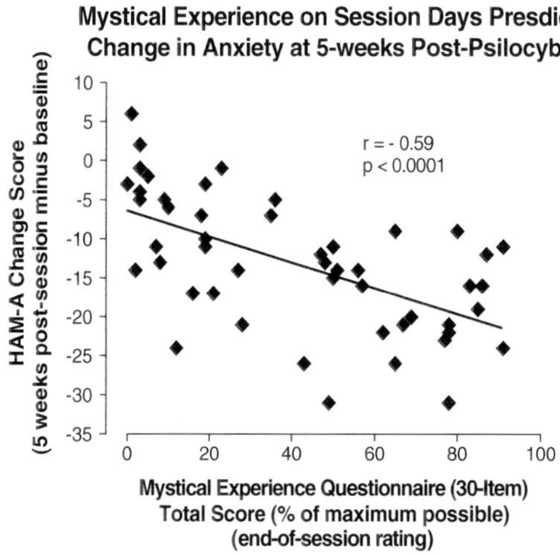

Fig. 6 Mystical experience on session days predicts change in anxiety at 5 weeks post-psilocybin. *X*-axis: total score on the 30-item version of the Mystical Experience Questionnaire (MEQ30) expressed as the percentage of the maximum possible score. *Y*-axis: difference in score on the Hamilton Anxiety Rating Scale between baseline and 5-week follow-up. Data points represent individual participants ($n = 51$). r = Pearson product moment correlation between HAM-A and MEQ30 scores. Data are from Griffiths et al. (2016)

7 Are Mystical Experiences Reducible to Neural Processes?

Whether experience itself is wholly reducible to neural processes is an open question. There are psychological abilities and processes, such as perception and cognition, that are amenable to neuroscientific investigation (Gazzaniga et al. 2014), and that could be considered the contents of consciousness (Chalmers 1995). Some theorists have argued that while abilities and processes such as these are amenable to investigation, the nature of experience itself is nonreducible and should be taken as a fundamental property of consciousness (Chalmers 1995). However, this position is not universally accepted (Dennett and Kinsbourne 1995). Both of these positions acknowledge that the nature of experience itself is difficult to approach from a reductive perspective, and the nature of mystical experience is no different.

Some of psychological processes, such as the experience of positive mood or alterations in the perception of time and space, are elements of mystical experiences which can be likely associated with primary neural processes. However, in the same way that memories or visual percepts do not wholly constitute consciousness itself but rather constitute the contents of consciousness, explanations of the individual neural elements of mystical experience may not provide a complete account of a mystical experience. Yet, there is still value in identifying and understanding neural and psychological processes that relate to mystical experience.

Investigation of mystical experiences may reveal valuable information about brain functioning. Much in the tradition of the brain lesion model of psychology and neuroscience, systematic perturbation of a neural system permits better understanding of that system. Thus, the study of the neural correlates of mystical experiences may lead to a better understanding of the possible brain mechanisms underlying self-referential, spatial, and temporal processing, as well as complex emotions such as reverence or sacredness. Further, the study of therapeutic outcomes of mystical experiences may advance the understanding of the neural basis of addiction and mood disorders.

8 Hallucinogens and Meditation as Tools to Investigate the Neural Correlates of Mystical Experiences

While little is currently known about the neural basis of discrete mystical experiences, we have a basic framework from which hypotheses can be generated. Stace's work provides a phenomenological model of mystical experiences. Griffiths and colleagues provide a pharmacological model for prospectively investigating mystical experiences that fit the Stace model. The reliability and pharmacological specificity of a selective 5-HT_{2A} receptor agonist (psilocybin) to occasion mystical experiences can be offered as initial evidence in favor of some neurobiological component of mystical experiences. Recent neuroimaging research with classic hallucinogens provides a basis for beginning to develop a neural model of mystical experiences.

Over the past two and a half decades, a wide range of neuroimaging methods, including molecular, magnetic, and electrocortical modalities, have been used to study the effects of classic hallucinogens in humans. Multiple neural processes in nearly every major cortical and subcortical division of the brain have been reported to be modulated by classic hallucinogens. Obviously, hallucinogen-related changes in these brain areas cannot be taken as absolute markers of mystical experience, since not all experiences with classic hallucinogens are mystical. Substances such as psilocybin are neither necessary nor sufficient for producing such experiences, and $5HT_{2A}$ receptor agonism may play only an initiatory role in the brain processes that account for or correlate with mystical experiences. Thus, the literature on the neural correlates of classic hallucinogens alone will be insufficient to provide a complete model of the neural basis of mystical experiences.

One approach to exploring the neural basis of mystical experience is to consider brain states produced by other approaches to occasioning mystical experiences, such as meditation practices. Although meditation encompasses a broad range of practices (Nash and Newberg 2013), a few specific practices have been the focus of brain imaging studies. These include focused awareness, open awareness or open monitoring (Lutz et al. 2008), and non-dual awareness (Josipovic 2010). Focused awareness practice typically involves attentional focus on an explicit object of consciousness (e.g., breath exhalation). In contrast, the intention of an open awareness practice is effortless sustaining of awareness without explicit selection of a discrete object of focus (Lutz et al. 2008). In both cases, a goal is to cultivate nonattachment to thoughts or distractions that may arise, with the further goal of stabilizing the mind. Such practices may ultimately lead to experiences of unity, with more specific objects of consciousness receding into the background. Non-dual awareness practices are a more subtle and perhaps a more direct technique of cultivating the experience of unity or pure awareness. In these practices, the object of attention is awareness itself or, as some have described it, awareness of awareness. The non-dual or transcendental unitive state that can arise from such practices appears to be descriptively identical to those described in peak mystical experiences, especially in terms of the dissolution of the conventional sense of personhood or the lack of differentiation between the sense of self and other (i.e., mystical unity).

9 Hallucinogens, Meditation, Mystical Experience, and the Default Mode Network

This section provides a review of hallucinogen studies and meditation studies that identify neural changes that may be relevant to mystical experiences—more specifically, changes in the default mode network (DMN) of the brain as they relate to the experiences of unity and transcendence of time and space, which are core dimensions of mystical experiences. The "default mode network" (DMN) consists of a number of brain areas in which task-related decreases in activity were reliability identified in early studies of human brain activity using positron emission tomography (PET) and functional magnetic resonance imaging (fMRI) (Raichle et al. 2001). The pattern of activity that is typically found in the DMN is hypothesized to reflect intrinsic patterns of communication in the brain (Raichle and Snyder 2007). Activity in the DMN (Power et al. 2011) (also called the "task-negative network") generally correlates with internally directed attention and is typically negatively correlated with activity in task-positive networks (i.e., networks that support externally directed attention and behavioral response) (Fox et al. 2005).

As is described in more detail below, the medial prefrontal cortex (MPFC), posterior cingulate cortex (PCC), and parahippocampal cortex (PHC), which densely express $5HT_{2A}$ receptor sites (the cellular targets of classic hallucinogens), display altered functioning in response to classic hallucinogens (Fig. 7, "Deactivations During Psilocybin Effects"). Activity and connectivity in these brain regions is also altered by acute and long-term meditation practice (Fig. 7, "Deactivations During Meditation"). It can be argued that alteration of neural activity within these brain regions is consistent with the decreases in self-referential processing (i.e., dissolution of a sense of self, likely related an experience of unity) that accompanies introvertive mystical experience. Neural functioning in the lateral default mode network, more specifically in the angular gyrus region of the inferior parietal lobule (IPL), is also implicated in meditation, self-transcendence, and

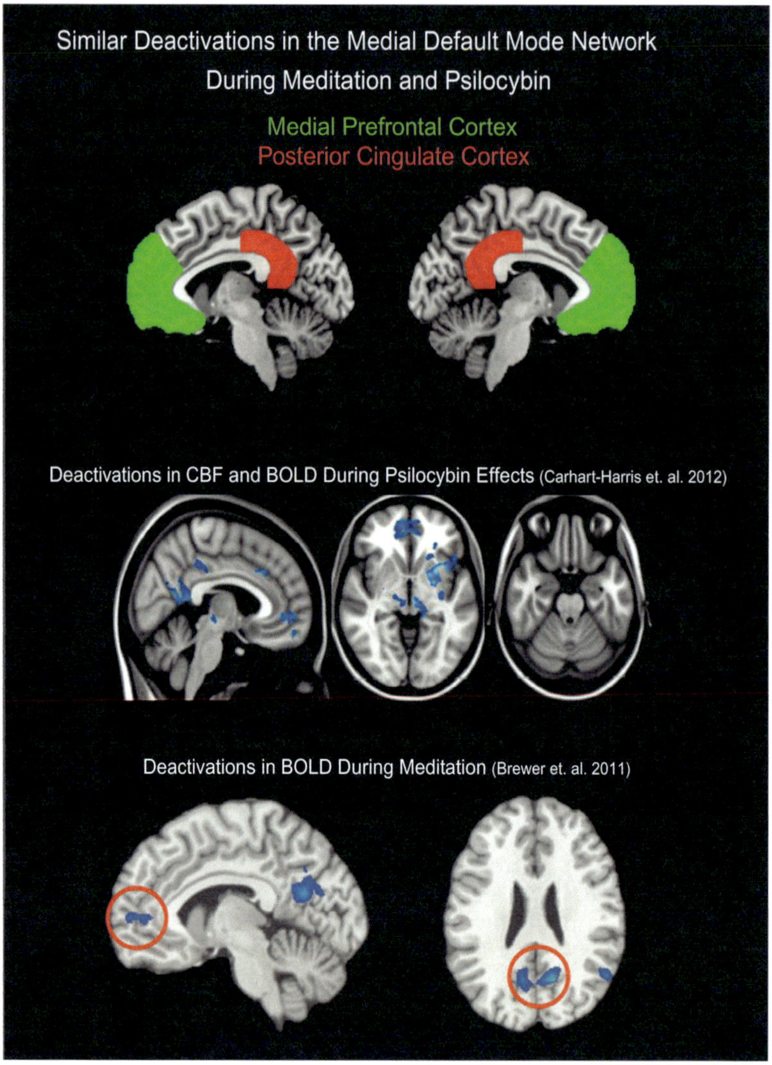

Fig. 7 Similar deactivations in the medial default mode network during meditation and experience with psilocybin. The upper portion of the figure illustrates approximate locations of the medial prefrontal cortex (labeled in *green*) and posterior cingulate cortex (labeled in *red*). The center portion of the figure (adapted from Carhart-Harris et al. 2014, Fig. 4, with permission), shows regions in the medial default mode network, including the medial prefrontal cortex and posterior cingulate, where both deactivation in blood-oxygenation-level-dependent (BOLD) data and decrease in cerebral blood flow (CBF) were observed after intravenous injection of psilocybin. The lower portion of the figure (adapted from Brewer et al. 2011, Fig. 1, with permission) shows regions in the medial default mode network, including the medial prefrontal cortex and posterior cingulate, where decreases in BOLD data were observed during meditation. The decreased activity within the medial prefrontal cortex and posterior cingulate which is observed after psilocybin and during meditation is consistent with decreased self-referential processing that accompanies introvertive mystical experience

experiences with classical hallucinogens. It is argued that change in neural activity in this region is consistent with spacelessness and timelessness that often accompanies introvertive mystical experience.

9.1 The Medial Default Mode Network

The medial prefrontal cortex (MPFC), the posterior cingulate cortex (PCC), and parahippocampal cortex (PHC) have been identified as major nodes of the medial aspect of the DMN (Fox et al. 2005). This has been confirmed in multiple large samples of volunteers (*N* between 180 and 1000), using a variety of analysis techniques (e.g., seed-based connectivity analysis, independent component analysis, graph-theoretic analysis, and clustering-based analysis) (Doucet et al. 2011; Power et al. 2011; Yeo et al. 2011). The psychological processes ascribed to the MPFC and PCC include various types of self-referential processing including mentalizing (i.e., thinking about your own or others' thoughts) (Gilbert et al. 2006), internal dialog (Northoff et al. 2006; Denny et al. 2012), self-related judgments (Northoff et al. 2006; Denny et al. 2012), and autobiographical memory retrieval (Svoboda et al. 2006). The process ascribed to the PHC is primarily the coding of episodic memory content, as opposed to semantic memory content which is coded by the perirhinal cortex (Ranganath and Ritchey 2012). The MPFC and PCC have been implicated in self-recognition and self-awareness in studies using tasks such as recognition of a face as either "you" or "not-you" (van Veluw and Chance 2014), tasks requiring the participant to identify adjectives as traits of themselves or others (Kelley et al. 2002), tasks assessing self-agency (Renes et al. 2014), and tasks that test false beliefs and theory of mind (van Veluw and Chance 2014). The PHC, and more broadly the hippocampal complex, has been implicated in supporting consciousness (Behrendt 2013) by supporting the moment-to-moment binding of sensory information into coherent memory representations, including those thought to be related to a sense of self. The PHC is involved in maintaining and recalling memories of self and self-relevant information.

Molecular (PET and single photo emission computed tomography, or SPECT) and fMRI (measuring blood-oxygenation-level-dependent, or BOLD, response) brain imaging methods have been used to show that the MPFC, PCC, and PHC are modulated by classic hallucinogens. Glucose metabolism after psilocybin ingestion (Vollenweider et al. 1997; Gouzoulis-Mayfrank et al. 1999) and cerebral blood flow after ingestion of either *ayahuasca* (Riba et al. 2006) or mescaline (Hermle et al. 1992) have been shown to increase in medial frontal areas that include or overlap with the MPFC. After intravenous administration of psilocybin or LSD, cerebral blood flow and BOLD activity in MPFC and PCC decreased substantially (Fig. 7, "Deactivations During Psilocybin Effects"), as did effective connectivity between these areas (Carhart-Harris et al. 2012, 2016). In the same studies, decreased cerebral blood flow correlated with increased intensity of drug effects and decreased sense of self (or increased "ego-dissolution"). The discrepancy between

Vollenweider, Riba, Gouzoulis-Mayfrank, and Hermle's work (showing increases in glucose metabolism and cerebral blood flow with classic hallucinogens) and the findings of Carhart-Harris and colleagues (showing decrease in BOLD activity and cerebral blood flow with classic hallucinogens) may be due to various factors including differing rates of drug CNS penetration, drug metabolic differences due to differing routes of administration (oral vs. intravenous, respectively), and differences in the specific neuronal process being measured by the different techniques (molecular techniques with PET and SPECT in the cases of Vollenweider, Riba, and Hermle, and magnetic resonance imaging techniques in the case of Carhart-Harris).

With injected psilocybin, decreased low-frequency power and long-range temporal correlations in BOLD data were observed in the PCC and MPFC, suggesting a breakdown in communication between these areas (Tagliazucchi et al. 2014). Increased spectral power and a greater diversity of functional connectivity was observed in the PHC after injected psilocybin, suggesting a substantial shift in communication involving the PHC (Tagliazucchi et al. 2014). After injected LSD, increased global connectivity was shown to correlate with increases in "ego dissolution" (Tagliazucchi et al. 2016) and decreases in mental time travel to the past (Speth et al. 2016). Both psilocybin (Lebedev et al. 2015) and LSD (Lebedev et al. 2016) were shown to increase global brain entropy. These findings all support a breakdown of long-distance communication in brain regions involved in the medial DMN.

Additional evidence for the breakdown of long-distance communication in the brain comes from studies using electroencephalography (EEG) or magnetoencephalography (MEG) that demonstrate changes in electrocortical activity induced by classic hallucinogens. Slow or low-frequency electrocortical oscillations (e.g., delta: 1–4 Hz; theta: 4–8 Hz, alpha: 8–16 Hz) are often shown to coordinate communication of distant neural areas, whereas faster or higher frequency oscillations (e.g., gamma: 32+ Hz) are often shown to coordinate local synchronization of a small number of neurons within limited populations (von Stein and Sarnthein 2000; Buzsaki et al. 2013). An absolute decrease in oscillatory power in all frequency bands was observed in frontal areas using EEG during the administration of ayahuasca (Riba et al. 2002), and with MEG during the administration of psilocybin (Muthukumaraswamy et al. 2013). With ayahuasca, this decrease was accompanied by an increase in the ratio of high to low-frequency oscillatory power (Riba et al. 2002), suggesting that low-frequency oscillatory power is more strongly decreased than high-frequency oscillatory power. With psilocybin, despite the decrease in power, stimulus-induced high-frequency oscillations were preserved (Muthukumaraswamy et al. 2013). Reduced synchronization of cortical oscillations in the PCC and PHC has been shown to correlate with a measure of spiritual experience during the effects of oral psilocybin (Kometer et al. 2015). Taken together, evidence in BOLD and electrocortical signals suggest an alteration of long-distance communication in the brain and a relative preservation of local and basic perceptual processing.

After psilocybin, oscillatory power decreases in all frequency bands in the PCC correlated with ratings of "disintegration of the self", with alpha power decreases correlating most strongly (Muthukumaraswamy et al. 2013). These power decreases in the PCC were consistent with the excitation of layer 5 pyramidal neurons, which densely express 5-HT_{2A} receptors, the primary site of action of classic hallucinogens. The negative correlation between alpha power and ratings of disintegration of self is notable because alpha power is positively correlated with perceptual framing, self-reflection, and introspection (Carhart-Harris et al. 2014). These findings may be relevant to mystical experiences insofar as a sense of disintegration of self is a primary aspect of the experience of introspective unity, a defining feature of the mystical experience.

With regard to meditation, amplitude of spontaneous fluctuations in the DMN has been shown to be reduced during mindfulness meditation (Berkovich-Ohana et al. 2016), and consistent deactivations of the MPFC, PCC, and IPL have been found during both focused awareness and open monitoring practices (Fox et al. 2016). Josipovic et al. (2011) demonstrated that while the typically observed negative correlations between the DMN (associated with internally directed attention) and task-positive networks (associated with externally directed attention) are maintained during a focused awareness practice and during a simple visual perception task, this negative correlation is significantly decreased during a non-dual awareness meditation practice, which may be associated with mystical-type experiences. Decreased negative correlation between the DMN and task-positive networks, as well as decreased activations and decreased spontaneous fluctuations in BOLD signal, are changes in brain dynamics that one might expect to accompany an experience of unity (a critical dimension of mystical experiences) in which representations of internal and external objects of consciousness blend together. Interestingly, decreased negative correlations between DMN and task-positive network areas in meditation are consistent with reports of increased resting-state functional connectivity between the DMN and various task-positive networks after intravenous psilocybin administration (Roseman et al. 2014), and increased global connectivity after LSD administration (Tagliazucchi et al. 2016). These decreases are also possibly consistent with both a breakdown of within-network functional connectivity in the DMN during the effects of psilocybin and LSD (i.e., between MPFC and PCC regions; Carhart-Harris et al. 2012, 2014, 2016; Tagliazucchi et al. 2014), increased brain entropy during the effects of psilocybin and LSD (Lebedev et al. 2015, 2016), and the development of new homological scaffolds in functional brain networks identified during the effects of psilocybin (Petri et al. 2014).

A very intriguing parallel between neuroimaging research on meditation (Brewer and Garrison 2014) and neuroimaging research with classic hallucinogens (Carhart-Harris et al. 2014) involves the functional relevance that is attributed to change in the activity of the PCC. The PCC is implicated in "internally directed cognition" (Leech and Sharp 2014), "'getting caught up in' one's experience" (Brewer et al. 2013), and "ego integrity" (Carhart-Harris et al. 2014). One could argue that activity in the PCC may reflect self-referential processing (i.e., an internal dialog or chatter) that is quieted by meditation training. This is supported by research demonstrating that mind-wandering is associated with PCC activity (Mason et al. 2007). Meditators with many years of experience in mindfulness practice showed decreased activity in the DMN (specifically the MPFC and PCC; Fig. 7, "Deactivations During Meditation") compared to non-meditators when data were collapsed across practices including focused awareness and open monitoring meditations (Brewer et al. 2011). Subsequent studies used a neuro-feedback paradigm to demonstrate that decreased activity in the PCC correlated with decreased mind-wandering during focused awareness practice (Garrison et al. 2013a, b). This is generally consistent with observations after intravenous psilocybin administration showing decreased low-frequency power and long-range temporal correlations in BOLD data in the PCC and MPFC (Tagliazucchi et al. 2014), substantially decreased effective connectivity in BOLD data between the PCC and MPFC (Carhart-Harris et al. 2012), and decreases in alpha oscillatory power in the PCC correlated with ratings of ego disintegration (Muthukumaraswamy et al. 2013). Considering all of the findings summarized above, it could be hypothesized that decreased activity and functional connectivity in the PCC and MPFC (medial nodes of the DMN) mediate aspects of the experience of introvertive unity (including decreased self-referential processing and a loss of sense of self) that is a key component of mystical experience, be that occasioned by classic hallucinogens or by meditation.

9.2 The Lateral Default Mode Network

The inferior parietal lobule (IPL) is one of the only lateral regions of the brain that is consistently identified as a node of the DMN across various techniques used to identify the DMN (Buckner et al. 2008). The IPL is a complex region that contains the anatomical locations of the angular gyrus (in the posterior IPL) and supramarginal gyrus (in the anterior IPL). The angular gyrus is variously implicated in spatial cognition (Amorapanth et al. 2009), semantics and memory (Humphreys and Lambon Ralph 2014), empathy, (Kubit and Jack 2013), and identifying and tracking the intentions of others (Scholz et al. 2009; Kubit and Jack 2013). The supramarginal gyrus is implicated more generally in categorical information processing (Amorapanth et al. 2009), phonological information processing, attentional processing (Scholz et al. 2009; Humphreys and Lambon Ralph 2014), target detection, and reorienting to salient stimuli (Kubit and Jack 2013). Distributed regions in the IPL have also been implicated in the processing of time and temporal relationships (Battelli et al. 2007; Bueti and Walsh 2009).

The angular gyrus is the region of the IPL most consistently included in the DMN, whereas the supramarginal gyrus most consistently overlaps with task-positive networks (Kubit and Jack 2013; Humphreys and Lambon Ralph 2014). The precise functional role of the angular gyrus may depend on the regions that are coactive with it in a given context (Seghier 2013). In the context of the DMN and self-referential processing, the angular gyrus may support spatial and temporal processing and the representation of the self as an individual in space and time.

With regard to meditation and mystical experience, d'Aquili and Newberg (2000) hypothesized that the experience of unity encountered in meditation and religious experiences is related to decreased activity in the posterior superior parietal lobe and IPL, consistent with alterations of the subjective experience of space and time. This relationship during meditation was confirmed using SPECT imaging in studies of Tibetan Buddhist meditators (Newberg et al. 2001). The IPL has also been implicated in the experience of unity outside of meditative practice. Patients undergoing parietal lobe surgery for resection of brain tumors completed a self-transcendence scale that included items that assess aspects of the mystical experience such as unity, timelessness, and spacelessness. The investigators reported that resection of the inferior posterior parietal lobes (the IPL in the left hemisphere, and specifically the angular gyrus in the right hemisphere) increased the experience of self-transcendence (Urgesi et al. 2010).

With regard to the classic hallucinogens, neuroimaging studies have reported changes in activity and functional connectivity of the IPL in response to these compounds. A decrease in regional cerebral blood flow in the inferior parietal cortex was identified after mescaline ingestion (Hermle et al. 1992), and an increase in glucose metabolism was identified in the lateral parietal cortex after psilocybin ingestion (Vollenweider et al. 1997). Decreased low-frequency power in BOLD signal in the angular gyrus but not the supramarginal gyrus (Tagliazucchi et al. 2014) and decreased functional connectivity between the PCC and the IPL (likely the angular gyrus, though exact coordinates were not provided) was demonstrated (Carhart-Harris et al. 2012) after psilocybin injection.

With regard to electrocortical findings, *ayahuasca* ingestion led to decreased current density in delta, alpha, and beta frequency bands localized to both the angular gyrus and the supramarginal gyrus (Riba et al. 2004). Psilocybin ingestion led to a decrease in pre-stimulus alpha band oscillations in parieto-occipital regions as detected using EEG (Kometer et al. 2013). Decreased power in delta, theta, alpha, and beta bands was detected with MEG over lateral parietal regions after psilocybin injection, specifically over the supramarginal gyrus, but the angular gyrus was not identified (Muthukumaraswamy et al. 2013).

Decreased activity in the IPL (and specifically in the angular gyrus), and decreased communication between the IPL and other areas involved in maintaining a sense of self (such as the PCC) are observed both in studies of meditation and in studies of classic hallucinogens. These observations are consistent with the experience of altered sense of time and space that has been reported with both meditation (d'Aquili and Newberg 2000) and classic hallucinogens (Griffiths et al. 2006). Considering the findings summarized above, it could be hypothesized that decreased activity and functional connectivity in the IPL (a lateral node of the DMN) mediates the experience of timelessness and spacelessness that accompanies introvertive mystical experience, be that occasioned by meditation or by classic hallucinogens.

10 Toward a Neural Model of Mystical Experiences

The experience of unity that is central to mystical experiences involves a decrease in self-referential processing. There is compelling evidence for a network of brain areas (i.e., the nodes of the DMN) that are involved in self-referential processing and maintenance of a sense of the self in space and time. Decreased activity in these areas has been observed using multiple imaging modalities, both after administration of classic hallucinogens and during meditation practices. After classic hallucinogens, BOLD and electrocortical data show decreased activity and low-frequency power localized to DMN areas (Riba et al. 2002, 2004; Carhart-Harris et al. 2012, 2016; Muthukumaraswamy et al. 2013), which may be related to a decrease in long-range neural communication between these and other brain areas. Alteration of the efficiency or fidelity of long-distance communication between nodes of the DMN after classic hallucinogens may be a neural mechanism underlying changes in activity in the DMN, and it may be a mechanistic change that is crucial to supporting mystical experiences.

Consistent with the general neural principle of low-frequency oscillations supporting long-range neural communication (von Stein and Sarnthein 2000; Buzsaki et al. 2013), decreased long-distance and increased short-distance connections may reduce "small world" features of brain networks and increase either "random graph" features or "regular lattice" features (Bullmore and Sporns 2009). In a "small world" network, most nodes only have a small number of connections to other nodes, and are generally not many connections away from any other node in the network. These features support efficient communication between nodes in large networks and they are nearly ubiquitous in natural complex systems (including healthy brains). Movement away from this small-world architecture (e.g., toward "random graph" architecture, in which each node has equal probability of being connected directly to every other node) would represent a notable departure from typical brain organization. This is consistent with idea that psychedelics induce disintegration and desegregation of functional brain networks (Carhart-Harris et al. 2016). The PCC, MPFC, and IPL can be considered connector hubs that support the small-world architecture of the brain; thus, breakdown of long-distance connections between these nodes, or decreased integration and segregation of these nodes of the DMN, that is observed with meditation and classic hallucinogens may be a marker of altered brain state that is consistent with mystical experiences.

Despite the lack of a ground truth against which to definitively validate a mystical experience, states engendered by classic hallucinogens and meditation share phenomenological descriptions that are consistent with mystical experiences, and it is tempting to interpret the neural correlates of hallucinogens and meditation as a model of the neural correlates of mystical experiences. While a number of observations have been made with classic hallucinogens and meditation that are consistent with aspects of mystical experiences, these observations have not been explicitly linked to "complete" mystical experiences. Thus, we conclude with the following hypotheses that are not specific not to meditation or experience with classic hallucinogens per se, but rather are specific to mystical experiences occasioned by meditation or classic hallucinogens: (1) activity within the medial nodes of the DMN will decrease, and communication between these nodes and other cortical targets in associative and sensory processing cortex will be fundamentally altered, in support of decreased self-referential processing during introvertive mystical experiences; (2) activity within the lateral nodes of the DMN will decrease, and communication between these nodes and other cortical targets will be fundamentally altered, in support of the experience of timelessness and spacelessness experienced during introvertive mystical experiences; (3) long-distance cortical communication will generally decrease, while local sensory and associative processing is maintained to some degree, during mystical experiences; (4) "small-world" properties of the brain will decrease during mystical experiences.

11 Conclusion

Naturally occurring mystical experiences (Stace 1960a) as well as the ceremonial use of classic hallucinogens (Schultes et al. 2001) have long histories that in some cases are complementary and may be intertwined (Osmond 1970; Dobkin De Rios 1984; Schultes et al. 2001). Certainly, not all experiences with classic hallucinogens are of the mystical type. However, with the proper setting, preparation, support, and dose of a classic hallucinogen, mystical experiences are occasioned at high probability (Griffiths et al. 2011). Post-session and follow-up ratings of mystical experience and positive impact of psilocybin session experiences are dose-dependent (Griffiths et al. 2011) and sustained at follow-up periods ranging from several weeks to 25 years (Pahnke 1963; Doblin 1991; Griffiths et al. 2006, 2008, 2011). Hallucinogen-occasioned mystical experiences are pharmacologically specific (i.e., occur at much higher frequency after a high dose of psilocybin than after placebo, low doses of psilocybin, methylphenidate, or nicotinic acid; Griffiths et al. 2006; Pahnke 1963) and not generally due to expectancy (i.e., demonstrated under conditions that obscured the range of drug conditions administered and not shown with placebo or a low dose of psilocybin Griffiths et al. 2006, 2011). Mystical experiences have a clear operational definition (MacLean et al. 2012; Barrett et al. 2015), and the positive outcomes associated with mystical experiences have been empirically demonstrated (Griffiths et al. 2006, 2008, 2011; Bogenschutz et al. 2015; Garcia-Romeu et al. 2015).

Although the most fundamental questions regarding mystical experiences presently evade a reductive neuroscientific explanation, analysis of the biological correlates suggestive of underlying mechanisms of mystical experiences is tractable. We have highlighted an intriguing overlap in neural findings on classic hallucinogens and neural findings on meditative practices that may occasion mystical experiences. More specifically, changes in activity, connectivity, and neural oscillatory processes in regions of the default mode network may underlie dimensions of mystical experience, especially decreased self-referential processing and altered sense of time and space that accompany introvertive mystical experiences. Further research with classic hallucinogens as experimental tools can be expected to provide significant insights into both the neural mechanisms and well as the biological and behavioral consequences of mystical experiences. Such information may have important implications for developing a science of moral and ethical behavior (Shermer 2015) as well as for developing novel therapeutic interventions for producing persisting positive behavioral and psychological changes.

References

Aaronson BS, Osmond H (1970) Introduction: psychedelics, technology, psychedelics. In: Aaronson BS, Osmond H (eds) Psychedelics: the uses and implications of hallucinogenic drugs. Anchor Books, New York, pp 3–20

Amorapanth PX, WIdick P, Chatterjee A (2009) The neural basis for spatial relations. J Cogn Neurosci 22(8):1739–1753

Anthony FV, Hermans CAM, Sterkens C (2010) A comparative study of mystical experience among Christian, Muslim, and Hindu students in Tamil Nadu, India. J Sci Study Relig 49(2):264–277

Barrett FS, Griffiths RR (2017) The factor structure of the Mystical Experience Questionnaire (MEQ): Reply to Bouso et al., 2016. Hum Psychopharm 32(1). doi:10.1002/hup.2564

Barrett FS, Johnson MJ, Griffiths RR (2015) Validation of the revised mystical experiences questionnaire in experimental sessions with psilocybin

Battelli L, Pascual-Leone A, Cavanagh P (2007) The 'when' pathway of the right parietal lobe. Trends Cogn Sci 11(5):204–210

Behrendt RP (2013) Hippocampus and consciousness. Rev Neuroscience 24(3):239–266

Berkovich-Ohana A, Harel M, Hahamy A, Arieli A, Malach R (2016) Alterations in task-induced activity and resting-state fluctuations in visual and DMN areas revealed in long-term meditators. NeuroImage 135:125–134

Bogenschutz MP, Johnson MW (2016) Classic hallucinogens in the treatment of addictions. Prog Neuro-psychoph 64:250–258

Bogenschutz MP, Pommy JM (2012) Therapeutic mechanisms of classic halluciongens in the treatment of addictions: from indirect evidence to testable hypotheses. Drug Test Anal 4(543–555):543

Bogenschutz MP, Forcehimes AA, Pommy JA, Wilcox CE, Barbosa PCR, Strassman RJ (2015) Psilocybin-assisted treatment for alcohol dependence: a proof-of-concept study. J Psychopharmacol

Bouso JC, Pedrero-Perez EJ, Gandy S, Alcazar-Corcoles MA (2016) Measuring the subjective: revisiting the psychometric properties of three rating scales that assess the acute effects of hallucinogens. Hum Psychopharm 31(5):356–372

Brewer JA, Worhunsky PD, Gray JR, Tang YY, Weber J, Kober H (2011) Meditation experience is associated with differences in default mode network activity and connectivity. Proc Natl Acad Sci U S A 108(50):20254–20259. doi:10.1073/pnas.1112029108

Brewer JA, Garrison KA, Whitfield-Gabrieli S (2013) What about the "self" is processed in the posterior cingulate cortex? Front Hum Neurosci 7:647. doi:10.3389/fnhum.2013.00647

Brewer JA, Garrison KA (2014) The posterior cingulate cortex as a plausible mechanistic target of meditation: findings from neuroimaging. Ann N Y Acad Sci 1307:19–27. doi:10.1111/nyas.12246

Buckner RL, Andrews-Hanna JR, Schacter DL (2008) The brain's default network: anatomy, function, and relevance to disease. Ann N Y Acad Sci 1124:1–38. doi:10.1196/annals.1440.011

Bueti D, Walsh V (2009) The parietal cortex and the representation of time, space, number and other magnitudes. Phil Trans R Soc B 364:1831–1840

Bullmore E, Sporns O (2009) Complex brain networks: graph theoretical analysis of structural and functional systems. Nat Rev Neurosci 10(3):186–198. doi:10.1038/nrn2575

Buzsaki G, Logothetis N, Singer W (2013) Scaling brain size, keeping timing: evolutionary preservation of brain rhythms. Neuron 80(3):751–764. doi:10.1016/j.neuron.2013.10.002

Carbonaro TM, Bradstreet MP, Barrett FS, MacLean KA, Jesse R, Johnson MW, Griffiths RR (2016) Survey study of challenging experiences after ingesting psilocybin mushrooms: acute and enduring positive and negative consequences. J Psychopharmacol 30(12):1268–1278

Carhart-Harris RL, Bolstridge M, Rucker J, Day CMJ, Erritzoe D, Kaelen M, Bloomfield M, Rickard JA, Forbes B, Feilding A, Taylor D, Pilling S, Curran HV, Nutt DJ (2016) Psilocybin with psychological support for treatment-resistant depression: an open-label feasibility study. Lancet Psychiatry [Online] May 17, 2016

Carhart-Harris RL, Erritzoe D, Williams T, Stone JM, Reed LJ, Colasanti A, Tyacke RJ, Leech R, Malizia AL, Murphy K, Hobden P, Evans J, Feilding A, Wise RG, Nutt DJ (2012) Neural correlates of the psychedelic state as determined by fMRI studies with psilocybin. Proc Natl Acad Sci U S A 109(6):2138–2143. doi:10.1073/pnas.1119598109

Carhart-Harris RL, Leech R, Hellyer PJ, Shanahan M, Feilding A, Tagliazucchi E, Chialvo DR, Nutt DJ (2014) The entropic brain: a theory of conscious states informed by neuroimaging research with psychedelic drugs. Front Hum Neurosci 8:20. doi:10.3389/fnhum.2014.00020

Chalmers DJ (1995) Facing up to the problem of consciousness. J Conscious Stud 2(3):200–219

Chen Z, Hood RWJ, Qi W, Watson PJ (2011a) Common core thesis and qualitative and quantitative analysis of mysticism in Chinese Buddhist Monks and Nuns. J Sci Study Relig 50 (4):654–670

Chen Z, Hood RWJ, Yang L, Watson PJ (2011b) Mystical experience among Tibetan Buddhists: the common core thesis revisited. J Sci Study Relig 50(2):328–338

Chen Z, Zhang Y, Hood RWJ, Watson PJ (2012) Mysticism in Chinese Christians and Non-Christians: measurement invariance of the mysticism scale and implications for the mean differences. Int J Psychol Relig 2(2):155–168

d'Aquili EG, Newberg AB (2000) The neuropsychology of aesthetic, spiritual, and mystical states. Zygon 35(1):39–51

Dennett DC, Kinsbourne M (1995) Time and the observer: the where and when of consciousness in the brain. Behav Brain Sci 15(2):183–247

Denny BT, Kober H, Wager TD, Ochsner KN (2012) A meta-analysis of functional neuroimaging studies of self- and other judgments reveals a spatial gradient for mentalizing in medial prefrontal cortex. J Cogn Neurosci 24(8):1742–1752. doi:10.1162/jocn_a_00233

Dittrich A (1998) The standardized psychometric assessment of altered states of consciousness (ASCs) in humans. Pharmacopsychiatry 31:80–84. doi:10.1055/s-2007-979351

Dittrich A, Lamparter D, Maurer M (2010) 5D-ASC: questionnaire for the assessment of altered states of consciousness. A short introduction. PSIN PLUS, Zurich, Switzerland

Dobkin de Rios M (1984) Visionary vine: hallucinogenic healing in the Peruvian Amazon. Waveland Press, Prospet Heights, Illinois

Dobkin de Rios M (1996) On "human pharmacology of Hoasca": a medical anthropology perspective. J Nerv Ment Dis 184(2):95–98

Doblin R (1991) Pahnke's "good friday experiment": a long-term follow-up and methodological critique. J Transpersonal Psychol 23(1):1–28

Doucet G, Naveau M, Petit L, Delcroix N, Zago L, Crivello F, Jobard G, Tzourio-Mazoyer N, Mazoyer B, Mellet E, Joliot M (2011) Brain activity at rest: a multiscale hierarchical functional organization. J Neurophysiol 105(6):2753–2763. doi:10.1152/jn.00895.2010

El-Seedi HR, De Smet PAGM, Beck O, Possnert G, Bruhn JG (2005) Prehistoric peyote use: alkaloid analysis and radiocarbon dating of archaeological specimens of Lophophora from Texas. J Enthnopharmacol 101:238–242

Fox KC, Dixon ML, Nijeboer S, Girn M, Floman JL, Lifshitz M, Ellamil M, Sedlmeier P, Christoff K (2016) Functional neuroanatomy of meditation: a review and meta-analysis of 78 functional neuroimaging investigations. Neurosci Biobehav R 65:208–228

Fox MD, Snyder AZ, Vincent JL, Corbetta M, Van Essen DC, Raichle ME (2005) The human brain is intrinsically organized into dynamic, anticorrelated functional networks. Proc Natl Acad Sci U S A 102(27):9673–9678. doi:10.1073/pnas.0504136102

Gabrielsson A (2010) Strong experiences with music. Oxford University Press, Handbook of Music and Emotion

Garcia-Romeu A, Griffiths RR, Johnson MW (2015) Psilocybin-occasioned mystical experiences in the treatment of tobacco addiction. Curr Drug Abuse Rev 7(3):157–164

Garrison KA, Santoyo JF, Davis JH, Thornhill TAt, Kerr CE, Brewer JA (2013a) Effortless awareness: using real time neurofeedback to investigate correlates of posterior cingulate cortex activity in meditators' self-report. Front Hum Neurosci 7:440. doi:10.3389/fnhum.2013.00440

Garrison KA, Scheinost D, Worhunsky PD, Elwafi HM, Thornhill TAt, Thompson E, Saron C, Desbordes G, Kober H, Hampson M, Gray JR, Constable RT, Papademetris X, Brewer JA (2013b) Real-time fMRI links subjective experience with brain activity during focused attention. Neuroimage 81:110–118. doi:10.1016/j.neuroimage.2013.05.030

Gazzaniga MS, Ivry RB, Mangun GR (2014) Cognitive neuroscience: the biology of the mind, 4th edn. W. W. Norton & Company, Inc., New York

Gilbert SJ, Spengler S, Simons JS, Steele JD, Lawrie SM, Frith CD, Burgess PW (2006) Functional specialization within rostral prefrontal cortex (area 10): a meta-analysis. J Cogn Neurosci 18(6):932–948. doi:10.1162/jocn.2006.18.6.932

Gouzoulis-Mayfrank E, Schreckenberger M, Sabri O, Arning C, Thelen B, Spitzer M, Kovar KA, Hermle L, Büll U, Sass H (1999) Neurometabolic effects of psilocybin, 3,4-methylenedioxyethylamphetamine (MDE) and d-methamphetamine in healthy volunteers. A double-blind, placebo-controlled PET study with [18F]FDG. Neuropsychopharmacology 20 (6):565–581. doi:10.1016/S0893-133X(98)00089-X

Griffiths RR, Richards WA, McCann U, Jesse R (2006) Psilocybin can occasion mystical-type experiences having substantial and sustained personal meaning and spiritual significance. Psychopharmacology (Berl) 187(3):268–283; discussion 284–292. doi:10.1007/s00213–006-0457-5

Griffiths RR, Richards W, Johnson M, McCann U, Jesse R (2008) Mystical-type experiences occasioned by psilocybin mediate the attribution of personal meaning and spiritual significance 14 months later. J Psychopharmacol 22(6):621–632. doi:10.1177/0269881108094300

Griffiths RR, Johnson MW, Richards WA, Richards BD, McCann U, Jesse R (2011) Psilocybin occasioned mystical-type experiences: immediate and persisting dose-related effects. Psychopharmacology 218(4):649–665. doi:10.1007/s00213-011-2358-5

Griffiths RR, Johnson MW, Carducci MA, Umbricht A, Richards WA, Richards BD, Cosimano MP, Klinedinst MA (2016) Psilocybin produces substantial and sustained decreases in depression and anxiety in patients with life-threatening cancer: a randomized double-blind trial. J Psychopharmcol 30(12):1181–1197

Grob CS, McKenna DJ, Callaway JC, Brito GS, Neves ES, Oberlaender G, Saede OL, Labigalini E, Tacla C, Miranda CT, Strassman RJ, Boone KB (1996) Human psychopharmacology of Hoasca, a plant hallucinogen used in ritual context in Brazil. J Nerv Ment Dis 184 (2):86–94

Grob CS, Danforth AL, Chopra GS, Hagerty M, McKay CR, Halberstadt AL, Greer GR (2011) Pilot study of psilocybin treatment of anxiety in patients with advanced-stage cancer. Arch Gen Psychiatry 68(1):71–78

Grof S, Grof C (2010) Holotropic breathwork: a new approach to self-exploration and therapy. State University of New York Press, Albany, New York

Guzmán G (1983) The genus psilocybe: a systematic revision of the known species including the history, distribution and chemistry of the hallucinogenic species. Nova Hedwigia Heft 74, Cramer, Vaduz, Germany

Guzmán G (2008) Hallucinogenic mushrooms in Mexico: an overview. Econ Bot 62(3):404–412

Halpern JH, Sherwood AR, Hudson JI, Yurgelun-Todd D, Pope HG Jr (2005) Psychological and cognitive effects of long-term peyote use among Native Americans. Biol Psychiatry 58(8):624–631. doi:10.1016/j.biopsych.2005.06.038

Hermle L, Fünfgeld M, Oepen G, Botsch H, Borchardt D, Gouzoulis E, Fehrenbach RA, Spitzer M (1992) Mescaline-induced psychopathological, neuropsychological, and neurometabolic effects in normal subjects: experimental psychosis as a tool for psychiatric research. Biol Psychiatry 32(11):976–991

Hood RWJ (1975) The construction and preliminary validation of a measure of reported mystical experience. J Sci Study of Relig 14:29–41

Hood RWJ, Morris RJ, Watson PJ (1990) Quasi-experimental elicitation of differential report of religious experience among intrinsic and indiscriminately pro-religious types. J Sci Study Relig 29(2):164–172

Hood RW, Williamson WP (2000) An empirical test of the unity thesis: the structure of mystical descriptors in various faith samples. J Psychol Christianity 19(3):232–244

Hood RWJ, Ghorbani N, Watson PJ, Ghramaleki AF, Bing MN, Davidson HK, Morris RJ, Williamson WP (2001) Dimensions of the mysticism scale: confirming the three-factor structure in the United States and Iran. J Sci Study Relig 40(4):691–705

Hood RWJ (2003) Conceptual and empirical consequences of the unity thesis. In: Belzen JA, Geels A (eds) Mysticism: a variety of psychological perspectives. Rodopi, New York

Hood RWJ (2006) The common core thesis in the study of mysticism. Where God and Science Meet, vol 3. Praeger, Westport, pp 119–138

Hood RWJ (2009) Mysticism. In: Hood RWJ, Hill PC, Spilka B (eds) The psychology of religion, 4th edn. The Guilford Press, New York

Humphreys GF, Lambon Ralph MA (2014) Fusion and fission of cognitive functions in the human parietal cortex. Cereb Cortex. doi:10.1093/cercor/bhu198

Huxley A (1947) The perennial philosophy. Chatto & Windus, London

Johnson MW, Garcia-Romeu A, Cosimano MP, Griffiths RR (2014) Pilot study of the 5-HT2AR agonist psilocybin in the treatment of tobacco addiction. J Psychopharmacol 28(11):983–992. doi:10.1177/0269881114548296

Josipovic Z (2010) Duality and nonduality in meditation research. Conscious Cogn 19:1119–1121

Josipovic Z, Dinstein I, Weber J, Heeger DJ (2011) Influence of meditation on anti-correlated networks in the brain. Front Hum Neurosci 5:183. doi:10.3389/fnhum.2011.00183

Josipovic Z (2014) Neural correlates of nondual awareness in meditation. Ann N Y Acad Sci 1307:9–18. doi:10.1111/nyas.12261

Katz ST (ed) (1978) Mysticism and philosophical analysis. Oxford University Press, New York

Katz ST (ed) (1983) Mysticism and religious traditions. Oxford University Press, New York

Kelley WM, Macrae CN, Wyland CL, Caglar S, Inati S, Heatherton TF (2002) Finding the self? An event-related fMRI study. J Cogn Neurosci 14(5):785–794. doi:10.1162/08989290260138672

Kometer M, Pokorny T, Seifritz E, Volleinweider FX (2015) Psilocybin-induced spiritual experiences and insightfulness are associated with synchronization of neuronal oscillations. Psychopharmacology 232(19):3663–3676

Kometer M, Schmidt A, Jancke L, Vollenweider FX (2013) Activation of serotonin 2A receptors underlies the psilocybin-induced effects on alpha oscillations, N170 visual-evoked potentials, and visual hallucinations. J Neurosci 33(25):10544–10551. doi:10.1523/JNEUROSCI.3007-12.2013

Kubit B, Jack AI (2013) Rethinking the role of the rTPJ in attention and social cognition in light of the opposing domains hypothesis: findings from an ALE-based meta-analysis and resting-state functional connectivity. Front Hum Neurosci 7:323. doi:10.3389/fnhum.2013.00323

Lebedev AV, Kaelen M, Lovden M, Nilsson J, Feilding A, Nutt DJ, Carhart-Harris RL (2016) LSD-induced entropic brain activity predicts subsequent personality change. Hum Brain Mapp 37(9):3203–3213

Lebedev AV, Lovden M, Rosenthal G, Feilding A, Nutt DJ, Carhart-Harris RL (2015) Finding the self by losing the self: neural correlates of ego-dissolution under psilocybin. Hum Brain Mapp 36(8):3137–3153

Leech R, Sharp DJ (2014) The role of the posterior cingulate cortex in cognition and disease. Brain 137(Pt 1):12–32. doi:10.1093/brain/awt162

Liechti ME, Dolder PC, Schmid Y (2016) Alterations of consciousness and mystical-type experiences after acute LSD in humans. Psychopharmacology

Lutz A, Slagter HA, Dunne JD, Davidson RJ (2008) Attention regulation and monitoring in meditation. Trends Cogn Sci 12(4):163–169

Lyvers M, Meester M (2012) Illicit use of LSD or psilocybin, but not MDMA or nonpsychedelic drugs, is associated with mystical experiences in a dose-dependent manner. J Psychoactive Drugs 44(5):410–417

MacLean KA, Leoutsakos JM, Johnson MW, Griffiths RR (2012) Factor analysis of the mystical experience questionnaire: a study of experiences occasioned by the hallucinogen psilocybin. J Sci Study Relig 51(4):721–737. doi:10.1111/j.1468-5906.2012.01685.x

Majic T, Schmidt TT, Gallinat J (2015) Peak experiences and the afterglow phenomenon: when and how do therapeutic effects of hallucinogens depend on psychedelic experiences? J Psychopharmacol. doi:10.1177/0269881114568040

Mason MF, Norton MI, Van Horn JD, Wegner DM, Grafton ST, Macrae CN (2007) Wandering minds: the default network and stimulus-independent thought. Science 315(5810):393–395. doi:10.1126/science.1131295

Miller WR, C'de Baca J (2001) Quantum change. Guilford Press, New York

Muthukumaraswamy SD, Carhart Harris RL, Moran RJ, Brookes MJ, Williams TM, Errtizoe D, Sessa B, Papadopoulos A, Bolstridge M, Singh KD, Feilding A, Friston KJ, Nutt DJ (2013) Broadband cortical desynchronization underlies the human psychedelic state. J Neurosci 33 (38):15171–15183. doi:10.1523/JNEUROSCI.2063-13.2013

Nash JD, Newberg A (2013) Toward a unifying taxonomy and definition for meditation. Front Psychol 4(806):1–18

Newberg A, Alavi A, Baime M, Pourdehnad M, Santanna J, d'Aquili E (2001) The measurement of regional cerebral blood flow during the complex cognitive task of meditation: a preliminary SPECT study. Psychiatry Res 106(2):113–122

Newberg AB, Iversen J (2003) The neural basis of the complex mental task of meditation: neurotransmitter and neurochemical considerations. [Review]. Med Hypotheses 61(2):282–291

Newberg AB, Wintering NA, Morgan D, Waldman MR (2006) The measurement of regional cerebral blood flow during glossolalia: a preliminary SPECT study. Psychiatry Res 148(1):67–71. doi:10.1016/j.pscychresns.2006.07.001

Newberg AB, Wintering N, Waldman MR, Amen D, Khalsa DS, Alavi A (2010) Cerebral blood flow differences between long-term meditators and non-meditators. Conscious Cogn 19 (4):899–905. doi:10.1016/j.concog.2010.05.003

Nichols DE (2004) Hallucinogens. Pharmacol Ther 101(2):131–181. doi:10.1016/j.pharmthera.2003.11.002

Nichols DE (2016) Psychedelics. Pharmacol Rev 68:264–355. doi:10.1124/pr.115.011478

Northoff G, Heinzel A, de Greck M, Bermpohl F, Dobrowolny H, Panksepp J (2006) Self-referential processing in our brain-a meta-analysis of imaging studies on the self. Neuroimage 31(1):440–457. doi:10.1016/j.neuroimage.2005.12.002

Nour MM, Evans L, Nutt D, Carhart-Harris RL (2016) Ego-dissolution and psychedelics: validation of the Ego-Dissolution Inventory (EDI). Front Hum Neurosci 10:269

Osmond H (1970) Peyote night. In: Aaronson B, Osmond H (eds) Psychedelics: the uses and implications of hallucinogenic drugs. Anchor Books, Garden City, pp 67–85

Pahnke WN (1963) Drugs and mysticism: an analysis of the relationship between psychedelic drugs and the mystical consciousness. Harvard University Press, Cambridge, MA

Pahnke WN (1967) LSD and religious experience. In: DeBold RC, Leaf RC (eds) LSD man & society. Wesleyan University Press, Middletown, CT

Pahnke WN (1969) Psychedelic drugs and mystical experience. Int Psychiatry Clin 5:149–162

Penman J, Becker J (2009) Religious ecstatics, "deep Listeners," and musical emotion. Empirical Musicol Rev 4(2):49–70

Petri G, Expert P, Turkheimer F, Carhart-Harris R, Nutt D, Hellyer PJ, Vaccarino F (2014) Homological scaffolds of brain functional networks. J R Soc Interface 11(101):20140873

Power JD, Cohen AL, Nelson SM, Wig GS, Barnes KA, Church JA, Vogel AG, Laumann TO, Miezin FM, Schlaggar BL, Petersen SE (2011) Functional network organization of the human brain. Neuron 72(4):665–678. doi:10.1016/j.neuron.2011.09.006

Proudfoot W (1985) Religious experience. University of California Press, Berkeley, CA

Raichle ME, MacLeod AM, Snyder AZ, Powers WJ, Gusnard DA, Shulman GL (2001) A default mode of brain function. Proc Natl Acad Sci 98(2):676–682

Raichle ME, Snyder AZ (2007) A default mode of brain function: A brief history of an evolving idea. NeuroImage 37:1083–1090

Ranganath C, Ritchey M (2012) Two cortical systems for memory-guided behaviour. Nat Rev Neurosci 13(10):713–726

Renes RA, van Haren NE, Aarts H, Vink M (2014) An exploratory fMRI study into inferences of self-agency. Soc Cogn Affect Neurosci. doi:10.1093/scan/nsu106

Riba J, Anderer P, Morte A, Urbano G, Jane F, Saletu B, Barbanoj MJ (2002) Topographic pharmaco-EEG mapping of the effects of the South American psychoactive beverage ayahuasca in healthy volunteers. Br J Clin Pharmacol 53(6):613–628

Riba J, Anderer P, Jane F, Saletu B, Barbanoj MJ (2004) Effects of the South American psychoactive beverage ayahuasca on regional brain electrical activity in humans: a functional neuroimaging study using low-resolution electromagnetic tomography. Neuropsychobiology 50(1):89–101. doi:10.1159/000077946

Riba J, Romero S, Grasa E, Mena E, Carrio I, Barbanoj MJ (2006) Increased frontal and paralimbic activation following ayahuasca, the pan-Amazonian inebriant. Psychopharmacology 186(1):93–98. doi:10.1007/s00213-006-0358-7

Richards WA, Rhead JC, DiLeo FB, Yensen R, Kurland AA (1977) The peak experience variable in DPT-assisted psychotherapy with cancer patients. J Psychedelic Drugs 9:1–10

Richards WA (2014) The rebirth of research with entheogens: lessons from the past and hypotheses for the future. In: Ellens JH (ed) Seeking the sacred with psychoactive substances: chemical paths to spirituality and to God, vol 2. Praeger, Santa Barbara, pp 1–14

Roberts TB (ed) (2001) Psychoactive sacramentals: essays on entheogens and religion. Council on Spiritual Practices, San Francisco

Roseman L, Leech R, Feilding A, Nutt DJ, Carhart-Harris RL (2014) The effects of psilocybin and MDMA on between-network resting state functional connectivity in healthy volunteers. Front Hum Neurosci 8:204. doi:10.3389/fnhum.2014.00204

Ross S, Bossis A, Guss J, Agin-Liebes G, Malone T, Cohen B, Mennenga SE, Belser A, Kalliontzi K, Babb J, Su Zhe, Corby P, Schmidt BL (2016) Rapid and sustained symptom reduction following psilocybin treatment for anxiety and depression in patients with life-threatening cancer: a randomized controlled trial. J Psychopharmcol 30(12):1165–1180

Scholz J, Triantafyllou C, Whitfield-Gabrieli S, Brown EN, Saxe R (2009) Distinct regions of right temporo-parietal junction are selective for theory of mind and exogenous attention. PLoS ONE 4(3):e4869. doi:10.1371/journal.pone.0004869

Schultes RE (1969) Hallucinogens of plant origin. Science 163:245–254

Schultes RE, Hofmann A, Rätsch C (2001) Plants of the Gods: their sacred, healing, and Hallucinogenic powers. Healing Arts Press

Seghier ML (2013) The angular gyrus: multiple functions and multiple subdivisions. Neuroscientist 19(1):43–61. doi:10.1177/1073858412440596

Sharf RH (1988) Experience. In: Taylor M (ed) Critical terms for religious studies. University of Chicago Press, Chicago, pp 94–116

Shermer M (2015) The Moral Arc: how science and reason lead humanity toward truth, justice, and freedom. Henry Holt and Co, New York

Smith H (2000) Cleansing the doors of perception: the religious significance of entheogenic plants and chemicals. Tarcher/Putnam, New York

Speth J, Speth C, Kaelen M, Schloerscheidt AM, Feilding A, Nutt DJ, Carhart-Harris RL (2016) Decreased mental time travel to the past correlates with default-mode network disintegration under lysergic acid diethylamide. J psychopharmacol 30(4):344–353

Stace WT (1960a) Mysticism and philosophy. MacMillan Press, New York

Stace WT (1960b). The teachings of the mystics. New American Library

Studerus E, Gamma A, Vollenweider FX (2010) Psychometric evaluation of the altered states of consciousness rating scale (OAV). PLoS ONE 5(8):e12412

Svoboda E, McKinnon MC, Levine B (2006) The functional neuroanatomy of autobiographical memory: a meta-analysis. Neuropsychologia 44(12):2189–2208. doi:10.1016/j.neuropsychologia.2006.05.023

Tagliazucchi E, Carhart-Harris R, Leech R, Nutt D, Chialvo DR (2014) Enhanced repertoire of brain dynamical states during the psychedelic experience. Hum Brain Mapp 35(11):5442–5456. doi:10.1002/hbm.22562

Tagliazucchi E, Roseman L, Kaelen M, Orban C, Muthukumaraswamy SD, Murphy K, Laufs H, Leech R, McGonigle J, Crossley N, Bullmore E, Williams T, Bolstridge M, Feilding A, Nutt DJ, Carhart-Harris R (2016) Increased global functional connectivity correlates with LSD-Induced Ego Dissolution. Curr biol 26(8):1043–1050

Tupper KW (2008) The globalization of ayahuasca: harm reduction or benefit maximization? Int J Drug Policy 19:297–303

Underhill E (1911 [2009]). Mysticism. Evinity Publishing

Urgesi C, Aglioti SM, Skrap M, Fabbro F (2010) The spiritual brain: selective cortical lesions modulate human self-transcendence. Neuron 65(3):309–319. doi:10.1016/j.neuron.2010.01.026

van Veluw SJ, Chance SA (2014) Differentiating between self and others: an ALE meta-analysis of fMRI studies of self-recognition and theory of mind. Brain Imaging Behav 8(1):24–38. doi:10.1007/s11682-013-9266-8

Vollenweider FX, Kometer M (2010) The neurobiology of psychedelic drugs: implications for the treatment of mood disorders. Nat Rev Neurosci 11:642–651

Vollenweider FX, Leenders KL, Scharfetter C, Maguire P, Stadelmann O, Angst J (1997) Positron emission tomography and fluorodeoxyglucose studies of metabolic hyperfrontality and psychopathology in the psilocybin model of psychosis. Neuropsychopharmacology 16(5):357–372. doi:10.1016/S0893-133X(96)00246-1

von Stein A, Sarnthein J (2000) Different frequencies for different scales of cortical integration: from local gamma to long range alpha/theta synchronization. Int J Psychophysiol 38(3):301–313

Wasson RG, Hofman A, Ruck CAP (1998) The road to eleusis: unveiling the secret of the mysteries. Hermes Press, Los Angeles

Watts A (1970) Psychedelics and religious experience. In: Aaronson B, Osmond H (eds) Psychedelics: the uses and implications of hallucinogenic drugss. Anchor Books, New York, pp 131–144

Westermeyer J (1988) The pursuit of intoxication: our 100 century-old romance with psychoactive substances. Am J Drug Alcohol Abuse 14(2):175–187

Wulff DM (1991) Psychology of religion; classic and contemporary views. Wiley, New York

Yeo BT, Krienen FM, Sepulcre J, Sabuncu MR, Lashkari D, Hollinshead M, Roffman JL, Smoller JW, Zöllei L, Polimeni JR, Fischl B, Liu H, Buckner RL (2011) The organization of the human cerebral cortex estimated by intrinsic functional connectivity. J Neurophysiol 106 (3):1125–1165. doi:10.1152/jn.00338.2011

Erratum to: Phenomenology, Structure, and Dynamic of Psychedelic States

Katrin H. Preller and Franz X. Vollenweider

Erratum to: Curr Topics Behav Neurosci, DOI 10.1007/7854_2016_459

The original version of the chapter was inadvertently published without using the revised manuscript for Chap. 459. The erratum chapter has been updated with the change.

The updated original online version of this chapter can be found at http://dx.doi.org/10.1007/7854_2016_459

K.H. Preller (✉) · F.X. Vollenweider (✉)
Neuropsychopharmacology and Brain Imaging Unit, Heffter Research Center Zurich Department of Psychiatry, Psychotherapy and Psychosomatics, Psychiatric University Hospital, University of Zurich, Lenggstrasse 31, 8032 Zurich, Switzerland
e-mail: preller@bli.uzh.ch

F.X. Vollenweider
e-mail: vollen@bli.uzh.ch

Curr Topics Behav Neurosci (2017) 36: E1
DOI 10.1007/7854_2017_477
Published Online: 15 July 2017